Microbial Processes and Products

METHODS IN BIOTECHNOLOGY™

John M. Walker, SERIES EDITOR

18. **Microbial Processes and Products**, edited by *José-Luis Barredo, 2005*
17. **Microbial Enzymes and Biotransformations**, edited by *José-Luis Barredo, 2005*
16. **Environmental Microbiology:** *Methods and Protocols*, edited by *John F. T. Spencer and Alicia L. Ragout de Spencer, 2004*
15. **Enzymes in Nonaqueous Solvents:** *Methods and Protocols*, edited by *Evgeny N. Vulfson, Peter J. Halling, and Herbert L. Holland, 2001*
14. **Food Microbiology Protocols**, edited by *J. F. T. Spencer and Alicia Leonor Ragout de Spencer, 2000*
13. **Supercritical Fluid Methods and Protocols**, edited by *John R. Williams and Anthony A. Clifford, 2000*
12. **Environmental Monitoring of Bacteria**, edited by *Clive Edwards, 1999*
11. **Aqueous Two-Phase Systems**, edited by *Rajni Hatti-Kaul, 2000*
10. **Carbohydrate Biotechnology Protocols**, edited by *Christopher Bucke, 1999*
 9. **Downstream Processing Methods**, edited by *Mohamed A. Desai, 2000*
 8. **Animal Cell Biotechnology**, edited by *Nigel Jenkins, 1999*
 7. **Affinity Biosensors:** *Techniques and Protocols*, edited by *Kim R. Rogers and Ashok Mulchandani, 1998*
 6. **Enzyme and Microbial Biosensors:** *Techniques and Protocols*, edited by *Ashok Mulchandani and Kim R. Rogers, 1998*
 5. **Biopesticides:** *Use and Delivery*, edited by *Franklin R. Hall and Julius J. Menn, 1999*
 4. **Natural Products Isolation**, edited by *Richard J. P. Cannell, 1998*
 3. **Recombinant Proteins from Plants:** *Production and Isolation of Clinically Useful Compounds*, edited by *Charles Cunningham and Andrew J. R. Porter, 1998*
 2. **Bioremediation Protocols**, edited by *David Sheehan, 1997*
 1. **Immobilization of Enzymes and Cells**, edited by *Gordon F. Bickerstaff, 1997*

Microbial Processes and Products

Edited by

José-Luis Barredo

R&D Biology
Antibióticos S. A., León, Spain

HUMANA PRESS ✳ TOTOWA, NEW JERSEY

© 2005 Humana Press Inc.
999 Riverview Drive, Suite 208
Totowa, New Jersey 07512

www.humanapress.com

This publication is printed on acid-free paper. ∞
ANSI Z39.48-1984 (American Standards Institute)

Permanence of Paper for Printed Library Materials.
Cover illustration: Figure 11A, B from Chapter 17,"Improved Polysaccharide Production Using Strain Improvement," by Thomas P. West.

Cover Design: Patricia F. Cleary

For additional copies, pricing for bulk purchases, and/or information about other Humana titles, contact Humana at the above address or at any of the following numbers: Tel.: 973-256-1699; Fax: 973-256-8341; E-mail: humana@humanapr.com; or visit our Website: www.humanapress.com

Printed in the United States of America. 10 9 8 7 6 5 4 3 2 1

Library of Congress Cataloging in Publication Data

e-ISBN: 1-59259-847-1

Microbial processes and products / edited by José-Luis Barredo.

    p. cm. -- (Methods in biotechnology ; 18)

  Includes bibliographical references and index.

  ISBN 1-58829-548-6 (alk. paper)

  1. Microbial biotechnology--Methodology. I. Barredo, José-Luis II. Series.

  TP248.27.M53M536 2004

  660.6'2--dc22

                                     2004048227

Preface

The development of biotechnology over the last 20 years, and particularly the use of recombinant DNA techniques, has rapidly expanded the opportunities for human benefits from living resources. Efforts to reduce pollution, prevent environmental damage, combat microbial infection, improve food production, and so on can each involve fermentation or the environmental release of microorganisms. Many products of fermentation technology, such as alcoholic beverages, bread, antibiotics, amino acids, vitamins, enzymes, and others, have been influenced by the progress of recombinant DNA techniques. The development of new products or the more efficient manufacturing of those already being produced often involve the use of microorganisms as cell factories for many productions and biotransformations.

Microbial Processes and Products is intended to provide practical experimental laboratory procedures for a wide range of processes and products mediated by microorganisms. Although not an exhaustive treatise, it provides a detailed "step-by-step" description of the most recent developments in such applied biotechnological processes. The detailed protocols we provide are cross-referenced in the Notes section, contain critical details, lists of problems and their troubleshooting, as well as safety recommendations that may not normally appear in journal articles and can be particularly useful for those unfamiliar with specific techniques.

The lead chapter of *Microbial Processes and Products* represents an overview on strain improvement programs and strategies to optimize fermentation processes. The remaining chapters detail comprehensive experimental methods for the optimal design of microbial metabolite production and for applying biotechnological processes to the manufacture of products used worldwide for human health, nutrition, and environmental protection, including semisynthetic derivatives of cephalosporins, erythromycin, antitumor compounds, plasmids for gene therapy and DNA vaccination, L-lysine, vitamins B_2 and B_{12}, the sweet-tasting protein thaumatin, the carotenoids β-carotene and astaxanthin, the polysaccharide gellan, and the bacteriocin-producing bacteria for sausage fermentation. Furthermore, the uses of the phenylacetyl-CoA catabolon for the enzymatic synthesis of penicillins, aromatic biotransformations, synthesis of new bioplastics, biosensor design, the synthesis of drug vehicles, and the development of a phosphatase encoding gene as a reporter and monitor gene expression are illustrated.

Additionally, *Microbial Processes and Products* offers techniques for analysis and quantification, including antimicrobial metabolites and carotenoids, volatile sulfur compounds, metabolic pathway fluxes, gene expression arrays, proteome analysis, methods to understand the mechanisms underlying bacterial modulation of the innate immune response, bioleaching activity, and microbial metal sulfide oxidation, and heavy metals remediation. Finally, three overview chapters on the transport of biological material, the deposit of biological material for patent purposes, and protection of biotechnological inventions are included.

Microbial Processes and Products has been written by outstanding experts in the field and provides a highly useful reference source for laboratory and industrial professionals, as well as for graduate students in a number of biological disciplines (biotechnology, microbiology, genetics, molecular biology) because of the uncommonly wide applicability of the procedures across the range of areas covered.

I am indebted to the authors who, in spite of their professional activities, agreed to participate in this book, to Dr. John Walker, Series Editor, for his encouragement and advice in reviewing the manuscripts, and to the rest of the staff of The Humana Press for their assistance in assembling this volume and their efforts in keeping this project on schedule. Last but not least, I warmly acknowledge my wife Natalia and our children Diego, José-Luis, Álvaro, and Gonzalo for their patience and support.

José-Luis Barredo

Contents

Preface v

Contributors xi

1. Development of Improved Strains and Optimization
   of Fermentation Processes
   Lei Han and Sarad R. Parekh 1

2. Experimental Design in Microbiology
   Guillaume E. Vanot and Michelle Sergent 25

3. Metabolic Engineering of *Acremonium chrysogenum*
   to Produce Deacetoxycephalosporin C and Bioconversion
   to 7-Aminodeacetoxycephalosporanic Acid
   Marta Rodríguez-Sáiz, Juan-Luis de la Fuente,
   and José-Luis Barredo 41

4. Production of Erythromycin
   With *Saccharopolyspora erythraea*
   Wolfgang Minas 65

5. The Phenylacetyl-CoA Catabolon: *Genetic, Biochemical,*
   and Biotechnological Approaches
   José M. Luengo, Belén García, Ángel Sandoval,
   Elsa Arias-Barrau, Sagrario Arias, Francisco Bermejo,
   and Elías R. Olivera 91

6. Recombinant Microorganisms for the Biosynthesis
   of Glycosylated Antitumor Compounds
   Carmen Méndez and José A. Salas 131

7. Assay Methods for Detection and Quantification
   of Antimicrobial Metabolites Produced
   by *Streptomyces clavuligerus*
   Paloma Liras and Juan F. Martín 149

8. Purification of Plasmid DNA Vectors Produced
   in *Escherichia coli* for Gene Therapy
   and DNA Vaccination Applications
   Maria Margarida Diogo, João António Queiroz,
   and Duarte Miguel F. Prazeres 165

9. Genome Breeding of an Amino Acid-Producing
   Corynebacterium glutamicum Mutant
   Masato Ikeda, Junko Ohnishi, and Satoshi Mitsuhashi 179

10. Metabolic Activity Profiling by ¹³C Tracer Experiments
    and Mass Spectrometry in *Corynebacterium glutamicum*
    Christoph Wittmann and Elmar Heinzle *191*

11. Protein and Vitamin Production in *Bacillus megaterium*
    Heiko Barg, Marco Malten, Martina Jahn, and Dieter Jahn *205*

12. Strategies for Large-Scale Production of Recombinant Proteins
    in Filamentous Fungi
    **Heidi Sisniega, José-Luis del Río, María-José Amaya,
    and Ignacio Faus** *225*

13. Strain and Culture Conditions Improvement
    for β-Carotene Production With *Mucor*
    **Enrique A. Iturriaga, Tamás Papp, Jesper Breum, José Arnau,
    and Arturo P. Eslava** *239*

14. Xanthophylls in Fungi: *Metabolic Engineering of the Astaxanthin
    Biosynthetic Pathway in* Xanthophyllomyces dendrorhous
    Hans Visser, Gerhard Sandmann, and Jan C. Verdoes *257*

15. Methodologies for the Analysis of Fungal Carotenoids
    Paul D. Fraser and Peter M. Bramley *273*

16. Insertional Mutagenesis in the Vitamin B₂ Producer
    Fungus *Ashbya gossypii*
    **María A. Santos, Laura Mateos, Karl-Peter Stahmann,
    and José-Luis Revuelta** *283*

17. Improved Polysaccharide Production Using Strain Improvement
    Thomas P. West *301*

18. Use of the *Morganella morganii phoC* Gene as Reporter
    in Bacterial and Yeast Hosts
    **Stefania Cresti, Cesira L. Galeotti, Serena Schippa,
    Gian Maria Rossolini, and Maria C. Thaller** *313*

19. Gene Expression Arrays in Food
    Bart Weimer, Yi Xie, Lan-szu Chou, and Adele Cutler *333*

20. Optimization of Proteome Analysis for Wine Yeast Strains
    **Tammi L. Olineka, Apostolos Spiropoulos, Paula A. Mara,
    and Linda F. Bisson** *345*

21. Bacteriocin-Producing Strains in a Meat Environment
    Frédéric Leroy and Luc De Vuyst *369*

22. In Vitro and In Vivo Interactions of Nonpathogenic Bacteria
    With Immunocompetent Cells
    **Eduardo J. Schiffrin, Nabila Ibnou-Zekri, Jean M. Ovigne,
    Thierry von der Weid, and Stephanie Blum** *381*

23. Volatile Sulfur Detection in Fermented Foods
   Bart Weimer and Ben Dias *397*

24. Bioleaching: *Analysis of Microbial Communities
      Dissolving Metal Sulfides*
   Axel Schippers and Klaus Bosecker *405*

25. Use of PhoN Phosphatase to Remediate Heavy Metals
   Marion Paterson-Beedle and Lynne E. Macaskie *413*

26. Safe Dispatch and Transport of Biological Material
   Vera Weihs and Christine Rohde *437*

27. How to Deposit Biological Material for Patent Purposes
   Vera Weihs *451*

28. Laws and Regulations for the Protection
      of Biotechnological Inventions
   Uwe Fitzner *465*

Index *495*

Contributors

MARÍA-JOSÉ AMAYA • *Grupo Uriach, Barcelona, Spain*

SAGRARIO ARIAS • *Departamento de Bioquímica y Biología Molecular, Universidad de León, Spain*

JOSÉ ARNAU • *Department of Fungal Biotechnology, Biotechnological Institute, Hørsholm, Denmark*

ELSA ARIAS-BARRAU • *Departamento de Bioquímica y Biología Molecular, Universidad de León, Spain*

HEIKO BARG • *Institut für Mikrobiologie, der Carolo-Wilhelmina-Universität, Braunschweig, Germany*

JOSÉ-LUIS BARREDO • *R&D Biology, Antibióticos S. A., León, Spain*

FRANCISCO BERMEJO • *Departamento de Química Orgánica, Universidad de Salamanca, Spain*

LINDA F. BISSON • *Department of Viticulture and Enology, University of California, Davis, CA*

STEPHANIE BLUM • *Nestlé Research Centre, Lausanne, Switzerland*

KLAUS BOSECKER • *Section Geomicrobiology, Federal Institute for Geosciences and Natural Resources, Hannover, Germany*

PETER M. BRAMLEY • *School of Biological Sciences, Royal Holloway, University of London, UK*

JESPER BREUM • *Department of Fungal Biotechnology, Biotechnological Institute, Hørsholm, Denmark*

LAN-SZU CHOU • *Center for Microbe Detection and Physiology, Utah State University, Logan, UT*

STEFANIA CRESTI • *Dipartimento di Biologia Molecolare, Università degli Studi di Siena, Italy*

ADELE CUTLER • *Department of Mathematics and Statistics, Utah State University, Logan, UT*

JUAN-LUIS DE LA FUENTE • *R&D Biology, Antibióticos S. A., León, Spain*

LUC DE VUYST • *Research Group of Industrial Microbiology, Fermentation Technology, and Downstream Processing, Department of Applied Biological Sciences, Vrije Universiteit Brussel, Brussels, Belgium*

JOSÉ-LUIS DEL RÍO • *Grupo Uriach, Barcelona, Spain*

BEN DIAS • *Kraft Foods Inc., Glenview, IL*

MARIA MARGARIDA DIOGO • *Center for Biological and Chemical Engineering, Instituto Superior Técnico, Lisbon, Portugal*

ARTURO P. ESLAVA • *Centro Hispano-Luso de Investigaciones Agrarias, Universidad de Salamanca, Spain*

IGNACIO FAUS • *Grupo Uriach, Barcelona, Spain*

UWE FITZNER • *Dres. Fitzner, Münch und Kluin, Ratingen, Germany*

PAUL D. FRASER • *School of Biological Sciences, Royal Holloway, University of London, UK*

CESIRA L. GALEOTTI • *I.R.I.S. Chiron s.r.l., Siena, Italy*

BELÉN GARCÍA • *Departamento de Bioquímica y Biología Molecular, Universidad de León, Spain*

LEI HAN • *Dow AgroSciences LLC, Indianapolis, IN*

ELMAR HEINZLE • *Biochemical Engineering Institute, Saarland University, Saarbrücken, Germany*

NABILA IBNOU-ZEKRI • *Nestlé Research Centre, Lausanne, Switzerland*

MASATO IKEDA • *Department of Bioscience and Biotechnology, Faculty of Agriculture, Shinshu University, Nagano, Japan*

ENRIQUE A. ITURRIAGA • *Area de Genética, Departmento de Microbiología y Genética, Universidad de Salamanca, Spain*

DIETER JAHN • *Institut für Mikrobiologie, Carolo-Wilhelmina-Universität, Braunschweig, Germany*

MARTINA JAHN • *Institut für Mikrobiologie, Carolo-Wilhelmina-Universität, Braunschweig, Germany*

FRÉDÉRIC LEROY • *Research Group of Industrial Microbiology, Fermentation Technology, and Downstream Processing, Department of Applied Biological Sciences, Vrije Universiteit Brussel, Brussels, Belgium*

PALOMA LIRAS • *Área de Microbiología, Departmento de Ecologia, Genética, y Microbiologia, Facultad de Ciencias Biológicas y Ambientales, Universidad de León, Spain*

JOSÉ M. LUENGO • *Departamento de Bioquímica y Biología Molecular, Universidad de León, Spain*

LYNNE E. MACASKIE • *School of Biosciences, The University of Birmingham, UK*

MARCO MALTEN • *Institut für Mikrobiologie, Carolo-Wilhelmina-Universität, Braunschweig, Germany*

PAULA A. MARA • *Department of Viticulture and Enology, University of California, Davis, CA*

JUAN F. MARTÍN • *Instituto de Biotecnología, León, Spain*

LAURA MATEOS • *Departamento de Microbiologia y Genetica, CSIC/ Universidad de Salamanca, Spain*

CARMEN MÉNDEZ • *Departamento de Biología Funcional e Instituto Universitario de Oncología del Principado de Asturias, Universidad de Oviedo, Spain*

WOLFGANG MINAS • *Anbics Management-Services AG, Zurich, Switzerland*

SATOSHI MITSUHASHI • *Tokyo Research Laboratories, Kyowa Hakko Kogyo Co., Ltd., Tokyo, Japan*

JUNKO OHNISHI • *Tokyo Research Laboratories, Kyowa Hakko Kogyo Co., Ltd., Tokyo, Japan*

TAMMI L. OLINEKA • *Department of Viticulture and Enology, University of California, Davis, CA*

ELÍAS R. OLIVERA • *Departamento de Bioquímica y Biología Molecular, Universidad de León, Spain*

JEAN M. OVIGNE • *Nestlé Research Centre, Lausanne, Switzerland*

TAMÁS PAPP • *Area de Genética, Departmento de Microbiología y Genética, Universidad de Salamanca, Spain*

SARAD PAREKH • *Dow AgroSciences LLC, Indianapolis, IN*

MARION PATERSON-BEEDLE • *School of Biosciences, The University of Birmingham, UK*

DUARTE MIGUEL F. PRAZERES • *Center for Biological and Chemical Engineering, Instituto Superior Técnico, Lisbon, Portugal*

JOÃO ANTÓNIO QUEIROZ • *Department of Chemistry, Universidade da Beira Interior, Covilhã, Portugal*

JOSÉ-LUIS REVUELTA • *Departamento de Microbiologia y Genetica, CSIC/ Universidad de Salamanca, Spain*

MARTA RODRÍGUEZ-SÁIZ • *R&D Biology, Antibióticos S. A., León, Spain*

CHRISTINE ROHDE • *DSMZ-Deutsche Sammlung von Mikroorganismen und Zellkulturen GmbH, Braunschweig, Germany*

GIAN MARIA ROSSOLINI • *Dipartimento di Biologia Molecolare, Università degli Studi di Siena, Italy*

JOSÉ A. SALAS • *Departamento de Biología Funcional e Instituto Universitario de Oncología del Principado de Asturias, Universidad de Oviedo, Spain*

GERHARD SANDMANN • *Botanical Institute, J.W. Goethe University, Frankfurt, Germany*

ÁNGEL SANDOVAL • *Departamento de Bioquímica y Biología Molecular, Universidad de León, Spain*

MARÍA A. SANTOS • *Departamento de Microbiologia y Genetica, CSIC/ Universidad de Salamanca, Spain*

EDUARDO J. SCHIFFRIN • *Nestlé Research Centre, Lausanne, Switzerland*

SERENA SCHIPPA • *Dipartimento di Scienze e di Sanità Pubblica, Università La Sapienza, Rome, Italy*

AXEL SCHIPPERS • *Section Geomicrobiology, Federal Institute for Geosciences and Natural Resources, Hannover, Germany*

MICHELLE SERGENT • *Laboratoire de Méthodologie de la Recherche Expérimentale, Faculté des Sciences et Techniques de Saint Jérôme, Marseille, France*

HEIDI SISNIEGA • *Grupo Uriach, Barcelona, Spain*

APOSTOLOS SPIROPOULOS • *Department of Viticulture and Enology, University of California, Davis, CA*

KARL-PETER STAHMANN • *Technische Mikrobiologie. FB Bio-, Chemie- und Verfahrenstechnik. Fachhochschule Lausitz, Senftenberg. Germany*

MARIA C. THALLER • *Dipartimento di Biologia, Università di Roma Tor Vergata, Roma, Italy*

GUILLAUME E. VANOT • *Faculté des Sciences et Techniques de Saint Jérôme, Marseille, France*

JAN C. VERDOES • *Dyadic Nederland BV, Zeist, The Netherlands*

HANS VISSER • *Section of Fungal Genomics, Wageningen University, The Netherlands*

THIERRY VON DER WEID • *Nestlé Research Centre, Lausanne, Switzerland*

VERA WEIHS • *DSMZ-Deutsche Sammlung von Mikroorganismen und Zellkulturen GmbH, Braunschweig, Germany*

BART WEIMER • *Center for Integrated BioSystems, Department of Nutrition and Food Sciences and Center for Microbe Detection and Physiology, Utah State University, Logan, UT*

THOMAS P. WEST • *Olson Biochemistry Laboratories, Department of Chemistry and Biochemistry, South Dakota State University, Brookings, SD*

CHRISTOPH WITTMANN • *Biochemical Engineering Institute, Saarland University, Saarbrücken, Germany*

YI XIE • *Center for Microbe Detection & Physiology, Department of Nutrition and Food Science, Utah State University, Logan, UT*

1

Development of Improved Strains and Optimization of Fermentation Processes

Lei Han and Sarad R. Parekh

Summary

Microbial strains overproducing commercially important metabolites are routinely obtained through mutagenesis and random screening and/or selection. Advances in recombinant DNA technology have made it possible for engineering improved microbial strains by specific addition or deletion of certain genes. The key to the genetic engineering approach, however, is the identification of genes controlling metabolite production. In recent years, innovative technologies have been developed to allow researchers to investigate the genetics and physiology of a microorganism on a global scale. Knowledge gained from these studies is beginning to modernize strain development processes. Fermentation processes must be constantly optimized in order to maximize the potential of each improved strain. This chapter reviews the various methods of developing improved strains and addresses the specific issues concerning each method. In addition, strategies commonly employed to optimize fermentation processes will be analyzed. Finally, new technologies and how they can help strain development and fermentation process optimization will be discussed.

Key Words: Strain improvement; fermentation; mutation; genetic engineering, scale-up.

1. Introduction

Microbial strain improvement involves the application of one or a combination of strategies that result in the development of new strains with desired phenotypes. The most commonly sought phenotype is increased metabolite production. Other desired phenotypes include reduced production of side products or ease of scalability at operation. Improving microbial strains for desired phenotypes has been the cornerstone of all commercial fermentation processes. Even today, much of the emphasis placed on improving microbial strains is the result of the diversity of the metabolites produced by microorganisms that have

From: *Methods in Biotechnology, Vol. 18: Microbial Processes and Products*
Edited by: J. L. Barredo © Humana Press Inc., Totowa, NJ

found novel applications in the food, chemical, agricultural, health care, and pharmaceutical industries.

Biologically active molecules synthesized by microorganisms isolated from nature are usually produced in extremely low quantities. The reason is that those molecules are either nonessential for the survival of the microorganisms or present in sufficient quantities to satisfy their primary needs. Low productivity of the metabolite of interest translates into high manufacturing cost per unit of the product *(1)*. Therefore, manufacturing of commercially important metabolites directly from microorganisms isolated from nature is often economically unfavorable. However, there are ways of enhancing productivity to make the production process economically viable. Lower fermentation manufacturing and capital costs can be achieved through improvements in the design of fermenters *(2)*. The greatest opportunity to lower manufacturing cost without significant capital investment, however, comes from microbial strains with increased productivity or the ability to utilize low-cost raw materials or some other beneficial traits *(3)*. Improved microbial strains can be obtained by using microbes isolated from nature as a starting point and by employing one or combinations of strategies.

Conventionally, strain improvement is achieved through random mutation and screening or selection, the so-called classical approach. This empirical method has been practiced for more than 50 yr and has a long history of success *(4)*. The best known example is the titer improvement achieved for penicillin *(5)*. In view of the long practice of the classical approach, it continues to be the primary strain improvement strategy of any newly established strain improvement program. One reason for continued interest in the classical approach is that this approach does not require prior knowledge of the metabolite biochemical pathway, regulation, or transport. Another reason is that the advancement made over the years in the precision and sensitivity of analytical instruments dramatically increases the reliability and sensitivity of detection. Furthermore, automation and miniaturization of screening processes have significantly reduced system variability and increased screening throughput.

In addition to the classical approach, there are targeted approaches for developing improved strains, including enrichment methods and genetic engineering. One significant advantage of the targeted approaches is that only a small number of strains have to be screened and evaluated when the rationale behind screening and genetic engineering is sound. Therefore, the key to the targeted approaches has been to identify the genes controlling metabolite production. Previously target gene identification was a time-consuming and painstaking task. Researchers were able to work only on a few leads at a time because of the complexity and difficulty involved in identifying the right gene target. However, with the aid of genomic sequencing, expression analysis, protein analysis, and metabolic flux analysis that offer the understanding of microbial physiology on a global scale, researchers are better equipped to identify potential targets. Ideally, the classical approach and targeted approaches are integrated to create a synergistic effect for rapid strain development.

Finally, because titer increments seen in small-scale fermentation might not scale to production fermenters, improved strains must be validated at a pilot scale designed to mimic fermenter conditions for production. In addition, this intermediate step allows engineers to discover potential problems associated with improved strains and affords the opportunity for further enhancing strain performance.

This chapter reviews the various methods used to improve microbial strains and fermentation processes and addresses the issues associated with each method. New technologies and their potential applications in strain development and process improvement will be discussed. The reader is also directed to the recent reviews concerning the classical approach, enrichment methods, and fermentation process improvement *(1,6–8)*.

2. Classical Approach

2.1. Process

The process of obtaining improved strains involves repetition of three steps: (1) mutation of a parent strain to introduce new genetic alteration in the genome, (2) random screening and assessment of mutagenesis survivors in small-scale fermentation vessels, and (3) quantification of the metabolite and identification of potentially improved strains. Potentially improved strains are evaluated again in multiple repetitions to confirm statistically an improvement over the parent strain. Each time an improved strain is identified and confirmed, it is used as a new parent strain in a new round of mutation and screening and/or selection (*see* **Fig. 1**).

2.1.1. Mutagenesis

The first key step of the classical approach is the generation of a mutant population. Alterations at the gene level in a microorganism are typically achieved by subjecting the organism to a mutagen. There are a variety of mutagens, which include X-ray, ultraviolet (UV) light, hydroxylamine, and *N*-methyl-*N'*-nitro-*N*-nitrosoguanidine *(7)*. Different mutagens have different mechanisms of action, including base transitions, base deletions, or base additions. It is essential at the

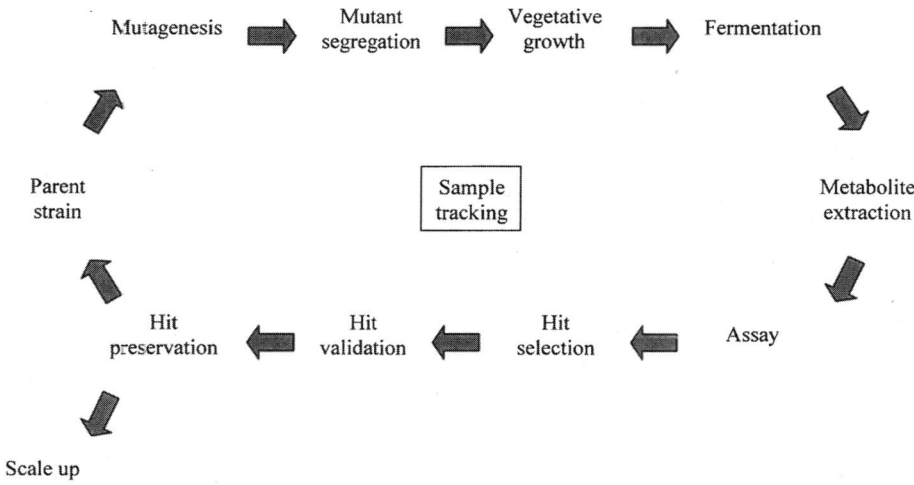

Fig. 1. Schematic representation of the classical strain improvement approach.

beginning of a strain improvement program that different mutagens are tested to identify the one that is most effective in generating the desired diversity in a mutation population. Sometimes, it might be necessary to switch to a different mutagen as strain improvement progresses. The key to the success in generating a diversified mutant population is to identify the ideal condition for mutation of the microbe *(9)*. "Undermutagenesis" leaves a large number of mutation survivors without genetic alteration and therefore results in screening a large number of strains identical to the parent. "Overmutagenesis," on the other hand, is more likely to produce a population in which survivors accumulate multiple mutations, significantly reducing the chance of finding desired mutants. Therefore, the aim of an ideal mutagenesis strategy is to identify the dose of the mutagen that maximizes the frequency of survivors that contain only one yield-affecting mutation.

2.1.2. Random Screening

Mutation survivors are segregated, followed by growth on either solid agar media or liquid media for the development of individual colonies. Single colonies are chosen randomly and fermented in small-scale fermentation vessels. One question concerning the classical approach is, how many colonies from a mutant population should be chosen for evaluation? In other words, how large a throughput is necessary to find improved mutants? Throughput typically is dictated by manpower, time, and equipment capacity. It has been demonstrated that throughput can be maximized by using a statistical approach to provide guidelines for efficient resource utilization *(3)*. When seeking mutants with increased productivity, it is more likely to find mutants with a small increment in productivity instead of a blockbuster high-producing mutant. However, in order to identify mutants with a small titer increment, the screening system must be sufficiently consistent and sensitive to reduce the chance of finding false mutants.

2.1.3. Quantification of Desired Products and Confirmation of Improved Strains

Once fermentation is complete, the metabolite is extracted from the fermentation culture and quantified. Mutants with increased metabolite productivity are considered potential hits and are re-evaluated in multiple replications to determine statistically if they are better than the parent strain. Mutants that demonstrate improvement during the confirmation process are used as parents for a new round of mutation and screening.

2.2. Assuring Successful Screening

The ability to identify improved strains using the classical approach depends primarily on the number of mutants screened and the variability of the screening system. Because desired mutations occur at a low frequency, the screening system must be designed to handle a large number of mutants. System variability dictates the ease of identifying potential improved strains and the number of replicates needed to confirm each improved strain *(6,10)*. In essence, the higher the system variability, the larger the throughput required to identify improved strains. Therefore, considerable effort is required to identify and eliminate factors contributing to variability.

2.3. New Tools and Applications

2.3.1. Automation

The rapid development of user-friendly robotics has led to the automation of single or multiple steps of high-throughput screening processes. The steps commonly subject to automation include (1) dispensing of media, (2) selection of individual colonies, (3) inoculation of individual colonies into vegetative media, (4) transfer of vegetative culture into screening media, (5) extraction of product, and (6) sample tracking. The advantage of automation is that it increases throughput and reduces the chance of human error and variability. In addition, automation facilitates rapid data capture that significantly improves the speed of downstream data processing. The success of automated programs requires skilled microbiologists, analytical specialists, and information specialists. In addition, constant monitoring and assessment are necessary to ensure that all aspects of the screening system are functioning properly. Obvious concerns regarding employment of an automated system are the high initial capital investment and the ongoing maintenance of equipment and software *(11)*. However, such investments are worthwhile if a long-term strain improvement program is projected and/or the possibility of using the same system for other products is predicted.

2.3.2. Miniaturization

Conventionally, small shake flasks have been the choice for screening improved strains because they are readily available in various sizes at low cost. One drawback of using flasks is that they usually require individual handling. Modules containing blocks of small bottles or culture tubes were widely accepted because of the ease of handling and sample tracking. However, these fermentation vessels require a working volume in the milliliter range. One way to make automated screening more cost-effective is to miniaturize fermentation vessels. In recent years, a variety of 96-well microtiter plates have become available to facilitate the cultivation of different microorganisms. The 96-well microtiter plates require a working volume only in the microliter range, which significantly reduces media volume for growth and metabolite production as well as buffer or solvent volume for metabolite extraction. The use of microtiter plates for screening tens of thousands or hundreds of thousands of individual mutants for improved strains has many advantages, including (1) higher throughput, (2) smaller footprint, (3) inexpensive vessel, and (4) lower operating cost. However, the use of microtiter plates for high-throughput fermentation is not problem-free. One of the concerns is the consistency of liquid dispensing. Because the overall volume is in the microliter range, small variations in liquid dispensing would significantly increase variability. Fortunately, robotic liquid-dispensing instruments now can handle small volumes reliably and consistently. Another concern with microtiter plates for fermentation is medium evaporation. Many fermentation screening processes require an incubation time of several days on a shaker. During this time, significant loss of volume can occur, especially when the starting volume is small. Humidity-controlled shakers capable of keeping evaporation to a minimum level and maintaining the same evaporation rate throughout the shaker have benefited many users.

2.3.3. Advancements in Analytical Capabilities

High-pressure liquid chromatography (HPLC) is most commonly used for metabolite quantification because of its sensitivity and specificity. The need for developing other analytical methods comes from the desire to eliminate solvent handling. One method that is gradually being adopted for analyzing fermentation broths *in situ* is near-infrared (NIR) spectroscopy. Drawbacks of the method are the lack of sensitivity as well as the difficulty of working with small volumes of fermentation broth. These weaknesses prevent the technology from being applied when metabolite titers are low and fermentation volumes are small. Another method uses mass spectroscopy. Goodacre et al. *(12)* demonstrated the possibility of combining pyrolysis mass spectroscopy (PyMS) with multivariate calibration and artificial neural networks (ANNs) for quantitative analysis of antibiotic titers in fermentation broths.

3. Selection of Improved Strains Using Enrichment Methods

Enrichment methods rely on the use of selection agents for identifying desired strains from an overwhelming number of unwanted mutants. The approach is powerful in that it increases screening efficiency by eliminating large numbers of undesirable mutants from the population, provided the logic behind the selection method is sound. Ideally, with the same or less throughput than needed for random screening, one can obtain improved strains with less labor. Enrichment methods do, however, have limitations. Some phenotypes might not be obtained using this approach and, sometimes, false-positive mutants are selected.

3.1. Rapid Detection Methods

The classical example of enrichment is the use of a "diagnostic" culture media for quick and semiquantitative measurement of antibiotic production. Usually, the survivors of a mutation treatment are grown on a solid agar medium followed by overlaying the agar-grown colonies with an indicator organism sensitive to the antibiotic. If a colony produces the antibiotic and secrete the antibiotic into the medium, there would be a clear zone, called a zone of inhibition, surrounding the colony, indicating the absence of growth of the indicator organism. In general, the more antibiotic secreted into the medium, the larger the inhibition zone surrounding the colony. This system is ideal for identifying high-producing strains when the overall antibiotic titer is still low but must be further modified to screen superior strains that produce large zones of inhibition. One drawback of this method is that the colonies producing the antibiotic under solid agar conditions might not produce the same compound when grown in liquid media. Similarly, colonies that do not produce the compound on solid media might be high-producing strains in liquid culture.

Another example of a classical enrichment method relies on the use of chromogenic agents. The agents, such as phenol red for an acid/base reaction or tetrazolium and methylene blue for oxidation/reduction reactions, are converted to a visible product by a specific biochemical reaction or change in the redox level in the media *(13)*. This leads to visual detection against the large background population of a specific strain having the desired biochemical activity.

3.2. Analog-Resistant Mutants

Microorganisms employ a variety of mechanisms to tightly control the synthesis of primary metabolites such as amino acids and nucleotides. Therefore, the metabolites are produced only in quantities sufficient for essential cellular activities. One of the mechanisms used to prevent unnecessary overproduction of essential metabolites is feedback regulation. There are two types of feedback regulation: One is feedback inhibition in which the end product of a biosynthetic pathway inhibits the action of an enzyme (usually the first) in the pathway; the second is feedback repression, which is inhibition of enzyme synthesis by the end product. Industrial microorganisms used for the production of primary metabolites such as amino acids are often developed to be less sensitive to feedback regulation by using analogs for selection. Such antimetabolites of amino acids and nucleotides often inhibit cell growth by interfering with the cell's machinery used for making those primary metabolites. Mutants deregulated in feedback control are resistant to the toxic effect of the analog and, typically, overproduce the metabolite to which the analog is antagonistic. Therefore, analog-resistant mutants overproduce the end product if (1) the toxicity is the result of the fact that the analog mimics the control properties of the end product and (2) the site of resistance of the resistant mutant is the site of control by the end product.

To isolate mutants deregulated in feedback control and, therefore, resistant to the analog, the minimum inhibitory concentration (MIC) of the analog must be determined *(14)*. This is usually achieved by plating the microorganism of interest on solid agar plates containing the toxic analog at various concentrations. The plates are incubated for several days at a temperature suitable for the organism before scoring the number of surviving colonies on each plate. The MIC is the lowest concentration of the analog that inhibits the growth of the microorganism. Once determined, the survivors of a mutagenesis treatment are plated on agar plates containing the analog at its MIC. Colonies that develop in the presence of the analog could be mutants that are deregulated in feedback control and, therefore, could overproduce the end product. The resistant colonies growing on analog plates must be screened again to confirm the resistance phenotype and ultimately confirmed for the production of the end product by fermentation. Mutants of *Bacillus, Streptomyces, Aspergillus,* and *Corynebacterium* overproducing amino acids, nucleotides, and vitamins have been successfully isolated using this approach *(15)*.

Use of analogs also has been adopted successfully for isolation of mutants overproducing secondary metabolites such as antibiotics. The logic behind the strategy is that precursors or intermediates of secondary metabolites are often the metabolites of primary metabolism whose synthesis is usually controlled by feedback regulation. Production of secondary metabolites could be improved by increasing the supply of the precursors or intermediates through the use of an analog. One of the first examples of the successful application of this approach is the isolation of pyrrolnitrin-overproducing *Pseudomonas fluorescens* strains through the use of tryptophan analogs *(16)*. In this case, tryptophan, whose synthesis is subject to feedback control, is a precursor of pyrrolnitrin. The analog-resistant mutants were resistant to feedback control by tryptophan and were pyrrolnitrin overproducers. Other examples include the use of a valine analog to isolate cephamycin-overproducing *Streptomyces lipmanii* strains *(17)* and a lysine analog to isolate β-lactam high producers *(18)*.

3.3. Auxotrophic Mutants

Pathways involved in the synthesis of many primary metabolites are interconnected, forming branched pathways. A typical example is the pathway of the aspartate-family amino acids, including lysine, methionine, and threonine. The lysine pathway constitutes one pathway branch that shares the common intermediate aspartyl semialdehyde with the second pathway branch involving methionine and threonine synthesis. Mutants blocked in one branch of the pathway often lead to accumulation of the end product of the other branch through redirection of the common intermediate. One example is the isolation of auxotrophic mutants of *Corynebacterium glutamicum* for lysine overproduction *(19,20)*. Because the end products of primary metabolism frequently are used as precursors for the synthesis of secondary metabolites, in theory one should be able to isolate auxotrophic mutants that overproduce the precursor and, ultimately, the secondary metabolite.

4. Genetic Engineering Approach to Creating Improved Strains

In addition to the above-described approaches, improved microbial strains now are routinely derived via genetic engineering. This approach offers several advantages, including (1) allowing specific control of beneficial genes as well as unwanted genes, (2) reducing the number of mutants screened, and (3) creating multiple beneficial mutations. Application of this approach, however, does require prior knowledge of the biochemical pathway. In addition, genetic tools must be available in order to manipulate the producing organism.

4.1. General Strategies

A number of strategies have proven successful in increasing the production of desired metabolites. They include (1) de-bottlenecking of rate-limiting steps, (2) eliminating feedback regulation, (3) manipulating regulatory gene(s), (4) perturbing central metabolism, (5) removing competing pathway(s), (6) reducing product toxicity, and (7) enhancing product transport.

4.1.1. Debottlenecking of Rate-Limiting Steps

Rate-limiting steps refer to the steps of a biosynthetic pathway that restrain the flow of intermediates and thereby limit the overall productivity of the final product. The classical way of identifying rate-limiting steps is by feeding pathway intermediates to the producing strain. If the intermediate is being imported into the cell but is not being converted to the final product, one or multiple steps between the intermediate and the final product are limiting. Once rate-limiting steps are identified, the genes encoding the limiting pathway enzymes are modified by either increasing gene dosage (amplification) or by placing the gene under the control of a stronger promoter. Ikeda et al. *(21)* successfully increased phenylalanine production in *C. glutamicum* by amplifying the genes encoding the rate-limiting enzymes in the phenylalanine pathway. Eggeling et al. *(22)* constructed a number of *C. glutamicum* mutant strains containing increased copy numbers of *dapA,* a gene encoding dihydrodipicolinate synthase at the branch point of the lysine and methionine/threonine pathway. Lysine production was higher in the mutant containing one extra copy of *dapA* than the wild-type strain and was highest in the mutant containing the highest copy number of the same gene. Kennedy and Turner *(23)* successfully increased penicillin production in *Aspergillus nidulans* by replacing the native promoter of δ-(L-α-aminoadipyl)-L-cysteinyl-D-valine (ACV) synthase gene with a strong inducible ethanol dehydrogenase promoter.

4.1.2. Eliminating Feedback Regulation

As discussed in the previous section, analogs are used routinely to relieve feedback regulation of many primary metabolites. Although this strategy is largely successful, it is recognized that subtle alteration of a task that an enzyme has evolved to perform is often difficult. Genetic engineering offers a promising alternative to overcome feedback regulation. An excellent example is isoleucine, an essential amino acid produced by *C. glutamicum* and used as a constituent of infusions and special dietary products. In this system, threonine dehydratase, the first committed enzyme in the isoleucine biosynthetic pathway, is sensitive to inhibition by the end product, isoleucine. Feedback inhibition is somewhat relieved through amplification of the native gene *(ilvA)* encoding threonine dehydratase *(24)*. However, when the *Escherichia coli* gene *(tcdB)*, encoding a threonine dehydratase insensitive to isoleucine feedback inhibition, was expressed in *C. glutamicum,* the isoleucine yield was significantly higher in this recombinant strain than the strain containing extra copies of the native *ilvA* gene.

4.1.3. Manipulating Regulatory Gene(s)

Regulatory genes encode proteins that control the expression of structural genes or other regulatory genes. There are two types of regulatory gene based on the effect they have on the genes they control. Positive regulators turn on the expression of the genes they control, whereas negative regulators repress the expression of the genes under their control. Regulatory genes control many biological processes such as the organism's developmental cycle and antibiotic production. Genes directly involved in antibiotic biosynthesis are often linked to form a gene cluster that usually contains pathway-specific regulatory gene(s). Such regulatory genes encode proteins that directly bind to the promoter region(s) of the biosynthetic gene(s) for that antibiotic. Such binding results in either an increase or decrease in the expression levels of the biosynthetic gene(s), which, in turn, either boost or hamper antibiotic production. Manipulation of pathway-specific regulatory gene(s) is relatively straightforward. Addition of extra copies of positive regulator(s) or disruption of negative regulator(s) often yields strains with increased production of the desired product. Some well-known examples of pathway-specific positive regulators include *actII–ORF4* of the actinorhodin pathway, *redD* of the undecylprodigiosin pathway, and *srmR* of the spiramycin pathway. Extra copies of the regulatory gene often cause overproduction of the corresponding antibiotic *(25)*. Some of the pathway-specific negative regulators in actinomycetes include *mmyR* of the methylenomycin pathway *(25)*, *tcmR* of the tetracenomycin pathway *(26)*, *jadR₂* of the jadomycin pathway *(27)*, and *dnrO* of the daunorubicin pathway *(28)*. Insertional inactivation or deletion of the genes resulted in increased production of the corresponding antibiotic.

Pathway-specific regulators are often controlled by higher-level regulators called global regulators, which coordinate many metabolic activities. Manipulation of global regulators also can be fruitful, although their identification and the understanding of their functions is time-consuming and labor-intensive. Tatarko and Romeo *(29)* engineered a high-phenylalanine-producing *E. coli* strain by disrupting the global regulatory gene, *csrA*. Disruption of this gene enhanced gluconeogenesis and decreased glycolysis, which, in turn, resulted in increased accumulation of phosphoenoylpyruvate (PEP), one of the two starting molecules needed for phenylalanine biosynthesis.

4.1.4. Perturbing Central Metabolism

Central metabolism provides primary metabolites and energy to support the survival of a microorganism. In addition, central metabolism contributes a small fraction of primary metabolites under special growth conditions to pathways that produce secondary metabolites such as antibiotics. In essence, central metabolism provides starting materials for the synthesis of a vast array of products of commercial importance. At some stages of strain improvement, it is necessary to tap into central metabolism for more of the starting materials when the pathway itself has reached maximum capacity. Perturbing central metabolism is complicated. Complications could arise from situations in which (1) disturbing the balance of metabolic activities is detrimental to the host or (2) the host is resistant to unnatural alterations imposed on it. Despite the difficulties and complications, manipulation of central metabolism has proven to be rewarding. Butler et al. *(30)* engineered a superior actinorhodin-producing strain by deleting the genes responsible for either of the first two steps in the pentose phosphate pathway. Peters-Wendisch et al. *(31)* constructed a lysine-overproducing strain by expressing at high levels the *pyc* gene encoding pyruvate carboxylase. Apparently, overexpression of the gene increased the availability of oxaloacetate, the precursor for the starting material (aspartate) of the lysine pathway. Increased availability of oxaloacetate, in turn, enhanced lysine production. Perturbing central metabolism might be better utilized for second-generation strain improvement in which the specific biosynthetic pathway for the desired product has already been optimized.

4.1.5. Blocking Competing Pathway(s)

Competing pathways refer to the pathways utilizing the precursor(s) and/or intermediate(s) of the desired pathway. Targeting competing pathways is generally considered after the biosynthetic pathway for the desired metabolite is optimized. Two approaches could be applied to remove competing pathways: (1) deletion of the entire competing pathway or (2) knocking out the function of the first gene of the competing pathway. Before applying the approaches, several factors must be taken into consideration. The most important factor is whether disruption of the competing pathway is lethal to the host. Many antibiotic producers often accumulate several compounds structurally related to the antibiotic of interest. Because these compounds do not usually possess biological activity, their synthesis is considered a waste of energy, precursors, and intermediates. In addition, the presence of these compounds could complicate downstream purification. In this case, knocking out the steps leading to the synthesis of the unwanted derivatives would simplify downstream processing and, more importantly, would redirect the precursors or intermediates toward the synthesis of the desired product. Backman et al. *(32)* engineered an elegant system in which tyrosine synthesis, competing for the common intermediate used for phenylalanine production, is interrupted during the phenylalanine production phase of the fermentation.

4.1.6. Overcoming Product Toxicity

Metabolites produced by microorganisms found in nature are usually made in small quantities. At such low levels, the producing organisms do not usually encounter the potential toxic effects of the metabolites. However, toxicity can become an issue, as the production level goes up and might hinder the efforts of isolating strains with enhanced productivity. The possible relationship between antibiotic resistance and productivity has been used to successfully develop improved strains by selecting mutants resistant to high levels of the antibiotic *(33)*.

4.1.7. Enhancing Product Transport

Increasing product transport across the cytoplasmic membrane is another important component of the strain improvement strategy. The presence of unusually high levels of any products in any organism is likely to alter the internal physiological environment and thereby prohibit further synthesis of the product because of feedback regulation. Most microorganisms producing metabolites of commercial interest have a means of transporting the products across the cell membrane. As biosynthetic capabilities become optimized, transport systems must be upgraded as well. Reinscheid et al. *(34)* showed that the intracellular concentration of threonine in *C. glutamicum* is proportional to the copy number of the deregulated threonine pathway genes while the extracellular concentration of the amino acid remained unchanged, indicating the importance of optimizing export machinery. Existence of an active transport system in *C. glutamicum* involved in lysine secretion recently was identified through cloning and characterization of the *lysE* gene *(35)*. The *lysE* knockout mutant lost the ability to secrete lysine, whereas the *lysE*-overexpressed mutant exported the amino acid at an accelerated rate.

4.2. Gene Transfer Methods

The most frequently used method to introduce DNA into microorganisms is by transformation. Transformation requires uptake of plasmid DNA by recipients in a physiological stage of competence, which usually occurs at a specific growth stage. However, DNA uptake based on naturally occurring competence is usually inefficient. Competence can be induced by treating bacterial cells with chemicals to facilitate DNA uptake. For *Escherichia,* the uptake of plasmid DNA is achieved when cells are first treated with calcium chloride or rubidium chloride *(36)*. Transformation of plasmid DNA into *Streptomyces,* the genus that produces the majority of the antibiotics, is more complicated than *Escherichia.* The process involves preparation of protoplasts by treating mycelia with lysozyme to remove most of the cell wall *(37)*. Protoplasts are mixed with plasmid DNA in the presence of polyethylene glycol (PEG) to promote the uptake of DNA. In some *Streptomyces* species, strong restriction barriers prohibit successful protoplasts transformation when double-stranded plasmid DNA is used. Sometimes, this problem can be overcome by the use of single-stranded plasmid DNA *(38)*.

Another method of introducing plasmid DNA into microorganisms is by conjugation. This method involves a donor strain that contains both the gene(s) of interest and the origin of transfer *(oriT)* on a plasmid and the genes encoding transfer functions on the chromosome. Upon brief contact between donor and recipient, DNA transfer occurs. After conjugation takes place, donor cells are eliminated with an antibiotic to which the recipient cells are resistant. Recipient cells containing the transferred plasmid are selected based on the antibiotic resistance marker carried by the plasmid. Successful intergeneric conjugation between *Escherichia* and *Streptomyces* was first demonstrated by Mazodier et al. *(39)*. Since then, the method has proven to be useful with a number of *Streptomyces* species and with various strains of other actinomycetes *(40–43)*. One advantage of this method is that it does not rely on the development of procedures for protoplast formation and regeneration of cell wall. In addition, this method offers the possibility of bypassing restriction barriers by transferring single-stranded DNA *(42)*.

Electroporation is an alternative method to the use of PEG-mediated DNA uptake. Recipient cells are made electrocompetent by subjecting the cells to ultrasound pulses in the presence of plasmid DNA. Transient pores are formed in the cell membrane as a result of the electroshock, thereby allowing DNA uptake. This method initially was shown to be highly effective in mediating DNA uptake by various plants and animal cells. Since then, there have been several reports documenting the application of this method to industrially important *Streptomyces* *(44,45)*.

4.3. Issues to Consider When Creating Genetically Modified Strains

4.3.1. Plasmid Instability

Plasmid instability is one of the concerns when constructing genetically modified strains for increased metabolite production. In general, the larger the plasmid, the less stable is the recombinant strain *(46)*. Other factors such as media composition and culture conditions also can contribute to plasmid instability. Loss of plasmid DNA containing the essential genetic information will lead to the loss of desired phenotypes. One way to ensure the stable presence of plasmid DNA is to maintain the recombinant strain under selective pressure. This can be achieved by including an antibiotic in the culture medium. This method is effective but often impractical, especially when the recombinant strain is to be grown in large fermenters for commercial production purposes. An alternative is the integration of the gene(s) of interest into the host's chromosome.

4.3.2. Foreign DNA

Foreign DNA refers to DNA originated from organisms other than the host organism itself. Concerns over the presence of foreign DNA in the final production strain include the potential impact of the DNA on the performance of the fermentation, the quality of the product, and the impact of accidental release of the recombinant strain into the environment. The most common source of foreign DNA is vector DNA. In this case, vector DNA can be eliminated by a double-crossover recombination event integrating only gene(s) of interest into the host chromosome, leaving vector DNA behind and eventually lost from the host organism. Another source of foreign DNA is the gene(s) of interest carried by vector DNA. Sometimes, a foreign gene is needed in order to introduce specific beneficial traits.

4.3.3. Antibiotic-Resistance Markers

Antibiotic-resistance markers are present in many vectors. Often, they are used to ensure that only the cells containing the vectors are selected and that the cultures used to synthesize the products of interest are pure. However, the DNA that confers resistance is foreign DNA. As discussed earlier, using an antibiotic-resistance marker to ensure a pure culture for production requires addition of large quantities of the antibiotic to large production fermenters, which usually is not desirable.

4.4. New Technologies

4.4.1. Genomic Approaches to Strain Improvement

Rapid advancement of sequencing technology makes it possible to unlock the genetic code of an organism of interest in a short period of time. The genetic code of the organism is assembled and putative open reading frames (ORFs) are determined using sophisticated computer programs. The function of each ORF is predicted based on sequence similarity to known proteins in the database. This wealth of information gives clues to the genes directly involved in the synthesis of the valuable product. The information also provides insight into the metabolic potential of the organism, such as pathways supplying precursors or cofactors required for the production of the metabolite of interest, pathways competing for valuable precursors or intermediates, and global regulatory networks that might control metabolite production. Genomic information is beginning to transform strain development efforts. Recently, Ohnishi et al. *(47)* reported the genetic engineering of a hyper-lysine-producing strain of *C. glutamicum.* The group identified the mutations carried by the lysine high-producing strain derived from the classical approach through comparative genomic analysis of the *C. glutamicum* wild-type strain and the lysine-producing strain. Introducing those mutations into the wild-type strain produced a recombinant strain superior compared to the classically derived lysine-producing strain both in lysine productivity and in growth.

Metabolic potential of an organism now can be investigated at a global scale using new technologies developed and optimized in recent years to examine gene expression, protein synthesis, and metabolite flux. The knowledge gained from these studies offers a window to "see" what is going on inside the organism. The knowledge can lead to the prediction of gene targets controlling metabolite production.

DNA microarray, a technology developed in the mid-1990s, is very powerful in that it allows simultaneous quantification of the expression levels of all genes in a given organism *(48–50)*. Technical details of the technology include array preparation by spotting DNA fragments representing genes of the organism, RNA purification and labeling, and hybridization. The technology can be used for strain comparison, identifying genes responding to nutritional or environmental stimulation, and detecting genes that are important for metabolite production.

Advanced technology has made it possible to simultaneously analyze all proteins of a given organism *(51–53)*. Like gene expression analysis, protein synthesis can be used to compare high- and low-producing strains. One advantage of protein analysis over gene expression analysis is that protein levels might be more indicative of the organism's physiology during fermentation.

In addition to gene expression and protein analysis, metabolic flux analysis is another way of investigating the physiology of a microorganism *(54,55)*. Integration of this approach will provide both a blueprint of the complex cellular metabolism of a microorganism and guidance toward rapid engineering of improved strains. Askenazi et al. *(56)* recently reported identification of key genetic components and physiological traits associated with the production of lovastatin and (+)-geodin in *Aspergillus terreus* by integrating gene expression analysis and metabolite-profiling methods.

4.4.2. DNA Shuffling

DNA shuffling, a technology introduced in 1994, is based on error-prone polymerase chain reaction (PCR) and random recombination of DNA fragments *(57)*. DNA shuffling could involve a single gene or multiple genes of the same family. Family gene shuffling is more powerful than shuffling of single genes because it takes advantage of the natural diversity that already exists within homologous genes *(58,59)*. DNA shuffling has been successfully applied for the improvement of numerous commercially important enzymes and proteins *(60)*. Genome shuffling, a recent breakthrough in shuffling technology, is a novel approach that combines the beneficial traits of multiple parental lines with the recombination of entire genomes *(61)*. Zhang et al. *(61)* demonstrated the utility of this approach in strain improvement through rapid generation of *Streptomyces fradiae* strains with increased tylosin titers.

5. Process Optimization and Scale-Up

5.1. Media Optimization

In most cases, a fermentation medium that supports the production of the metabolite of interest is first developed based on information available in the literature, the researchers' past experiences, and trial and error. In industrial settings, this medium becomes the starting point for further optimization. The primary drive here is to increase product titer and to improve the economy of a fermentation process. A secondary drive includes the development of a simplified metabolite-purification process.

The basic nutrients for the growth of microorganisms include carbon, nitrogen, and minerals. Depending on the organism and the fermenter conditions used for growth and metabolite production, additional nutrients such as amino acids, vitamins, nucleotides, or even specialty chemicals might be required. There is an array of carbon and nitrogen sources available from which to choose. Some are in relatively pure form, whereas others occur in complex forms such as the byproducts of the food and agricultural industries. Pure or high-quality raw materials, although expensive, are more consistent from batch to batch and vendor to vendor. The byproducts of the food and agricultural industries are often less consistent in quality. Pure or higher-quality carbon and nitrogen sources are generally used for high-value products such as therapeutic proteins, whereas the less expensive raw materials are used for the production of low-cost and high-volume commodity items such as organic acids or bulk chemicals. When developing fermentation media, one must consider cost, availability, and lot consistency. In addition, fermentation media that facilitate downstream processing have an added advantage.

Once a fermentation medium is developed, the next phase is to optimize the media composition to achieve the best possible benefit. This task can be tedious and labor-intensive. The traditional method of media optimization is to alter one ingredient at a time until its optimum concentration is identified while the remaining ingredients are held constant. Recently, new techniques such as statistical experimental design, genetic algorithms (GAs), and particle swam optimization (PSO) have become available to assist fermentation scientists. These tools and techniques have been extensively reviewed by Strobel and Sullivan *(62)*, Parekh et al. *(7)*, and Conner *(8)*.

5.2. Fermentation Process Scale-Up

The performance of improved strains selected from a screening process using small-scale fermentation vessels must be validated in pilot-scale fermenters. Once validated, the improved strains will be used for commercial production. The process of transferring a laboratory fermentation process to an industrial operation is referred to as scale-up. The goal of scale-up is to recreate in large production fermenters the optimal conditions in which the improved strains are cultivated in the laboratory. This is accomplished by controlling environmental conditions, including power input, mixing ability, oxygen transfer, shear stress, heat transfer, media sterilization, and seed culture preparation.

Historically, scale-up of improved strain candidates has centered on maintaining the same physical environment for the growing cells as in small fermentation vessels. This goal, however, is difficult to achieve because of the geometry of large-scale fermenters having low surface-to-volume ratios compared to model fermentation vessels *(7)*. Nevertheless, scale-up might be successful by maintaining (1) constant impeller tip speed to achieve equal shear stress, (2) constant agitation power per unit volume of broth, (3) constant mixing time, and (4) constant oxygen mass transfer *(7)*. Shear stress is an important issue when cultivating filamentous microorganisms such as the antibiotic-producing actinomycetes and fungi in large fermenters. Filamentous microorganisms are more likely to suffer damage from shear stress that, in turn, results in decreased product yield. One way to reduce shear stress is to culture those microorganisms in fermenters equipped with large-diameter impellers or multiple impellers while maintaining adequate dissolved oxygen levels at low agitation speed. In small-scale fermentation vessels, rapid mixing of culture broth creates a homogeneous environment in which nutrient concentration, dissolved oxygen, and shear are identical throughout the vessel. In large fermenters, however, a homogeneous environment is difficult to attain, especially when dealing with viscous cultures of filamentous microorganisms. This issue can be addressed by isolating morphological mutants with shorter mycelial fragments.

In a production facility, sterilization of media takes place in the fermenters and usually takes more time because of the large volume compared to the time used to sterilize the same media for small-scale fermentation vessels. The prolonged sterilization time can cause media to caramelize and/or media components to degrade. When that happens, growth and product yield is often significantly impacted.

Microorganisms are usually grown to reach a certain cell mass and physiological status, as a "seed culture" preparation, before being transferred to production media. Seed culture preparation often involves a single seed stage when small fermentation vessels are used for metabolite production. However, when cultivating in large fermenters, multiple seed stages are often necessary to generate enough cell mass. Cell mass is only one of the parameters indicating whether a seed culture is ready for production. Additional parameters include the physiological status of the culture and the pH of the seed culture.

6. Conclusion

Microbial strain improvement is essential for the commercialization of beneficial compounds such as antibiotics and amino acids. The classical approach and enrichment methods were once the only methods available for developing improved strains. With the advancement in recombinant DNA technology and the development of new technologies to facilitate studying microbial physiology, improved strains are being developed using the genetic engineering approach. Integration of various approaches will offer the best chance for rapid development of improved strains. Optimization of fermentation processes continues to play an essential role in maximizing the potential of each improved strain.

Acknowledgment

The authors wish to thank Professor Arnold Demain for critical review of the manuscript.

References

1. Parekh, S. (1999) Strain improvement, in *Encyclopedia of Microbiology* (Lederberg, J. S., ed.), Academic, San Diego, CA, Vol. 4, pp. 6170–6197.
2. Doran, P. (1995) Reactor engineering, in *Bioprocess Engineering Principles* (Doran, P., ed.), Academic, London, pp. 333–391.
3. Stanbury, P. F., Whitaker, A., and Hall, S. J. (1995) Fermentation economics, in *Principles of Fermentation Technology* (Stanbury, P. F., Whitaker, A., and Hall, S., eds.), Butterworth–Heinemann, Oxford, pp. 331–349.
4. Queener, S. W. and Lively, D. H. (1986) Screening and selection for strain improvement, in *Manual of Industrial Microbiology and Biotechnology* (Demain, A. L. and Solomon, N. A., eds.), ASM, Washington, DC, pp. 155–169.
5. Elander, R. P. (2002) University of Wisconsin contributions to the early development of penicillin and cephalosporin antibiotics. *SIM News* **52,** 270–278.
6. Vinci, V. A. and Byng, G. (1999) Strain improvement by non-recombinant methods, in *Manual of Industrial Microbiology and Biotechnology* (Demain, A. L. and Davis, J. E., eds.), ASM, Washington, DC, pp. 103–113.
7. Parekh, S., Vinci, V. A., and Strobel, R. J. (2000) Improvement of microbial strains and fermentation processes. *Appl. Microbiol. Biotechnol.* **54,** 287–301.
8. Conner, N. (2003) Culture medium optimization and scale-up for microbial fermentations, in *Handbook of Industrial Cell Culture: Mammalian, Microbial and Plant Cells* (Vinci, V. A. and Parekh, S., eds.), Humana, Totowa, NJ, pp. 171–193.
9. Baltz, R. H. (1986) Mutagenesis in *Streptomyces* spp., in *Manual of Industrial Microbiology and Biotechnology* (Demain, A. L. and Solomon, N. A., eds.), ASM, Washington, DC, pp. 184–190.
10. Rowlands, R. T. (1984) Industrial strain improvement: mutagenesis and random screening procedures. *Enzyme Microbiol. Technol.* **6,** 3–10.

11. Nolan, R. (1986) Automation system in strain improvement, in *Overproduction of Microbial Metabolites: Strain Improvement and Process Control Strategies* (Vanek, Z. and Hostalek, Z., eds.), Butterworth, Boston, pp. 215–230.
12. Goodacre, R., Trew, S., Wrigley-Jones, C., et al. (1994) Rapid screening for metabolite overproduction in fermentor broths using pyrolysis mass spectrometry with multivariate calibration and artificial neural networks. *Biotechnol. Bioeng.* **44,** 1205–1216.
13. Elander, R. P. and Vournakis, J. N. (1986) Genetic aspects of overproduction of antibiotics and other secondary metabolites, in *Overproduction of Microbial Metabolites: Strain Improvement and Process Control Strategies* (Vanek, Z. and Hostalek, Z., eds.), Butterworth, Boston, pp. 63–80.
14. Sermonti, G. (ed.) (1969) *Genetics of Antibiotic Producing Organisms,* Wiley–Interscience, London.
15. Wang, D. I. C., Cooney, C. L., Demain, A. L., Dunhill, P., Humphrey, A. E., and Lilly, M. D. (eds.) (1979) Biosynthesis of primary metabolites, in *Fermentation and Enzyme Technology,* Wiley and Sons, New York, pp. 14–25.
16. Elander, R. P., Mabe, J. A., Hamill, R. L., and Gorman, M. (1971) Biosynthesis of pyrrolnitrins by analogue-resistant mutants of *Pseudomonas fluorescens. Folia Microbiol.* **16,** 156–165.
17. Godfrey, O. W. (1973) Isolation of regulatory mutants of the aspartic and pyruvic acid families and their effect on antibiotic production in *Streptomyces lipmanii. Antimicrob. Agents Chemother.* **4,** 73–79.
18. Elander, R. P. (1989) Bioprocess technology in industrial fungi, in *Fermentation Process Development of Industrial Organisms* (Neway, J. O , ed.), Marcel Dekker, New York, pp. 169–220.
19. Nakayama, K., Kitada, S., Sato, Z., and Kinoshita, K. (1961) Induction of nutritional mutants of glutamic acid bacteria and their amino acid accumulation. *J. Gen. Appl. Microbiol.* **7,** 41–51.
20. Kinoshita, S. and Nakayama, K. (1978) Amino acids, in *Economic Microbiology: Primary Products of Metabolism* (Rose, A. H., ed.), Academic, New York, pp. 210–262.
21. Ikeda, M., Ozaki, A., and Katsumata, R. (1993) Phenylalanine production by metabolically engineered *Corynebacterium glutamicum* with the *pheA* gene of *Escherichia coli. Appl. Microbiol. Biotechnol.* **39,** 318–323.
22. Eggeling, L., Oberle, S., and Sahm, H. (1998) Improved L-lysine yield with *Corynebacterium glutamicum:* use of *dapA* resulting in increased flux combined with growth limitation. *Appl. Microbiol. Biotechnol.* **49,** 24–30.
23. Kennedy, J. and Turner, G. (1996) Delta-(L-alpha-aminoadipyl)-L-cysteinyl-D-valine synthetase is a rate limiting enzyme for penicillin production in *Aspergillus nidulans. Mol. Gen. Genet.* **253,** 189–197.
24. Guillouet, S., Rodal, A. A., An, G., Lessard, P. A., and Sinskey, A. J. (1999) Expression of the *Escherichia coli* catabolic threonine dehydratase in *Corynebacterium glutamicum* and its effect on isoleucine production. *Appl. Environ. Microbiol.* **65,** 3100–3107.

25. Champness, W. (1999) Cloning and analysis of regulatory genes involved in *Streptomyces* secondary metabolite biosynthesis, in *Manual of Industrial Microbiology and Biotechnology* (Demain, A. L. and Davies, J. E., eds.), ASM, Washington, DC, pp. 725–739.

26. Guilfoile, P. G. and Hutchinson, C. R. (1992) Sequence and transcriptional analysis of the *Streptomyces glaucescens tcmAR* tetracenomycin C resistance and repressor gene loci. *J. Bacteriol.* **174,** 3651–3658.

27. Yang, K., Han, L., and Vining, L. C. (1995) Regulation of jadomycin B production in *Streptomyces venezuelae* ISP5230: involvement of a repressor gene, *jadR₂*. *J. Bacteriol.* **177,** 6111–6117.

28. Otten, S. L., Ferguson, J., and Hutchinson, C. R. (1995) Regulation of daunorubicin production in *Streptomyces peucetius* by the *dnrR₂* locus. *J. Bacteriol.* **177,** 1216–1224.

29. Tatarko, M. and Romeo, T. (2001) Disruption of a global regulatory gene to enhance central carbon flux into phenylalanine biosynthesis in *Escherichia coli*. *Curr. Microbiol.* **43,** 26–32.

30. Butler, M. J., Bruheim, P., Jovetic, S., Marinelli, F., Postma, P. W., and Bibb, M. J. (2002) Engineering of primary carbon metabolism for improved antibiotic production in *Streptomyces lividans. Appl. Environ. Microbiol.* **68,** 4731–4739.

31. Peters-Wendisch, P. G., Schiel, B., Wendisch, V. F., et al. (2001) Pyruvate carboxylase is a major bottleneck for glutamate and lysine production by *Corynebacterium glutamicum. J. Mol. Microbiol. Biotechnol.* **3,** 295–300.

32. Backman, K., O'Connor, M. J., Maruya, A., et al. (1990) Genetic engineering of metabolic pathways applied to the production of phenylalanine. *Ann. NY Acad. Sci.* **589,** 16–24.

33. Woodruff, H. B. (1966) The physiology of antibiotic production: role of producing organisms. *Symp. Soc. Gen. Microbiol.* **16,** 22–46.

34. Reinscheid, D. J., Kronemeyer, W., Eggeling, L. Eikmanns, B. J., and Sahm, H. (1994) Stable expression of *hom-1-thrB* in *Corynebacterium glutamicum* and its effect on the carbon flux to threonine and related amino acids. *Appl. Environ. Microbiol.* **60,** 126–132.

35. Vrljic, M., Sahm, H., and Eggeling, L. (1996) A new type of transporter with a new type of cellular function: L-lysine export from *Corynebacterium glutamicum. Mol. Microbiol.* **22,** 815–826.

36. Sambrook, J., Fritsch, E. F., and Maniatis, T. (eds.) (1989) *Molecular Cloning: A Laboratory Manual,* Cold Spring Harbor Laboratory, Cold Spring Harbor, NY.

37. Kieser, T., Bibb, M. J., Buttner, M. J., Chater, K. F., and Hopwood, D. A. (eds.) (2000) *Practical Streptomyces Genetics,* The John Innes Foundation, Norwich, UK.

38. Oh, S. H. and Chater, K. F. (1997) Denaturation of circular or linear DNA facilitates targeted integrative transformation of *Streptomyces coelicolor* A3(2): possible relevance to other organisms. *J. Bacteriol.* **179,** 122–127.

39. Mazodier, P., Petter, R., and Thompson, C. J. (1989) Intergeneric conjugation between *Escherichia coli* and *Streptomyces* species. *J. Bacteriol.* **171,** 3583–3585.

40. Bierman, M., Logan, R., O'Brien, K., Seno, E. T., Rao, R. N., and Schoner, B. E. (1992) Plasmid cloning vectors for the conjugal transfer of DNA from *Escherichia coli* to *Streptomyces* spp. *Gene* **116,** 43–49.

41. Sun, Y., Zhou, X., Liu, J., et al. (2002) *"Streptomyces nanchangensis,"* a producer of the insecticidal polyethe antibiotic nanchangmycin and the antiparasitic macrolide meilingmycin, contains multiple polyketide gene clusters. *Microbiology* **148,** 361–371.

42. Matsushima, P., Broughton, M. C., Turner, J. R., and Baltz, R. H. (1994) Conjugal transfer of cosmid DNA from *Escherichia coli to Saccharopolyspora spinosa:* effects of chromosomal insertions on macrolide A83543 production. *Gene* **146,** 39–45.

43. Paranthaman, S. and Dharmalingam, K. (2003) Intergeneric conjugation in *Streptomyces peucetius* and *Streptomyces* sp. strain C5: chromosomal integration and expression of recombinant plasmids carrying the *chiC* gene. *J. Bacteriol.* **69,** 84–91.

44. Pigac, J. and Schrempf, H. (1995) A simple and rapid method of transformation of *Streptomyces rimosus* R6 and other streptomycetes by electroporation. *Appl. Environ. Microbiol.* **61,** 352–356.

45. Tyurin, M., Starodubtseva, L., Kudryavtseva, H., Voeykova, T., and Livshits, V. (1995) Electrotransformation of germinating spores of *Streptomyces* spp. *Biotech. Techniques* **9,** 737–740.

46. Leer, R. J., van Luijk, N., Posno, M., and Pouwels, P. H. (1992) Structural and functional analysis of two cryptic plasmids from *Lactobacillus pentosus* MD353 and *Lactobacillus plantarum* ATCC 8014. *Mol. Gen. Genet.* **234,** 265–274.

47. Ohnishi, J., Mitsuhashi, S., Hayashi, M., et al. (2002) A novel methodology employing *Corynebacterium glutamicum* genome information to generate a new L-lysine-producing mutant. *Appl. Microbiol. Biotechnol.* **58,** 217–223.

48. Lashkari, D. A., DeRisi, J. L., McCusker, J. H., et al. (1997) Yeast microarrays for genome wide parallel genetic and gene expression analysis. *Proc. Natl. Acad. Sci. USA* **94,** 13,057–13,062.

49. Tao, H., Bausch, C., Richmond, C., Blattner, F. R., and Conway, T. (1999) Functional genomics: expression analysis of *Escherichia coli* growing on minimal and rich media. *J. Bacteriol.* **181,** 6425–6440.

50. Price, C. W., Fawcett, P., Ceremonie, H., Su, N., Murphy, C. K., and Youngman, P. (2001) Genome-wide analysis of the general stress response in *Bacillus subtilis*. *Mol. Microbiol.* **41,** 757–774.

51. Pandey, A. and Mann, M. (2000) Proteomics to study genes and genomes. *Nature* **405,** 837–846.

52. VanBogelen, R. A., Schiller, E. E., Thomas, J. D., and Neidhardt, F. C. (1999) Diagnosis of cellular states of microbial organisms using proteomics. *Electrophoresis* **20,** 2149–2159.

53. Vohradsky, J., Li, X. M., and Thompson, C. J. (1997) Identification of procaryotic developmental stages by statistical analyses of two-dimensional gel patterns. *Electrophoresis* **18,** 1418–1428.

54. Schilling, C. H., Edwards, J. S., and Palsson, B. O. (1999) Toward metabolic phenomics: analysis of genomic data using flux balances. *Biotechnol. Prog.* **15,** 288–295.

55. Schilling, C. H., Edwards, J. S., Letscher, D., and Palsson, B. O. (2000–2001) Combining pathway analysis with flux balance analysis for the comprehensive study of metabolic systems. *Biotechnol. Bioeng.* **71,** 286–306.

56. Askenazi, M., Driggers, E. M., Holtzman, D. A., et al. (2003) Integrating transcriptional and metabolite profiles to direct the engineering of lovastatin-producing fungal strains. *Nat. Biotechnol.* **21,** 150–156.

57. Stemmer, W. P. (1994) Rapid evolution of a protein *in vitro* by DNA shuffling. *Nature* **370,** 389–391.

58. Crameri, A., Raillard, S. A., Bermudez, E., and Stemmer, W. P. (1998) DNA shuffling of a family of genes from diverse species accelerates directed evolution. *Nature* **391,** 288–291.

59. Ness, J. E., Welch, M., Giver, L., et al. (1999) DNA shuffling of subgenomic sequences of subtilisin. *Nat. Biotechnol.* **17,** 893–896.

60. del Cardayre, S. and Powell, K. (2003) DNA shuffling for whole cell engineering, in *Handbook of Industrial Cell Culture: Mammalian, Microbial, and Plant Cells* (Vinci, V. A. and Parekh, S., eds.), Humana, Totowa, NJ, pp. 465–481.

61. Zhang, Y. X., Perry, K., Vinci, V. A., Powell, K., Stemmer, W. P., and del Cardayre, S. B. (2002) Genome shuffling leads to rapid phenotypic improvement in bacteria. *Nature* **415,** 644–646.

62. Strobel, R. J. and Sullivan, G. R. (1999) Experimental design for improvement of fermentations, in *Manual of Industrial Microbiology and Biotechnology* (Demain, A. L. and Davies, J. E., eds.), ASM, Washington, DC, pp. 80–93.

2

Experimental Design in Microbiology

Guillaume E. Vanot and Michelle Sergent

Summary

The field of predictive microbiology is rapidly widening and ranges from bioreactor engineering to the modeling of foodstuff degradation and contamination. In this chapter, we point out how important the choice of experimental conditions is. We also briefly describe some tools (methodology of experimental research) that can be used to design optimal experimental strategies with respect to the study's aim. Emphasis is given on the optimization of microbial metabolite production, with an example showing the screening of factors and another illustrating the use of the response surface methodology.

Key Words: Predictive microbiology; methodology; experimental research; optimal design; response surface methodology; screening of factors, experimentation plan.

1. Introduction

Experiments are essential to scientific exploration and are still highly favored by researchers. Any work relying on a large number of them is normally considered of good standard. The methods of study and analysis of the physical, chemical, and biological phenomena have recently been rapidly improving, thanks to increasingly refined techniques and equipment (often linked to a computer in charge of part of the data analysis) and the development of mathematical and statistical methods for the analysis and processing of numerical data (factorial analysis, pattern recognition, classification, etc.).

By increasing the number of sensors, recorders, analyzers, and so forth, it is now possible to have an almost unlimited quantity of information concerning the study of a phenomenon. It is understandable that the researcher wants to gather as much "material" as possible during his experimentation even if this means delaying (sometimes forever) the analysis of these data. However, for some time, the tendency has been to reduce the number of experiments mainly because of their cost. Nevertheless, a few researchers point out that a plethora of results

From: *Methods in Biotechnology, Vol. 18: Microbial Processes and Products*
Edited by: J. L. Barredo © Humana Press Inc., Totowa, NJ

does not guarantee valuable scientific information. Significant improvement has taken place in the field of methodology of experimental research, but, to date, it has not succeeded in convincing a wide audience of experimental scientists. Two important facts are often overlooked. First, as powerful as the hardware and software might be, one cannot extract more information than contained in the experimental data. Second, in spite of the increase in the number of measurements, it is not rare that essential information is lacking.

It is difficult to persuade researchers that the result of an experiment contains no information and that all information is contained in the chosen experimental conditions. For example, with an infinite number of experiments carried out at two precise points, it is only possible to study a linear first-order model. The information quality does not depend on the number of experiments. The planning of the experiments is of the highest importance, although often neglected. In most cases, a classical approach is needed: The researcher makes assumptions from which he then deduces consequences. If the necessary information required to check these assumptions is not available, he must undertake the necessary experiments. The aim of the "methodology of experimental research" (*see* **Note 1**) is to control, describe, foresee, or explain the phenomenon under study.

In **Subheading 3.,** we give a brief overview of the ERM (Experimental Research Methodology) and illustrate how it can be used with the example of microbial metabolite production optimization.

2. Materials

NEMRODW software (Mathieu D., Nony J. and Phan Tan Luu R., New Efficient Methodology for Research Using Optimal Design, Windows version; LPRAI, Marseille, France) has been used to study the given examples and create the corresponding input tables and output figures.

3. Methods

3.1. Methodology of Experimental Research

We propose a methodological approach that emphasizes the importance of planning the experiments rather than running the experiments. Planning is necessary not only during the initial stages but also throughout all the research process. New information obtained from optimally designed experiments must be used to redefine the experimental strategy if required. The different steps of this procedure are as follows:

1. Clear definition of the problem being studied: proposed targets, consequences of a wrong decision, budget (in time, cost, means, etc.).
2. Compilation of the current local and bibliographical knowledge. If some necessary information is not available, experiments must be undertaken. Therefore, a complete and precise list of the factors likely to be influential, the responses, and the con-

straints must be established. The area of the experimental domain in which the missing information is to be sought has to be defined. It is referred to as the experimental domain of interest.

3. Setting up an experimental strategy (or experimental design) (i.e., to choose the experiments to be carried out according to the defined targets, the means available, and the desired information). The researcher seeks a relationship of cause and effect between some parameters of the phenomenon (called factors) (*see* **Note 2**), which are supposed to influence the behavior of the phenomenon and other parameters (called responses) (*see* **Note 3**) that define the result of the phenomenon. The planning of experiments consists in forcing the factors (input) to vary in a precise way, measuring the induced variations of the answers (output) and then deducing the relationships between causes and consequences.

4. Carrying out the experiments that will give us the values of the studied responses.

5. Deduction of the answers to the questions either directly or with the help of a mathematical model.

There are many different types of study, depending on the proposed targets. This implies different experimental strategies. We will give an example of two of them with the adequate experimental design.

3.2. Exploratory Research

At the beginning of a new study, the researcher usually does not know much about the phenomenon. He sometimes even ignores if he will be able to reproduce it. Therefore, he undertakes preliminary experiments to ensure that he has control over the phenomenon, to choose the favorable experimental fields and responses, and to check the reproducibility. These experiments are usually done without planning. However, there are simple methods to perform the exploratory research in a more organized way with grids, saturated designs of experiments, random research, and so forth.

3.3. Screening of Factors

This step is often done after the exploratory research. The researcher quickly picks the factors potentially influential in the chosen experimental field. Because of the belief that an increasing number of factors increases the number of experiments exponentially, a high number of factors is often reduced to a number that seldom exceeds three or four. This reduction is artificial and relies only on laboratory practices and the researcher's feeling (i.e., they retain the factors they like and reject those they do not). For a more scientific screening, there are methodological techniques with a low number of experiments. If the number of factors is very high (50–10,000), group screening or sequential bifurcation or supersaturated designs can be used. If it is lower, symmetrical and asymmetrical fractional factorial designs can be used. In any case, each factor is weighed, which allows choosing the most important ones (and not the preferred ones) for a later

Table 1
Factors and Experimental Domain

	Factors	Unit	Level (–)	Level (+)
X_1	Glucose concentration	g/L	1	5
X_2	Initial pH	—	5	7
X_3	Inoculum size	Spores/mL	10^6	10^8
X_4	Agitation rate	rpm	100	200
X_5	Inducer added	—	Yes	No
X_6	Nitrogen source	—	Corn steep	Peptone
X_7	Temperature	°C	20	30

Table 2
Hadamard Matrix

X_1	X_2	X_3	X_4	X_5	X_6	X_7
+	+	+	–	+	–	–
–	+	+	+	–	+	–
–	–	+	+	+	–	+
+	–	–	+	+	+	–
–	+	–	–	+	+	+
+	–	+	–	–	+	+
+	+	–	+	–	–	+
–	–	–	–	–	–	–

and more precise analysis. This approach is very effective but not often used mainly because researchers are afraid of working with more than three factors.

Screening designs can be written $s_1{}^{k_1}s_2{}^{k_2} \cdots s_1{}^{k_i}//N$, where s_i is the number of levels (i.e., values) of k_i factors and N is the minimal number of experiments required (*see* **Note 4**). If several factors have a different number of levels, the design is asymmetrical, whereas if k factors have the same number s of levels (i.e., $s^k//N$), it is symmetrical. Optimal experimental designs such as Addelman's (*1*) or Hadamard's also known as Plackett and Bluman (*2*) can be used. A Hadamard design is a symmetrical design for two-level factors, and for the design to be optimal, it requires that N be a multiple of 4.

This design was used in a build-up example to screen seven factors supposed to have an influence on the lipase production of a *Penicillium* sp. (*see* **Table 1**). The response studied is the enzymatic activity of the culture broth after 150 h (in U/mL). The factors are transposed in the Hadamard matrix (*see* **Table 2**), giving the datasheet of the experiments to run (*see* **Table 3**). In a screening

Table 3
Experimental Datasheet

Run	Glucose conc.	Initial pH	Inoculum size	Agitation rate	Inducer	Nitrogen source	Temp.	Enzymatic activity
1	5	7.0	10^8	100	No	Corn steep	20	61
2	1	7.0	10^8	200	Yes	Peptone	20	284
3	1	5.0	10^8	200	No	Corn steep	30	23
4	5	5.0	10^6	200	No	Peptone	20	69
5	1	7.0	10^6	100	No	Peptone	30	238
6	5	5.0	10^8	100	Yes	Peptone	30	236
7	5	7.0	10^6	200	Yes	Corn steep	30	105
8	1	5.0	10^6	100	Yes	Corn steep	20	167

Table 4
Calculation of the Estimation of b_i Effects

b_i	Value
b_0	147.87
b_1	−30.12
b_2	24.12
b_3	3.12
b_4	−27.62
b_5	−50.12
b_6	58.87
b_7	2.62

study, the effects are supposed to be additive; this implies that the relationship between the experimental responses and the studied variables is a first-order polynomial model with values of $X_i = \pm 1$:

$$\eta \ (\text{response}) = \beta + \beta_1 X_1 + \beta_2 X_2 + \beta_3 X_3 + \beta_4 X_4 + \beta_5 X_5 + \beta_6 X_6 + \beta_7 X_7$$

From the experimental data, the estimations of the β_i effects can be calculated (*see* **Table 4**). These values can then be represented with various diagrams such as a bar chart (*see* **Fig. 1**). This study clearly highlights two very influent key factors: the nature of the nitrogen source and the presence of the inducer. Glucose concentration, agitation rate, and initial pH are less influent. Inoculum size and temperature are without any significant effect on the response in the chosen experimental field. Other graphical tools can be used to analyze the screening results, such as the Pareto technique or normal and half-normal plot charts.

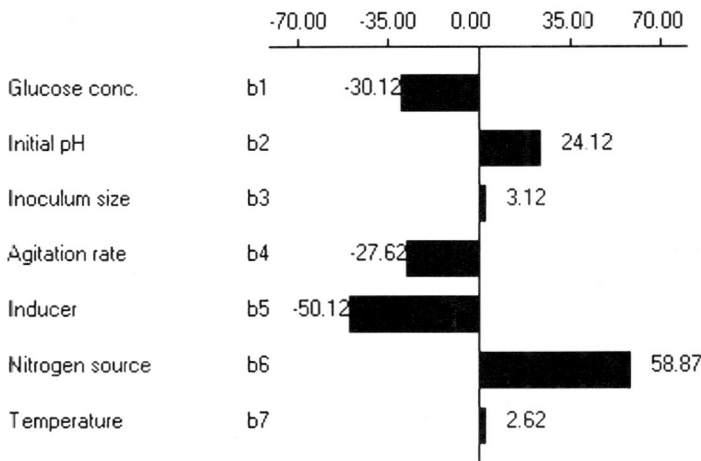

Fig. 1. Bar chart of the factor screening results.

3.4. Quantitative Studies of Factors

Once the influent factors are known, they can be studied more precisely. The hypothesis of additive effects used in the screening of factors is abandoned. The old strategy, which consists in studying a factor at a time while the others are maintained at a fixed value, requires many experiments and does not take into account the possible interactions (*see* **Note 5**). Information obtained is incomplete and might not allow the solving of the studied problem.

There are two kinds of interaction: those that we can postulate (of which we want to know their possible existence importance) and those of which we are unaware and wish to discover. To study them, the desired information is well defined and factorial experiment designs, whether symmetrical or asymmetrical, complete, or fractional, are used *(3)*. These experiment designs have all the desired qualities and especially "sequentiality." A complete factorial design is, by definition, composed of all the possible combinations of the levels of each factor. The number N of different combinations is equal to the product of the number of levels (i.e., $N = s_1 \, s_2 \, s_3 \ldots s_k$). These designs, especially the two-level factorial ones, are very often used and they have many applications.

3.5. Quantitative Studies of Responses

In many applications, the interest does not lie in studying the effects of the factors or the importance of the interactions but in knowing how one or several measured characteristics (responses) behave in a well-defined experimental domain. It is then possible to seek the optimum of one or more experimental responses without having to undertake a great number of experiments. Whatever the domain of application, the objective is to find a region of the experimental domain in which all of the studied properties meet the desired constraints as closely as possible. This region is called the area of acceptable compromise. To find the relationship between the factors and the responses, the phenomenon studied is simplified by mathematical modeling. Depending on the problem studied, the model could be linear or nonlinear, a differential equation, and so forth. This part of the methodological tool is called "response surface methodology" *(4–6)*. Experimentation will determine the values of the mathematical model's coefficients. However, prediction is of little value if one does not know how precise it is. The precision of the model depends on the precision of its coefficients. However, the quality (variance) of these depends only on the experimental measuring accuracy (experimental or residual variance), the structure of the experimental design, and the postulated mathematical model; it is completely independent of the experimental results. Therefore, the quality of an experimental design can be either the precision of the prediction in a given experimental field or the precision with which the coefficients are known. These models must be good representations of the experimental response within the domain of interest, and if this condition is fulfilled, they must give an estimation of the response that is acceptable, qualitywise. It is considered acceptable if it can be compared (for a given experimental point) to the quality obtained by running the experiment. Any model type can be chosen, provided it possesses the two aforementioned properties. Polynomial models are very often used because of their simplicity and their sequential approach: Once the model and a corresponding optimal design are chosen, it is tested for validity. If it is valid (i.e., it is a good representation of the phenomenon), the response can be calculated for every point of the experimental domain.

The most commonly used models are first- and second-order polynomial models and the corresponding classical optimal designs are as follows:

For first-order models:

1. Hadamard or Plackett and Burman experimental designs
2. Full (2^k) or fractional (2^{k-r}) factorial experimental designs
3. Equiradial experimental designs
4. Simplex experimental designs

For second-order models:

1. Composite experimental designs
2. Doehlert uniform shell designs
3. Equiradial experimental designs

Table 5
Factors and Experimental Domain

Factor	Unit	Center	Variation step
Nitrogen conc.	%	3.0	2.0
Inducer conc.	%	0.50	0.46
pH	U	6.0	2.0

Table 6
Doehlert Matrix

Run	X_1	X_2	X_3	Y_1
1	1.0000	0.0000	0.0000	298.00
2	−1.0000	0.0000	0.0000	260.00
3	0.5000	0.8660	0.0000	290.00
4	−0.5000	−0.8660	0.0000	248.00
5	0.5000	−0.8660	0.0000	269.00
6	−0.5000	0.8660	0.0000	278.00
7	0.5000	0.2887	0.8165	274.00
8	−0.5000	−0.2887	−0.8165	267.00
9	0.5000	−0.2887	−0.8165	290.00
10	0.0000	0.5774	−0.8165	284.00
11	−0.5000	0.2887	0.8165	268.00
12	0.0000	−0.5774	0.8165	264.00
13	0.0000	0.0000	0.0000	285.00
14	0.0000	0.0000	0.0000	288.00
15	0.0000	0.0000	0.0000	284.00

4. Box and Behnken experimental designs
5. Hybrid experimental designs
6. Hoke experimental designs

Here, we give an example of a Doehlert (7) uniform shell design. The settings are the same as in the screening example. The three most influent factors (*see* **Table 5**) were retained in order to maximize lipase production (experimental response). The Doehlert matrix (*see* **Table 6**) was used to create the experimental datasheet (*see* **Table 7**). Experiment 13 is done in triplicate to check reproducibility. The resulting coefficients are shown in (**Table 8**). With the complete equation, response surfaces can be drawn (*see* **Figs. 2–4**). From **Fig. 2,** it can be seen that the response is minimal when both concentrations are low. When the inducer concentration rises from 0.04 to 0.50, the response rap-

Table 7
Experimental Datasheet

Run	Nitrogen concentration (%)	Inducer concentration (%)	pH	Enzymatic activity (UI/mL)
1	5.0	0.50	6.0	298.00
2	1.0	0.50	6.0	260.00
3	4.0	0.90	6.0	290.00
4	2.0	0.10	6.0	248.00
5	4.0	0.10	6.0	269.00
6	2.0	0.90	6.0	278.00
7	4.0	0.63	7.6	274.00
8	2.0	0.37	4.4	267.00
9	4.0	0.37	4.4	290.00
10	3.0	0.77	4.4	284.00
11	2.0	0.63	7.6	268.00
12	3.0	0.23	7.6	264.00
13	3.0	0.50	6.0	285.00
14	3.0	0.50	6.0	288.00
15	3.0	0.50	6.0	284.00

Table 8
Model Coefficients

Coefficient	Value
b_0	285.667
b_1	17.250
b_2	12.846
b_3	−7.144
b_{11}	−6.667
b_{22}	−17.001
b_{33}	−10.832
b_{12}	−5.196
b_{13}	−8.573
b_{23}	−2.594

idly increases; whereas from 0.50 upward, it does not change significantly. When the nitrogen concentration increases, so does the response. From **Fig. 3,** we can see a response enhance with the increase of nitrogen concentration if the initial pH is under 6.0. If the pH is higher than this limit, the response decreases.

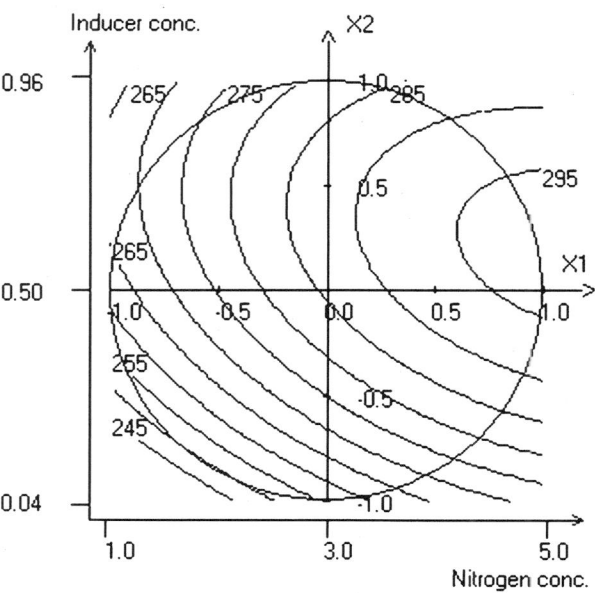

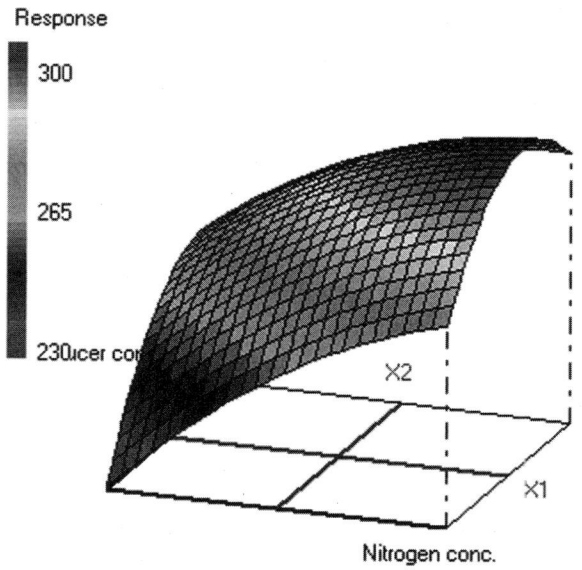

Fig. 2. A two-dimensional (2D) and three-dimensional (3D) graphical study of the response variation in the plane: nitrogen concentration and inducer concentration. Fixed factor: pH = 6.0.

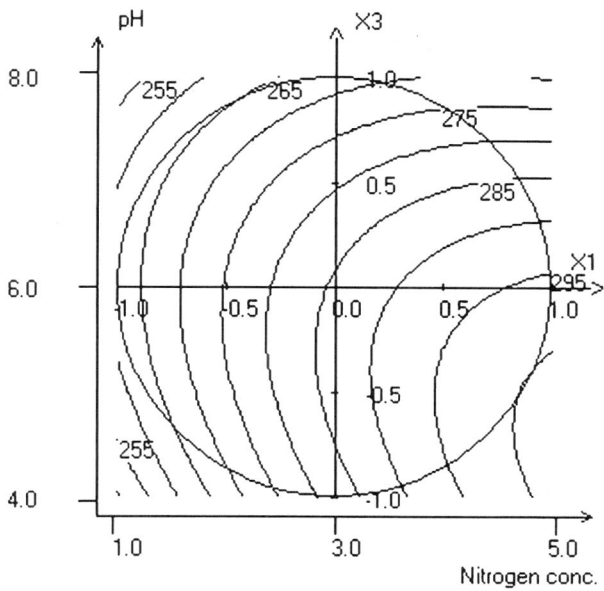

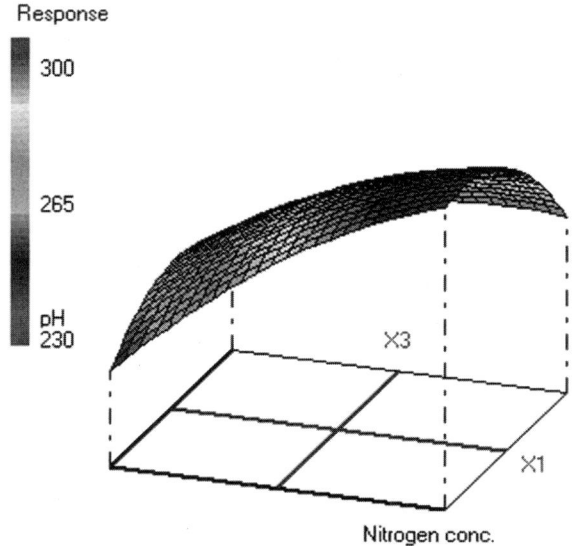

Fig. 3. A 2D and 3D graphical study of the response variation in the plane: nitrogen concentration and pH. Fixed factor: inducer concentration = 0.50 %.

Finally, **Fig. 4** shows that the response is high for inducer concentrations above 0.04 but no longer rises when it equals 0.50. The maximum response is obtained for inducer concentration above 0.50 and pH under 6.0.

3.6. Mixtures

In many industries, a great number of products are obtained by mixing two or more components or ingredients. The properties of the final product depend on the proportion of each component in the studied mixture. In the case of mixtures, the factors are the proportions of each component. They have two significant characteristics: (1) Their total amount is equal to one and they are, thus, not independent and (2) their values are dimensionless numbers, perfectly comparable.

These constraints on the values that components can take account for the fact that the mixtures cannot be treated as usual. Considering these constraints, any variation of the proportion of a component causes a variation of the proportions of the other components. The problems involving mixtures have two main differences: the experimental domain (a regular polyhedron of dimension [q–1] for a mixture of q components) and the form of the mathematical model *(8)*. For this type of study, specific experimental designs are available, such as Scheffe's *(9)* simplex lattice designs or particular optimal designs when the components are under constraints.

3.7. Particular Experimental Designs

The traditional experimental designs presented cover a significant share of the experimenter needs. There are, however, many circumstances under which these designs are inapplicable:

1. Nonsymmetrical experimental domains. This is a very significant limitation to the use of the traditional experimental designs that are usable only in the case of a symmetrical experimental domain. In certain cases, the experimental domains limited by technological or economic constraints that give it a nonsymmetrical form, some combinations of factors leading to expensive, dangerous, or impossible experiments, and discontinuities can be feared. It can also happen that some experimental fields are discrete and have a reduced number of possible experiments.
2. A fixed number of experiments. It can happen for economic reasons that the number of experiments is limited. This number is seldom in agreement with that of a traditional experiment design.
3. Unspecified linear mathematical model. The traditional experiment matrices are designed to study well-defined linear models and do not allow, economically, the study of a particular linear model postulated according to existing information.
4. Complement of a design. It is extremely rare that previous experiments can be reused. The traditional experimental designs, because they are rigid and pre-

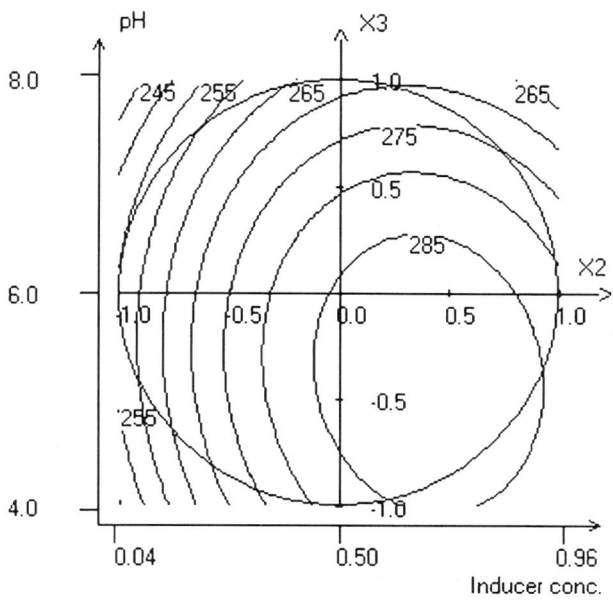

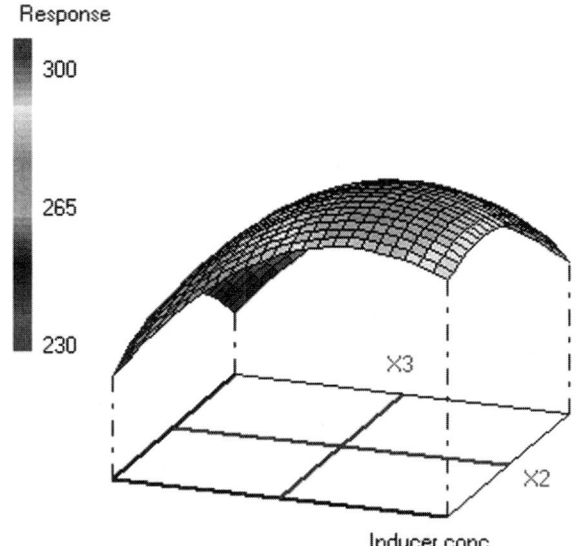

Fig. 4. A 2D and 3D graphical study of the response variation in the plane: inducer concentration and pH. Fixed factor: nitrogen concentration = 3.0 %.

established, usually do not allow modifications (such as adding new factors), especially if they were not planned in advance.

5. Repair of an experimental design. Even if none of the above cases prevents the construction of a traditional experimental design, new difficulties can emerge. We can imagine many situations in which the selected experimental design does not work. For example, if one or more experiments are impossible to perform in the course of experimentation, the experimental results will be incomplete and will not satisfy the studied objectives. The missing information will have to be acquired by adding one or more experiments.

In conclusion, we currently have new, powerful, and very flexible tools that make it possible to build, according to the problem arising, the most economic and most informative experimental strategies by taking account of the reality of the studied problem. The application field of this methodology is very broad because it includes not only the traditional applied sciences (physics, chemistry, biology, etc.), but also, for example, social sciences or economy and various situations of simulation (the experiment being taken in the broad sense).

4. Notes

1. For a long time, the expression "experimental design" has been used to describe a group of hitherto well-known experimental strategies for variance analysis such as factorial designs, Latin squares, Greco-Latin squares, and so forth. In 1970, when we began our research in this field, in order to avoid the ambiguous use of the overworked expression "experimental design" and to underline the extreme importance of the planning stage, we decided to use the expression "methodology of experimental research." "Methodology" describes all the methods and tools that can be used to define the studied problem, undertake the necessary experiments, and exploit the results. This also includes the concept of "experimental strategy," where "experimental" indicates experimentation is the only way of obtaining the as-yet unavailable information.

2. The factors are the causes, either supposed or certain, responsible for the studied phenomenon. They can be controlled or not (noise factors), but only those we can control are taken into account. They can be quantitative or qualitative, continuous or discontinuous.

3. An experimental response is a measurable change observed when the factors vary. A phenomenon can be described by several responses. Problems can arise during interpretation of the results if the response can take only discrete values.

4. With a design including k factors and s levels, the number of coefficients to calculate is $p = \Sigma_{i=1}^{k} (s_i - 1)$ and, therefore, the minimal number of experiments N to undertake is:

$$N \geq 1 + \Sigma_{i=1}^{k} (s_i - 1) \tag{1}$$

5. An interaction effect between two factors means that the effect of one of them depends on the value of the other one.

References

1. Addelman, S. (1962) Orthogonal main-effects for asymmetrical factorial experiments. *Technometrics,* **4,** 21–46.
2. Plackett, R. L. and Burman, J. P. (1946) The design of optimum multifactorial experiments. *Biometrica* **33,** 305–332.
3. Box, G. E. P., Hunter, W. G., and Hunter, J. S. (eds.) (1978) *Statistics for Experimenters,* Wiley, New York.
4. Box, G. E. P. and Draper, N. R. (eds.) (1987) *Model Building and Response Surfaces,* Wiley, New York.
5. Montgomery, D. C. (1991) *Design and Analysis of Experiments,* Wiley, New York.
6. Cochran, W. G. and Cox, G. M. (1950) *Experimental Designs,* Wiley, New York.
7. Doehlert, D. H. (1970) Uniform shell designs, *Appl. Statist.* **19,** 231–239.
8. Cornell, J. A. (1990) *Experiments with Mixtures,* Wiley, New York.
9. Scheffe, H. (1958) Experiments with mixtures. *J. R. Statist. Soc. B* **20,** 344–360.

3

Metabolic Engineering of *Acremonium chrysogenum* to Produce Deacetoxycephalosporin C and Bioconversion to 7-Aminodeacetoxycephalosporanic Acid

Marta Rodríguez-Sáiz, Juan-Luis de la Fuente, and José-Luis Barredo

Summary

7-Aminodeacetoxycephalosporanic acid (7-ADCA) and 7-aminocephalosporanic acid (7-ACA) are the starting materials for the production of all clinically important semisynthetic derivatives of cephalosporins. Whereas 7-ADCA is conventionally produced from penicillin by a synthetic chemical method, here we describe an alternative bioprocess for its production. The method is based on the disruption by one-step replacement of the *cefEF* gene, encoding the bifunctional expandase/hydroxylase activity, of a cephalosporin C-producing strain of *Acremonium chrysogenum*. Subsequently, cloning and expression of the *cefE* gene from *Streptomyces clavuligerus* in the *A. chrysogenum* disrupted transformant yield recombinant strains producing deacetoxycephalosporin C (DAOC). DAOC production level is almost equivalent to the total β-lactams biosynthesized by the parental strain. DAOC deacylation is carried out by two final enzymatic bioconversions catalyzed by D-amino acid oxidase (DAO) and glutaryl acylase (GLA), yielding 7-ADCA.

Key Words: *Acremonium chrysogenum;* cephalosporin; deacetoxycephalosporin C; deacetylcephalosporin C; DAOC; DAC; 7-ADCA; *cefEF; cefE;* D-amino acid oxidase; glutaryl acylase.

1. Introduction

Cephalosporins, penicillins, and their semisynthetic derivatives are antibiotics of extensive clinical use, obtained from intermediate compounds produced by fermentation of microorganisms. Specifically, cephalosporins show a broad-spectrum activity against Gram-positive and Gram-negative bacteria, and, in some cases, even against anaerobes. Furthermore, they have a high degree of resistance to staphylococcal penicillinase, as well as a very favorable adverse effects profile, which confers an outstanding value from a clinical and industrial point of view.

From: *Methods in Biotechnology, Vol. 18: Microbial Processes and Products*
Edited by: J. L. Barredo © Humana Press Inc., Totowa, NJ

Fig. 1. Cephalosporin C biosynthetic pathway of *A. chrysogenum*. Enzyme activities are indicated in uppercase bold letters: ACVS (ACV synthetase), IPNS (isopenicillin N synthase), IPNE (isopenicillin N epimerase), DAOCS/DACS (deacetoxycephalosporin C synthase/deacetylcephalosporin synthase), DAC-AT (deacetylcephalosporin C acetyltransferase). The corresponding genes are shown in italics in parentheses. Whereas IPNE is encoded by a single gene *(cefD)* in *Streptomyces*, two genes *(cefD1* and *cefD2)* are involved in *A. chrysogenum*. The *cefEF* gene encodes an enzyme with both DAOCS/DACS activities in *A. chrysogenum*. Prokaryotic species possess separate *cefE* and *cefF* genes and enzymes and need the activity of lysine ε-aminotransferase for the biosynthesis of α-aminoadipic acid. In filamentous fungi, α-aminoadipic acid is a precursor of lysine biosynthesis.

Cephalosporins are chemically characterized by a cephem nucleus composed of a β-lactam ring fused to a dihydrothiazine ring. The cephalosporin C biosynthetic pathway *(see* **Fig. 1**) has been investigated in-depth in *Acremonium chrysogenum,* the fungus chosen for its industrial production. The route begins with the condensation of L-cysteine and L-α-aminoadipic acid to form the dipeptide L-α-aminoadipyl-L-cysteine (AC). Then, L-valine is epimerized to D-valine, activated, and condensed with AC to form the tripeptide δ-(L-α-aminoadipyl)-L-cysteinyl-D-valine (ACV) *(1,2)*. The first two reactions are catalyzed by the enzymatic activity ACV synthetase (ACVS), encoded by the *pcbAB* gene *(3)*. Further cyclization of ACV to isopenicillin N (IPN) is carried out by the isopenicillin N synthase (IPNS or "cyclase"), encoded by the *pcbC* gene *(4)*. The isopenicillin N epimerase activity (IPNE) of *A. chrysogenum,* responsible for the epimerization of the L-α-aminoadipic side chain of IPN to the D-configuration of penicillin N (PN), has been recently identified as a two-step reaction encoded by the *cefD1* and *cefD2* genes *(5)*.

The ring expansion of PN to form deacetoxycephalosporin C (DAOC) is catalyzed by an α-ketoglutarate-dependent dioxygenase named DAOC synthase (DAOCS or "expandase"). Although DAOCS and the following enzyme of the pathway, deacetylcephalosporin C synthase (DACS), are encoded by different genes in bacteria *(cefE* and *cefF,* respectively), both activities are present as a single protein encoded by the *cefEF* gene in *A. chrysogenum (6,7)*. Finally, the methoxy group of the deacetylcephalosporin C (DAC) located at C3 is acetylated to give cephalosporin C. This final step is catalyzed by a DAC acetyltransferase encoded by the *cefG* gene *(8)*.

Currently, the clinically important semisynthetic derivatives of cephalosporins are manufactured from 7-aminodeacetoxycephalosporanic acid (7-ADCA) and

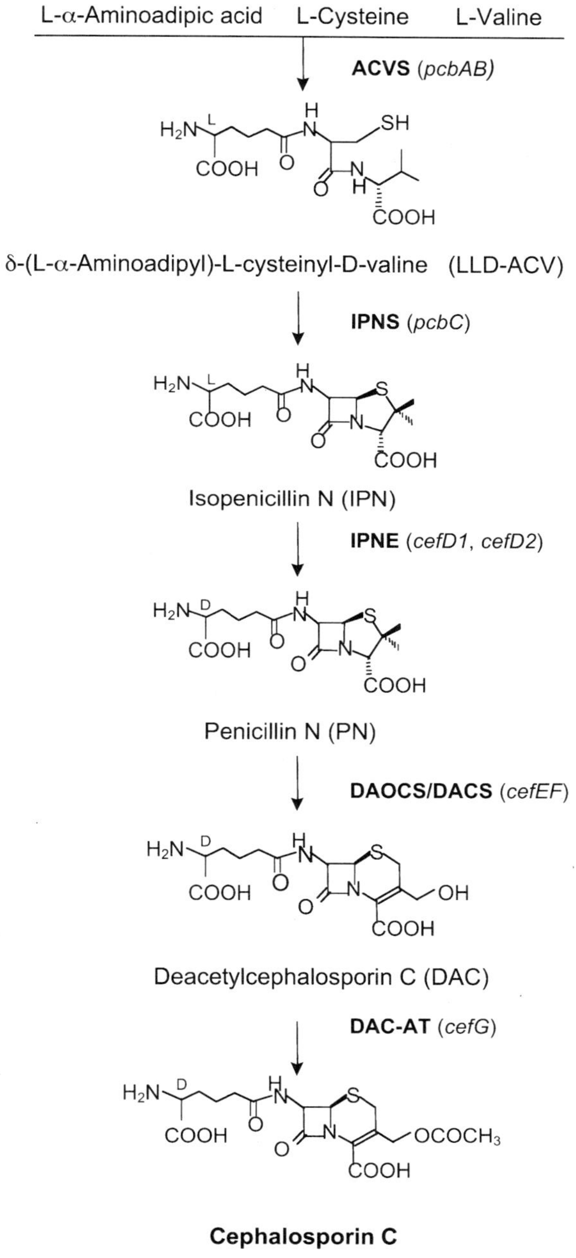

L-α-Aminoadipic acid   L-Cysteine   L-Valine

ACVS (*pcbAB*)

δ-(L-α-Aminoadipyl)-L-cysteinyl-D-valine   (LLD-ACV)

IPNS (*pcbC*)

Isopenicillin N (IPN)

IPNE (*cefD1, cefD2*)

Penicillin N (PN)

DAOCS/DACS (*cefEF*)

Deacetylcephalosporin C (DAC)

DAC-AT (*cefG*)

Cephalosporin C

Fig. 1.

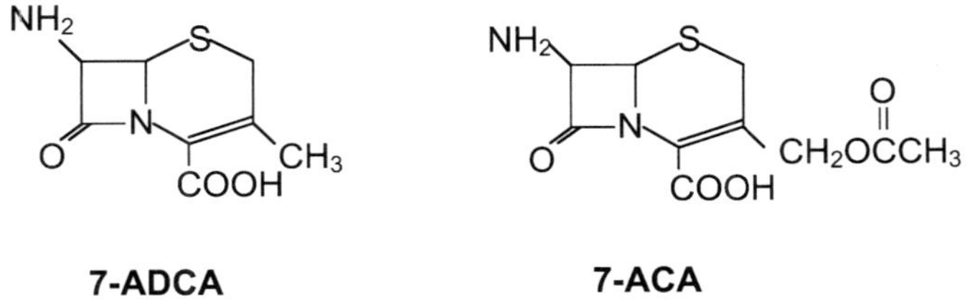

7-ADCA   **7-ACA**

Fig. 2. Chemical formula of 7-ADCA and 7-ACA, the most important intermediates for the production of cephalosporin derivatives.

7-aminocephalosporanic acid (7-ACA) (*see* **Fig. 2**) by side-chain substitutions. The classical process to produce 7-ADCA is based on a chemical ring expansion of the penicillin G produced by *P. chrysogenum* fermentation. This is a multistep and complex process, which requires expensive reagents and the previous purification of a highly impure starting material before chemical treatment. Additionally, significant quantities of process by-products can cause environmental problems.

The development of recombinant DNA techniques and their application to industrial microorganisms has allowed the design of biosynthetic pathways, giving rise to new molecules and to new processes *(9)*. In this way, metabolic engineering to develop fermentative processes alternative to the chemical pathway to 7-ADCA was attempted by several research groups without successful results *(10)*. Good results were obtained by engineering the penicillin pathway in industrial strains of *P. chrysogenum*. Thus, the transformation of *P. chrysogenum* with the *cefE* gene (DAOCS) from *S. clavuligerus,* expressed under the control of promoter and terminator sequences of the *pcbC* gene of *P. chrysogenum,* originated in vivo enzymatic ring expansion of penicillin G, achieving up to 70-fold higher DAOCS activity (equivalent to industrial strains of *A. chrysogenum*) *(11)*. Moreover, strains of *P. chrysogenum* have been engineered by expressing the *cefD* gene (IPNE) of *S. lipmanii* and the *cefE* gene (DAOCS) of *S. clavuligerus,* obtaining low levels of DAOC (corresponding to expanded IPN), but ring expansion of penicillin G was not observed *(10,12)*.

The fermentation of industrial strains of *P. chrysogenum* carrying the *cefE* gene of *S. clavuligerus* by feeding adipic acid as a side chain precursor (instead of phenylacetic or phenoxyacetic acid) originated the in vivo expansion of the biosynthesized adipyl-6-APA into adipyl-7-ADCA *(13)*. Significant amounts of this product (up to 17% of the penicillin V synthesized by the untransformed *P. chrysogenum* strain fed with phenoxyacetate) were produced by selected transformants. Final removal of the D-α-aminoadipyl side chain to obtain 7-ADCA was efficiently accomplished by using a glutaryl acylase (GLA) from *Pseudomonas (14,15)*. Nevertheless, although this process presents high production efficiencies, the cost of adipic acid as side-chain precursor as well as the presence of contaminants as adipyl-6-APA, adipyl-7-ACA, and adipyl-7-ADAC, which must be removed before the final acylase treatment, could decrease the industrial benefits of the process.

One of the problems found in the design of a competitive recombinant 7-ADCA-producing strain is the narrow substrate specificity shown by DAOCS, with no detectable activity on inexpensive available substrates such as penicillin G or V *(16,17)* produced by standard fermentation. The development of engineered DAOCSs with modified substrate specificity might be a good option to improve conversion efficiency.

In this chapter, we report the construction of a DAOC-producing strain of *A. chrysogenum*. This intermediate can be subsequently converted into 7-ADCA by a two-step enzymatic bioconversion based on the enzymes D-amino acid oxidase (DAO) *(18,19)* and glutaryl acylase (GLA) *(20–22)*. The strategy is based on the *cefEF* gene disruption of a cephalosporin C-producing strain of *A. chrysogenum,* obtaining transformants able to accumulate PN. This inactivation is carried out by transformation with a disruption plasmid using the one-step gene disruption technique *(23)*, which allows the selection of genetically stable disrupted transformants. Those transformants accumulating large amounts of PN are selected by bioassay against the β-lactam-supersensitive strain *E. coli* ESS22-31 *(24)* and further fermented in a flask to confirm their production level. Subsequent expression of the *cefE* gene from *S. clavuligerus* in the selected PN-producing strain causes in vivo expansion of PN into DAOC (*see* **Fig. 1**). Transformants obtained are tested by a new bioassay against *E. coli* ESS22-31, but in the presence of a penicillinase. In this case, unexpanded PN is destroyed, being able to select those transformants producing cephalosporins because they originate inhibition halos. Further flask fermentations allow the selection of the best producers and the analysis of the pathway intermediates accumulated. In contrast to the data reported for recombinant strains of *P. chrysogenum* expressing ring-expansion activity, no detectable contamination with other cephalosporin intermediates occurred *(25)*.

DAOC constitutes the starting material for 7-ADCA production using a two-step enzymatic bioconversion based on DAO *(18,19)* and GLA *(20–22)* (*see* **Fig. 3**). DAO catalyzes the oxidative deamination of DAOC into ketoadipyl-7-ADCA [7-β-(5-carboxy-5-oxopentanamido)-deacetoxycephalosporanic acid]. The side product of the reaction, hydrogen peroxide, promotes the decarboxylation of the ketoacid intermediate to glutaryl-7-ADCA [7-β-(4-carboxybutanamido)-

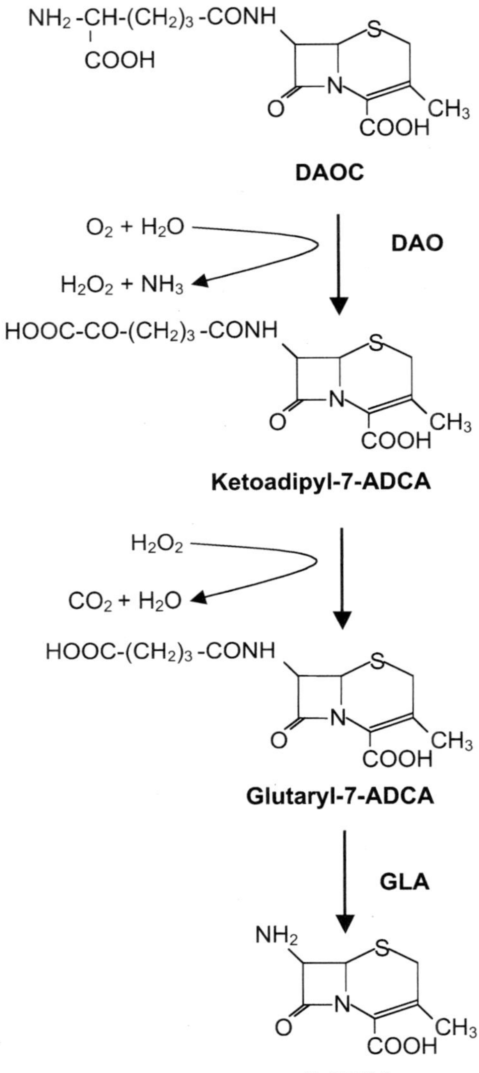

Fig. 3. Two-step enzymatic bioconversion of DAOC into 7-ADCA based on DAO and GLA activities.

deacetoxycephalosporanic acid], which is then hydrolyzed by GLA to 7-ADCA. In this way, the entire synthesis of 7-ADCA is carried out using a green route that avoids chemical steps and constitutes a very efficient bioprocess both from industrial and environmental points of view.

2. Materials

2.1. Construction of Plasmids for Disruption and Heterologous Expression in A. chrysogenum

1. Plasmid pBC KS (+) (Stratagene, La Jolla, CA).
2. Plasmid pBluescript I KS (+) (Stratagene).
3. Plasmid pAN7.1 *(26)*.
4. Plasmid pALfleo7 *(27)*.
5. Restriction enzymes, T4 DNA ligase, DNA polymerase I large (Klenow) fragment, Pfu Turbo DNA polymerase (Stratagene).
6. *Escherichia coli* DH5α *(28)*.
7. Luria–Bertani (LB) medium: 10 g/L bacto-tryptone, 5 g/L bacto-yeast extract, 10 g/L NaCl, and 20 g/L agar. Adjust to pH 7.0 with 5 N NaOH. Autoclave at 121°C for 15 min.
8. Chloramphenicol.
9. Ampicillin.
10. Agarose gel electrophoresis equipment.
11. Polymerase chain reaction (PCR) equipment.
12. Electroporation equipment.
13. Oligonucleotide 1: 5′-GATCAGTGAGAGTCCATGGACACGACGGTGCCC-3′.
14. Oligonucleotide 2: 5′-GGGCACCGTCGTGTCCATGGACTCTCACTGATC-3′.
15. Phenol–chloroform–isoamylalcohol (25 : 24 : 1) (v/v).
16. TE buffer: 10 mM Tris-HCl, pH 8.0, and 1 mM EDTA.
17. IPTG (isopropyl-β-D-thio-galactopyranoside).
18. X-Gal (5-bromo-4-chloro-3-indolyl-β-D-thiogalactopyranoside).
19. Qiaex II Gel Extraction System (Qiagen, Hilden, Germany).
20. Qiagen Plasmid Midi Kit (Qiagen).
21. QuickChange Kit (Stratagene).
22. Refrigerated centrifuges.
23. Micropipets.

2.2. Transformation of A. chrysogenum Strains With pALC73 and pALC88

1. Strain of *A. chrysogenum* producing cephalosporin C *(3,5)*.
2. Plasmid pALC73 *(25)*.
3. Plasmid pALC88 *(25)*.
4. LePage–Campbell sporulation medium: 1 g/L glucose, 2 g/L yeast extract, 1.5 g/L NaCl, 10 g/L CaCl$_2$, and 22.5 g/L agar. Adjust to pH 6.8 with 1 N NaOH. Autoclave at 121°C for 15 min *(29)*.
5. MMC medium: 31.6 g/L sucrose, 2.2 g/L glucose, 0.5 g/L corn step solid, 7.5 g/L L-asparagine, 0.22 g/L ammonium acetate, 15 g/L KH$_2$PO$_4$, 21 g/L K$_2$HPO$_4$, 0.75 g/L Na$_2$SO$_4$, 0.18 g/L MgSO$_4$ · 7H$_2$O, 0.06 g/L CaCl$_2$, and 1 mL/L salts solution. Adjust to pH 7.0 with 1N NaOH. Autoclave at 121°C for 15 min *(30)*.
6. Salts solution: 15 g/L Fe(NH$_4$)$_2$(SO$_4$)$_2$ · 6H$_2$O, 3 g/L MnSO$_4$ · 4H$_2$O, 3 g/L ZnSO$_4$ · 7H$_2$O, and 0.8 g/L CuSO$_4$ · 5H$_2$O. Autoclave at 121°C for 15 min.

7. TPC buffer: 50 mM potassium phosphate buffer, pH 7.0, and 0.8 M NaCl. After sterilization by autoclave at 121°C for 15 min, add 20 mM MgSO$_4$.
8. DTT (1,4 dithio-DL-threitol).
9. Caylase C3: enzymatic mixture from *Tolypocladium geodes* (Cayla, Tolouse, France).
10. Nytal filter 30-μm pore (Swiss silk bolting; Zurich, Switzerland).
11. NCM buffer: 10 mM 3-morpholinopropanesulfonic acid (MOPS), pH 7.0, 0.8 M NaCl, and 50 mM CaCl$_2$. Autoclave at 121°C for 15 min.
12. CCM buffer: 10 mM MOPS, pH 7.0, 18 % PEG-8000, and 50 mM CaCl$_2$. Autoclave at 121°C for 15 min.
13. Counting chamber.
14. TSAS medium (TSA-sucrose): 3 g/L soy peptone, 16 g/L casein peptone, 2.5 g/L glucose, 6 g/L NaCl, 2.5 g/L K$_2$HPO$_4$, 103 g/L sucrose, and 20 g/L agar. Adjust to pH 7.0. Autoclave at 121°C for 15 min *(30)*.
15. Phleomycin (Cayla).
16. Hygromycin (Sigma, St. Louis, MO).
17. Slant tubes (200 × 30 mm).
18. Orbital shaker.
19. Water bath.
20. Laminar airflow cabinet.
21. Centrifuge.

2.3. Selection by Bioassay of A. chrysogenum *Transformants Producing* PN or DAOC

1. *Escherichia coli* ESS22-31 *(24)*.
2. Bacto Penase (Difco Laboratories; Detroit, MI).
3. Penicillin G (Antibióticos S. A., León, Spain).
4. Stainless-steel cylinders for agar plugs.
5. Bioassay dishes 245 × 245 × 25 mm (Nalge Nunc International, Hereford, UK).
6. Spectrophotometer.
7. Icubator.

2.4. A. chrysogenum *Flask Fermentation*

1. *A. chrysogenum* Δ*cefEF*-T1, a PN-producing strain lacking the enzymatic activities DAOCS/DACS obtained from the parental *A. chrysogenum* by inactivation of the *cefEF* gene with pALC73.
2. *A. chrysogenum* Δ*cefEF*-T1/*cefE*-T1, a DAOC-producing strain lacking the enzymatic activities DAOCS/DACS but expressing the DAOCS activity encoded by the *cefE* gene of *S. clavuligerus*. It was obtained from *A. chrysogenum* Δ*cefEF*-T1 by transformation with pALC88.
3. *A. chrysogenum* inoculum medium: 30 g/L sucrose, 1.5 g/L CaCO$_3$, 5 g/L corn steep solid, and 15 g/L beef extract. Distribute 50-mL aliquots in 250-mL flasks. Autoclave at 121°C for 15 min *(31)*.

4. *A. chrysogenum* fermentation medium: 36 g/L sucrose, 27 g/L glucose, 3.2 g/L DL-methionine, 12 g/L asparagine, 0.16 g/L $Fe(NH_4)_2(SO_4)_2 \cdot 6H_2O$, 15 g/L K_2HPO_4, 15 g/L KH_2PO_4, 1.6 g/L $Na_2SO_4 \cdot 10H_2O$, 0.34 g/L $MgSO_4 \cdot 7H_2O$, 0.028 g/L $ZnSO_4 \cdot 7H_2O$, 0.028 g/L $MnSO_4 \cdot H_2O$, 0.0073 g/L $CuSO_4 \cdot 5H_2O$, and 0.072 g/L $CaCl_2 \cdot 2H_2O$. Adjust to pH 7.4 with 1 *N* NaOH. Distribute 50-mL aliquots in 500-mL flasks. Autoclave at 121°C for 15 min *(31)*.
5. Cryogenic vials, 1.8 mL (Nalge Nunc International).
6. pH meter.
7. Microwave.

2.5. Cephalosporins and PN Quantification by HPLC

1. Mobile phase: ammonium formate, pH 3.3, methanol, and tetrahydrofurane (99.85 : 0.1 : 0.05).
2. Reverse-phase high-performance liquid chromatography (HPLC) column Nucleosil C_{18}-10 µm (250 × 4.6 mm) (Phenomenex, Torrance, CA).
3. HPLC equipment.

2.6. Bioconversion of DAOC Into 7-ADCA

1. *Rhodotorula gracilis* ATCC 26217.
2. YM medium: 3 g/L yeast extract, 3 g/L malt extract, 5 g/L peptone, 10 g/L glucose, and 15 g/L agar. Autoclave at 121°C for 15 min.
3. *R. gracilis* DAO production medium: 0.5 g/L NaCl, 1.5 g/L K_2HPO_4, 1 g/L $MgSO_4 \cdot 7H_2O$, 0.25 g/L $CaCl_2$, 0.002 g/L $ZnSO_4$, 0.003 g/L $FeCl_3$, 25 g/L glucose, and 7 g/L D-alanine. Adjust to pH 5.6 with 2 *N* H_2SO_4 and distribute 100-mL aliquots in 500-mL flasks. Autoclave at 121°C for 15 min *(22)*.
4. *E. coli* ATCC 9637/pJC200 *(21)*. *E. coli* ATCC 9637 is a non-β-lactamase-producing strain. Plasmid pJC200 contains the gene encoding GLA from *Acinetobacter* sp. ATCC 53891 expressed under the control of the *tac* promoter and the chloramphenicol resistance gene as selection marker.
5. *E. coli* GLA inoculum medium: 12 g/L bacto-tryptone, 24 g/L yeast extract, 4 g/L glycerol, 2.31 g/L KH_2PO_4, and 12.54 g/L K_2HPO_4. Distribute 100-mL aliquots in 500-mL flasks. Autoclave at 121°C for 15 min. Add 30 µg/mL chloramphenicol *(22)*.
6. *E. coli* GLA fermentation medium: 6 g/L acid sodium glutamate, 3 g/L yeast extract, 20 g/L collagen hydrolysates, 3 g/L corn steep liquor, 1.1 g/L KH_2PO_4, and 6.2 g/L K_2HPO_4. Distribute 100-mL aliquots in 500-mL flasks. Autoclave at 121°C for 15 min. Add 30 µg/mL chloramphenicol and 6 g/L glucose *(22)*.
7. Standard buffer: 25 m*M* phosphate buffer, pH 8.0.
8. HPLC standards: PN, DAOC, DAC, cephalosporin C, ketoadipyl-7-ADCA, glutaryl-7-ADCA, 7-ADCA.
9. Bioreactor with oxygen input.
10. Filtration membrane, 30-kDa cutoff (Amicon, Bedford, MA).
11. Separating funnel.
12. Paper Whatman 3MM (Whatman, Kent, UK).
13. Slants tubes (200 × 18 mm).

3. Methods

This section describes the construction of a strain of *A. chrysogenum* able to produce DAOC and further bioconversion of DAOC into 7-ADCA. The methods outline (1) the construction of the plasmids for disruption and heterologous expression in *A. chrysogenum*, (2) the transformation of *A. chrysogenum* for gene disruption and gene expression, (3) the selection by bioassay of transformants producing PN or DAOC, (4) the fermentation of the selected transformants, (5) the cephalosporins and PN quantification by HPLC, and (6) the two-step bioconversion of DAOC into 7-ADCA.

3.1. Construction of Plasmids for Disruption and Heterologous Expression in A. chrysogenum

The construction of a disruption plasmid to inactivate the *cefEF* gene of *A. chrysogenum* is described in **Subheading 3.1.1.**, and the subcloning of the *cefE* gene of *S. clavuligerus* under the control of the *pcbC* promoter (P*pcbC*) for its heterologous expression in *cefEF*-disrupted mutants of *A. chrysogenum* is detailed in **Subheading 3.1.2.** DNA manipulations were performed according to standard procedures *(32)* and are not described here in detail because of space limitations.

3.1.1. Construction of pALC73, a cefEF Disruption Plasmid

The inactivation of the *cefEF* gene of *A. chrysogenum* was approached with the one-step gene disruption technique *(23)*, which allows the selection of genetically stable disrupted transformants. The disruption plasmid includes a selection marker suitable for *A. chrysogenum* inserted into the *cefEF* gene sequence, having at both flanking edges enough *cefEF* gene sequence to allow the single crossover. In this way, the hygromycin-resistance expression cassette was chosen as the selection marker. It is constituted by the P*gpdA* (glyceraldehyde-3-phosphate dehydrogenase promoter) from *A. nidulans*, the hygromycin resistance gene *(hyg^R)* from *E. coli*, and the T*trpC* (tryptophan C terminator) from *A. nidulans* *(26)*. First, a 7.2-kb *Bam*HI fragment containing the *cefEF* and *cefG* genes of *A. chrysogenum* was subcloned into the previously modified plasmid pBC KS (+) (*Xho*I and *Sal*I single sites were removed by double digestion and subsequent ligation), yielding pALC72. Subsequently, a *Bgl*II–*Xba*I DNA fragment from pAN7.1 *(26)* containing the described hygromycin resistance cassette was purified from agarose gel using Qiaex II, filled in with DNA polymerase I large (Klenow) fragment and subcloned into the single *Xho*I site of pALC72 (located inside of the *cefEF* gene), previously filled in with Klenow, yielding pALC73 (*see* **Fig. 4**). This plasmid contains the hygromycin-resistance cassette flanked by 2.2-kb and 5.0-kb DNA fragments including *cefEF* and *cefG* genes.

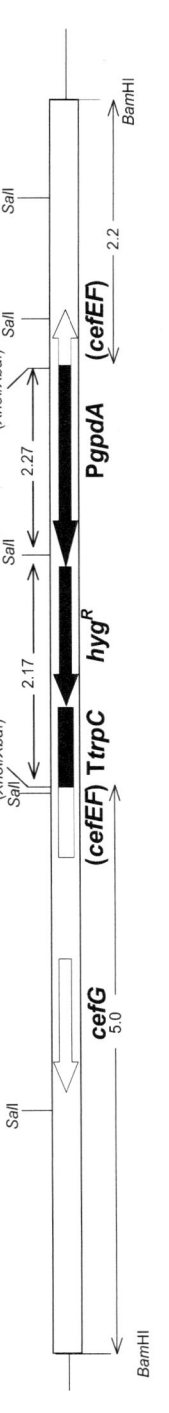

Fig. 4. Genetic construction for the disruption of the *cefEF* gene included in the plasmid pALC73. Sizes are indicated in kilobases.

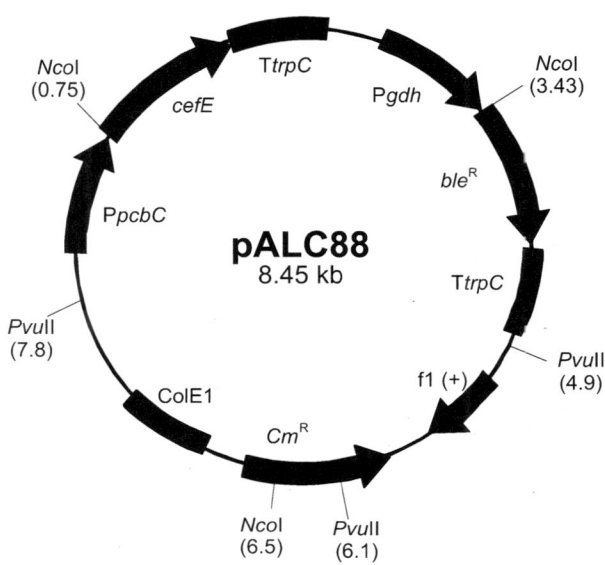

Fig. 5. Genetic map of the plasmid pALC88 used to introduce ring-expansion activity (DAOCS) into *A. chrysogenum* Δ*cefEF*-T1. The *cefE* gene from *S. clavuligerus* is expressed under the control of P*pcbC*. Phleomycin (ble^R) is used for selection in *A. chrysogenum* and chloramphenicol (Cm^R) in *E. coli*.

3.1.2. Construction of pALC88, a cefE Gene Expression Plasmid

Plasmid pALC88 (*see* **Fig. 5**) was constructed to express the *cefE* gene of *S. clavuligerus* in *A. chrysogenum*. P*pcbC* from *P. chrysogenum* was used to improve its transcription level. The 1.6-kb *Bam*HI–*Kpn*I fragment including the *cefE* gene *(33)* was subcloned into pBluescript I KS (+), yielding pALC82. Afterward, a *Nco*I site was created in the ATG starting codon of the *cefE* gene to allow subcloning at the *Nco*I site located at the 3′ end of P*pcbC*. Site-directed mutagenesis was done using the QuickChange kit and two complementary oligonucleotides corresponding to opposite strands, including the *Nco*I site (oligonucleotides 1 and 2). PCR amplification reactions were done with *Pfu* Turbo DNA polymerase in a GeneAmp PCR System 2400 for 15 cycles: 95°C, 30 s (60 s first cycle); 55°C, 60 s; 68°C, 12 min (10 min last cycle). The resulting plasmid, pALC83, was partially sequenced to confirm the presence of the *Nco*I site and the absence of additional single mutations. Subsequently, pALC83 was sequentially digested with *Kpn*I, filled in with Klenow, and digested with *Nco*I, yielding a 0.95-kb (*Nco*I-blunt) fragment that was subcloned under the control of P*pcbC* to obtain pALC87. After *Pvu*II digestion of pALC87, a 3.0-kb fragment containing the *cefE* gene expression cassette was purified with Qiaex II and subcloned into the *Eco*RV site of pALfleo7 *(27)* to give pALC88 (*see* **Fig. 5**).

3.2. Transformation of A. chrysogenum Strains With pALC73 and pALC88

Transformation of *A. chrysogenum* was carried out as previously described *(30)*, with some modifications. The formation of *A. chrysogenum* protoplasts is described in **Subheading 3.2.1.**, whereas the transformation reactions are reported in **Subheading 3.2.2.** The transformation protocol is equivalent for the selection of PN- or DAOC-producing strains, but using different selection markers: 5–10 µg/mL hygromycin for pALC73 (PN) and 5 µg/mL phleomycin for pALC88 (DAOC).

3.2.1. Protoplasts Formation

1. Pour about 25 mL of LePage medium into slants and store at room temperature until the condensed water present in the slants is eliminated. Seed six slants from a frozen vial of *A. chrysogenum* with a sterile pipet, and incubate at 28°C for 4–6 d (*see* **Note 1**).
2. Suspend each slant in 8 mL of deionized water, briefly scraping with a sterile pipet, and then transfer this suspension sequentially to the remaining slants until all of them have been harvested (*see* **Note 2**).
3. Seed the biomass suspension in a 500-mL flask with 100 mL of defined medium MMC and incubate for 24–30 h at 28°C and 250 rpm in an orbital shaker.
4. Harvest the mycelium through a sterile Nytal filter (30 µm pore) and wash with 3–5 vol of 9 g/L NaCl.
5. Carefully dry the mycelium sandwiched between sterile filter papers to eliminate the excess of the liquid (*see* **Note 3**).
6. Suspend the mycelium in 100 mL of TPC buffer supplemented with 100 m*M* DTT and incubate for 2 h at 28°C and 250 rpm in an orbital shaker (*see* **Note 4**).
7. Collect the mycelium as in **step 4.**
8. Dissolve 200 mg of Caylase C3 in 40 mL of TPC buffer. Once mixed homogenously, suspend the harvested mycelium in this solution and incubate in a 500-mL flask for 3 h at 28°C in an orbital shaker with gentle agitation (75–100 rpm). Protoplast release is checked by microscopic examination (*see* **Note 5**).
9. To eliminate the mycelium debris, filter the protoplast suspension in sterile conditions through a sterile Nytal filter (30 µm pore). Collect the suspension in 10-mL sterile polypropylene tubes.
10. Centrifuge the filtered protoplasts suspension at 600*g* and room temperature for 4 min.
11. Discard the supernatant and wash the protoplast pellet with 10 mL of 0.8 *M* NaCl. Centrifuge again at 600*g* and room temperature for 4 min (*see* **Note 6**).

12. Wash again twice more as indicated in the **step 11.**
13. Wash the pellet with 10 mL of NCM buffer and centrifuge at 600g and room temperature for 4 min.
14. Suspend the pellet in approx 5 vol of NCM buffer. Determine protoplast concentration by microscopic counting with a counting chamber and adjust the suspension to (1–5) × 10^8 protoplasts/mL (*see* **Note 7**).
15. Add CCM buffer (1/10 of the total volume) and keep the protoplasts suspension in an ice-water bath until use (*see* **Note 8**).

3.2.2. Transformation and Plating

1. Mix gently 100 µL of protoplasts suspension and 1–10 µg (10–15 µL) of plasmid DNA in a 10-mL polypropylene tube (*see* **Note 9**).
2. Incubate the transformation reactions in a water-ice bath for 20 min.
3. Add 500 µL of CCM buffer and incubate at room temperature for 20 min.
4. Add 600 µL of NCM buffer and mix gently until homogenization. Keep at room temperature.
5. Prepare small Petri dishes (50 mm in diameter) containing 5 mL of TSAS medium supplemented with the suitable antibiotic: 10 µg/mL hygromycin for pALC73 or 5 µg/mL phleomycin for pALC88. Likewise, prepare at least two control plates without antibiotic (*see* **Note 10**).
6. Mix each transformation reaction with 5 mL of top TSAS medium at 50°C, add the antibiotic (amount necessary for a final concentration of 10 µg/mL hygromycin or 5 µg/mL phleomycin), and spread onto one Petri dish (*see* **Notes 11** and **12**).
7. Once the top layer is solidified, incubate the plates at 28°C for 6–8 d.

3.3. Selection by Bioassay of A. chrysogenum Transformants Producing PN or DAOC

Production of PN or DAOC by the transformants was initially tested by bioassay against the β-lactam-supersensitive strain *E. coli* ESS22-31 *(24)*. Bacto Penase, a narrow spectrum β-lactamase that destroys the β-lactam ring present in penicillins (but not in cephalosporins), is fed to the bioassay plates to differentiate between penicillins and cephalosporins. Only those strains producing cephalosporins are able to originate growth-inhibition halos in the presence of Bacto Penase.

For the selection of disrupted transformants blocked in cephalosporin biosynthesis, two simultaneous bioassays (with and without penicillinase) are conducted. The development of a growth-inhibition halo in the absence of penicillinase (but not in its presence) indicates the accumulation of PN and, consequently, *cefEF* gene disruption (*see* **Fig. 1**). Best producers of PN selected by bioassay are then fermented in flask to determine their production level by HPLC and chemical structure by nuclear magnetic resonance (NMR). In the same way, the best DAOC producers are selected by the higher growth-inhibition halos in the presence of penicillinase. Further flask fermentation is used to define DAOC production by HPLC and chemical structure by NMR. The following steps outline the bioassay procedure:

1. Seed the selected transformants on Le Page–Campbell sporulation medium and incubated for 7 d at 28°C (*see* **Note 13**).
2. Inoculate 100 mL of LB liquid medium with an isolated colony of *E. coli* ESS22-31 and incubate for 18–24 h at 37°C and 250 rpm. Measure the OD_{660} and adjust to 1.65 with LB (*see* **Note 14**).
3. Mix 500 µL of *E. coli* ESS22-31 with 50 mL of LB agar maintained at 50°C and spread onto a 245-mm × 245-mm bioassay plate. Plates are prepared in the same way but adding 50,000 IU/mL Bacto Penase to detect cephalosporin intermediates specifically.
4. Take plugs of the transformants from Le Page–Campbell plates and place onto the bioassay plates.
5. Keep the plates at 4°C for 1 h to facilitate the diffusion of the β-lactam antibiotics present in the plug and then incubate at 37°C for 24 h.
6. Determine the diameter of the growth-inhibition halo formed and select the higher producers.
7. Seed slants of the selected transformants from Le–Page Campbell plates.
8. Determine antibiotic production by flask fermentation.

3.4. A. chrysogenum *Flask Fermentation*

The steps described in **Subheadings 3.4.1.** and **3.4.2.** outline the procedure for flask fermentation of *A. chrysogenum.*

3.4.1. Strain Maintenance and Preservation

1. Inoculate 200 × 30-mm slants of Le Page–Campbell sporulation medium with 0.1 mL of a frozen vial.
2. Incubate for 7 d at 28°C.
3. Store these slants at 4°C up to 1 mo.
4. For long-term storage, suspend one slant in 10 mL of 10% glycerol and store 1-mL aliquots in 1.8-mL cryogenic vials at –80°C (*see* **Note 15**).

3.4.2. Flask Fermentation

1. Suspend one slant in 5 mL of 9 g/L NaCl.
2. Seed the vegetative phase: 50 mL of *A. chrysogenum* inoculum medium in 250-mL flasks with 1.7 mL each.
3. Incubate at 28°C and 250 rpm for 48 h in an orbital shaker (*see* **Note 16**).
4. Seed the production phase: 30 mL of *A. chrysogenum* fermentation medium in 500-mL flasks with 1 mL of vegetative each.
5. Incubate at 25°C and 250 rpm for 7 d in an orbital shaker (*see* **Note 16**).
6. Take a 2-mL sample for β-lactam antibiotics quantification.

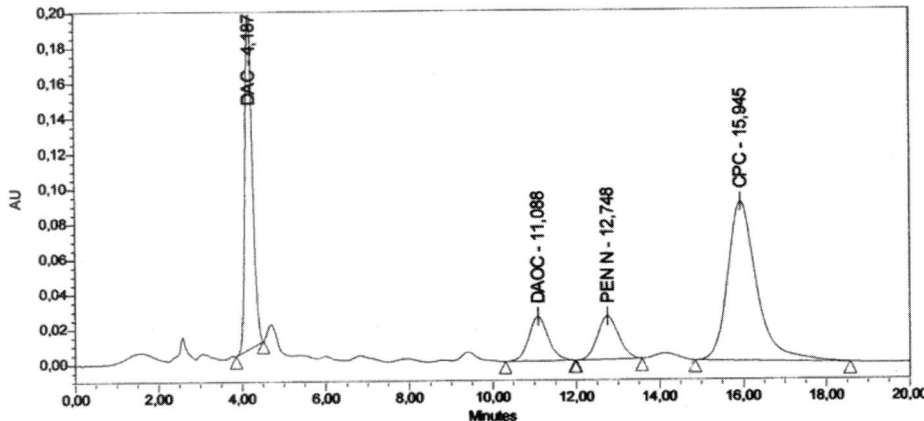

Fig. 6. HPLC profile of deacetylcephalosporin C (DAC), deacetoxycephalosporin C (DAOC), penicillin N (PN), and cephalosporin C (CPC) standards.

3.5. Cephalosporins and PN Quantification by HPLC

3.5.1. Mobile-Phase Preparation

1. Mix 2 mL of 99% formic acid with 2 L of deionized water.
2. Adjust to pH 3.3 with 30% ammonium hydroxide.
3. Add 2 mL of methanol and 1 mL of tetrahydrofuran.

3.5.2. Samples Preparation

4. Centrifuge 2 mL of culture broth at 15,000g for 5 min to remove the mycelium.
5. Mix 1 mL of supernatant with 9 mL deionized water.
6. Filtrate through 0.45 μm syringe filter unit and fill HPLC vials (*see* **Note 17**).

3.5.3. Chromatographic Performance

1. Adjust column temperature to 30°C.
2. Adjust mobile-phase flow to 2.3 mL/min in an isocratic elution mode.
3. Inject samples of 50 μL and run for 30 min.
4. Monitor OD_{214} for penicillins and OD_{254} for cephalosporins.

After resolving the typical HPLC profile of DAC, DAOC, PN, and cephalosporin C standards (*see* **Fig. 6**), β-lactam antibiotic production was determined for the parental strain *A. chrysogenum* and the transformants *A. chrysogenum* Δ*cefEF*-T1 and *A. chrysogenum* Δ*cefEF*-T1/*cefE*-T1 (*see* **Table 1**).

Table 1
β-Lactam Antibiotic Production by the Parental Strain *A. chrysogenum,*
the Transformant Δ*cefEF*-T1 Lacking the Enzymatic Activities DAOCS/DACS,
and the transformant Δ*cefEF*-T1/*cefE*-T1 Lacking the Enzymatic Activities
DAOCS/DACS But Expressing the DAOCS *(cefE)* From *S. clavuligerus*

Strain	PN	DAOC	DAC	CPC
A. chrysogenum	16.5	1.9	6.3	75.2
A. chrysogenum Δ*cefEF*-T1	99.5	0.0	0.5	0.0
A. chrysogenum Δ*cefEF*-T1/*cefE*-T1	3.7	95.9	0.3	0.1

Note: Average of penicillin N (PN), deacetoxycephalosporin C (DAOC), deacetyl-cephalosporin C (DAC), and cephalosporin C (CPC) production from three independent trials is shown as percentage of total β-lactams.

3.6. Bioconversion of DAOC Into 7-ADCA

The two-step enzymatic bioconversion of DAOC into glutaryl-7-ADCA begins with a reaction catalyzed the DAO activity of *R. gracilis* ATCC 26217 *(18,34)*. This reaction transforms DAOC into a mixture of glutaryl-7-ADCA and ketoadipyl-7-ADCA. The residual ketoadipyl-7-ADCA is chemically oxidized by hydrogen peroxide into glutaryl-7-ADCA. The second step involves the use of the GLA activity from *E. coli* ATCC 9637/pJC200 *(21,35)*. This reaction converts glutaryl-7-ADCA into 7-ADCA.

3.6.1. First Step: Bioconversion of DAOC Into Glutaryl-7-ADCA

3.6.1.1. *R. gracilis* Maintenance and Preservation

1. Inoculate 200 × 18-mm slants of YM medium with one colony suspended in 0.1 mL of 9 g/L NaCl.
2. Incubate for 7 d at 30°C.
3. Store at 4°C up to 1 mo.
4. For long-term storage, suspend one slant in 10 mL of 20% glycerol. Distribute 1-mL aliquots in 1.8-mL vials and store at −80°C (*see* **Note 15**).

3.6.1.2. *R. gracilis* Fermentation

1. Suspend one slant in 5 mL of 9 g/L NaCl.
2. Seed the vegetative phase: 100 mL of *R. gracilis* DAO production medium in a 500-mL flask with 0.5 mL of suspension.
3. Incubate at 30°C and 250 rpm for 48 h in an orbital shaker (*see* **Note 16**).
4. Seed 10 flasks of production phase: 100 mL of *R. gracilis* DAO production medium in 500-mL flasks with 0.5 mL of vegetative each.
5. Incubate at 30°C and 250 rpm for 48 h in an orbital shaker (*see* **Note 16**).

6. Determine DAO activity of the fermentation broth (*see* **Subheading 3.6.1.3.** and **Note 18**).
7. Centrifuge at 5000*g* and 4°C for 5 min. Discard the supernatant and keep the biomass for further assays (*see* **Note 19**).

3.6.1.3. DAO ASSAY

1. The reaction mixture is carried out in 1 mL and includes 25 m*M* D-phenylglycine, 50 m*M* sodium phosphate buffer, pH 8.0, 1 µ*M* FAD, and DAO solution.
2. Incubate at 30°C monitoring the increment of OD_{252} (*see* **Note 20**).

One unit (U) of DAO activity is the amount of enzyme able to transform 1 µmol substrate/min *(36)*.

3.6.1.4. BIOCONVERSION OF DAOC INTO GLUTARYL-7-ADCA

3. Harvest 1 L of *A. chrysogenum* Δ*cefEF*-T1/*cefE*-T1 fermentation broth and determine DAOC production.
4. Centrifuge at 2000*g* and room temperature for 15 min.
5. Pour the supernatant in a chemical reactor. Adjust to pH 7.5 and 25°C. Set the feeding of 1 L oxygen per minute through the sparger located at the bottom. Mix with gentle stirring.
6. Add *R. gracilis* cell paste and keep at pH 7.5 by feeding 5% ammonia for 2 h (*see* **Note 21**).
7. Filter through a 30-kDa cutoff filtration unit to remove solids.
8. Add 10 mL of 3.5% hydrogen peroxide per liter and incubate at 25°C for 15 min.
9. Add 0.5 g sodium pyruvate per liter.

After this treatment, all ketoadipyl-7-ADCA is transformed into glutaryl-7-ADCA, providing a solution close to 1% glutaryl-7-ADCA.

3.6.1.5. GLUTARYL-7-ADCA EXTRACTION

The glutaryl-7-ADCA is first extracted at acid pH to an organic phase and then re-extracted to water at alkaline pH. The method is as follows.
1. While stirring, add 1 vol of isobutyl acetate and chill to 4°C.
2. Adjust to pH 1.5 with 2 *N* H_2SO_4.
3. Centrifuge at 5000*g* and 4°C for 5 min. Recover the organic phase.
4. Add, with stirring, 0.5 vol of 25 m*M* phosphate buffer, pH 8.0. Adjust to pH 8.0 with 2 *N* KOH.
5. Centrifuge at 5000*g* and 4°C for 5 min. Recover the watery phase with a separating funnel.

3.6.2. Second Step: Bioconversion of Glutaryl-7-ADCA Into 7-ADCA

3.6.2.1. E. COLI MAINTENANCE AND PRESERVATION

1. Inoculate 200 × 18-mm slants of LB medium with one colony suspended in 0.1 mL of 9 g/L NaCl.
2. Incubate for 24 h at 28°C.

3. Store at 4°C up to 2 wk.
4. For long-term storage, suspend one slant in 10 mL of 10% glycerol. Distribute 1-mL aliquots in 1.8-mL cryogenic vials and store at –80°C (*see* **Note 15**).

3.6.2.2. *E. COLI* FERMENTATION

1. Suspend one slant with 5 mL of 9 g/L NaCl.
2. Seed the vegetative phase: 100 mL of *E. coli* GLA inoculum medium in a 500-mL flask with 0.5 mL of suspension.
3. Incubate at 28°C and 250 rpm until OD_{600} 15–20 (i.e., around 24 h) (*see* **Note 16**).
4. Seed 10 flasks of production phase: 100 mL of *E. coli* GLA production medium in 500-mL flasks with 2 mL of vegetative each.
5. Incubate at 23°C and 250 rpm for 72 h in an orbital shaker. The culture broth should get pH 8.0 and OD_{600} 20–25 (*see* **Note 16**).
6. Determine GLA activity of the fermentation broth (*see* **Subheading 3.6.2.3.** and **Note 22**).
7. Centrifuge at 5000*g* and 4°C for 5 min and discard the supernatant.
8. Suspend the biomass pellet in 9 g/L NaCl, centrifuge again at 5000*g* and 4°C for 5 min, and recover the cell paste (*see* **Note 19**).

3.6.2.3. GLA ASSAY

GLA activity is evaluated by measuring the rate of hydrolysis of glutaryl-7-ADCA into 7-ADCA. 7-ADCA is determined spectrophotometrically at 410 nm against a standard curve, by measuring the yellow color (Schiff's base) that forms with the reagent *p*-dimethylaminobenzaldehyde *(37)*. One unit (U) of GLA is defined as the quantity of enzyme which under the conditions of the method produces 1 µmol of 7-ADCA per minute.

1. The reaction is carried out in 1 mL and contains 0.5% glutaryl-7-ADCA, 10 m*M* sodium phosphate buffer, pH 8.0, and GLA solution.
2. Incubate at 37°C for 10 min.
3. Stop the reaction by adding 3 mL of 20% glacial acetic acid.
4. Spin for 10 min at 11,200*g*.
5. Mix 3 mL of supernatant with 0.5 mL of 0.5% *p*-dimethylaminobenzaldehyde dissolved in methanol.
6. Keep for 10 min at 25°C and determine OD_{415} against a blank without enzyme.

3.6.2.4. BIOCONVERSION OF GLUTARYL-7-ADCA INTO 7-ADCA

1. Mix by gentle stirring 5 g of glutaryl-7-ADCA in 300 mL of 0.1 *M* sodium phosphate buffer, pH 8.0, at 25°C.
2. Add *E. coli* cell paste and keep at pH 8.0 by feeding 5% ammonia for 2 h (*see* **Note 23**).
3. Filter through 30-kDa cutoff filtration unit to remove solids.

After this treatment, the conversion yield of glutaryl-7-ADCA into 7-ADCA is near 85%.

3.6.2.5. 7-ADCA Extraction

1. While stirring, add 4 g of charcoal and incubate at room temperature for 20 min.
2. Filter through Whatman 3MM paper to remove charcoal.
3. Adjust to pH 4.5 with $6N$ H_2SO_4 and incubate with gentle stirring for 15 min to initiate 7-ADCA crystallization.
4. Chill to 5–10°C and adjust to pH 3.8 with $6N$ H_2SO_4. Incubate for 2 h.
5. Filter through Whatman 3MM paper to recover the precipitate and wash it with 25 mL of water.

4. Notes

1. The presence of condensed water into the slants hinders the proper growth of the fungus. It is important store the slants at room temperature for 2–3 d before being seeded.
2. A good way to collect the biomass from all the slants in a volume as low as possible is to suspend the first slant in MMC medium and then transfer this suspension to the second, and so on. The final suspension is used to seed the inoculum. In this way, you can use higher volumes of suspension avoiding the dilution of the medium.
3. An excessive drying of the harvested mycelium hinders its further suspension in TPC buffer.
4. Dithiothreitol reduces the disulfide bonds of the fungus cellular wall, facilitating protoplast release.
5. The time necessary for the maximum release of protoplasts must be determined for each strain. *A. chrysogenum* wild-type strains frequently have a cellular wall stronger than overproducing strains, which, as result of the mutagenic treatment, are more susceptible to the lytic enzymes. In general, treatment for 2–3 h is enough for release up to 80% of the protoplasts.
6. The pellet must be homogeneously suspended to improve the washing procedure. Sometimes, the protoplasts form lumps difficult to disaggregate, indicating the presence of mycelial debris that can negatively affect the transformation efficiency.
7. From this step, you must take special care with protoplasts manipulation, because of their fragile physiological state. A vigorous agitation can damage protoplasts, decreasing noteworthy transformation efficiency. When possible, you must avoid the use of micropipets to mix the components, because they create an aspiration effect that can produce protoplasts lysis. Likewise, protoplast suspension must be kept in an ice-water bath while the counting is being done.
8. The CCM buffer is very viscous as a result of the presence of PEG, and its mixture with protoplasts suspension is difficult. To help this operation, invert the tube gently several times until a homogeneous mixture is obtained.
9. In addition to the transformation reactions, it is necessary to carry out negative control reactions without plasmid DNA. In this case, plasmid DNA is substituted by an equivalent volume of deionized water. These control reactions must be seed in TSAS plates (1) without the antibiotic chosen for the selection of the transformants to check protoplasts regeneration and (2) with the antibiotic to verify the sensibility of the strain and the absence of background growth. A deficient growth in the

plates without antibiotic would point to protoplast damage, yielding low transformation efficiency. The presence of background in the plates supplemented with antibiotic indicates insufficient dosage or heat inactivation.

10. Prepare control plates while the process for protoplasts formation is running and keep them drying in a laminar airflow cabinet. Likewise, melt top TSAS medium, pour into 10-mL polypropylene tubes, and keep at 50°C in a water bath until use.

11. Temperatures higher than 50°C would cause protoplast damage and antibiotic deactivation, whereas lower temperatures would lead to a partial solidification of the medium, preventing the homogeneous spread onto the plates. The antibiotic is fed with a micropipet near the bottom of the tube containing the prewarmed medium, whereas the transformation reaction is placed in the top, avoiding the direct contact between protoplasts and antibiotic. Mix gently by inversion of the tube several times and spread onto one plate before the solidification of the medium.

12. When the disrupted strain *A. chrysogenum* Δ*cefEF*-T1 (resistant to hygromycin) is transformed with pALC88 (confers resistance to phleomycin), both antibiotics must be present in the selection medium. Nevertheless, once integrated into the chromosome, the plasmid become very stable, being unlikely to be lost in absence of selective pressure.

13. The transformants selected by hygromycin or phleomycin resistance are seeded in individual Petri plates (50 mm in diameter) with a sterile platinum loop, spreading the mycelium into a square region of about 3–4 cm^2 ("patch"). The loop must be sterilized by heating before the seed of each transformant. Once grown, an agar plug is taken from the patch with a sterilized stainless-steel cylinder and placed in the bioassay plate. The plates with the patches are stored at 4°C for further use. The best producers selected by bioassay are picked up from the patch and seeded in slants. Slants are incubated at 28°C for 7 d, and their biomass is (1) stored in 10% glycerol at –80°C and/or (2) used for flask fermentation.

14. The culture of *E. coli* grown to OD$_{660}$ 1.65 can be stored as frozen aliquots (–80°C) in 20% glycerol for their use in further bioassays. These aliquots are prepared in 1.8-mL cryogenic vials by mixing 1 vol of cell suspension with 1 vol of 40% glycerol. Frozen aliquots are thawed at room temperature, and 1 mL of suspension is used to inoculate 100 mL LB medium.

15. To improve the viability of the frozen vial, incubate the mixture of protector agent (glycerol) and microorganism for 30 min at room temperature before freezing and then freeze slowly. Thaw must be done as quickly as possible.

16. High relative humidity (near 80%) is necessary to avoid medium evaporation. The orbit diameter of the orbital shaker should be 5 cm.

17. Antibiotics in solution deactivate quickly. Analyze samples after they are taken; do not store.

18. It is not necessary to disrupt *R. gracilis* cells for DAO activity determination. Typical results are nearly 4000 U/L.

19. Cell paste can be frozen at –20°C and preserved until use.

20. The DAO reaction is carried out in a water bath. The DAO solution is the last component added.

21. Four hundred units of DAO are enough to convert 5 g of DAOC in 2 h at 25°C. At this step, the yield of DAOC transformation into glutaryl-7-ADCA is near 90%, with a relevant 5–8% of ketoadipyl-7-ADCA. This ketoadipyl-7-ADCA is converted into glutaryl-7-ADCA by a further treatment involving hydrogen peroxide and sodium pyruvate.
22. It is not necessary to disrupt *E. coli* cells for GLA activity determination. Typical results are nearly 2000 U/L.
23. Two thousand units of GLA are enough to convert 5 g of glutaryl-7-ADCA in 2 h at 25°C.

Acknowledgment

The authors are indebted to the personnel of Área de Biotecnología for their excellent technical assistance.

References

1. Banko, G., Wolfe, S., and Demain, A. L. (1986) Cell-free synthesis of delta-(L-alpha-aminoadipyl)-L-cysteine, the first intermediate of penicillin and cephalosporin biosynthesis. *Biochem. Biophys. Res. Commun.* **137,** 528–535.
2. Banko, G., Demain, A. L., and Wolfe, S. (1987) δ-(L-α-aminoadipyl)-L-cysteinyl-D-valine synthetase (ACV synthetase): a multifunctional enzyme with broad substrate specificity for the synthesis of penicillins and cephalosporins precursors. *J. Am. Chem. Soc.* **109,** 2858–2860.
3. Gutiérrez, S., Díez, B., Montenegro, E., and Martín, J. F. (1991) Characterization of the *Cephalosporium acremonium pcbAB* gene encoding α-aminoadipyl-cysteinyl-valine synthetase, a large multidomain peptide synthetase: linkage to the *pcbC* gene as a cluster of early cephalosporin–biosynthetic genes and evidence of multiple functional domains. *J. Bacteriol.* **173,** 2354–2365.
4. Samson, S. M., Belagaje, R., Blankenship, D. T., et al. (1985) Isolation, sequence determination and expression in *Escherichia coli* of the isopenicillin N synthetase gene from *Cephalosporium acremonium. Nature* **318,** 191–194.
5. Ullán, R. V., Casqueiro, J., Bañuelos, O., Fernández, F. J., Gutiérrez, S., and Martín, J. F. (2002) A novel epimerization system in fungal secondary metabolism involved in the conversion of isopenicillin N into penicillin N in *Acremonium chrysogenum. J. Biol. Chem.* **277,** 46,216–46,225.
6. Scheidegger, A., Kuenzi, M. T., and Nüesch, J. (1984) Partial purification and catalytic properties of a bifunctional enzyme in the biosynthetic pathway of β-lactams in *Cephalosporium acremonium. J. Antibiot.* **37,** 522–531.
7. Samson, S. M., Dotzlaf, J. E., Slisz, M. L., et al. (1987) Cloning and expression of the fungal expandase\hydroxylase gene involved in cephalosporin biosynthesis. *Biotechnology* **5,** 1207–1214.
8. Gutiérrez, S., Velasco, J., Fernández, F. J., and Martín, J. F. (1992) The *cefG* gene of *Cephalosporium acremonium* is linked to the *cefEF* gene and encodes a deacetylcephalosporin C acetyltransferase closely related to homoserine *O*-acetyltransferase. *J. Bacteriol.* **174,** 3056–3064.

9. Díez, B., Mellado, E., Rodríguez, M., Fouces, R., and Barredo, J. L. (1997) Recombinant microorganisms for the industrial production of antibiotics. *Biotechnol. Bioeng.* **55,** 216–226.

10. Cantwell, C. A., Beckmann, R., Whiteman, P., Queener, S. W., and Abraham, E. P. (1992) Isolation of deacetoxycephalosporin C from fermentation broths of *Penicillium chrysogenum* transformants: construction of a new fungal biosynthetic pathway. *Proc. R. Soc., London B* **248,** 283–289.

11. Queener, S. W., Beckmann, R. J., Cantwell, C. A., et al. (1994) Improved expression of a hybrid *Streptomyces clavuligerus cef*E gene in *Penicillium chrysogenum.* *Ann. NY Acad. Sci.* **721,** 178–193.

12. Beckman, R., Cantwell, C. A., Whiteman, P., Queener, S. W., and Abraham, E. P. (1993) Production of deacetoxycephalosporin C by transformants of *Penicillium chrysogenum:* antibiotic biosynthetic pathway engineering, in *Industrial Microorganisms: Basic and Applied Molecular Genetics* (Baltz, R. H., Hegeman, G. D., and Skatrud, P. L., eds.), American Society for Microbiology, Washington DC, pp. 177–182.

13. Crawford, L., Stepan, A. M., McAda, P. C., et al. (1995) Production of cephalosporin intermediates by feeding adipic acid to recombinant *Penicillium chrysogenum* strains expressing ring expansion activity. *Biotechnology* **13,** 58–62.

14. Shibuya, Y., Matsumoto, K., and Fujii, T. (1985) Isolation and properties of 7β-(4-carboxybutanamido) cephalosporanic acid acylase-producing bacteria. *Agric. Biol. Chem.* **45,** 1561–1567.

15. Li, Y., Jiang, W., Yang, Y., Zhao, G., and Wang, E. (1998) Overproduction and purification of glutaryl 7-amino cephalosporanic acid acylase. *Protein Express Purif.* **12,** 233–238.

16. Dotzlaf, J. E. and Yeh, W. K. (1987) Copurification and characterization of deacetoxycephalosporin C synthase/hydroxylase from *Cephalosporium acremonium. J. Bacteriol.* **169,** 1611–1618.

17. Maeda, K., Luengo, J. M., Ferrero, O., et al. (1995) The substrate specificity of deacetoxycephalosporin C synthase ("expandase") of *Streptomyces clavuligerus* is extremely narrow. *Enzyme Microb. Technol.* **17,** 231–234.

18. Alonso, J., Barredo, J. L., Díez, B., et al. (1998) D-Amino-acid oxidase gene from *Rhodotorula gracilis (Rhodosporidium toruloides)* ATCC 26217. *Microbiology* **144,** 1095–1101.

19. Alonso, J., Barredo J. L., Armisén, P., et al. (1999) Engineering the D-amino-acid oxidase from *Trigonopsis variabilis* to facilitate its overproduction in *Escherichia coli* and its downstream processing by tailor-made metal chelate supports. *Enzyme Microb. Technol.* **25,** 88–95.

20. Matsuda, A. and Komatsu, K. I. (1985) Molecular cloning and structure of the gene for 7β-(4-carboxybutanamido) cephalosporanic acid acylase from a *Pseudomonas* strain. *J. Bacteriol.* **163,** 1222–1228.

21. Croux, C., Costa, J., Barredo, J. L., and Salto, F. (1994) Process for the enzymatic preparation of 7-aminocephalosporanic acid. US patent 05354667.

22. Cambiaghi, S., Tomaselli, S., and Verga, R. (1995) Enzymatic process for preparing 7-aminocephalosporanic acid and derivatives. US patent 5424196.
23. Rothstein, R. J. (1983) One-step gene disruption in yeast. *Methods Enzymol.* **101**, 202–211.
24. Shen, Y. Q., Wolfe, S., and Demain, A. L. (1986) Levels of isopenicillin N synthetase and deacetoxycephalosporin C synthetase in *Cephalosporium acremonium* producing high and low levels of cephalosporin C. *Biotechnology* **4**, 61–64.
25. Velasco, J., Adrio, J. L., Moreno, M. A., Díez, B., Soler, G., and Barredo, J. L. (2000) Environmentally safe production of 7-aminodeacetoxycephalosporanic acid (7-ADCA) using recombinant strains of *Acremonium chrysogenum. Nature Biotechnol.* **18**, 857–861.
26. Punt, P. J., Oliver, R. P., Dingemanse, M. A., Pouwels, P. H., and van den Hondel, C. A. M. J. J. (1987) Transformation of *Aspergillus* based on the hygromycin B resistance marker from *Escherichia coli. Gene* **56**, 117–124.
27. Díez, B., Mellado, E., Rodríguez, M., Bernasconi, E., and Barredo, J. L. (1999) The NADP-dependent glutamate dehydrogenase gene from *Penicillium chrysogenum* and the construction of expression vectors for filamentous fungi. *Appl. Microbiol. Biotechnol.* **52**, 196–207.
28. Hanahan, D. (1985) Techniques for transformation of *E. coli,* in *DNA Cloning: A Practical Approach* (Glover, D. M., ed.), IRL Press, Oxford, pp. 109–135.
29. LePage, G. A. and Campbell, E. (1946) Preparation of streptomycin. *J. Biol. Chem.* **162**, 163–171.
30. Queener, S. W., Ingolia, T. D., Skatrud, P. L., Chapman, J. L., and Kaster, K. R. (1985) A system for genetic transformation of *Cephalosporium acremonium,* in *Microbiology—1985* (Lieve, L., ed.), American Society of Microbiology, Washington, DC, pp. 468–472.
31. Shen, Y. Q., Wolfe, S., and Demain, A. L. (1986) Levels of isopenicillin N synthetase and deacetoxycephalosporin C synthetase in *Cephalosporium acremonium* producing high and low levels of cephalosporin C. *Biotechnology* **4**, 61–64.
32. Sambrook, J., Fritsch, E. F., and Maniatis, T. (1989) *Molecular Cloning: A Laboratory Manual.* Cold Spring Harbor Laboratory Press, Cold Spring Harbor, NY.
33. Kovacevic, S., Tobin, M. B., and Miller, J. R. (1990) The β-lactam biosynthesis genes for isopenicillin N epimerase and deacetoxycephalosporin C synthetase are expressed from a single transcript in *Streptomyces clavuligerus. J. Bacteriol.* **172**, 3952–3958.
34. Simonetta, M. P., Vanoni, M. A., and Curti, B. (1982) D-Amino acid oxidase activity in the yeast *Rhodotorula gracilis. Microbiol. Lett.* **15**, 27–31.
35. Armisén, P., Mateo, C., Cortés, E., et al. (1999) Selective adsorption of poly-His tagged glutaryl acylase on tailor-made metal chelate supports. *J. Chromatogr.* **848**, 61–70.
36. Fonda. M. L. and Anderson, B. M. (1967) D-Amino acid oxidase. Spectrophotometric studies. *J. Biol. Chem.* **242**, 3957–3962.
37. Balasingham, K., Warburton, D., Dunnill, P., and Lilly, M. D. (1972) The isolation and kinetics of penicillin amidase from *Escherichia coli. Biochem. Biophys. Acta* **276**, 250–256.

4

Production of Erythromycin With *Saccharopolyspora erythraea*

Wolfgang Minas

Summary

Actinomycetes are among the most fascinating microorganisms. Their developmental life cycle with its morphological and physiological differentiation and the rich repertoire of secondary metabolites (about 70–80% of bioactive secondary metabolites are being produced by actinomycetes) have resulted in a large research community studying these microbes. Of particular interest are actinomycetes for large-scale industrial production of bioactive molecules, such as polyketides, representing a diverse class of such compounds, including molecules with anticancer activity (e.g., mithramycin, daunorubicin, doxorubicin), antibacterials (e.g., erythromycin and derivatives or tetracyclines), antiparasitics (avermectins), or immunosuppressants (rapamycin). The biotechnological process for erythromycin production is described.

Key Words: Actinomycetes; polyketide; macrolide; erythromycin.

1. Introduction

Actinomycetes, and members of the genus *Streptomyces* in particular, are ubiquitous in nature and have been an excellent source for biologically active secondary metabolites. **Table 1** was recently compiled, listing some 62 useful secondary metabolites out of the over 9000 biologically active molecules isolated so far from actinomycetes, which account for two-thirds of the known antibiotics made by microorganisms and about 60% of all secondary metabolites with biological activities other than antibiotic. In both cases, nearly 80% of the molecules are made by members of the genus *Streptomyces* (*1*). Secondary metabolites are, by definition, superfluous to the metabolic activities that are needed for the growth of an organism. They can accumulate in substantial quantities and they are often excreted into the medium. Their function in nature, in most instances, cannot be defined clearly. Some molecules are designed as competitive weapons; others are implicated in metal transport or

From: *Methods in Biotechnology, Vol. 18: Microbial Processes and Products*
Edited by: J. L. Barredo © Humana Press Inc., Totowa, NJ

Table 1
List of Useful Actinomycete Antibiotics and Their Classification and Use

Antibiotic	Producer	Chemical class[a]	Target[b]	Application
Actinomycin D	*Streptomyces* spp.	Peptide	Transcription	Antitumor
Antimycin A	*Streptomyces* spp.	Macrolide	Cytochrome system	Telocidal
Avermectin	*S. avermitilis*	Macrolide (PK)	Chloride ion channels	Antiparasitic
Bambermycin	*S. bambergiensis*	Substituted amino-glycosides (complex of at least four moenomycins)	Peptidoglycan	Growth promoting
Bialaphos	*S. hygroscopicus*	Peptide	Glutamine synthetase	Herbicidal
Bleomycin	*S. verticillus*	Glycopeptide	DNA strand breakage	Antitumor
Candicidin	*S. griseus*	Polyene macrolide (PK)	Membrane (pore former)	Antifungal
Cephamycin C	*Nocardia lactamdurans* (and others)	β-Lactam	Peptidoglycan	Antibacterial
Chloramphenicol	*S. venezuelae*	N-Dichloracyl phenylpropanoid	R	Antibacterial
Chlorotetracycline	*S. aureofaciens*	Tetracycline (PK)	R	Antibacterial
Clavulanic acid	*S. clavuligerus*	β-lactam	β-Lactamase inhibitor	Combined with a β-lactam as anti-bacterial
Cycloserine	*S. orchidaceus*	Substituted cyclic peptide	Peptidoglycan	Antibacterial
Daptomycin	*S roseosporus*	Lipopeptide	Lipoteichoic acid?	Antibacterial
Daunorubicin (daunomycin)	*S. peucetius*	Anthracycline (PK)	DNA intercalation	Antitumor
Desferrioxamine	*S. pilosus*	Peptide	Iron chelating	Iron purging in iron overload

66

Doxorubicin (adiamycin)	S. peucetius var. caesius	Anthracycline (PK)	DNA intercalation	Antitumor
Erythromycin	S. erythraea	Macrolide (PK)	R	Antibacterial
FK506 (tacrolimus)	S. hygroscopicus	Macrolide (PK)	Binds to FK protein	Immuno-suppressant
Fortimicin	Micromonospora olivoasterospora	Aminoglycoside	R	Antibacterial
Fosfomycin	Steptonyces spp.	Phosphoric acid	Peptidoglycan	Antibacterial
Gentamycin	Micromonospora spp.	Aminoglycoside	R	Antibacterial
Hygromycin B	S. hygroscopicus	Substituted aminoglycoside	R	Antihelminitic
Kanamycin	S. kanamyceticus	Aminoglycoside	R	Antibacterial
Lasalocid	S. lasaliensis	Polyether (PK)	Membrane (ionophore)	Anticoccidial; growth promoting
Lincomycin	S. lincolnensis	Sugar-amide	R	Antibacterial
Milbemycin	S. hygroscopicus	Macrolide (PK)	Chloride ion channels	Antiparasitic
Mithramycin	S. argillaceus	Aureolic acid	DNA alkylation	Antitumor
Mitomycin C	S. caespitosus, S. verticillatus	Benzoquinone	DNA crosslinking	Antitumor
Monensin	S. cinnamonensis	Polyether (PK)	Membrane (ionophore)	Anticoccidial; growth promoting
Natamycin	S. nataensis	Tetraene polyene (PK)	Membrane (pore former)	Antifungal
Neomycin	S. fradiae	Aminoglycoside	R	Antibacterial
Nikkomycin	S. tendae	Nucleoside	Chitin biosynthesis	Antifungal; insecticidal

(continued)

Table 1 (continued)
List of Useful Actinomycete Antibiotics and Their Classification and Use

Antibiotic	Producer	Chemical class[a]	Target[b]	Application
Nocardicin	*Nocardia uniformis*	β-Lactam	Peptidoglycan	Antibacterial
Nosiheptide	*S. actuosus*	Thiopeptide	R	Growth promoting
Novobiocin	*S. niveus*	Coumerin glycoside	DNA gyrase (GyrB-subunit)	Antibacterial
Nystatin	*S. noursei*	Polyene macrolide (PK)	Membrane (pore former)	Antifungal
Oleandromycin	*S. antibioticus*	Macrolide (PK)	R	Antibacterial
Oxytetracycline	*S. rimosus*	Tetracycline (PK)	R	Antibacterial
Paromomycin	*S. rimosus forma paromomycinus*	Aminoglycoside	R	Antiamoebal
Phleomycin	*S. verticillus*	Glycopeptide	DNA strand breakage	Antitumor
Polyoxins	*S. cacaoi var. asoensis*	Nucleoside–peptide	Chitin biosynthesis	Antifungal (plant protection)
Pristinamycin	*S. pristinaespiralis*	Peptidic macrolactone + polyunsaturated macrolactone (PK)	R	Antibacterial
Puromycin	*S. alboniger*	Purine nucleoside	R	Research
Rapamycin	*S. hygroscopicus*	Macrolide (PK)	Binds FK protein	Immunosuppressant
Rifamycin	*Amycolatopsis mediterranei*	Ansamycin (PK)	RNA polymerase	Antibacterial
Ristocetin	*Nocardia lrruida*	Glycopeptide	Peptidoglycan	Antibacterial
Salinomycin promoting	*S. albus*	Polyether (PK)	Membrane (ionophore)	Anticoccidial; growth
Spectinomycin	*S. spectabilis*	Aminocyclitol	R	Antibacterial

68

Name	Organism	Type	Target	Use
Spinosyns	*S. spinosa*	Tetracyclic mocrolide (PK)	Unknown	Insecticidal
Spiramycin	*S. ambofaciens*	Macrolide (PK)	R	Antibacterial
Streptogramins	*S. graminofaciens*	Macrocyclic lactons	R	Antibacterial
Streptomycin	*S. griseus*	Aminoglycoside	R	Antibacterial
Streptothricin	*S. lavendulae*	N-Glycoside	R	Growth promoting; plant protection
Teichoplanin	*Actinoplanes teichomyceticus*	Glycoprotein	Peptidoglycan	Antibacterial
Tetracycline	*S. aureofaciens*	Tetracycline (PK)	R	Antibacterial
Thienamycin	*S. cattleya*	β-Lactam	Peptidoglycan	Antibacterial
Thiostrepton	*S. azureus*	Thiopeptide	R	Growth promoting
Tobramycin	*S. tenebrarius*	Aminoglycoside	R	Antibacterial
Tylosin	*S. fradiae*	Macrolide (PK)	R	Growth promoting
Validamycin	*S. hygroscopicus*	Aminoglycoside	R	Plant protection
Vancomycin	*Amycolatopsis orientalis*	Glycopeptide	Peptidoglycan	Antibacterial
Virginamycin	*S. virginiae*	Macrocyclic lactone (PK) + macrocyclic peptidolactone	R	Growth promoting

[a] PK = polyketide.
[b] R = binds to ribosomes and thus inhibits protein synthesis.
Source: **ref. 1**.

might act as stimulants for symbiosis, sexual hormones, or effectors for differentiation *(2)*.

Among these secondary metabolites, the polyketides are a particularly interesting large group of structurally and chemically diverse molecules that exhibit different important biological activities, including anticancer (e.g., mithramycin, daunorubicin, doxorubicin), antibacterial (e.g., erythromycin and derivatives or tetracyclines), antiparasitic (avermectins), or immunosuppressive (rapamycin) properties (*see* **ref. 3** for a review). Most of the known polyketides are produced by *Actinomycetes,* but other organisms such as plants and fungi have been shown to be polyketide producers as well *(4–6)*.

Polyketides are built from short carboxylic acid units (primarily acetate, propionate, butyrate) that are activated as coenzyme A thioesters for decarboxylative (Claisen) polycondensations. The enzymes involved are polyketide synthases (PKS). Two types of polyketide biosynthetic system have been described. Type I modular PKSs *(7,8)* are large multifunctional proteins (modules) that harbor a distinct enzyme activity for every step catalyzed and function largely nonreiteratively. Type II PKSs *(7)* are multiprotein complexes that carry a single set of reiteratively used activities and consist of several largely monofunctional proteins. Hence, the synthesis of the polyketides is similar to fatty acid synthesis in eukaryotes (PKS type I) and prokaryotes (PKS type II). The structural diversity of the polyketides arises from the PKS-encoded variability in the chain length of the backbone and from the subsequent reactions such as ketoreductions, ring closures, aromatizations, hydroxylations, methylations, dimerizations, glycosylations, and so forth. Among *Actinomycetes,* type I enzymes are exemplified by the PKS for erythromycin biosynthesis by *Saccharopolyspora erythraea (9)*, whereas type II PKSs are responsible for the synthesis of a variety of aromatic polyketides such as actinorhodin *(10)*.

The structural diversity of the polyketides is also reflected in the diversity of their producers. It is generally accepted that an imbalance in the primary metabolism and the resulting stress induces the production of secondary metabolites *(11–13)*. Other stress factors or even bacterial communication via A-factor autoinducers can also induce or contribute to the induction of secondary metabolism *(14,15)*. The details, however, vary widely between the producing strains.

The macrolide antibiotic erythromycin has been chosen for a detailed description of its production. Erythromycin is made from *Saccharopolyspora erythraea* (formerly *Streptomyces erythreus*). *S. erythraea* was first described in 1919 by Waksman *(16)*. In 1952, the antibiotic activity was reported in a different strain, *S. erythraea* NRRL2338, derived from this isolate *(17)*. Strain NRRL 2338 produced about 0.25–1 g erythromycin per liter, depending on the fermentation conditions. This strain can be viewed as the ancestor of most current production strains. It was surprising that 50 yr since fermentation data was first published, fermentation yields had increased by little more the 10 times to around 10–13 g/L of the final broth. The biosynthesis is encoded by the *ery* genes clustered on an approximate 60-kb DNA fragment. The biosynthesis of erythromycin has been described elsewhere *(9)* and is summarized in **Fig. 1.** The fermentation will yield a mixture of erythromycin A, B, C, and D, of which erythromycin A is the only desired compound. The other compounds represent partially hydroxylated or methylated intermediates. A good production strain is characterized by a high titer of erythromycin A, at least 8–10 g/L, combined with a >90% content of erythromycin A. Erythromycin was marketed first by Lilly (1952). In the 1970s, Abbott emerged as the main producer and still holds the largest share of the erythromycin market in both amount and value. The annual production of erythromycin is about 4000 tons. Although some of it is being used as erythromycin (1000 tons), most of it is chemically converted into azithromycin (1500 tons), clarithromycin (1500 tons), and roxithromycin (400 tons).

This chapter will describe the fermentation of *S. erythraea* and the principle steps for the isolation of erythromycin. Furthermore, a brief description is given for the derivatizations of erythromycin A into the aforementioned products. Some descriptions, in particular the media composition and the feed profiles, are strain-specific. Nevertheless, most steps and notes will generally apply. In particular, some to the relatively simple methodology employed in the industrial setting will complement the more sophisticated techniques often used in the academic environment.

2. Materials

2.1. Strain Selection, Maintenance, and Fermentation

1. *Saccharopolyspora erythraea* NRRL, 2338 (NRRL, Peoria, IL), *S. erythraea* sp. (industrial strain), or *S. erythraea::vhb* (recombinant strain) *(18,19)* (*see* **Note 1**).
2. Tryptic soy broth (TSB) (Oxoid, Basingstoke, UK) solidified with 2% agar (bacto agar) (Difco Laboratories, Detroit, MI).
3. M1 agar medium: 5 g/L anhydrous glucose, 5 g/L tryptone (Oxoid, Basingstoke, UK), 0.5 g/L betaine-HCl, 5 g/L ARGO corn starch (CPC International Inc., Englewood Cliffs, NJ), 1 g/L corn steep liquor C*Plus F15855 (Cargill Cerestar BVBA, Mechelen, Belgium), 200 mg/µL $MgSO_4 \cdot 7H_2O$, 2 mg/L $ZnSO_4 \cdot 7H_2O$, 0.8 mg/L $CuSO_4 \cdot 5H_2O$, 0.2 mg/L $CoCl_2 \cdot 6H_2O$, 4 mg/L $FeSO_4 \cdot 7H_2O$, 80 mg/L $CaCl_2 \cdot 6H_2O$, 10 g/L NaCl, 150 mg/L KH_2PO_4, and 20 g/L agar (Difco Laboratories) *(19)* (*see* **Note 2**).
4. CMAE-1 medium: 25 g/L oat meal, 100 g/L fresh pealed, sliced, and boiled potatoes (no sprouts), 0.25 g/L L-arginine, and 25 g/L agar. pH adjusted with HCl or NaOH to 6.85 and sterilized for 20 min at 121°C.
5. V1 medium: 16 g/L ARGO corn starch (CPC International Inc., Englewood Cliffs, NJ), 10 g/L dextrin D-2256 (Sigma, St. Louis, MO), 15 g/L soybean flour 32H0411 (Sigma), 2.5 g/L NaCl, 5 mL/L corn steep liquor C*Plus F15855 (Cargill Cerestar

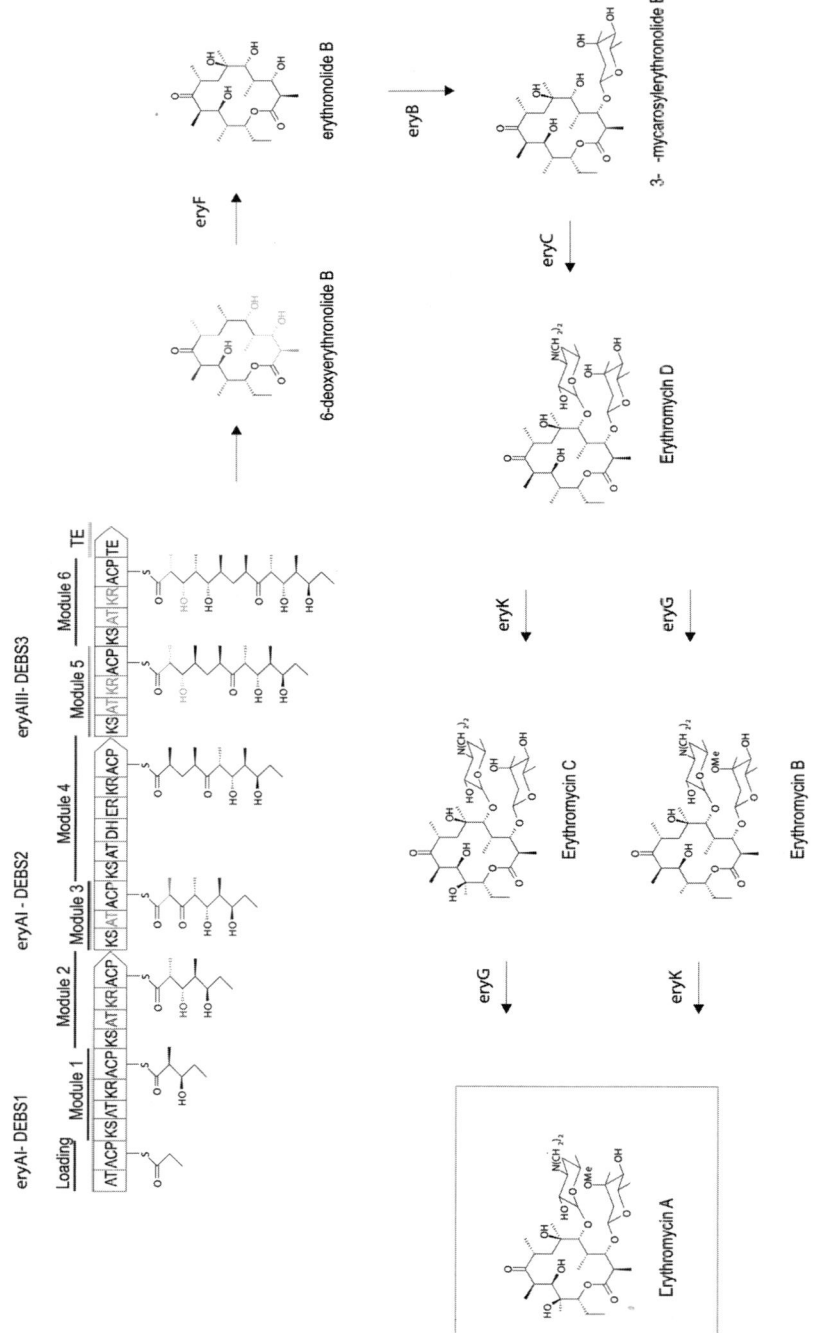

Fig. 1.

BVBA, Mechelen, Belgium), 1 g/L $(NH_4)_2SO_4$, 6 mL/L pure soybean oil #01190 (Nef Lebensmittel AG, Zürich, Switzerland), and 4 g/L $CaCO_3$ (Sigma, St. Louis, MO). The pH was adjusted to 6.5 before autoclaving *(19)* (*see* **Notes 2** and **3**).

6. V2 medium: 18 g/L corn starch, 12 g/L dextrin, 15 g/L soybean flour, 3 g/L NaCl, 6 mL/L corn steep liquor C*Plus F15855 (Cargill Cerestar BVBA, Mechelen, Belgium), 1.2 g/L $(NH_4)_2SO_4$, 6 mL/L soybean oil, and 5 g/L $CaCO_3$. The pH was adjusted to pH 6.8 after the medium was autoclaved *(19)* (*see* **Notes 2** and **3**).

7. F1 medium: 35 g/L ARGO corn starch (CPC International Inc., Englewood Cliffs, N.J.), 32 g/L dextrin, 33 g/L soybean flour, 7 g/L NaCl, 20 mL/L corn steep liquor C*Plus F15855 (Cargill Cerestar BVBA, Mechelen, Belgium), 2 g/L $(NH_4)_2SO_4$, 6 mL/L soybean oil, and 8 g/L $CaCO_3$. The pH of the sterile F1 medium was adjusted to pH 6.5 *(19)* (*see* **Notes 2–5**).

8. Sterile antifoam, Mazu DF 204 (PPG Ouvrie, Lesquin, France), was added prior to sterilization (approx 0.5 mL/L) and during cultivations as needed.

9. Soy bean oil, fresh, preferably nonrefined (e.g., no. 01190) (Nef Lebensmittel AG, Zürich, Switzerland) (*see* **Note 2**).

10. Dextrin D-2256 (Sigma, St. Louis, MO) (*see* **Note 5**).

11. Methylen blue reagent: 1% methylen blue stock solution (prepared in ethanol) in water (1 : 100).

12. YSI 2700 Biochemistry analyzer fitted with the 2730 Monitor and Control Module (YSI Inc., Yellow Springs, OH).

13. Cross-flow filter assembly (BioEngineering, Wald, CH) fitted with a 0.2-μm IRIS 6502 membrane (Rhodia Orelis, formerly Rhone-Poulenc Tech-Sep, Miribel, France).

14. Bio-Rad Protein Assay Kit (Bio-Rad Laboratories, Hercules, CA) or Folin reagent procedure after alkaline lysis of the culture, using bovine serum albumin as a standard *(20,21)*.

15. Preweighed glass fiber filters GF/C (Whatman, Maidstone, UK) and filter holder.

16. Perlite filter aid.

Fig. 1. Erythromycin biosynthesis. The three genes *eryAI–III* encode for the multidomain proteins DEBS1–3. The first protein, DEBS1, consist of a loading module with an actyltransferase (AT) and acyl carrier protein (ACP) domain, followed by module 1 with the ketosynthase (KS), AT, ketoreductase (KR), and an ACP domain, followed by module 2 with KS, AT, KR, and ACP domains. The second protein, DEBS2, consists of two modules, with a KS, AT, ACP domain and a KS, AT, dehydrogenase (DH), enoylreductase (ER), KR, ACP domain, respectively. The third protein, DEBS3, also consists of two modules and a thioesterase (TE) domain to release the polyketide from the module. The growing chain is indicated below the proteins. The genes involved in the hydroxylations and glycosylations and methylation are indicated by their name: eryF, 6-deoxyerythromolid B hydroxylase; eryB, locus for L-mycarose synthesis and transferase; eryC, locus for D-desosamine synthesis and transferase; eryK, erythromycin DIB hydroxylase; eryG, *o*-methyltransferase

17. Test tubes for agar slants with 15-mm diameter and 15- to 20-cm length.
18. Benchtop centrifuge and graduated centrifuge tubes.
19. Orbital shaker incubator for 500-mL Erlenmeyer flasks with 250 rpm and a 5-cm stroke and cooling facility.
20. Spectrophotometer.
21. Microscope.
22. Lab- or pilot-scale bioreactors equipped with gas flow control, a regulation of dissolved oxygen and pH, and the auxiliary equipment to implement the three feed streams in the main fermentation, e.g. 5-L LH 210 bioreactor (Adaptive Biosystems, Luton, UK), 20-L Infors ISF200 bioreactor (Infors AG, Bottmingen, Switzerland).
23. Equipment for on-line measurement of glucose and for off-gas analysis (optional), e.g., mass spectrometer VG Prima 600, VG Gas Analysis Systems, Middlewich, UK).
24. Viscometer (optional).

2.2. Isolation of Erythromycin

Refrigerated centrifuge generating $2000g$ to $3000g$.

2.3. Assay for Erythromycin Purity and Activity

1. *Micrococcus luteus* ATCC 9341 (ATCC, Manassas, VA) was used as challenge strain in the erythromycin bioassay.
2. *M. lutens* growth medium: Luria broth (LB) (Difco Laboratories, Detroit, MI).
3. Bioassay medium: Tryptic soy broth (TSB) (Oxoid, Basingstoke, UK) supplemented with 2 g/L glucose, and 2% agar.
4. Antibiotic test disks 6.6 mm in diameter (Difco Laboratories, Detroit, MI).
5. Erythromycin A.
6. Erythromycin B and erythromycin C (U.S. Pharmacopeia, Rockville, MD).
7. $1M$ phosphate buffer.
8. Chloroform : octan 70 : 30 (v/v).
9. Rotatory evaporator connected to vacuum.
10. High-performance liquid chromatography (HPLC) system equipped with an ultraviolet (UV) detector set to 215 nm and a µBondapack C_{18} (300 × 3.9 mm inner diameter) column (Waters, Milford, MA).
11. Mobile phase: acetonitrile–methanol–0.2 M ammonium acetate–water (45 : 10 : 10 : 35).

3. Methods

The methods described outline the selection and maintenance of high producing strains of *S. erythraea* and the fermentation of the strain (*see* **Subheading 3.1.**) (both summarized in **Fig. 2**), the extraction of the crude erythromycin (*see* **Subheading 3.2.**), and the assays for erythromycin A purity and activity (*see* **Subheading 3.3.**). **Subheading 3.4.** provides a brief look at the conversion of erythromycin into its semisynthetic derivatives.

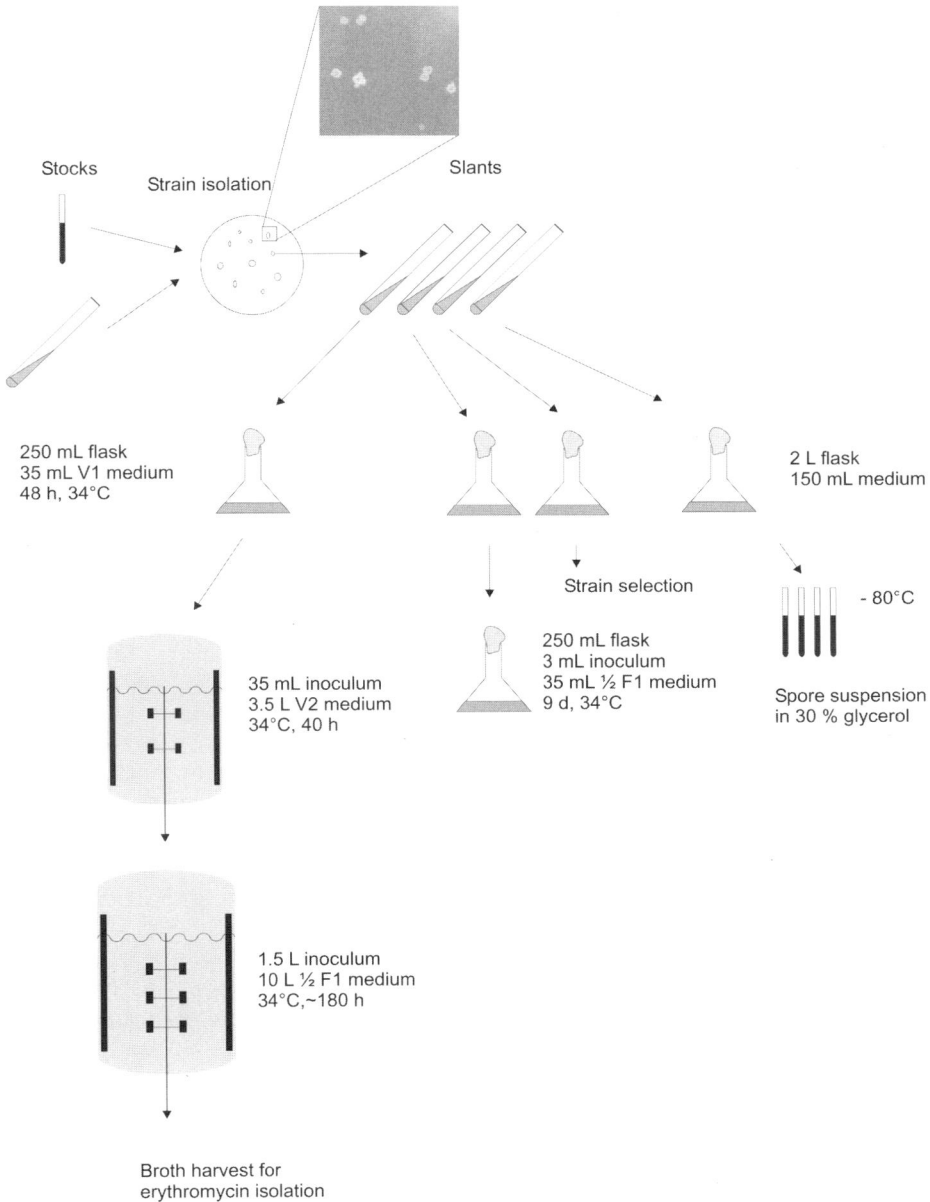

Stocks

Strain isolation

Slants

250 mL flask
35 mL V1 medium
48 h, 34°C

2 L flask
150 mL medium

Strain selection

- 80°C

35 mL inoculum
3.5 L V2 medium
34°C, 40 h

250 mL flask
3 mL inoculum
35 mL ½ F1 medium
9 d, 34°C

Spore suspension
in 30 % glycerol

1.5 L inoculum
10 L ½ F1 medium
34°C,~180 h

Broth harvest for
erythromycin isolation

Fig. 2. Overview of strain maintenance, strain selection, and fermentation processes.

A

B

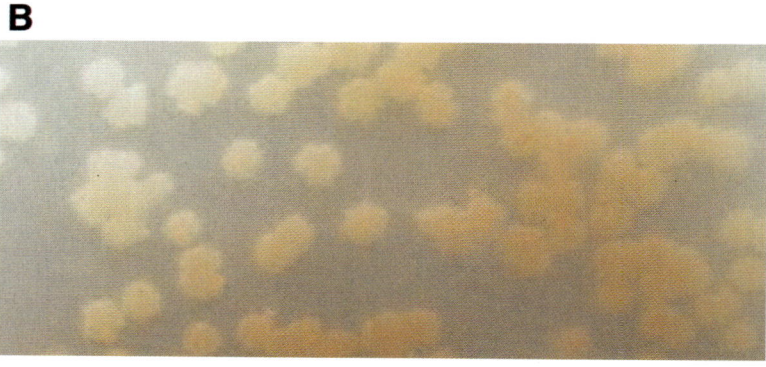

Fig. 3. Colony morphology of *S. erythraea:* (**A**) bottom view of a plate and (**B**) top view of a plate.

3.1. Strain Selection, Maintenance, and Fermentation

Genetic variability in high-producing strains makes necessary a continuous effort to reisolate high-producing strains with the desired low spectrum of side products, mainly erythromycin B and C (*see* **Note 1**).

3.1.1. Selection of a High-Yielding S. erythraea Strain

1. Upon receiving a strain, typically as a slant culture, lyophilized stock, or spore suspension, the first cultivation is done using the V1 medium in a 500-mL Erlenmeyer flask filled with 30 mL medium (*see* **Notes 2** and **3**).
2. After 48 h incubation, dilute the cell suspension and plate for single colonies onto either M1 or CMAE-1 medium. Because of differences between strains, the best medium for sporulation needs to be determined experimentally.
3. After 10–14 d of incubation at 34°C and approx 60% humidity, visually inspect sporulated colonies for variations in colony morphology. Typically, colonies grew

starlike to 3 mm in diameter with a pinkish gray color and a light brown spore pigment (*see* **Fig. 3**). The bottom of the colonies appeared deep brown.

4. Isolate individual colonies of different morphologies and use them for the preparation of agar slant cultures (*see* **Subheading 3.1.2.**).

3.1.2. Maintenance of S. erythraea

1. Prepare slants in wide test tubes (15 mm in diameter) filled with 8 mL M1 or CMAE-1 agar. The medium is allowed to solidify with a slope.
2. Use a single colony, picked from a plate, to inoculate the surface of two to three slants.
3. Incubate the cultures for 10–14 d at 34°C until sporulated. These slants can be stored for up to 1 mo at 4°C.
4. Prepare spore suspensions by adding 5 mL of sterile water to the slant and gently scraping off the surface using a sterile pipet.
5. Add sterile glycerol to 20% and store 1.5-mL aliquots at –80°C. Glycerol stocks can be used within 1 yr without any effect on viability or erythromycin production.
6. Inoculate a 30-mL seed culture in V1 medium with 1 mL of the fresh spore suspension or 1.5 mL frozen glycerol stocks and incubate for 40 h at 34°C with 250 rpm agitation in an orbitary shaker (*see* **Notes 6–8**).
7. Inoculate 27 mL half-strength F1 production medium with 3 mL of the seed culture (*see* **Notes 2** and **3**).
8. Run cultivations for erythromycin production in this medium for 9 d with daily addition of soybean oil (0.2 mL, d 0–6) and *n*-propanol (0.1 mL, d 0–5; 0.15 mL, d 6–9).
9. Weigh the shake flasks daily and add sterile water to compensate for evaporation if necessary.
10. Take samples, filter, and store for determination of the erythromycin titer (*see* **Subheading 3.3.**).
11. Choose the best performing isolates for the preparation of frozen stocks.

3.1.3. S. erythraea Fermentation

Cultivations were performed in three stages.

1. Grow the first-stage seed culture in 35 mL of V1 medium.
2. After 48 h, use this culture to inoculate 3.5 L of V2 medium. This second-stage seed cultivation is performed in a 5-L bioreactor equipped with a pitched blade turbine. The agitation speed is set to 800 rpm, the airflow rate is set to 0.6 volume/volume/minute (vvm), the temperature is controlled at 34°C, and the dissolved oxygen tension (DOT), pH, and redox potential profiles are monitored. CO_2 and O_2 concentrations in the exhaust gas are monitored on-line with a mass spectrometer (VG Prima 600). Growth of the culture is monitored by determination of the packed mycelium volume (PMV) centrifuging 10 mL whole broth 10 min at 3000g and determination of the percentage of the packed biomass. Alternatively, cell dry weight is determined by filtering 10 mL broth through preweighed glass fiber filters, washing once with 1 vol of water, and drying the filters for 24 h at 80°C prior to weighing (*see* **Notes 3, 5, 9,** and **10**).

3. After about 40 h, 1.5 L of culture broth is transferred to a 20-L bioreactor equipped with two Rushton turbines and containing 10 L of half-strength F1 production medium (see **Notes 3, 5,** and **9–13**). The proper transfer time from the V2 medium into the productive fermentation is determined by methylen blue reduction assay. Briefly, 4 mL freshly drawn sample are mixed with 1 mL methylen blue reagent. After mixing, the time needed for reduction of the color is registered. A reduction time of 1–2 min is characteristic for a healthy, well-grown, and ready for transfer culture. Overgrown cultures as well as cultures that did not grow well are characterized by reduction times of more than 2.5 min are not used. Cultivation conditions for the productive stage are set as follows: The temperature is set to 34°C; the pH is controlled with H_2SO_4 so as not to exceed 7.2, the agitation speed is initially set to 700 rpm and controlled by the DOT signal to increase to a maximum of 900 rpm if the DOT fell below 45% air saturation, the airflow rate is set to 0.37 vvm for the first 12 h and then changed to 0.83 vvm, which fell to 0.7 vvm as the feed was added to the reactor volume, and the pressure is set to 0.1 bar gage (overpressure). Feeding rates and durations are: 2.4 mL/L/d n-propanol from 12 to 160 h, 4.8 mL/L/d soybean oil from 25 h until the end of the cultivation, and 48 mL/L/d 15% dextrin from 30 to 90 h. The low flow rates are realized by activating the feed pumps once per minute to add the proper amounts of feed compounds. Redox potential and CO_2 and O_2 concentrations in the exhaust gas are monitored on-line, as described earlier. In addition, free glucose is measured hourly with a YSI 2700 Biochemistry analyzer fitted with the 2730 Monitor and Control Module. Samples are drawn aseptically through a cross-flow filter assembly fitted with a 0.2-μm IRIS 6502 membrane. Daily samples of 50 or 100 mL are drawn to determine the erythromycin titer, PMV, or cell dry weight and microscopic inspection for both possible contamination and mycelial morphology. One-milliliter aliquots are 0.2-μm filtered for HPLC analysis and the samples stored at –20°C until the analysis.

Figure 4 shows the summary of two productive fermentations using two different strains. The horizontal bars in the center indicate the duration of the respective feeds. Because glucose will repress erythromycin production, a close control of the dextrin feed is advised to avoid high free-glucose concentration (more than 1 g/L).

The first and greatest instance of free glucose, about 1 g/L, occurred during the first 12 h after inoculation, which is likely the result of the induction of the amylase and release of glucose from starch. A second peak is observed after 38 h following the start of the dextrin feed, and a third is observed at the end of the dextrin feed, when the metabolic activity shifted toward oil consumption.

The oil feed was also adjusted to the consumption by visually inspecting the samples. A major accumulation of oil droplets was circumvented by reducing the feed rate.

The erythromycin production rates were calculated for the time interval from 48 to 144 h and found to be 57.5 mg/L/h for the improved strain, compared to

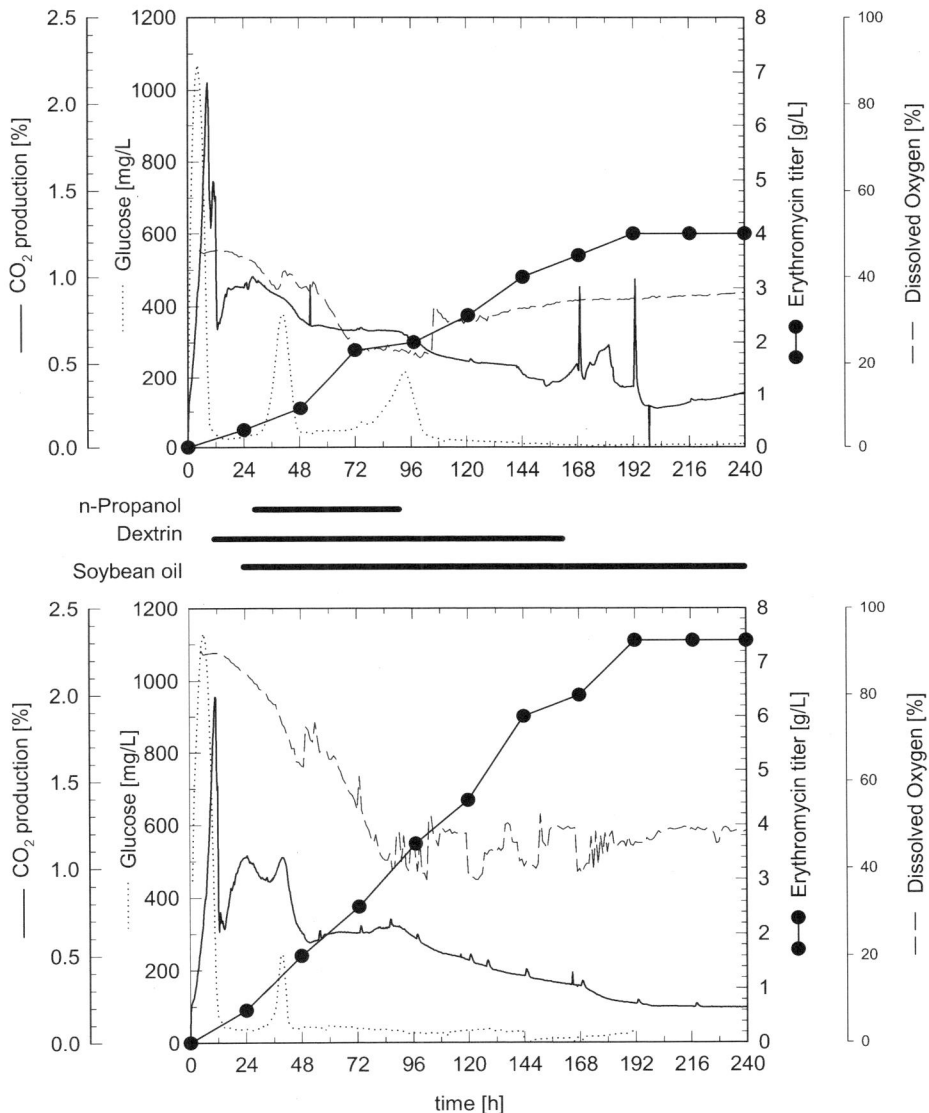

Fig. 4. On-line data for dissolved oxygen, CO_2 production, and glucose and off-line data for erythromycin titer of two productive fermentations of two strains are shown. The top panel is a recombinant strain harboring the hemoglobin-encoding gene from *Vitreoscilla* (**19**); the lower panel was the original strain. The feed intervals are indicated in the middle.

Table 2
Summary of Data on Biomass (Total Protein) and Erythromycin Yield
of the Two Cultivations Shown in Fig. 4

Strain	Time (h)	Total protein (g/L)	Erythromycin titer (g/L) in bioassay	Erythromycin yield on biomass (g/g)
S. erythraea::vhb	168	2.6	6.7	2.6
(improved recombinant)	216	3.2	7.4	2.3
S. erythraea spp.	168	4.1	3.6	0.9
	216	4.2	4.0	1.0

Source: **ref. *19*.**

only 24.3 mg/L/h for the original production strain. The higher erythromycin production together with 24–40% lower biomass concentration obtained with the improved *S. erythraea* strain resulted in an about 2.5-fold higher overall specific erythromycin yield (grams of erythromycin/g of protein) in this recombinant strain (*see* **Table 2**). While the biosynthesis rate in *S. erythraea* spp. dropped after about 90 h into the cultivation from initial high rate to a reduced rate of 1.9 mg/L/h, the erythromycin production rate of the improved recombinant strain *S. erythraea::vhb* remained at the high rate of 4.2 mg/L/h (*see* **Fig. 4**). The combination of faster erythromycin production with continuation of production to higher titers by *S. erythraea::vhb* ensures even greater advantages in overall space-time yield of the process. For example, by harvesting the *S. erythraea::vhb* after 144 h, a space-time yield of 1.1 g of erythromycin/L/d is achieved, which is 100% greater than the space-time yield achieved by *S. erythraea* spp. fermentation (0.56 g of erythromycin/L/d, harvesting at the point of maximum titer after 192 h).

The HPLC analysis of broth samples revealed that the ratios of erythromycin A (the active compound) to erythromycin B and C were identical for the two strains. The distribution was 75–80% erythromycin A, about 15% erythromycin C, and the rest being erythromycin B (data not shown). Microscopic analysis of the samples revealed that, starting at about d 8 (192 h), a progressive fragmentation of the mycelium occurred. This disintegration was closer studied in a separate cultivation (*see* **Fig. 5**), showing that broth viscosity, PMV, and cell dry weight correlated well. The increase in total protein determined after 192 h coincided with the maximum erythromycin titer and indicated a protein release because of the disintegrations of the mycelium, observed in the microscopic analysis of the samples.

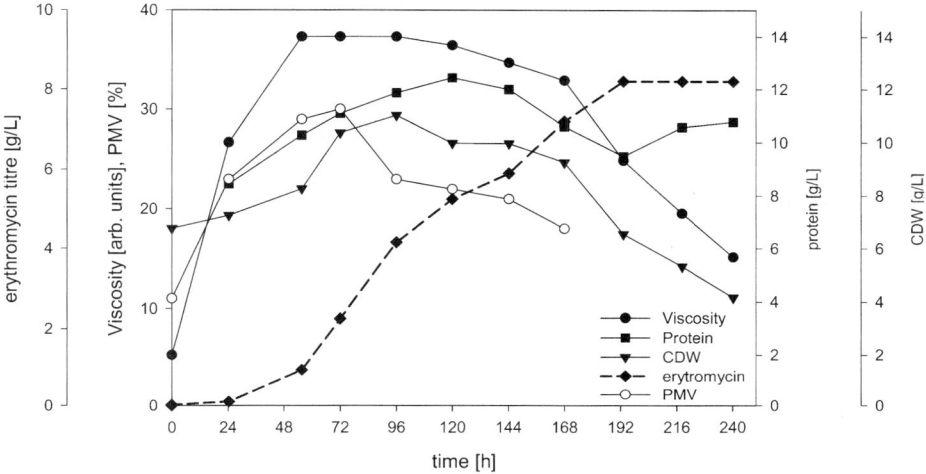

Fig. 5. Correlation among broth viscosity, PMV, cell dry weight, erythromycin titer, and protein content in productive cultivations. Details in the text.

3.2. Isolation of Erythromycin

3.2.1. Isolation of the Erythromycin–Isothiocyanate

1. Adjust the pH of approx 5 kg of whole broth with 20% NaOH to 8.5.
2. Remove the biomass by first centrifugation with 3000*g* for 30 min at 4°C. The supernatant is further clarified by filtration through a glass fiber filter (Whatmann, GF/C).
3. Adjust the pH to 9.5 ± 0.1.
4. Add 1.5% methylisobutylketone and stir the mixture for 10 min at 4°C.
5. Stop stirring and separate phases by centrifugation (5 min with 2000*g*).
6. Collect the organic phase and determine the erythromycin content by HPLC (*see* **Subheading 3.3.2.**).
7. Heat the organic phase to 30°C and add 0.63 L of a 33% Na isothiocyanate solution to the organic phase for each gram of erythromycin.
8. Adjust the pH to 5.8 with 10% acetic acid under mild agitation.
9. Cool the mixture to 10°C and incubate for 3 h under mild agitation for the crystallization of the erythromycin–isothiocyanate. Monitor the pH and adjust to 5.8 as needed.
10. Filter the suspension and wash the white filter cake with distilled water.
11. Dry overnight at 50°C under vacuum.

The weight yield throughout these processes is approx 83%.

3.2.2. Preparing the Erythromycin Base

Erythromycin–isothiocyanate, wet from the filter, is converted into erythromycin crude base.

1. Resuspend about 1.3 kg of the filter cake in 4.6 kg of dichloromethane.
2. Under mild stirring at 33°C, add 0.6 kg of a 15% NaOH solution. The final pH should be between 9.5 and 10.0.
3. Add 3 g of each filter aid (Perlite) and activated carbon and filter the suspension.
4. Wash with a small quantity of dichloromethane.
5. Extract the organic phase (filtrate) with 2 L deionized water for 30 min at 33°C under slow stirring.
6. After phase separation, the aqueous phase is discarded and the organic phase reduced to 1.5 L under vacuum. The resulting increase in erythromycin concentration beyond its solubility induces precipitation of the macrolide.
7. Maintain slow agitation and low temperature (5°C) for 3–5 h for complete crystallization.
8. Recover the precipitated erythromycin by centrifugation or filtration, wash with 200 mL dichloromethane, and then dry at 50°C. About 700 g crude erythromycin base are obtained.
9. Resuspend the crude base in 3.6 L deionized water and heat to 50–60°C.
10. Add 150 mg lauryl sulfate.
11. Centrifuge or filter and wash with deinonized water (60°C) to obtain around 685 g purified erythromycin base.

Erythromycin base can be converted further into the different esters, salts, or semisynthetic derivates, some of which are briefly listed in **Subheading 3.4.**

3.3. Assay for Erythromycin Purity and Activity

3.3.1. Bioassay for Erythromycin Activity

The titers of erythromycin produced in a shake flask or bioreactor cultivations of *S. erythraea* were determined using a conventional bioassay with commercially available erythromycin as a standard.

1. Pour portions (35 mL) of TSB into Petri dishes (12×12 cm).
2. Once the medium is solidified, add a second layer consisting of 35 mL TSB medium containing 35 µL of a *Micrococcus luteus* overnight culture (grown 18 h at 30°C in LB).
3. Determine the erythromycin titer by pipetting 10 µL culture supernatant (or appropriate dilutions in methanol) on antibiotic test disks placed onto the solidified test plates. After 48 h incubation at 30°C, the growth-inhibition zones of *M. luteus* are measured and the erythromycin titer (g/L) is calculated using a standard curve.

3.3.2. HPLC Assay for Erythromycin Purity

An HPLC method based on the method described by Tsuji and Goetz **(22)** was used to determine the amount of the erythromycins A, B, and C in the broth. Prior to HPLC analysis, the erythromycins had to be extracted from the whole broth.

1. Place 20 mL broth into a 100-mL volumetric flask, add 1 vol of water, and stir the mixture for 2–3 min.
2. Filter the suspension through a paper filter or milk cloth.
3. Transfer 20 mL of the filtrate into a separatory funnel.
4. Adjust the pH by addition of 5 mL of 1 M phosphate buffer pH 10.0.
5. Add 40 mL solvent (chloroform : octane, 70 : 30 [v/v]) and place the funnel on a shaker
6. Agitate for 15 min and then allow phases to separate.
7. Collect 10 mL of the lower organic phase and dry under vacuum in a rotary evaporator at 60°C.
8. Dissolve the residue in 1 mL mobile phase (acetonitrile–methanol–0.2 M ammonium acetate–water (45 : 10 : 10 : 35), adjust pH with hydrochloric acid to 7–7.8, filter, and use for HPLC analysis. Fifty-microliters of broth extract aliquots are analyzed. Erythromycin A, B, and C concentrations were determined taking into account a 10-fold dilution during extraction.
9. Prepare erythromycin standards by accurately weighting 8–9 mg erythromycin reference standards into a 10-mL volumetric flask. Just prior to the analysis, the mobile phase is added to volume; sonicate to facilitate solubilization. Calibration curves are prepared for erythromycins A, B and C, respectively, using the peak heights.
10. Perform HPLC analysis on a system equipped with a UV detector set to 215 nm and a μBondapack C_{18} (300 × 3.9 mm inner diameter) column. The mobile phase (filtered and degassed) is pumped at a flow rate of 1 mL/min, and the column was operated at 40°C.

3.4. Synthesis of Semisynthetic Erythromycin Derivates

3.4.1. Synthesis of Azithromycin

Pliva published a series of process patents between 1968 and 1981 describing the synthetic steps needed to transform erythromycin to azithromycin. The process, summarized in **Fig. 6A,** includes an oxamination *(23)*, followed by a Beckmann rearrangement of erythromycine A oxime and the reduction of 11-aza-erythromycin A *(24)*, and the selective 11-*N*-methylation *(24,25)*. Some newer patents describe the preparation of novel crystalline forms that are claimed to avoid the aforementioned patents. The overall yields are at 40%, requiring 2.5 kg erythromycin base per kilogram of azithromycin.

3.4.2. Synthesis of Clarithromycin

This process is summarized in **Fig. 6B.** All synthesis of clarithromycin starts from *O,N*-dibenzylcarbonyl-des-3′-*N*-methylerythromycin A. This intermediate is produced by the reaction of erythromycin A with benzyl chloroformate. As well as protecting the 3′ nitrogen as a benzyloxycarbonyl derivate, it also protects the 2′-hydroxyl group. This reaction was discovered by Flynn and co-workers *(26)*. The following reactions with hydroxylamine hydrochloride

A

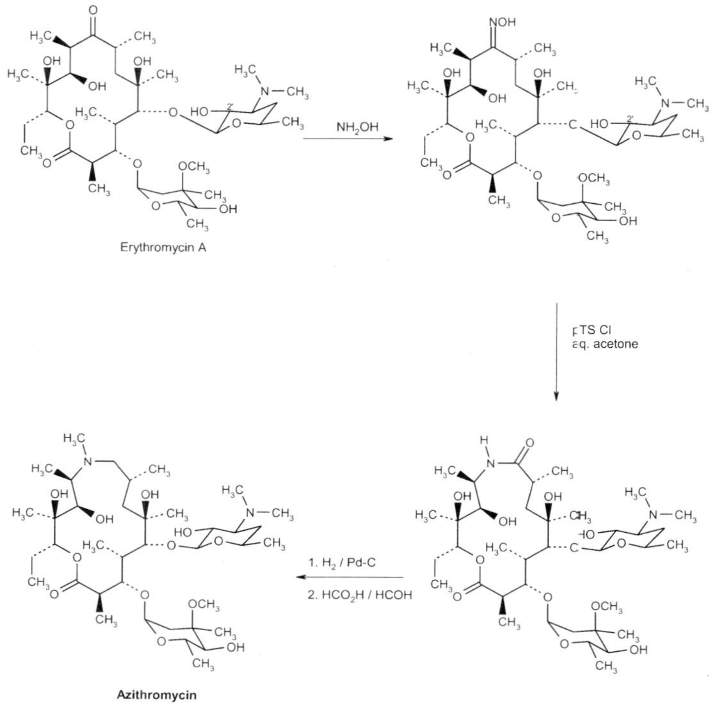

Fig. 6. Routes to semisynthetic erythromycin derivates: (**A**) azithromycin. *(Figure continues on following two pages.)*

and sodium acetate in methanol yield the oxime, which is reacted with *o*-chlorobenzyl chloride and sodium hydride in dimethylformamide (DMF) yielding the *o*-chlorobenzyloxime. Methylation with methyl iodine and potassium hydroxide in tetrahydrofuran (THF) and dimethyl sulfoxide (DMSO) produces the corresponding 6-*O*-methyl ether, which is then deprotected (debenzylation with hydrogen and palladium on charcoal) and remethylated on the 3' nitrogen with formaldehyde in refluxing formic acid to yield the 9-oxime of chalrithromycin. After deoximation with ethanolic bisulfite, clarithromycin is obtained in an overall yield of some 30%, roughly requiring 3 kg erythromycin base for 1 kg clarithromycin.

3.4.3. Synthesis of Roxithromycin

Roxithromycin is prepared from erythromycin A oxime (*see* **Subheading 3.4.1.**) and methoxyethylmethyl chloride (MEM chloride), which alkylates the

B

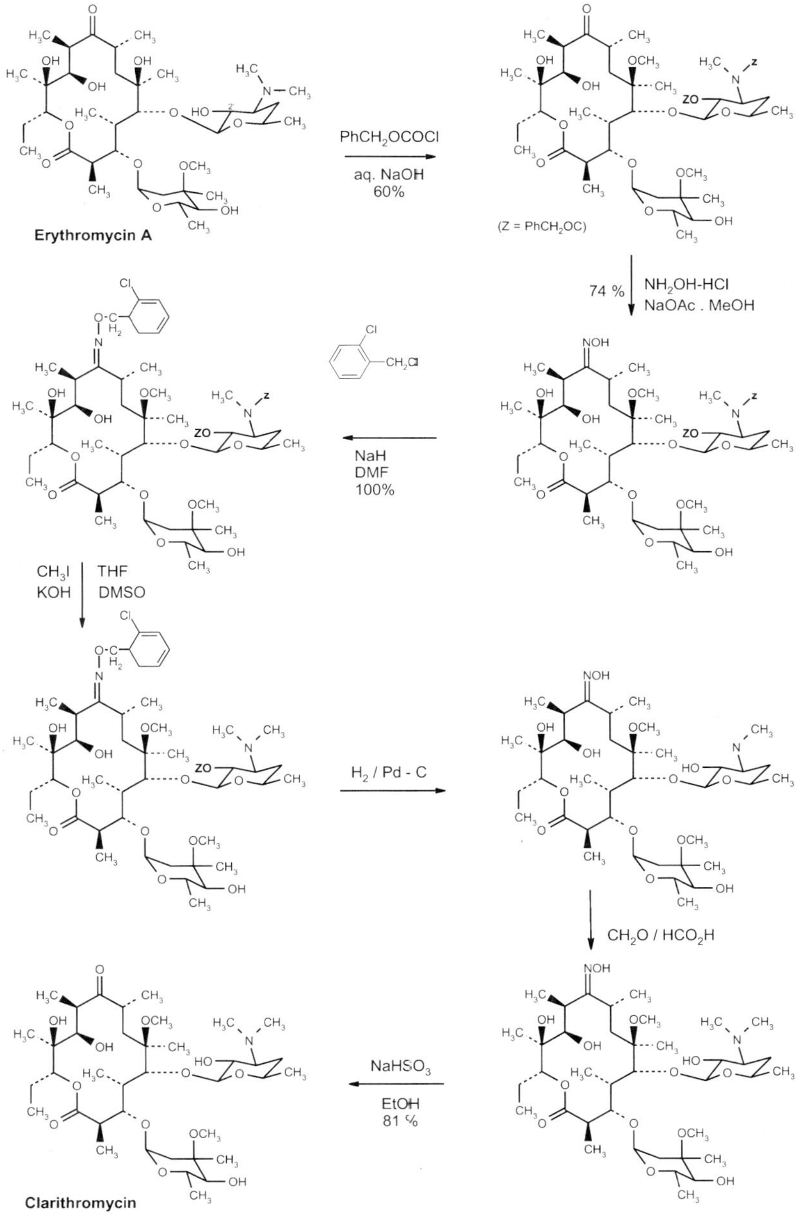

Fig. 6. *(continued)* **(B)** clarithromycin.

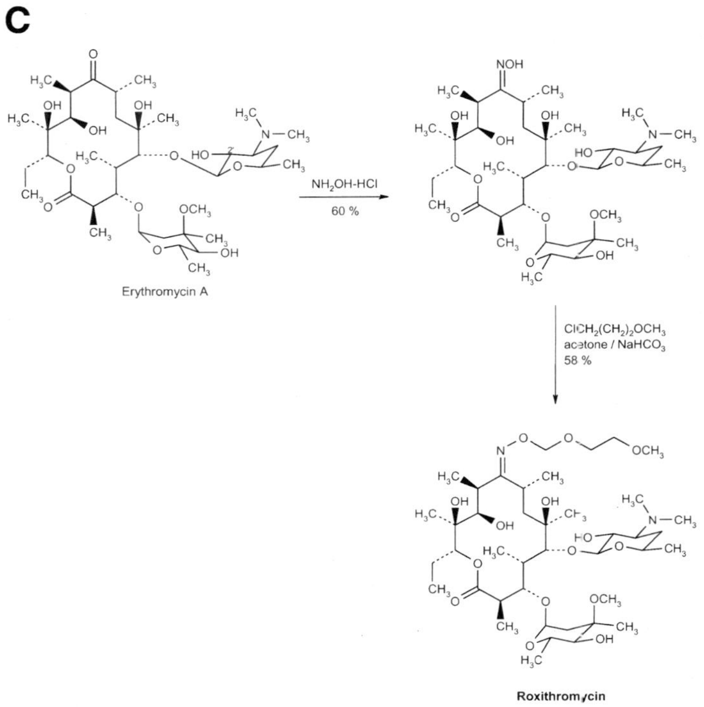

C

Fig. 6. *(continued)* **(C)** roxithromycin.

oxime hydroxyl specifically (*see* **Fig. 6C**). The process has been patented by Roussel–UCALF *(27)*.

1. Sodium bicarbonate and MEM chloride are added to an acetone solution of erythromycin A oxime and the suspension is refluxed for several hours under nitrogen. Extra equivalents of MEM chloride are added during the reaction.
2. After removal of the acetone, the whole is taken up in aqueous sodium bicarbonate and the oily residue is separated and recrystallized from acetone in the presence of kieselgur and charcoal.
3. Three crops of roxithomycin are collected and the crude product (68% yield) is recrystallized again giving purified product (58% yield).

Overall molar yield is about 40%, roughly requiring 2 kg erythromycin base for 1 kg roxithromycin

4. Notes

1. For projects with an applied and industrial background, it should be noted that a decent industrial strain produces 90–95% erythromycin A. the rest being ery-

thromycins B and C. This should be documented and tested prior to any work, in particular prior to any genetic engineering approach.

2. As it is not always possible to obtain the listed media ingredients because certain commodities from the United States are not allowed into Europe and probably other countries, corn and soy products from local distributors need to be tested. The soy oil should be crude, nonrefined, and fresh. This oil will turn rancid (odor) after a period of storage and should be replaced. Corn steep liquor is available in two qualities: incubated (recommended for this application) and nonincubated. Whereas the former is characterized by a completed lactic acid fermentation that stabilizes the product, the latter might undergo changes during storage. In either case, the slurry should be well resuspended prior to use. Soy bean flour is available in different particle sizes and fat content. The finer material is likely to cause increased foaming.

3. Media containing soybean flour started foaming at around 70°C. This could soil the exhaust air filter that is then blocked. If excessive foaming is observed and the addition of a few drops of antifoam does not solve the problem, coarser milled flour can be used. Alternatively, V2 medium was sterilized in the bioreactor without soybean flour. The latter was autoclaved separately as slurry and added aseptically to the sterile bioreactor.

4. In lab-scale fermentations, the original F1 medium was used at half concentration only.

5. All salts and dextrin were autoclaved separately and were added aseptically to the sterilized bioreactor that contained all of the other components.

6. Erlenmeyer flasks should be equipped with a long and wide neck and one baffle to ensure ventilation and oxygen transfer, but preventing excessive splashing and wetting of the covers.

7. Erlenmeyer flasks should be closed with a sandwich made of a layer gauze, one layer cotton, and a layer of gauze. The cover should be secured with two elastic rubber bands. This ensures a good air exchange in and out of the bottle.

8. Shaker incubators with a 5-cm stroke are recommended to provide sufficient aeration. The shaker should also be equipped with a cooling option, as the growth is very exothermic. A humidifier is recommended to prevent excessive loss of water by evaporation during the 8–9 d of cultivation.

9. Because of the high portion of undissolved matter, sterilization times for bioreactors were increased to 50 min at 121°C.

10. Dried filters for dry weight determinations and dried biomass are very hygroscopic and needed to be stored dry. Filters with or without biomass were dried at 80°C for 24 h and then stored in a dissector until just prior to weighting.

11. Soy oil was sterilized with about 10% water added into the bottle.

12. The dextrin feed needs agitation to prevent sedimentation.

13. The oil feed was adjusted such that no oil was floating on top of the samples.

Acknowledgment

We thank Mike Barber for his comments and suggestions.

References

1. Kieser, T., Bibb, J., Buttner, M. J., Chaiter, K. F., and Hopwood, D. A. (eds.) (2000) *Practial* Streptomyces *Genetics,* The John Innes Foundation, Norwich, UK.
2. Demain, A. L. (1989) Function of secondary metabolites, in *Genetics and Molecular Biology of Industrial Microorganisms* (Hershberger, C. L., Queener, S. W., and Hegemann, eds.), American Society for Microbiology, Washington DC, pp. 1–11.
3. Vining, L. C. and Stuttad, C. (eds.) (1995) *Genetics and Biochemistry of Antibiotic Production,* Butterworth–Heinemann, Boston, MA.
4. Dimroth, P., Walter, H., and Lynen, F. (1970) Biosynthesis of 6-methylsalicylic acid. *Eur. J. Biochem.* **13,** 98–110.
5. O'Hagan, D. (ed.) (1991) *The Polyketide Metabolites.* Ellis Horwood, London.
6. Reimold, U., Kröger, M., Kreuzaler, F., and Hahlbrock, K. (1983) Coding and 3′ non-coding nucleotide sequence of chalcone synthase mRNA and assignment of amino acid sequence of the enzyme. *EMBO J.* **2,** 1801–1805.
7. Katz, L. and Donadio, S. (1993) Polyketide synthesis: prospects for hybrid antibiotics. *Annu. Rev. Microbiol.* **47,** 875–912.
8. Katz, L. (1997) Manipulation of modular polyketide synthases. *Chem. Rev.* **97,** 2557–2576.
9. Staunton, J. and Wilkinson, B. (1997) Biosynthesis of erythromycin and rapamycin. *Chem. Rev.* **97,** 2611–2630.
10. Malpartida, F. and Hopwood, D. A. (1986) Physical and genetic characterisation of the gene cluster for the antibiotic actinorhodin in *Streptomyces coelicolor* A3(2). *Mol. Gen. Genet.* **205,** 66–73.
11. Viollier, P. H. Minas, W., Dale, G. E., Folcher, M., and Thompson, C. J. (2001) Roles of aconitase in growth, metabolism, and morphological differentiation of *Streptomyces coelicolor. J. Bacteriol.* **183,** 3193–3203.
12. Kelemen, G. H. Viollier, P. H., Tenor, J., Marri, L., Buttner, M. J., and Thompson, C. J. (2001) A connection between stress and development in the multicellular prokaryote *Streptomyces coelicolor* A3(2). *Mol. Microbiol.* **40,** 804–814.
13. Vohradsky, J. Li, X., Dale, G., et al. (2000) Developmental control of stress stimulons in *Streptomyces coelicolor* revealed by statistical analyses of global gene expression patterns. *J. Bacteriol.* **182,** 4979–4986.
14. Yamazaki, H., Ohnishi, Y., and Horinouchi, S. (2000) An A-factor-dependent extracytoplasmic function sigma factor (σ^{AdsA}) that is essential for morphological development in *Streptomyces griseus. J. Bacteriol.* **182,** 4596–4605.
15. Ueda, K., Kawai, S., Ogawa, H.-O., et al. (2000) Wide distribution of interspecific stimulatory events on antibiotic production and sporulation among *Streptomyces* species. *J. Antibiot.* **53,** 979–982.
16. Waksman, S. A. (ed.) (1947) *Microbial Antagonisms and Antibiotic Substances.* Commonwealth Fund, New York.
17. Bunch, R. L., and McGuire, J. M. (1953) US patent 2653899 (to Eli Lilly Co.).
18. Brunker, P., Minas, W., Kallio, P. T., and Bailey, J. E. (1998) Genetic engineering of an industrial strain of *Saccharopolyspora erythraea* for stable expression of the *Vitreoscilla* haemoglobin gene *(vhb). Microbiology* **144(Pt. 9),** 2441–2448.

19. Minas, W., Brünker, P., Kallio, P. T., and Bailey, J. E. (1998) Improved erythromycin production in a genetically engineered industrial strain of *Saccharopolyspora erythraea*. *Biotechnol. Prog.* **14,** 561–566.

20. Ausubel, F., Brent, R., Kingston, R. E., et al. (eds.) (1989) *Short Protocols in Molecular Biology,* Wiley & Sons, New York, USA.

21. Gerhardt, B., Murray, R. G. E., Wood, W. A., and Krieg, N. R. (eds.) (1994) *Manual of the Methods for General Bacteriology,* American Society for Microbiology, Washington, DC.

22. Tsuji, K. and Goetz, J. F. (1978) High-performance liquid chromatographic determination of erythromycin. *J. Chromatogr.* **147,** 359–367.

23. Tamburasov, Z. and Djokic, S. (1969) British patent GB 1100504 (to Pliva Pharm & Chem Works.).

24. Djokic, S. and Kobrehel, G. (1985) British patent GB 2094293 (to Pliva Pharm & Chem Works.).

25. Bright, G. M. (1984) European patent EP0101186 (to Pfizer).

26. Flynn, E. H., Murphy, H. W., and McMahon, R. E. (1955) Erythromycin. II. Des-*N*-methylerythromycin and *N*-Methyl-C14-erythromycin. *J. Am. Chem. Soc.* **77,** 3104–3106.

27. Gouin d'Ambrieres, S., Lutz, A., and J.-C., G. (1982) French patent FR2473525 (to Roussel UCLAF).

5

The Phenylacetyl-CoA Catabolon

Genetic, Biochemical, and Biotechnological Approaches

José M. Luengo, Belén García, Ángel Sandoval, Elsa Arias-Barrau, Sagrario Arias, Francisco Bermejo, and Elías R. Olivera

Summary

The phenylacetyl-CoA catabolon is a complex degradative unit integrated by several catabolic pathways that transform different unrelated aromatic compounds (i.e., phenylacetic acid [PhAc], phenylacetaldehyde, styrene, tropic acid, 2-phenylethylamine, 2-phenylethanol, ethylbenzene, penylacetylamides and other esters, *n*-phenylalkanoates with an even number of carbon atoms) into general metabolites. This supracatabolic unit contains (1) a central pathway (catabolon core or route of convergence) that catalyzes the conversion of PhAc via phenylacetyl-CoA, into TCA intermediates, and (2) different peripheral routes (upper pathways) that are specifically involved in the conversion of different molecules into phenylacetic acid or into phenylacetyl-CoA. The phenylacetyl-CoA catabolon, which has important biotechnological applications (enzymatic synthesis of penicillins, aromatic biotransformations, synthesis of new bioplastics, biosensor design, synthesis of drug vehicles, etc.), has been described in different nonphylogenetically related microbes, reinforcing the metabolic and industrial importance of this complex pathway.

Key Words: Phenylacetic acid; aromatics; catabolon; phenylacetyl-CoA; phenylalkanoates; bioplastics; PHAs.

1. Introduction

The bacterial pathways involved in the degradation and assimilation of many different aromatics has been studied in different microbes for decades *(1)*. Most of these compounds are catabolized through different unrelated routes that facilitate their mineralization *(2)*. Although considerable knowledge about these catabolic pathways has been accrued *(1–3)*, some of them either remain to be fully elucidated or have only been discovered recently. One such case is the phenylacetate pathway *(4)*. We have previously shown that the aerobic degradation of phenylacetic acid (PhAc) in *Pseudomonas putida* U is carried out by

From: *Methods in Biotechnology, Vol. 18: Microbial Processes and Products*
Edited by: J. L. Barredo © Humana Press Inc., Totowa, NJ

a novel pathway that involves the participation of 13 catabolic enzymes (PaaABCEFGHIJKLMN) organized in the following 5 functional units: (1) a transport system integrated by a permease (PaaL) and a porine (PaaM), which catalyzes the uptake of PhAc *(4,5)*; (2) an acyl-CoA-activating enzyme (PaaF), which converts PhAc into phenylacetyl-CoA (PhAc-CoA) *(6,7)*; (3) a ring-hydroxylation complex (PaaGHIJK), which catalyzes the hydroxylation of PhAc-CoA to 2-hydroxy-phenylacetyl-CoA (2-OH-PhAc-CoA); (4) a ring-opening enzyme (PaaN), which transforms 2-OH-PhAc-CoA into an aliphatic intermediate; and (5) a β-oxidation system integrated by two enoyl-CoA hydratases (PaaA and PaaB), a 3-OH-acyl-CoA dehydrogenase (PaaC), and a ketothiolase (PaaE) *(4)*, which finally converts this compound into succinyl-CoA (*see* **Fig. 1**). The identification of the end product (succinyl-CoA) has been confirmed by using *P. putida* mutants handicapped in the transformation of succinyl-CoA via succinate to fumarate. These mutants were also unable to catabolize PhAc. Ismail and co-workers in *Escherichia coli (8)* have recently proposed a similar mechanism. Additionally, a further two regulatory proteins (PaaX and PaaY) are involved in the control of the flux of intermediates through the pathway *(4,9)*.

All of these enzymes are encoded by 15 genes organized in 5 operons: 3 grouping the catabolic genes (*paaABCEF, paaGHIJK,* and *paaLMN*) and 2 regulatory ones (*paaX* and *paaY*), which are translated divergently *(4)* (*see* **Fig. 2**).

We have also shown that this catabolic pathway is widely distributed among many different and phylogenetically unrelated microbes *(3)*, highlighting the importance of this pathway for study of the genetic, enzymatic, and regulatory modifications that this pathway has undergone along evolution. Moreover, we have shown that this pathway is the central route of a more complex supracatabolic unit (the phenylacetyl-CoA catabolon) integrated by different assimilatory routes (upper pathways) involved in the degradation of several aromatic compounds that generates PhAc or phenylacetyl-CoA as final products *(3)* (*see* **Fig. 3**). Thus, with the term "catabolon," we defined a complex catabolic unit integrated by different degradative pathways involved in the degradation of structurally related molecules, which converge in a common catabolite (hence, the term "catabolon") and which are or could be regulated in a coordinated fashion.

Taking into account the broad biotechnological and industrial applications of the phenylacetyl-CoA catabolon *(3,10–12)*, here we describe some methods that can be routinely employed for genetic, biochemical, and biotechnological studies of this complex functional unit. Genetic approaches are directed to cloning the whole pathway (some of them addressing the operon encoding the enzymes involved in a particular function), to the overexpression of a required protein, to either the disruption or the deletion of a particular gene, and to the isolation of different mutants lacking a particular function. Biochemical assays

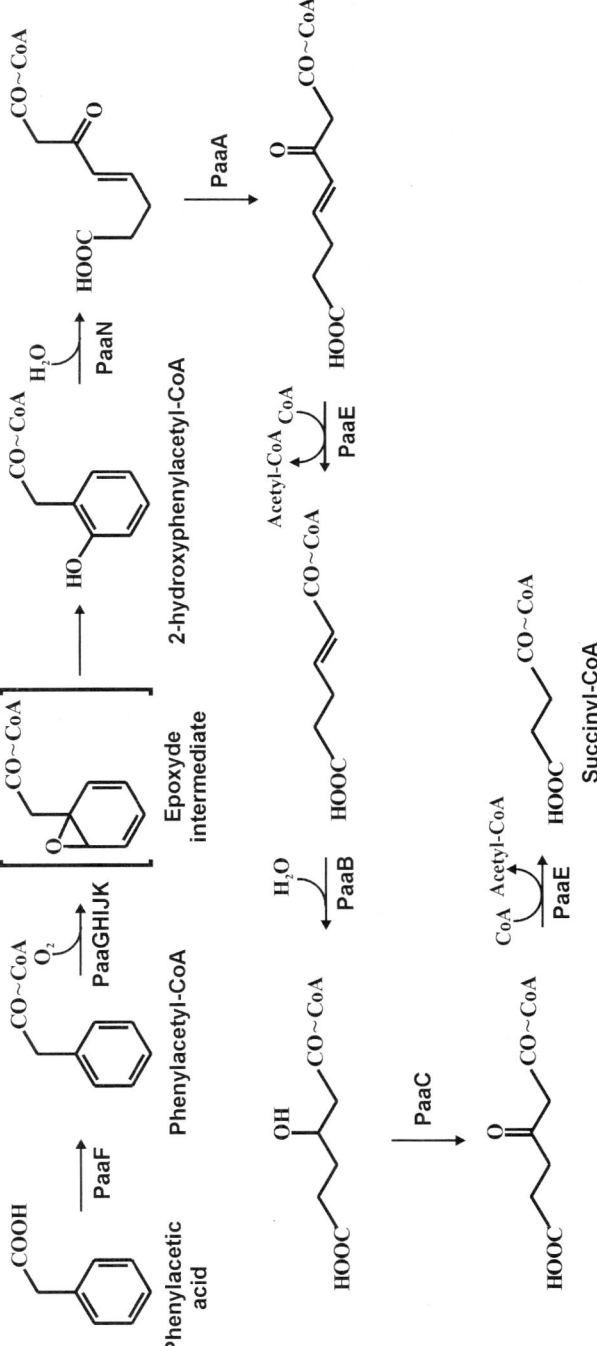

Fig. 1. Hypothetical pathway involved in the aerobic catabolism of PhAc.

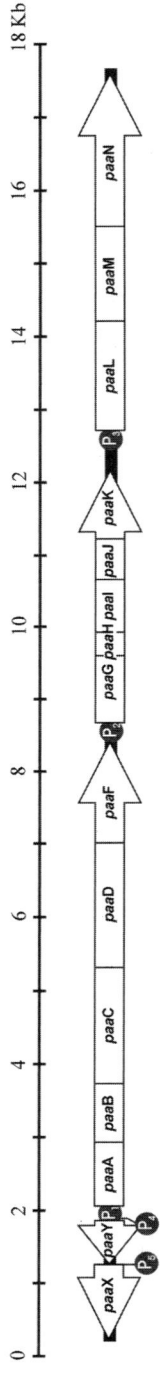

Fig. 2. Genetic organization of the PhAc catabolic pathway.

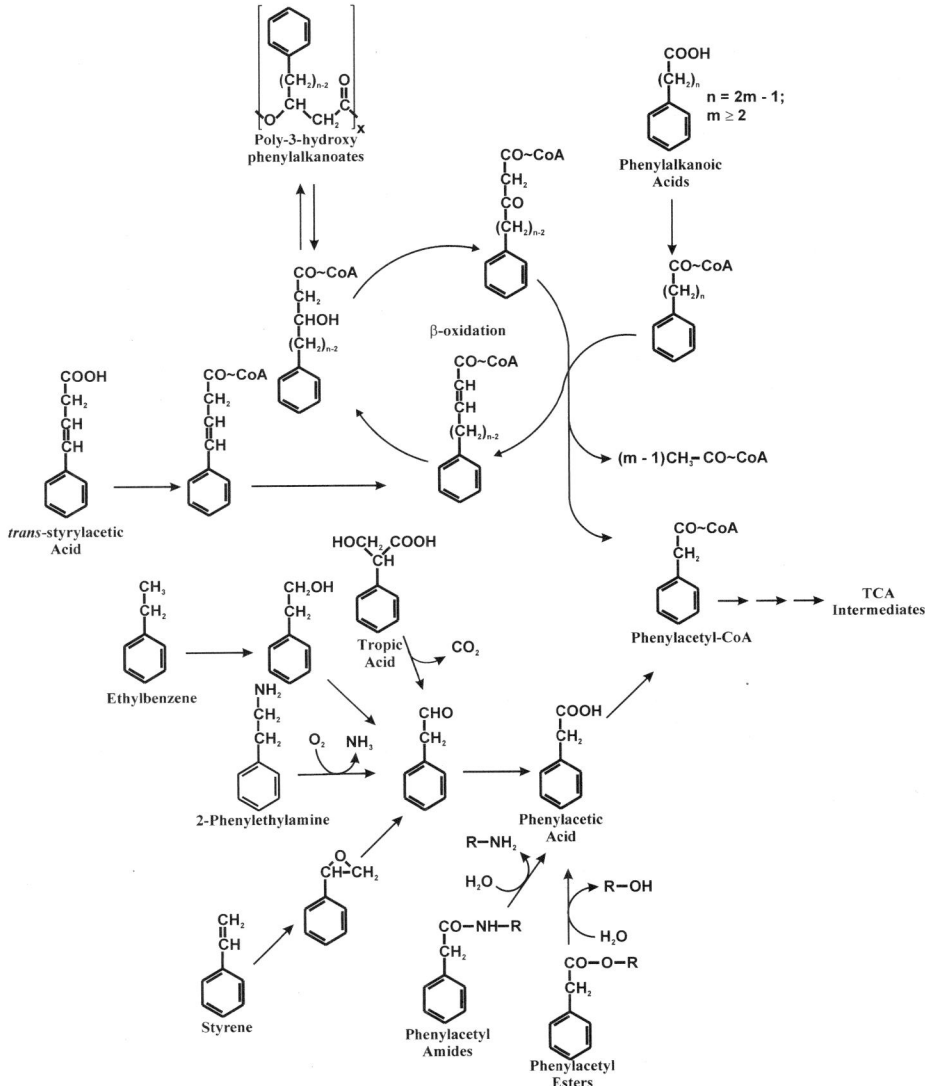

Fig. 3. Schematic representation of the phenylacetyl-CoA catabolon.

describe useful methods to properly evaluate in vitro or in vivo some of the enzymes belonging either to the central pathway or corresponding to the upper pathways. Finally, some procedures related to the biotechnological applications of the phenylacetyl-CoA catabolon, as well as general analytical procedures (with a specific use), are also reported.

2. Materials

1. pJQ200KS vector. Gm^R *(13)*.
2. pK18::*mob*. Km^R *(14)*.
3. pBC and pBluescript plasmids (Stratagene, La Jolla, CA).
4. The QIA*expressionist* Kit (Qiagen, Hilden, Germany).
5. *E. coli* strains DH5α, XL1-Blue, and HB101 (pGS9) *(4)*.
6. *Pseudomonas putida* U strain *(4)*.
7. Oligonucleotide primer D3017: 5′-GATCCGGCGATCATGGGTTA-3′.
8. Oligonucleotide primer D3350: 5′-CGCGTTCAAAGACAGTGGTGA-3′.
9. Oligonucleotide primer A19X: 5′-GAA<u>TCTAGA</u>GCCATCACGGTTAAG-GCTCTTG-3′. The *Xba*I restriction site is underlined.
10. Oligonucleotide primer A302P: 5′-ACT<u>CTGCAG</u>TCTGCGAAGGCGCTGAA-GAT-3′. The *Pst*I restriction site is underlined.
11. Oligonucleotide primer COMPFADB: 5′-TGACGCCAAGCCATACTTCCG-3′.
12. Thermocycler.
13. *Micrococcus luteus* ATCC 9341 *(7)*.
14. Antibiotics: ampicillin (Ap), kanamycin (Km), rifampicin (Rf), chloramphenicol (Cm), gentamicin (Gm), tetracycline (Tc).
15. Restriction enzymes, T4 DNA ligase, *Taq* polymerase, Pfu *(Pyrococcus furiosus)*.
16. Agarose and DNA sequencing equipment.
17. IPTG (isopropyl-β-D-thio-galactopyranoside).
18. X-Gal (5-bromo-4-chloro-3-indolyl-β-D-thio-galactopyranoside).
19. Sodium dodecyl sulfate–polyacrylamide gel electrophoresis (SDS-PAGE) equipment.
20. Loading buffer: 96.25 µL of a saturated solution of sucrose, 93.75 µL of 0.5 M Tris-HCl, pH 6.8, 10 µL of bromophenol blue (1% w/v), 50 µL of β-mercaptoethanol, and 250 µL of SDS (10% w/v).
21. Ferric chloride reagent (Lipmann reactive): 0.37 M ferric chloride, 20 mM trichloroacetic acid, and 0.66 M hydrochloric acid *(15)*.
22. [1-^{14}C]-PhAc (3.4 mCi/mmol) (Sigma, St. Louis, MO)
23. Millipore HAWPO 2500 filters (pore size, 0.45 µm) (Millipore Ibérica SA, Madrid, Spain)
24. Scintillation fluid (Dupont, Les Ulis Cedex A, France).
25. Spectrophotometer VIS/UV UV-120-02 (Shimadzu, Kyoto, Japan).
26. Spectrophotofluorimeter UV-RF 510 (Shimadzu).
27. Cell-shocking buffer: 20% (w/v) sucrose, 0.2 M Tris-HCl, pH 7.2, and 0.005 M EDTA.
28. Trypticase soy agar (Difco, Detroit, MI).
29. Hydroxylamine solution pH 8.0: 1 mL of 5 M hydroxylamine hydrochloride, 250 µL of distilled water, and 1.25 mL of 4 M KOH.
30. Ferric chloride reagent: 0.37 M ferric chloride, 20 mM trichloroacetic acid, and 0.66 M hydrochloric acid *(15)*.
31. Disruption buffer: 0.5 M Tris-HCl, pH 8.2, 5 mM mercaptoethanol, 4 mM EDTA, and 1 mM phenylmethylsulphonyl fluoride (PMSF).
32. Glass beads, Ballotini, 0.17–0.18 mm in diameter.

33. Braun MSK mechanical disintegrator (Braun, Melsungen, Germany).
34. Protoplasts preparation lysis buffer: 25 mM phosphate buffer, pH 5.8, 0.35 M KCl Novozym 234 (Novo Industri, Bagsvaerd, Denmark), and 0.5 µg/µL.
35. Protoplast washing solution: 0.7 M KCl, 50 mM CaCl$_2$, and 10 mM MOPS buffer, pH 5.8.
36. Enzyme cocktail: 30 µL adenylate kinase, 15 µL pyruvate kinase, and 4 µL lactate dehydrogenase in a total volume of 1 mL of 50 mM Tris-HCl buffer, pH 8.0.
37. Indole solution: prepare 100 mM indole in dimethylformamide and dilute it with 50 mM potassium phosphate to reach a final concentration of 0.25 mM indole.
38. Granules buffer: 100 mM Tris-HCl, pH 8.0, 250 mM sucrose, 0.5 mM EDTA, and 0.25 mM MgCl$_2$.
39. Tris-NaCl buffer (TBS) (DiaPharma, West Chester, OH).
40. Antibodies: rabbit anti-PHA polimerase C1 from *Pseudomonas putida* U.
41. Adenylate kinase; pyruvate kinase and lactate dehydrogenase.
42. Protran nitrocellulose membranes (Schleicher & Schuell, Dassel/Relliehausen, Germany).
43. Bio-Rad Transblot SD semidry electrophoretic transfer cell (Bio-Rad, Hercules, CA).
44. Chemiluminescence kit reagents (Amersham Life Science, Barcelona, Spain).
45. 3-Hydroxyoctanoate, acyl Coenzyme A synthetase from *Pseudomonas fragi,* and CoA (Sigma).
46. High-performance liquid chromatography (HPLC) equipment SP8800 (Spectra Physics, Mountain View, CA).
47. Variable-wavelength UV/VIS detector 2487 (Waters, Milford, MA).
48. Millennium software V 3.2 (Waters).
49. Microparticulate reversed-phase column, particle size = 5 µm, pore size = 100 nm, Nucleosil C-18, 4.6 (inner diameter) by 250 nm (Phenomenex Laboratories, Torrance, CA).
50. HPLC equipment Chromatograph Series 3B (Perkin-Elmer, Norwalk, CT).
51. LC-75 Spectrophotometric Detector (Perkin-Elmer).
52. Chromatography Data Station 10 B (Sigma).
53. µBondapack C-18 column, 30 cm (Waters).
54. Preparative HPLC equipment (Labomatic Instruments, Weil am Rhein, Germany).
55. Preparative C8-spherisorb-RP column (Bischoff Chromatography, Leonberg, Germany).
56. Buffer A: 90% of 50 mM NH$_4$Ac, pH 8.0, and 10% acetonitrile.
57. Buffer B: 60% of 50 mM NH$_4$Ac, pH 8.0, and 40% acetonitrile.
58. Silica-Gel plates HPTLC Fertigplatten Kieselgel 60, 10×10 cm (Merck & Co, Whitehouse Station, NJ).
59. Rotary evaporator.
60. Bovine serum albumin (BSA).
61. Liquid scintillation counter 1800 LS (Beckman).
62. Luria–Bertani medium (LB): 10 g/L bacto-tryptone, 5 g/L bacto-yeast extract, 10 g/L NaCl, and 20 g/L agar. Adjust to pH 7.0 with 5 N NaOH. Autoclave at 121°C for 15 min.

63. MM chemically defined medium: 13.6 g/L potassium phosphate, 2 g/L ammonium sulfate, 0.25 g/L magnesium sulfate, and 0.5 mg/L iron(II) sulfate. Carbon source at convenience. Adjust to pH 7.0 with 5 N KOH. Autoclave at 121°C for 15 min (*6*).
64. M9 chemically defined medium (g/L): Na_2HPO_4, 6; KH_2PO_4, 3; NaCl, 0.5; NH_4Cl, 1; HBO_3, 7.5×10^{-5}; $ZnCl_2$, 1.25×10^{-5}; $MnCl_2\cdot4H_2O$, 7.5×10^{-6}; $CoCl_2$, 5×10^{-5}; $CuCl_2\cdot2H_2O$, 2.5×10^{-6}; $NiCl_2\cdot6H_2O$, 5×10^{-6}; $NaMO_4\cdot2H_2O$, 7.5×10^{-6}; and thiamine, 0.042 (*6*).
65. DNA ladder, 1–30 kb (Invitrogene, Carlsbad, CA).
66. Oligonucleotide PCLA: 5′-GGCGCAAGGGTGACAA-3′.
67. Oligonucleotide PCLB: 5′-ATCTGGGTCGGGAACAC-3′.
68. Ultra-Turrax T25 (IKA Labortechnik, Staufen, Germany).

3. Methods

The methods described outline (1) the genetic approach, (2) biochemical studies, (3) biotechnological applications, and (4) some general analytical procedures for the study of the phenylacetyl-CoA catabolon.

3.1. Genetic Methods

3.1.1. Isolation of Mutants Altered in Some of the Genes Belonging to the Phenylacetyl-CoA Catabolon

In this subsection, we describe some of the methods routinely used to obtain different mutants that allow study of the phenylacetyl-CoA catabolon. Thus, these procedures outline the selection of different mutants, using (1) the transposon Tn5, (2) the plasmid pK18::*mob* (homologous recombination), or (3) the plasmid pJQ200KS (gene deletion).

3.1.1.1. Tn5 Transposon Mutagenesis

This method is used to obtain mutants affected in some of the functions required for the degradation of all the aromatic compounds that are catabolized through the phenylacetyl-CoA catabolon. The bacterial strains used are *E. coli* HB101 (pGS9) (donor of the transposon) (Cm^R, Km^R), and *P. putida* U (the recipient strain) (Rf^R) (*16*). The procedure followed is summarized in **Fig. 4** and the structure of the transposon is schematized in **Fig. 5**.

1. Incubate *E. coli* HB101 (pGS9) in 3 mL of LB medium supplemented with kanamycin (Km) (25 µg/mL) and Cm (30 µg/mL) at 37°C for 8 h. At the same time, seed *P. putida* in 5 mL of chemically defined medium (MM or M9) (*6*) supplemented with rifampicin (Rf) (20 µg/mL) and cultured for the same time (8 h) at 30°C.
2. Collect two aliquots (0.7 mL) from each culture, mix them in a sterile Eppendorf tube (1.5 mL), and centrifuge the bacteria (10,000*g* for 5 min).
3. Resuspend the pellet in 0.5 mL sterile LB. Wash it and centrifuge again as reported in **step 2**.

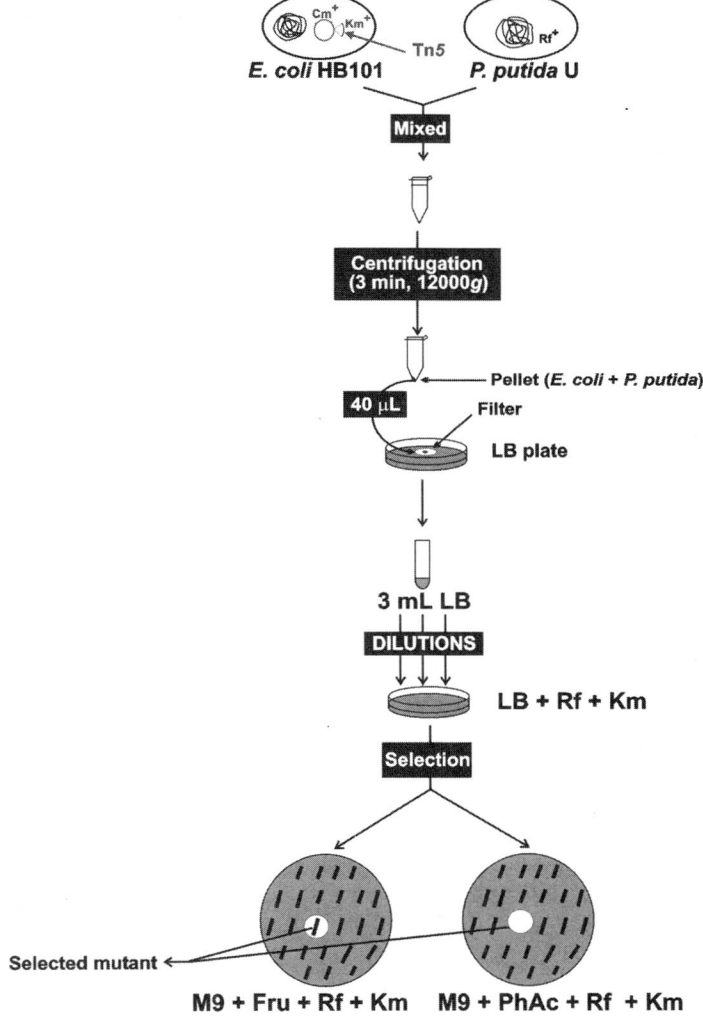

Fig. 4. Schematic representation of the Tn5 mutagenesis procedure.

4. Resuspend the mixed cultures in 30–50 µL of LB and apply the bacteria onto a sterile Millipore bacterial filter (0.45 µm pore size) that has been previously placed on the surface of a plate containing solid LB.
5. Incubate the plate containing the filter, holding the mixed cultures of *E. coli* and *P. putida* at 30°C for 12 h.

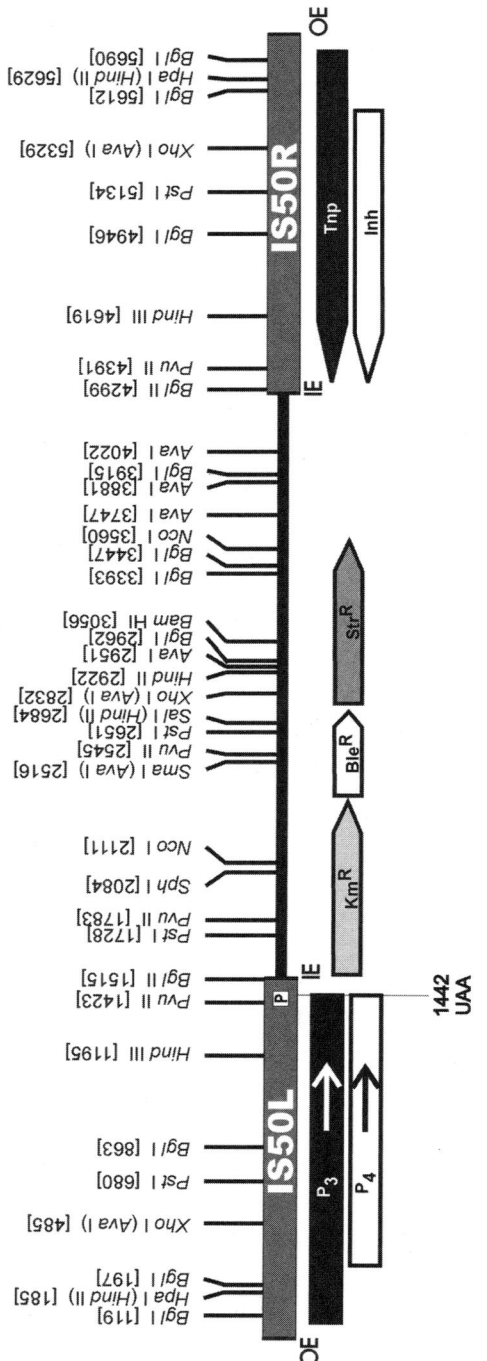

Fig. 5. Genetic organization of the transposon Tn5.

100

6. Once the mixed bacterial populations have grown over the filter, place them in a test tube containing 3 mL of LB without antibiotics and shake until most of the bacteria have been resuspended from the filter.

7. Dilute the bacterial suspension suitably and seed 100 µL of each dilution onto LB plates supplemented with Km and Rf (at the same concentrations indicated in **step 1**). Parental strains are unable to grow on these plates because they do not have both antibiotic resistances. The only bacteria able to grow will be *Pseudomonas* transconjugants that, in addition to R_f resistance, have incorporated the tranposon Tn5, which confers the recipient bacteria Km resistance.

8. After 48 h of incubation at 30°C, select transconjugants as a function of their ability to grow or not in chemically defined medium (solid M9) containing Km, Rf (as selection markers), and PhAc (as the sole carbon source).

3.1.1.2. GENE DISRUPTION BY RECOMBINATION

Using a specific mutagenesis procedure based on homologous recombination, it is possible to analyze the function of a particular gene *(4,14)*. The procedure followed is summarized in **Fig. 6.**

1. First, an internal fragment (300–900 bp, depending on the gene size) of the gene to be disrupted must be obtained. This can be isolated by either digestion with the appropriate restriction endonucleases or by the polymerase chain reaction (PCR) using specific oligonucleotides.

2. Clone the fragment in the plasmid pK18::*mob* (a conjugative plasmid that is replicative in *E. coli* but not in *P. putida* U).

3. Transform *P. putida* U with this construction by triparental conjugation *(17)* (*see* **Note 1**).

During the selection process, the presence of Km in the medium forces the recipient strain to insert the plasmid, which confers antibiotic resistance to the bacterial genome. This process is performed by homologous recombination between the internal fragment of the target gene that was cloned in the pK18:*mob* and the chromosomal copy. By following this method, it is possible to isolate mutants of any of the genes encoding the enzymes involved in a particular metabolic route (*see* **Note 2**).

3.1.1.3. GENE DELETION

Taking into account that some of the mutants obtained by gene disruption are unstable, in order to study a particular function, to clarify the role of a specific enzyme, or to obtain catabolic intermediates belonging to the phenylacetyl-CoA catabolon, we suggest the selection of mutants in which the desired gene has been eliminated. In such cases, the method followed for producing deletions in the *P. putida* U genome is that described by Donnenberg et al. *(18)* using plasmid pJQ200KS *(13)*.

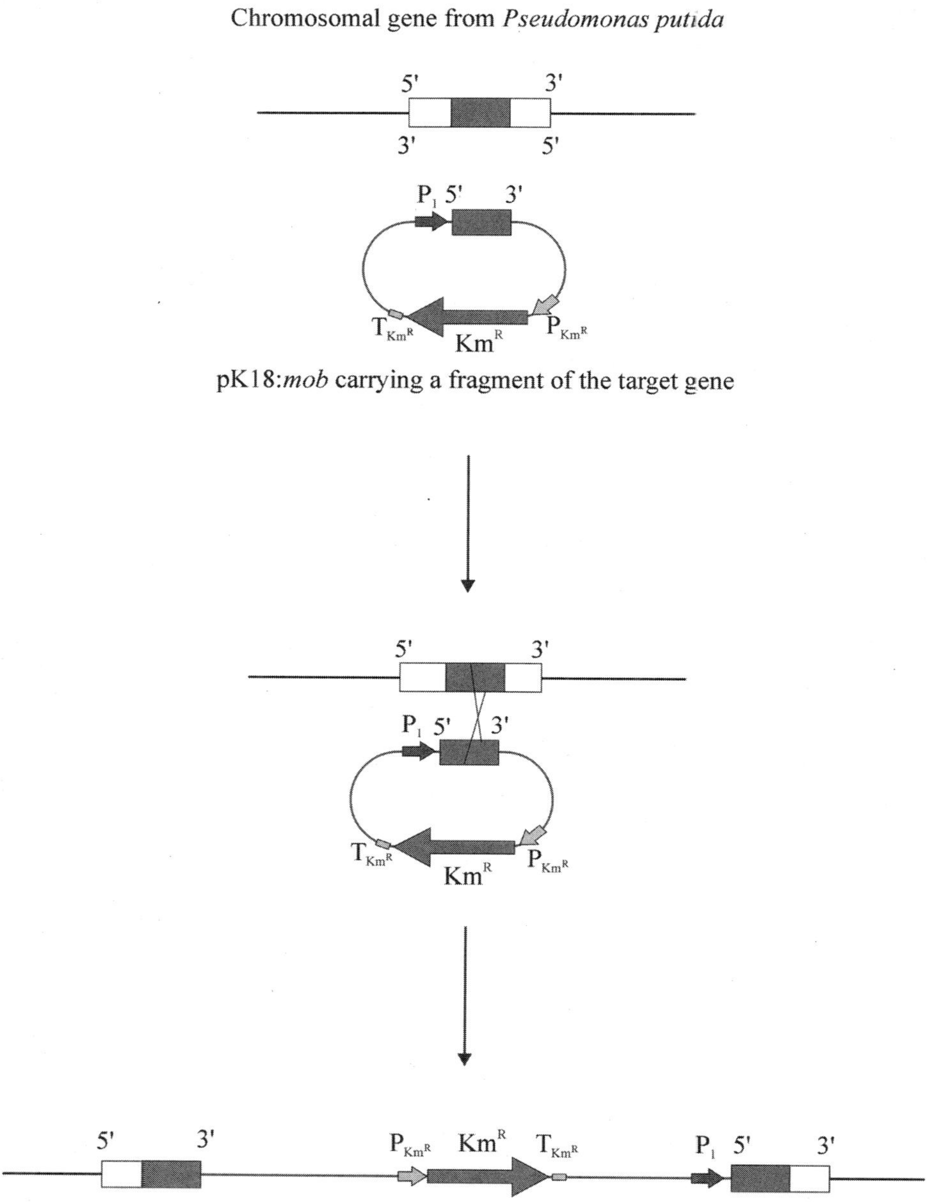

Chromosomal gene from *Pseudomonas putida*
5'
3'
3'
5'
P1 5' 3'
TKmR
KmR
PKmR
pK18:*mob* carrying a fragment of the target gene
5'
3'
P1 5' 3'
TKmR
KmR
PKmR
5' 3'
PKmR KmR TKmR
P1 5' 3'

Fig. 6. Procedure of gene disruption by homologous recombination.

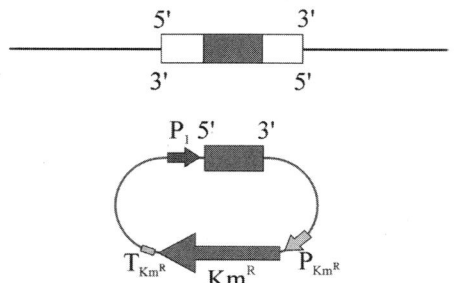

1. Clone two fragments of DNA placed at both sides of the gene (or the cluster) to be deleted in plasmid pJQ200KS (Gm^R). It is necessary that in the plasmid they maintain the same orientation as they show in the bacterial genome.
2. Transform *E. coli* DH10B with plasmid pJQ200KS containing the two fragments. The recombinant bacteria are employed to transform *P. putida* U by triparental mating (*see* **Subheading 3.1.1.**). The selective pressure caused by the presence of gentamicin (Gm) promotes a recombination event between one of the two fragments cloned in the plasmid and the homologous region found in the bacterial genome.
3. Select the recombinant strains as a function of their ability to grow in LB supplemented with Rf and Gm. Only the strains in which recombination has taken place are able to grow in this medium.

Plasmid pJQ200KS harbors the *sacB* gene *(13,18)*, which encodes an enzyme that polymerizes sucrose in the periplasmic space, causing the death of the recombinant strain cultured in media containing this disaccharide. Thus, in order to survive, the recombinant strains cultured in LB plus 10% sucrose force a second recombination event that eliminates the inner fragment of the target gene and the plasmid. The result of this second event (in 50% of the colonies analyzed) is the isolation of mutants that have undergone a deletion in the required gene (*see* **Note 3**).

This procedure can be used to eliminate some of the genes belonging to different routes involved in the catabolism of aromatics, facilitating the study of a particular function or contributing to the accumulation of some intermediate. Especially interesting is the deletion of the *fadBA* genes (those involved in the β-oxidation of aliphatic and aromatic acyl-CoA derivatives) because this leads to the accumulation of new bioplastic with broad applications. The deletion procedure is the following.

1. Clone into pGEM-T Easy a 300-bp fragment obtained from the *P. putida* genome by PCR using the oligonucleotides D3017 and D3350. This construction is named pGEM-D.
2. Digest this plasmid with *Eco*RI and clone the 300-bp fragment obtained into the pBluescript vector (previously cut with the same enzyme and treated with alkaline phosphatase). This plasmid is called pBluescript-D.
3. Release the 300-bp fragment from pBluescript-D with *Xho*I and *Pst*I and clone it into the pJQ200KS vector (pJQ200-D).
4. Amplify a second 300-bp fragment with the oligonucleotides A19X, which introduces an *Xba*I restriction site (TCTAGA) at the 5′ end, and A302P with a *Pst*I site (CTGCAG) also at the 5′ end.
5. Digest the PCR fragment with these two endonucleases (*Xba*I and *Pst*I) and clone the restriction product into pJQ200-D.
6. To be sure that the deletion event has indeed taken place, analyze the mutants by PCR with the oligonucleotides D3350 and COMFADB (they are located outside

Fig. 7. Genetic construction showing the cloning of *paaF* into the pQE-32 plasmid.

the deletion area). If deletion has occurred, a fragment of 800 bp will be obtained. In another case, a fragment of 3500 bp will be amplified.

3.1.2. Hyperexpression of the Gene Encoding the Phenylacetyl-CoA Ligase (paaF)

Phenylacetyl-CoA ligase is the first enzyme of the central catabolic pathway involved in the aerobic catabolism of PhAc *(4)*. This protein, which catalyzes the activation of PhAc to PhAc-CoA in the presence of ATP, Mg^{2+}, and CoA *(6)*, has important biotechnological applications (*see* **Subheadings 3.3.1.–3.3.3.**). For this reason, the collection of large amounts of this enzyme is needed to approach different biochemical and biotechnological studies. In order to achieve a high level of expression of the phenylacetyl-CoA ligase gene *(paaF)*, the QIAexpressionist Kit can be used. This procedure involves (1) the cloning of the required gene into the appropriate pQE vector and (2) hyperexpression of the selected gene.

3.1.2.1. CLONING THE REQUIRED GENE INTO THE PQE VECTOR

This kit provides a variety of vectors (pQE vectors) that allow a high-level expression of 6× His-tagged proteins in *E. coli* based on the T5 promoter transcription–translation system *(19)*. The ribosome-binding site (Shine–Dalgarno sequence) and the ATG initiation codon must be removed from the fragment to be inserted into these vectors, because internal starts from control sequences provided by the inserted fragment itself will result in the expression of proteins that lack the 6× His tag and, therefore, cannot be purified. For the overexpression of phenylacetyl-CoA ligase, the initial ATG present in the *paaF* gene is replaced by ATC (generating a *Bam*HI restriction site useful for cloning this gene into the pQE-32 multicloning site) in order to maintain the phase and allowing the synthesis of a fusion protein with the 6× His tag (*see* **Fig. 7**).

3.1.2.2. GENE HYPEREXPRESSION

The genetic construction containing the *paaF* gene in the pQE-32 vector is maintained in *E. coli* XL1-Blue (Tc^R), an ampicillin-sensitive strain that harbors the lacI^q gene on the F-factor (episome). This strain produces very high levels of the *lac* repressor and, hence, the PaaF protein is not expressed.

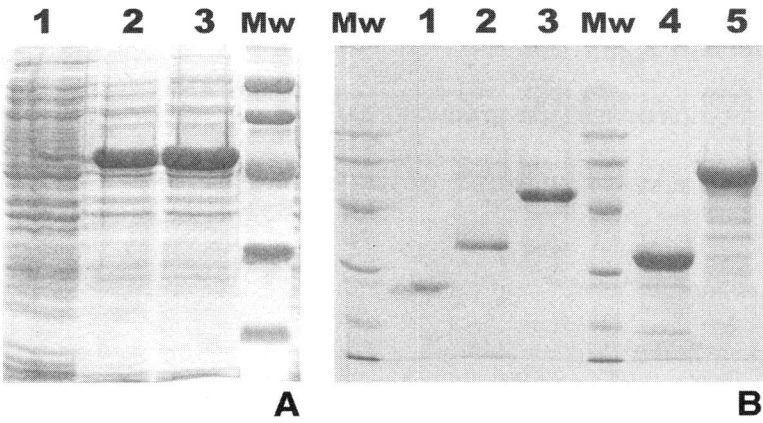

Fig. 8. SDS–10% PAGE to check overexpression of different proteins using a pQ32 vector containing **(A)** *paaF* gene (encoding phenylacetyl-CoA ligase) at different times of induction: (1) 0 min, (2) 120 min, and (3) 240 min, **(B)** *paaY* (1), *paaX* (2), *paaF* (3), *phaZ* (4), and *paaC1* (5) after 240 min of induction. Mw-molecular mass standards (94, 67, 43, 29, and 20 kDa, respectively).

1. Culture the recombinant bacteria in 500-mL Erlenmeyer flasks containing 100 mL of LB medium supplemented with 100 μL of ampicillin (Ap, 100 μg/μL) and 150 μL of tetracycline (Tc, 25 μg/μL).
2. Incubate the cultures at 250 rpm, 37°C, until they reach an OD_{540} of 0.5.
3. At this moment, after taking an initial sample of 1 mL, add 100 μL of 1 *M* IPTG to each flask to induce PaaF expression.
4. Take samples (1 mL) every hour up to a maximum of 6 h and analyze them by SDS gel electrophoresis (*see* **Fig. 8A**) to evaluate paaF expression.
5. Centrifuge the samples (10,000*g;* 10 min.), resuspend them in 50 μL of loading buffer and incubate at 95°C (2 min.). Later, analyse 15 μL of the crude cell lysate by SDS-PAGE.

By following a similar procedure, other catabolic enzymes (PaaGHIJK, PaaX, PhaC1, PhaC2, PhaZ) or regulatory proteins (PaaX and PaaY) belonging to the phenylacetyl-CoA catabolon can also be overexpressed in *E. coli* XL1-Blue (Tc^R) (*see* **Fig. 8A,B**) (*see* **Note 4**).

3.1.3. Directed Mutagenesis: Involvement of Some Cysteine Residues in PhAc-CoA Ligase Activity

Phenylacetyl-CoA ligase (PCL) catalyzes the activation of PhAc to PhAc-CoA in the presence of ATP, Mg^{2+}, and CoA. It has previously been suggested that formation of the phenylacetyl-CoA thioester would involve the participation

Table 1
Analysis of PCL Activity in Different Recombinant Phenylacetyl-CoA Ligases in Which a Residue of Cysteine Has Been Mutated

Strain	Oligonucleotide sequence	PCL activity
Wild-type	TTT gCC gAA **TgC** ggC gCC CAC CCC	+
C52T	TTT gCA gAA *ACC* gg*T* gCC CAC CCA	
Wild-type	gAC CTC AC**g** **TgC** CTg gAA gAC CTg	–
C61S	gAC CTC AC*T* AgT CTg gA*g* gAC CTg	
Wild-type	CgC CTg gg**C** **TgT** ACg gTA ATC CCg	–
C161S	CgC CTg gg*A* T*CC* *AC*g gTA ATC CCg	
Wild-type	ATg gAA **TgC** ATC gAA ACC AAg gAC	–
C256S	ATg gAA T*CT* ATC gAA ACC AAg gAC	
Wild-type	gAg **TTg** CgT gCg gAg **TgC** CAg CAC	+
C385S	gAg *CTC* CgT gCg gAg T*CC* CAg CAT	
Wild-type	CAg gCT **TgC** ggC ACg CTC AAg CgT	+
C419S	CAg gCT T*CC* ggC ACg CTC AAg CgT	
Wild-type	AAg gCg **TgC** CAC gTg TAC gAC AAA	–
C430S	AAg gCg *AgC* CAC gTg TAC gAC AAA	

Note: The modified sequence is indicated in **bold**.

of some cysteine residues that seem to be essential for the formation of the enzyme–substrate complex *(6,20)*. The functional role of the cysteine residues present in PCL can be established by replacing all of them (one by one) by other analogous residues (serine or threonine). Thus, using directed mutagenesis, the influence of the seven cysteines present in PCL in the catalytic process can be established. Using this procedure, seven mutants—each containing a different recombinant protein modified in a single residue of cysteine—can be obtained. The procedure followed is now summarized.

1. Design two oligonucleotides for each cysteine (inverse and complementary) that harbor the desired changes (a codon of serine or threonine instead of one of cysteine) as well as a restriction site, which allows the linking the two fragments to be amplified (*see* **Table 1**).
2. PCR-amplify the fragment containing the *paaF* gene mutated using a plasmid that contains the *paaF* gene and the specific oligonucleotides designed as a template.
3. Join the two amplified fragments and clone the mutated *paaF* gene into a vector (pBluescript, pUC18, pGEM-T Easy, etc.).
4. Analyze the phenylacetyl-CoA ligase activity of all the recombinant strains containing a mutated copy of *paaF* (*see* **Note 5**).

This procedure could be used to modify other amino acids in this enzyme and to study their involvement in the catalytic process.

3.1.4. Cloning of Different Pathways Belonging to the Phenylacetyl-CoA Catabolon

In this subsection, we describe an interesting procedure that is useful for cloning all of the genes encoding the two most important pathways of the phenylacetyl-CoA catabolon: the central pathway (the catabolon core) and the route involved in the biosynthesis and catabolism of poly-3-hydroxyalkanoates (PHAs pathway). The method followed, based on a single recombination event, seems to be a useful strategy that can be used to isolate many other linked sets of genes. This procedure requires the following steps:

1. Analysis of the restriction map of the whole pathway and search for a restriction site located outside the cluster. In the case of the PhAc cluster *(paa)*, the selected restriction endonuclease site was *Bam*HI, which only exists outside the *paa* genes and is also present at the multicloring site (polylinker) of plasmid pK18::*mob*.
2. Insert into the plasmid pK18::*mob* a gene (or a DNA sequence) located just before or after the cluster to be cloned (depending on the location of the endonuclease restriction enzyme that is going to be used). If the restriction site is located downstream of the pathway, the gene that should be cloned in pK18::*mob* would have been either the one encoding the first enzyme of the route or another one placed upstream of it, and vice versa when the restriction site is located upstream of the pathway. To clone the *paa* pathway, a transcriptional activator *(OrfA)* located just upstream from the *paa* route was cloned in the *Pst*I site of the pK18::*mob* multicloning site, so the *Bam*HI stays on the right of the cloned gene (henceforth called pK18-*OrfA*).
3. Transform this construction into *E. coli* DH10B and select the transformants that harbor the desired insert (this plasmid allows blue/white selection).
4. Force homologous recombination (through the cloned gene in pK18::*mob*) with the parental strain (in our case, *P. putida* U) and select the transconjugants as a function of their ability to grow in LB medium supplemented with Km and Rf. Only strains in which a homologous recombination event has occurred will be able to grow in media containing both antibiotics. The transconjugants obtained after the insertion of the pK18-*OrfA* construction should have a copy of the *A* gene followed by the pK18::*mob,* another *A* gene, and the whole *paa* cluster (*see* **Fig. 9A**).
5. Extract genomic DNA from the transconjugants and digest it with the selected endonuclease (in our case, *Bam*HI).
6. Analyze the digested products in an agarose gel (0.5% w/v) in order to select a band with a size corresponding to the whole cluster plus the pK18::*mob* size (in our case 23,198 bp—the *paa* pathway—plus 3700 bp—the plasmid pK18::*mob*).
7. Cut the selected band of DNA and extract it from the agarose using any of the many commercial kits available. Ligate the DNA fragment to reconstitute the plasmid containing all the genes encoding the required pathway inside the multicloning site (plasmid pK*paa*).
8. Transform any of the *E. coli* strains usually employed in molecular biology (we have used *E. coli* DH10B) and select the transformants by growing them either in

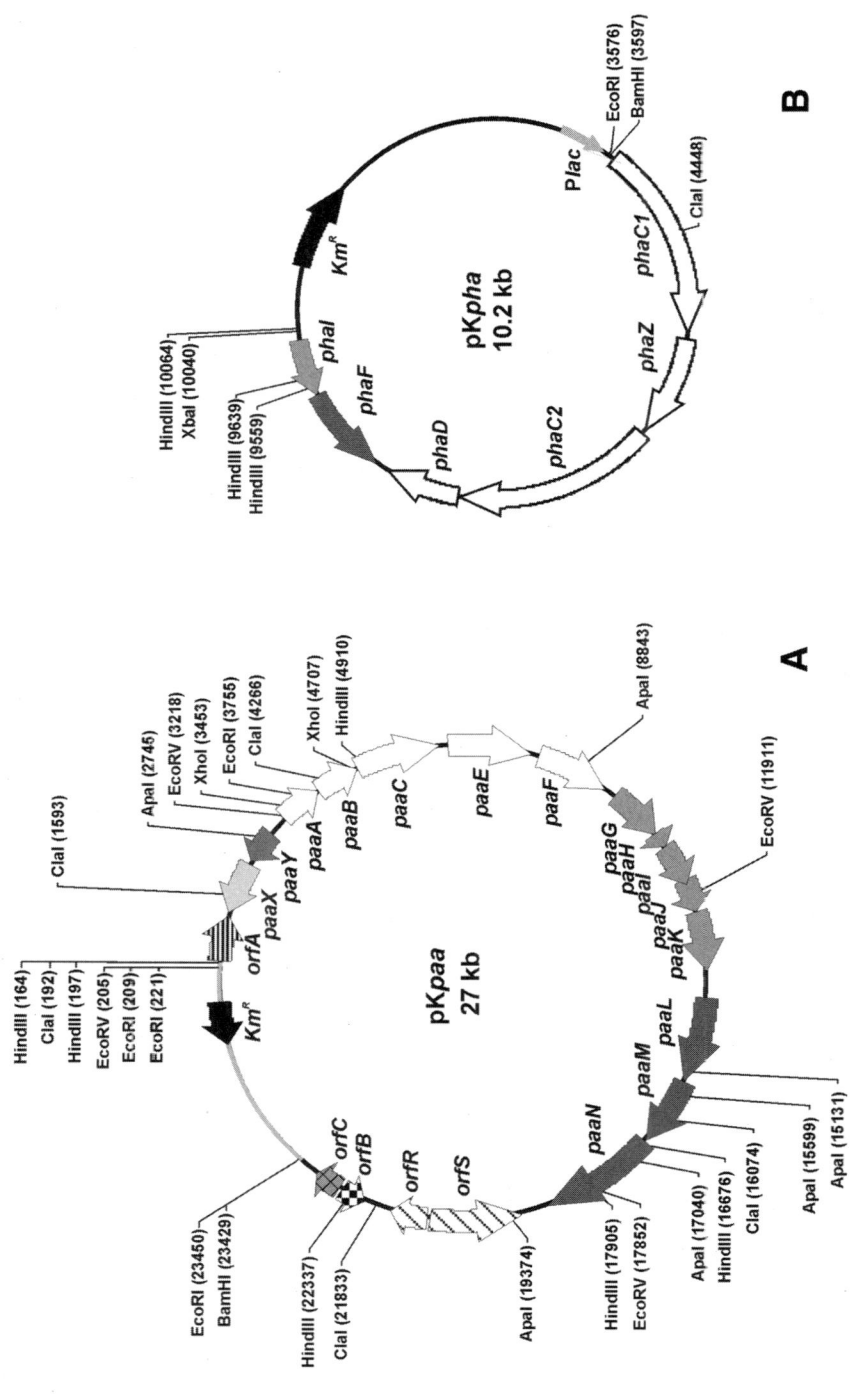

Fig. 9. Cloning into a plasmid of the genes encoding (**A**) the phenylacetic catabolic pathway and (**B**) the poly-3-hydroxyalkanoates metabolic pathway.

LB–Km or in MM containing Km and PhAc as the sole carbon source (when strains without auxotrophies are used).

By following a similar procedure, the genes encoding the polyhydroxyalkanoate pathway (PHA route) *(21)* have been cloned and expressed in different microbes (*see* **Fig. 9B**). In sum, this procedure could facilitate the cloning of the genes encoding many different pathways as long as they are clustered or not too distant from one to another. Furthermore, the cloning and expression of a pathway into a replicative or integrative plasmid could help to facilitate genetic and biochemical studies, to isolate some catabolic intermediates or biosynthetic precursors, or to enhance the metabolic versatility of different microbes. Therefore, this method could be used as a tool in evolutionary studies, because many routes might have been easily transferred to phylogenetically unrelated microbes.

3.2. Biochemical Assays and Related Procedures

3.2.1. Enzymes or Enzymatic Systems Belonging to the Central Pathway

In this subsection, we describe some of the procedures followed to evaluate properly certain enzymes belonging either to the central route or to the peripheral ones of the phenylacetyl-CoA catabolon.

3.2.1.1. ASSAY OF PHAC UPTAKE IN *P. PUTIDA* U

1. Inoculate *P. putida* U in 500-mL Erlenmeyer flasks containing 100 mL of a chemically defined medium (MM) with PhAc as the sole carbon source.
2. Incubate the bacteria in a rotary shaker (250 rpm at 30°C) until OD_{540} reaches a value of 0.14 (diluted 1/10, absolute value = 1.4).
3. Take aliquots of the culture (20 mL), harvest the bacteria by centrifugation ($10,000g$ for 10 min), wash the cells twice with sterile distilled water, centrifuge again, and resuspend the precipitate in 50 mM phosphate buffer, pH 7.0
4. Place aliquots of 1.4 mL (1.4×10^9 bacteria) in 25-mL Erlenmeyer flasks and preincubate them at 30°C for 3 min in a thermostatically controlled bath (160 strokes/min).
5. Add 21.4 μM PhAc containing 9.9 μM 1-^{14}C-labeled PhAc.
6. Incubate the bacterial suspension for different periods of time (usually 1 min) and halt PhAc uptake by the addition of 10 vol of water.
7. Quickly filter the bacteria through Millipore HAWPO 2500 filters (0.45 μm pore size), and wash with 2×10 mL of sterile distilled water.
8. Place the filters in 10 mL of scintillation fluid until they have dissolved and count the radioactivity incorporated by the bacteria on a liquid scintillation counter. PhAc uptake ability should be given as picomoles of PhAc incorporated in 1 min, bearing in mind that 21.4 μM correspond to 10^5 counts per minute (cpm).

PhAc uptake can be also determined in bacteria grown in solid medium. In this case, 25 g/L of agar is added to MM and, once sterilized, this is distributed in plates (9 cm in diameter) containing 28 mL. Plates are seeded with *P. putida*

U and incubated at 30°C for 24 h or the time required in each set of experiments. Bacteria are resuspended in 50 mM phosphate buffer, pH 7.0, to an OD$_{540}$ of 1.0 (10^9 bacteria/mL) and PhAc uptake *(5)* is measured as indicated earlier.

3.2.1.1.1. Determination of the PhAc Transport System Half-Life. *P. putida* U is cultured in liquid MM at 30°C, and at 10 h, or at the required time, protein synthesis was stopped by adding 75 μg of chloramphenicol/mL. From this time to 15 h, the PhAc transport system (PATS) was measured at 30-min intervals *(5)*.

3.2.1.1.2. Cell-Shocking Procedure. Osmotic shock in *P. putida* U is carried by following the procedure described by Neu and Heppel, modified by us *(22)*.

1. A 50-mL portion of bacterial suspension grown in MM for 19 h is centrifuged (10,000g, 10 min).
2. Wash the bacteria with sterile distilled water and suspend in 50 mL of cell-shocking buffer.
3. Incubate the bacteria in a rotary shaker (250 rpm) at 30°C for 5 min.
4. Harvest the precipitate by centrifugation (10,000g, 10 min) in a cold room (4°C).
5. Remove the supernatant fluid and mix the well-drained pellet rapidly with 50 mL of cold sterile distilled water (4°C) containing 0.5 M MgCl$_2$.
6. Incubate the suspension in an ice bath (using a thermostatically controlled bath at 160 strokes/min) for 10 min, centrifuge it (10,000g, 10 min), and remove the supernatant fluid. Resuspend cells in 50 mM phosphate buffer, pH 7.0, adjust them to an OD$_{540}$ of 1.0, and use these bacteria for the uptake experiments. Control cells are treated in a similar way, but without osmotic shock (*see* **Note 6**).

3.2.1.1.3. PATS Induction Experiments

1. Culture *P. putida* U in MM in which the PhAc had been replaced by an alternative growth substrate (5 mM): octanoic acid, succinic acid, benzoic acid, 4-hydroxy-PhAc or other hydroxyderivatives, several phenylderivatives, glycerol, acetate, or glucose and other sugars (*see* **Note 7**).
2. Monitor the appearance of PATS at intervals.

3.2.1.2. COLORIMETRIC DETERMINATION OF PHENYLACETYL-COA LIGASE

3.2.1.2.1. Cell-free Extracts Preparation. Crude extracts containing phenylacetyl-CoA ligase (PCL) activity are prepared as follows:

1. Culture *P. putida* U in MM for 14 h.
2. Collect the bacteria by centrifugation (10,000g, 10 min at 2°C).
3. Wash the bacteria with sterile saline solution and resuspend them in disruption buffer.
4. Disrupt the bacteria (28 g wet weight/115 mL of buffer) with 100 g of glass beads using a Braun MSK mechanical disintegrator.

5. Remove cell debris by centrifugation at 17,000*g*, 10 min at 2°C.
6. Eliminate the pellet and the supernatant obtained can be used to evaluate the PCL activity (*see* **Note 8**).

PhAc-CoA ligase activity can be evaluated by measuring the rate of formation of phenylacetylhydroxamate in the presence of ATP, CoA, PhAc, and neutral hydroxylamine, as has been reported for other CoA-activating enzymes *(6,7,23)*. The procedure is as follows.

1. Reaction mixture composition: 12.5 µL of 0.2 *M* $MgCl_2$, 50 µL of 0.1 *M* ATP, 30 µL of 20 m*M* CoA, 30 µL of 0.2 *M* potassium phenylacetate, and 50 µL of hydroxylamine solution. All the substrates except $MgCl_2$ are dissolved in 50 m*M* Tris-HCl and adjusted to pH 8.2. When required, PhAc or ATP are replaced by other molecules. In standard tests, potassium phenylacetate, ATP, $MgCl_2$, or CoA are omitted.
2. Equilibrate the temperature of the reaction in a water bath at 30°C (5 min). Add 100 µL of enzyme solution (about 500 µg of protein), prepared as indicated in **Subheading 3.2.1.2.1.**, and incubate the assay for 20 min (or the required time).
3. Stop the reactions by adding 450 µL of the ferric chloride reagent and keep on ice for 30 min.
4. Centrifuge in an Eppendorf microcentrifuge for 2 min (10,000*g*) and measure the red-purple color generated at 540 nm in a Shimadzu UV-120-02 spectrophotometer. The extinction coefficient of phenylacetylhydroxamate under these conditions is 0.9 mM^{-1} cm^{-1}.

One unit (U) of enzyme activity is defined as the catalytic activity leading to the formation of 1 nmol phenylacetylhydroxamate in 1 min. Specific activity is given as units per milligram of protein. Protein determination is usually performed using the method of Bradford *(24)*, using bovine serum albumin as the standard. Buffers identical to those containing the protein samples are used as blanks.

3.2.1.3. FLUORIMETRIC DETERMINATION OF PHENYLACETYL-CoA LIGASE

This procedure is an alternative assay for phenylacetyl-CoA ligase. The method, which is faster and more sensitive (it allows the quantification of very low concentrations of phenylacetyl-CoA [PhAc-CoA]), involves the participation of three different enzymes (adenylate kinase, E.C. 2.7.4.3; pyruvate kinase, E.C. 2.7.1.40; and lactate dehydrogenase, E.C. 1.1.1.27), which are coupled in vitro with PCL (E.C. 6.2.1.) *(25)*. Each molecule of PhAc-CoA formed requires the hydrolysis of a molecule of ATP, generating 1 molecule of AMP, which is measured indirectly by following the oxidation of NADH at OD_{340}.

1. Mix 600 µL of 50 m*M* Tris-HCl buffer, pH 8.0, 100 µL of enzyme cocktail, 100 µL of 0.1 *M* KCl, 100 µL of 0.1 *M* phosphoenolpyruvate, 12.5 µL of 0.1 *M* $MgCl_2$, 50 µL of 0.1 *M* ATP, 100 µL of 3.5 m*M* NADH (or 10 µL when the spectrophotofluorimetric assay was used), 30 µL of 0.02 *M* CoA, 30 µL of 0.2 *M* potassium

phenylacetate, and 100 μL of enzyme solution (10 μg of pure PCL). Control reactions contain all of the substrates except phenylacetate.

2. Start the reactions by the addition of phenylacetate and incubate at 30°C.
3. Measure reactions on a Shimadzu UV-260 spectrophotometer or a Shimadzu UV-RF 510 spectrophotofluorimeter (excitation at 340 nm and analysis of the emitted light at 465 nm). The determination of PCL activity by the spectrophotofluorimetric procedure is about 10-fold more sensitive than the spectrophotometric one (*see* **Note 9**).

Enzyme activity is given in units (U), where 1 U is defined as the amount of enzyme that, under standard assay conditions, oxidizes 2 nmol of NADH per minute. Specific activity is given as U/mg protein. Protein is measured by the method of Bradford using bovine serum albumin as standard *(24)*.

3.2.1.4. BIOCHEMICAL ANALYSIS OF THE PHAC-COA LIGASE-PHAC-COA HYDROXYLATION COMPLEX (PAAFGHIJK ENZYMATIC SYSTEM)

An in vitro assay for the enzymatic complex involved in the hydroxylation of phenylacetyl-CoA to 2-hydroxyPhAc has not yet been developed. However, it is possible to analyze the reaction products by cloning together the *paaFGHIJK* genes into a plasmid, transforming a selected bacterium and, once expressed, analyzing in vivo the biotransformation of certain aromatic compounds structurally related to the natural substrate (PhAc). The procedure for the in vivo assay is the following:

1. PCR amplify the genes *paaF* and *paaGHIJK* from either *P. putida* DNA or from a pK*paa* plasmid (*see* **Subheading 3.1.4.** and **Note 10**) or from any construction containing all of these genes, using specific oligonucleotides designed with the required restriction sites.
2. Digest the two fragments and clone the cluster (paaFGHIJK) into the plasmid pK18::*mob*.
3. Transform different strains of *E. coli* (DH10B, W14, etc.).
4. Incubate the recombinant bacteria for different times (24–96 h) in different media supplemented with PhAc or with different structural analogs (containing 2-OH-PhAc, 3-OH-PhAc, 4-OH-PhAc, 2,5-diOH-PhAc, 3,4-OH-PhAc, benzoic acid, or *n*-phenylalkanoic acid [10 < *n* > 2 carbon atoms]) (*see* **Note 11**).
5. Analyze the aromatic products accumulated in the culture broths by HPLC or by thin-layer chromatography (*see* **Subheadings 3.4.1.** and **3.4.2.**).

3.2.2. Enzymes Belonging to the Upper Pathways

3.2.2.1. STYRENE MONOOXYGENASE ASSAY

This enzyme catalyzes the first step of the catabolic pathway involved in the catabolism of styrene *(26)*.

1. Inoculate *P. putida* in 500-mL Erlenmeyer flasks containing 100 mL of a chemically defined medium (MM) with styrene (gas phase) as the sole carbon source.

2. Incubate the bacteria in a rotary shaker (250 rpm, at 30°C) until OD_{540} reaches a value of 0.1 (diluted 1/10, absolute value = 1.0).

3. Collect the bacteria by centrifugation (10,000g, 10 min, 2°C).

4. Wash the bacteria with 50 mM potassium phosphate buffer, pH 7.0, twice, centrifuge them, and resuspend the precipitate in the volume of buffer required to concentrate the bacterial suspension threefold for the determination of specific rates of indigo formation.

5. Add aliquots (100 μL of concentrated bacteria) to 400 μL of 50 mM potassium phosphate buffer, pH 7.0, containing 0.25 mM indole in a 1.5-mL polypropylene microcentrifuge tube, and preincubate them for 10 min at 30°C.

6. Incubate the samples in a vigorously shaken water bath at 30°C.

7. Take vials for analysis every 5 min over a 30-min period and centrifuge them (10,000g, 2 min).

8. Eliminate the supernatant and then resuspend the bacterial pellet in 1 mL of dimethylformamide.

9. Incubate this suspension (with shaking) at 25°C for 10 min to ensure the removal and the dissolution of cell-associated indigo *(26)*.

10. Centrifuge the tubes to remove cell debris.

11. Measure OD_{610} of the supernatant.

The specific rate of indigo production is calculated from the increase in absorbance over time. The molar extinction coefficient for indigo at 610 nm is 15,900 L/mol/cm *(26)*.

3.2.2.2. PHENYLACETALDEHYDE DEHYDROGENASE ASSAY

Phenylacetaldehyde dehydrogenase is a key enzyme in the catabolism of different aromatic compounds that are degraded through the phenylacetyl-CoA catabolon. Thus, this enzyme is directly involved in the catabolism of styrene, ethylbenzene, 2-phenylethanol, 2-phenylethylamine, and tropic acid, as well as phenylacetaldehyde *(3)*.

Phenylacetaldehyde dehydrogenase activity is measured in cell-free extracts of *P. putida* U grown in MM + 5 mM 2-phenylethylamine by following the formation of NADH (increase in OD_{340}) after the incubation of cell-free extracts (obtained as indicated in **Subheading 3.2.1.2.1.**) with phenylacetaldehyde and NAD$^+$ *(27)*.

The standard assay mixture (1 mL) contains 50 μL of 2 mM phenylacetaldehyde, 35 μL of 5 mM NAD$^+$, 200 μL of 0.5 M phosphate potassium buffer, pH 8.9, and 990 μL of sterile distilled water. The reaction is initiated with the addition of 100 μL of enzyme solution (0.5 mg of protein). The increase in OD_{340} is measured in a Shimadzu UV-120-02 spectrophotometer.

A unit of phenylacetaldehyde dehydrogenase is defined as the amount of enzyme that produces 1 μmole of NADH in 1 min at 25°C. Specific activity is expressed as the number of units per milligram of protein.

3.2.2.3. PHA P‌OLYMERASE A‌SSAY

The biosynthesis of poly-3-hydroxyalkanoates in *P. putida* U involves the participation of two different polymerases that catalyze the condensation of several 3-hydroxy-acyl-CoA derivatives into a complex structure (PHA granule), which is accumulated intracellularly *(9,11,12)*. Moreover, two additional proteins (phasins F and I) seem to be associated with the granules, and it has been assumed that these contribute to granule formation as well as to the acquisition of a correct macromolecular structure *(21)*.

In this subsection, we analyze the procedure followed to (1) isolate the PHA granules, (2) analyze the granule-bound proteins, (3) synthesize *R/S*-3-hydroxyoctanoyl Coenzyme A, and (4) assay PHA polymerase *(28)*.

3.2.2.3.1. Isolation of PHA Granules

1. Culture *P. putida* U in MM containing alkanoates and/or phenylalkanoates as a carbon source in order to induce the synthesis and accumulation of PHAs (visualized by direct microscopic observation of the bacteria stained with safranine).
2. Harvest the bacteria by centrifugation (20,000*g*, 30 min, 2°C) when most of their cytoplasm is occupied by PHA granules.
3. Resuspend the pellets in granules buffer up to a final concentration of 6 mg cell dry weight/mL.
4. Incubate for 1 h on ice and harvest the bacteria by centrifugation.
5. Resuspend them in water (up to a final concentration of 6 mg cell dry weight/mL) and disrupt them by sonication (or using a French press).
6. Layer the broken cells (5 mL) on a 20% (w/v) sucrose solution (dissolved in 50 m*M* Tris-HCl, pH 8.0) and ultracentrifuge them (110,000*g*, 4°C, 3 h).
7. Collect the PHA granules settled on top of the sucrose layer and wash twice with 50 m*M* Tris-HCl, pH 8.0.
8. Resuspend the final PHA pellet in 1 volume of 50 m*M* Tris-HCl, pH 8.0 *(28)*.

3.2.2.3.2. Analysis of Granule-Associated Proteins and Protein Determination of PHA Granules. Proteins are recovered from the granules by SDS-PAGE following the procedure indicated here *(29)*.

1. Mix samples of purified granules (*see* **Subheading 3.2.2.3.1.**) 1 : 1 (v/v) with loading buffer, boil them for 10 min, and analyze the samples by sodium dodecyl sulfate-polyacrylamide gel electrophoresis (SDS-PAGE).
2. Stain the proteins with Coomassie brilliant blue, R-250.
3. Evaluate the protein content in the PHA granules by densitometric scanning of SDS-PAGE gels (they are compared with known quantities of molecular mass standards).
4. When Western blots are required, transfer the proteins from unstained gels to 0.45-mm-pore-diameter Protran nitrocellulose membranes, using a Bio-Rad Transblot SD semidry electrophoretic transfer cell (1.5 h at 100 V, 10 mA). Wash the mem-

brane three times in Tris-buffered saline (TBS, pH 8.0) and treat with TBS with 3% milk powder to block nonspecific binding.

5. Apply primary antibody (rabbit anti-PHA polymerase C1 ([1 : 250]) in fresh blocking solution and incubate at 25°C for 2 h.
6. Rinse the membranes three times in TBS, add horseradish peroxidase-conjugated donkey anti-rabbit antibody (1 : 5000) or sheep anti-mouse antibody (1 : 10,000) in fresh blocking solution, and incubate at 25°C for 1 h.
7. Wash the membranes again in TBS (three times for 5 min each) and detect the antibodies after incubating with enhanced chemiluminescence kit reagents by exposure of the membrane to autoradiograms.
8. Estimate the quantity of polymerase C1 (or other proteins such as phasins F and I) by comparison to known amounts of a polymerase C1 (or of the protein studied) standard *(29)*.

3.2.2.3.3. Synthesis and Purification of R/S-3-Hydroxyoctanoyl Coenzyme A.

The synthesis of this thioester is performed by following the procedure reported by de Roo et al. *(28)*.

1. Synthesize *R/S*-3-hydroxyoctanoyl-CoA (3-OH-Oc-CoA) by enzymatic conversion of 3-hydroxyoctanoate into its CoA thioester through a reaction catalyzed by the acyl-CoA synthetase (10 μg) from *P. fragi* or by the acyl-CoA synthetase involved in the activation of medium-chain-length fatty acid in *P. putida* U. The reaction is performed as is indicated in the reaction of PaaF (*see* **Subheading 3.2.1.2.**), but now replacing PhAc by 10 m*M* of 3-hydroxyoctanoate (*see* **Note 12**).
2. Separate 3-OH-Oc-CoA from the other reaction components using a preparative C8-spherisorb-RP column, together with a preparative HPLC system. Purification is successful using a two-step elution procedure under isocratic conditions starting with buffer A for 10 min and subsequent elution of 3-OH-CoA with buffer B for 20 min.
3. Evaporate acetonitrile and ammonium acetate by freeze-drying. The purified substrate is kept at –20°C.

Using a similar protocol, many other thioesters can be synthesized (*see* **Note 11**).

3.2.2.3.4. Assay of PHA Polymerase.

This reaction is based on the measurement of the CoA residue released during PHA formation. It is evaluated using the reagent 5,5′-dithiobis(2-nitrobenzoic acid) (DTNB) *(30)*, according to the method reported by de Roo et al. *(28)*.

1. The reaction mixture (500 μL) contains 1 m*M* of 3-OH-Oc-CoA and 2.5 mg/mL of PHA granules resuspended in 100 m*M* Tris-HCl, pH 8.0.
2. Aliquots (100 μL) are taken at intervals and the reaction is rapidly stopped in them by adding 60 μL of trichloroacetic acid (1% w/v).
3. The PHA granules are eliminated by centrifugation (20,000*g*, 10 min) and a sample of the supernatant (100 μL) is mixed with 100 μL of 2 m*M* DTNB dissolved in 100 m*M* Tris-HCl, pH 8.2, in a microtiter plate.
4. Measure OD_{412} ($\varepsilon = 13.7$ mM^{-1} cm^{-1}).

One unit (U) is defined as the quantity of polymerase required to release 1 μmol of CoA/min. Specific activity is given as U/mg protein.

3.3. Biotechnological Applications

3.3.1. Coupled Assay of PA-CoA Ligase From P. putida U and acyl-CoA : 6-APA Acyltransferase From Penicillium chrysogenum

By coupling the enzymes acyl-CoA : 6-aminopenicillanic acid (isopenicillin N) acyltransferase from *P. chrysogenum* and phenylacetyl-CoA ligase from *P. putida,* it is possible to reproduce in vitro the last step involved in the biosynthesis of different hydrophobic penicillins *(31,32)*.

1. The reaction mixture contains 12.5 μL of 0.2 *M* MgCl₂, 50 μL of 0.1 *M* ATP, 30 μL of 20 m*M* CoA, 30 μL of 0.2 *M* potassium phenylacetate (or the corresponding side-chain precursor), 30 μL of 0.3 m*M* 6-aminopenicillanic acid (6-APA) (at this concentration, 6-APA does not inhibit the growth of *Micrococcus luteus*), 10 μL of 20 m*M* DTT, 100 μL of PA-CoA ligase (2 μg of pure protein), and 100 μL of purified acyl-CoA : 6-APA acyltransferase (8 μg of protein).
2. All the substrates except MgCl₂ are dissolved in 50 m*M* Tris-HCl and adjusted to pH 8.2.
3. Incubations are carried out at 30°C for 1 h (or the required time) and halted by the addition of 362.5 μL of methanol. When required, *E. coli* penicillin acylase (4 IU) or β-lactamase (10 μL) are added to the reactions and incubated for an additional 10 min before adding methanol.
4. The antibiotics generated are measured by bioassay against *Micrococcus luteus* ATCC 9341 *(7,33)*
5. Control reactions are carried out under the same conditions without PhAc or CoA.

3.3.1.1. Chemical Synthesis of Octanoyl, Heptanoyl, and Hexanoyl Penicillins

Octanoyl, heptanoyl, and hexanoyl penicillins and other noncommercial hydrophobic penicillins are routinely used as standards in many experiments. For this reason, they must be synthesized or extracted from the fermentations carried out in the absence of PhAc. Taking into account that—usually—very pure material is required, here we describe a method to synthesize these penicillins. It is based on the acylation of the amino group of 6-APA with the acid moiety of different acyl chlorides.

1. Dissolve 54 g of 6-APA in 1.2 L of 10% (w/v) aqueous NaHCO₃.
2. Dissolve 55 g of octanoyl, heptanoyl or hexanoyl chloride in 100 mL of dry acetone.
3. Add the acetone solution dropwise to the 6-APA solution and stir the mixture for 20 min at 16°C.
4. Evaporate the acetone under reduced pressure.
5. Add ammonium sulfate (45% w/v) to each aqueous solution until a pale white solid precipitates.

6. Filter the product (penicillin DF or K, or heptanoylpenicillin), wash it with dry acetone, dry the powder in an oven (28°C overnight), and, finally, identify the penicillin by HPLC (*see* **Subheading 3.4.1.**) *(34)*.

3.3.2. Expression of paaF Gene From P. putida in P. chrysogenum

The activation of PhAc to PhAc-CoA, a precursor of benzylpenicillin, is a critical step in the biosynthesis of this β-lactam antibiotic *(35)*. Although all the enzymes involved in the specific pathway of penicillin G (L-α-aminoadipoyl-L-cysteinyl-D-valine synthetase, isopenicillin N synthase, and acyl-CoA : 6-APA-isopenicillin *N*-acyltransferase) have been purified and assayed in vitro, phenylacetyl-CoA has never been purified or characterized *(3,35)*. Furthermore, it has been suggested that the activation of PhAc to PhAc-CoA is either a critical regulatory point in the control of the benzylpenicillin biosynthetic rate *(36)* or that this step is catalyzed by a different enzyme *(36,37)*. Accordingly, the phenylacetyl-CoA ligase from *P. putida* has been purified *(6)*, cloned, and expressed into *P. chrysogenum* to study its effect on the yields of penicillin G production *(7)*. Genetic engineering in the fungus is carried out according to the protocols reported previously *(38)*.

1. Culture *P. chrysogenum* Wis 54-1255 in minimal medium as previously reported *(7,38)*.
2. Collect the mycelia by centrifugation and resuspend it (100 mg wet weight/mL) in lysis buffer containing Novozym 234.
3. Incubate the suspension at 28°C with slow agitation for 2–3 h. Remove the undigested mycelium by filtration through 30-μm-pore-size filters.
4. Wash the protoplasts with 0.7 *M* KCl and once with a solution containing 0.7 *M* KCl, 50 m*M* CaCl$_2$, and 10 m*M* MOPS, pH 5.8.
5. Resuspend the protoplasts in the same solution (0.7 *M* KCl, 50 m*M* CaCl$_2$, and 10 m*M* MOPS, pH 5.8.) at a final concentration of about 10^8 protoplasts/mL.
6. Transform 50 μL of the protoplasts suspension with a plasmid (1–5 μg of DNA) derived from pBC (Stratagene) containing the following genetic information: (1) a gene that confers the fungus resistance to the antibiotic fleomycin *(ble^r)* *(39)* and that is under the control of the promoter of *P. chrysogenum* glutamate dehydrogenase (P$_{gdh}$), (2) the promoter of the *pcbAB* gene, which encodes ACVS in *P. chrysogenum* (P$_{pcbAB}$), (3) the *paaF* gene, isolated from *P. putida* U, and (4) the terminator of the *trpC* gene (T$_{trpC}$) of *P. chrysogenum*. The construction designated pALPs9 is represented schematically in **Fig. 10.** It can be seen that *paaF* is under the control of the promoter of the *pcbAB* gene of *P. chrysogenum*.
7. Select fungal transformants expressing *ble^r* and analyze the presence of *paaF* gene by PCR amplification of an internal fragment (651 bp) located between the synthetic oligonucleotides PCLA and PCLB. In the controls (untransformed strains of *P. chrysogenum*) or in transformants containing the same construction without the *paaF* gene, no amplification should be observed because this sequence is not present in the genome of the fungus.

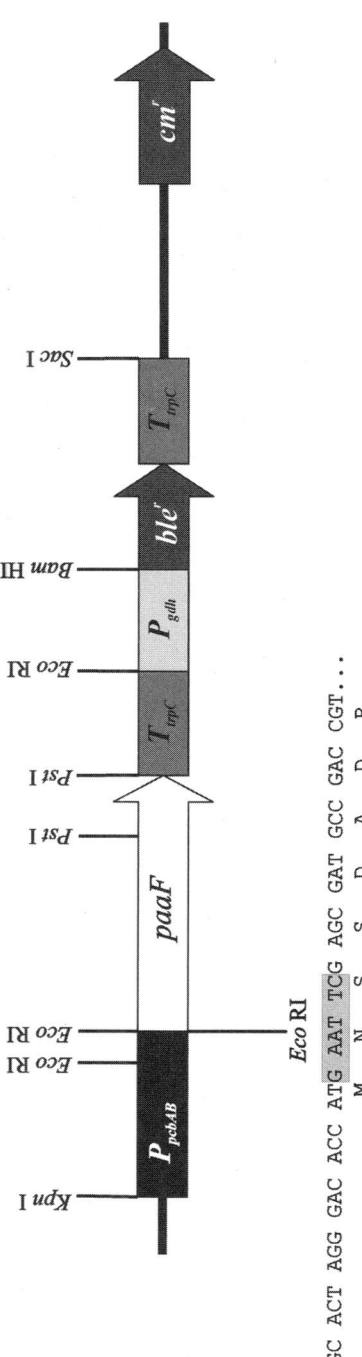

AGC ACT AGG GAC ACC ATG AAT TCG AGC GAT GCC GAC CGT...

M   N   S   S   D   A   D   R...

PaaF (*P. putida* U)

pALPs9
(9.0 kb)

Fig. 10. Schematic representation of the construction used to transform *Penicillium chrysogenum*. P$_{pcbAB}$, promoter of the *pcbAB* gene that encodes ACVS in *Penicillium chrysogenum*; *paaF*, *P. putida* phenylacetyl-CoA ligase gene; T$_{trpC}$, terminator of the *trpC* gene of *P. chrysogenum*; P$_{gdh}$, promoter of *P. chrysogenum* glutamate dehydrogenase; *ble^r*, fleomycin-resistance gene; *cm^r*, chloramphenicol-resistance gene.

8. Assay phenylacetyl-CoA ligase activity (*see* **Subheading 3.2.1.2.**) in cell-free extracts of the transformants in which the fragment of 621 bp (internal zone of *paaF*) has been amplified.

3.3.3. Production of 2-HydroxyPhAc

2-Hydroxy-phenylacetate is an aromatic compound that can be used as starting material to synthesize many other organic molecules with different environmental or industrial applications *(3)*. Thus, its microbial collection (by biotransformation from a direct precursor such as PhAc) could facilitate the synthesis of many different interesting and broadly used products.

Here, we describe a biotransformation procedure that permits efficient synthesis of this molecule:

1. Digest the plasmid pK*paa* (which harbors the whole *paa* pathway) with *Apa*I and *Eco*RI.
2. Load this sample into an agarose gel in order to isolate a 6288- bp band that contains the end of the *paaF* gene, the complete oxidation complex, and part of the *paaL* gene (*see* **Fig. 2**).
3. Clean the agarose band with any of the commercial kits used for DNA extraction from agarose gels.
4. Clone this band into pBluescript previously digested with *Apa*I and *Eco*RI.
5. Transform *E. coli* DH10B with this construction and select the recombinant strains in LB plus ampicillin, IPTG, and X-Gal (the plasmid allows blue/white selection). The construction contained in these transformants is called pBS*Cox*.

To include the gene encoding the phenylacetyl-CoA ligase (paaF) in the pBS*Cox*, the following steps must be performed:

1. Digest pK*paa* with *Apa*I and *Kpn*I.
2. Load the digestion product onto an agarose gel and rescue a 1182-bp fragment that contains the end of the *paaE* gene and a fragment of the *paaF* gene (from the ATG to the *Apa* I restriction site). Clean the selected band as indicated earlier.
3. Clone this fragment in the pBS*Cox* construction. The plasmid obtained, called pBS*Cox-paaF*, contains the pBluescript plasmid, the *paaF* gene, and all of the genes encoding the whole hydroxylation complex *(paaGHIJK)*.
4. Transform *E. coli* DH10B with pBS*Cox-paaF* and select transformants as reported in **step 5.**
5. Incubate this bacterium in MM containing glycerol (1% w/v) as the carbon source and PhAc (0.14% w/v) as the source of 2-OH-PhAc for 48–60 h.
6. Analyze the culture broths by HPLC to measure the quantity of 2-OH-PhAc accumulated.

3.3.4. PHA Production

3.3.4.1. OVERPRODUCTION OF PLASTIC POLYMERS

Among the different strategies followed to increase the production of bioplastics (PHAs) in *P. putida* U, one of the most efficient is the deletion of some

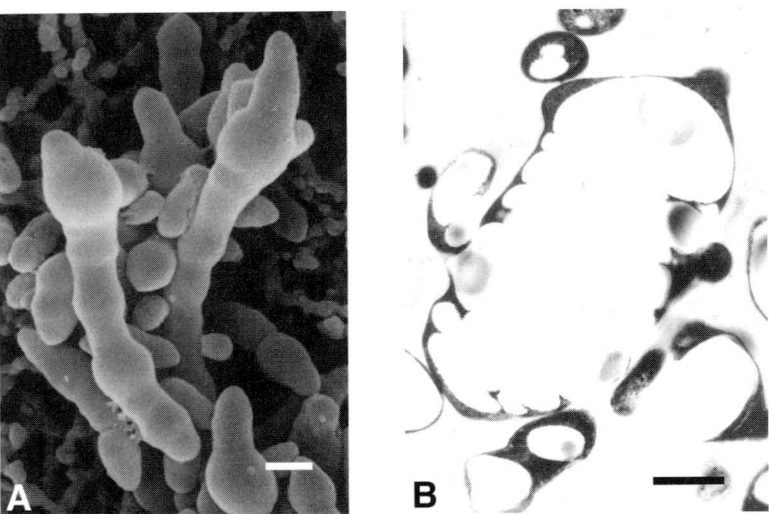

Fig. 11. Electron micrographs. Scanning (**A**) and transmission (**B**) microphotographs of *P. putida* U Δ*fadBA,* β-oxidation mutant cultured in chemically define solid medium (MM) containing *n*-phenylalkanoates. Bars correspond to 1 μm.

of the genes belonging to the β-oxidation pathway *(12)*. This method allows the elimination of the two genes *(fadB* and *fadA)* encoding five enzymatic activities (enoyl-CoA-hydratase, 3-OH-acyl-CoA-dehydrogenase, cis-δ³-trans-δ²-enoyl-CoA-isomerase, 3-OH-acyl-CoA-epimerase, and 3-ketoacyl-CoA-thiolase) and the collection of mutants showing an abnormal morphology overproduce aliphatic and aromatic bioplastics (most than 90% of the bacterial cytoplasm is occupied by PHAs granules) *(see* **Fig. 11**).

3.3.4.2. EXTRACTION OF PHAS

Study of the physicochemical characteristics and the chemical, clinical, pharmacological or industrial applications of PHAs require their isolation from the bacteria. In this subsection, we describe a useful procedure that affords efficient extraction. The current procedure is a slight modification of the method described by Lageveen et al. *(40)* and must be done as follows:

1. Inoculate *P. putida* Δ*fadBA* (10^{10} bacteria) in 2-L Erlenmeyer flasks containing 500 mL of chemically defined medium (MM) supplemented with 20 m*M* of 8-phenyloctanoic acid (or any other carbon source that provides 3-hydroxyphenylalkanoates, 3-hydroxyalkanoates, or combinations thereof).
2. Incubate in a rotary shaker at 250 rpm for 24–60 h at 30°C.

3. Harvest cells by centrifugation in 250-mL tubes at 10,000*g* for 7 min.
4. Freeze the pellet at –20°C.
5. Lyophilize the frozen cells overnight.
6. Weigh the bacterial dry powder obtained.
7. Break up the dry pellet with a mortar.
8. Add 100 mL chloroform per gram of dry cells.
9. Extract at constant flow for 3 h (boiling temperature of chloroform is 65°C).
10. After cooling, filter and obtain the eluate where the polymer is dissolved.
11. Evaporate off the chloroform at 35°C in a rotary evaporator, until a dark viscous liquid appears (the polymer).
12. Resuspend the polymer in 3 mL chloroform and, later, add 30 mL methanol.
13. Introduce this in a rotary evaporator (just rolling) until the polymer sticks on the glass surface.
14. Eliminate the excess liquid, resuspend the polymer in 3 mL chloroform, and spread on a glass plate (previously weighed).
15. Allow the chloroform to evaporate off (overnight) and calculate the weight of the polymer obtained (difference between the full-plate weight and the empty-plate weight).
16. Refer polymer production to grams of cell dry weight.

3.3.4.3. Synthesis of PHAs Microspheres as Drug Vehicles

Most bioplastics accumulated by *P. putida* mutants have important applications. One of them is the use of these polyesters to build vehicles for trapping drugs or pharmacological products *(10,11)*.

The general procedure followed is based on the collection of microspheres at atmospheric pressure by means of a solvent evaporation method *(41–43)*.

1. Dissolve 0.5 g of polymer purified as indicated in **Subheading 3.3.4.2** in 12 mL of dichloromethane at room temperature.
2. Emulsify the organic phase in 100 mL of aqueous phase containing 2% (w/v) polyvinyl alcohol (PVA) under high stirring (Ultra-Turrax) conditions at 8000 rpm for 30 min.
3. To complete the evaporation of dichloromethane, stir the solution magnetically for 6 h.

3.4. General Analytical Procedures

3.4.1. Analysis of PhAc, Penicillins, and Related Compounds by HPLC

Some of the assays reported earlier, as well as other biochemical procedures related to the study of the phenylacetyl-CoA catabolon, require the determination of the rate of utilization of the different carbon sources, the identification of the catabolic intermediates accumulated by the recombinant bacteria, or the analysis of the products generated in the enzymatic reactions or in the biotransformation processes. In most of these cases, an HPLC apparatus equipped with a variable-wavelength UV/vis detector, Millennium software, and a microparticulate reversed-phase column Nucleosil C-18 were used. In this subsection, we describe the method used to analyze PhAc and other structural analogs.

1. Take the samples (50 µL) to be analyzed from the culture broth, enzymatic reaction, cell-free extracts, and so forth.
2. Centrifuge the sample (10,000g 5 min), filter through a Millipore filter (0.45 µm), and inject it into the HPLC apparatus. The mobile phase used is 0.2 M KH_2PO_4 (pH 4.0) (isocratic conditions), the flow rate 1 mL/min, and the column temperature is 30°C. The eluate is monitored at 254 and 280 nm. Under these conditions, the retention times (Rt) of L-tyrosine (L-Tyr), 2,5-dihydroxyphenylacetic acid (2,5-OH-PhAc, homogentisic acid), 3,4-dihydroxyphenylacetic acid (3,4-OH-PhAc), 4-hydroxyphenylacetic acid (4-OH-PhAc), 3-hydroxyphenylacetic acid (3-OH-PhAc), 2-hydroxyphenylacetic acid (2-OH-PhAc), and PhAc are 3, 8, 15, 29, 31, 39, and 56 min, respectively (*see* **Note 13**).

When hydrophobic penicillins are analyzed by HPLC, the HPLC equipment used is a Chromatograph Series 3B with a LC-75 spectrophotometric detector, a Chromatography Data Station 10 B, and a µBondapack C-18 column of 30 cm. The mobile phase used to separate benzylpenicillin, hexanoylpenicillin, and heptanoylpenicillin is 83% (w/v) of 0.1 M sodium acetate–acetic acid, pH 4.5, and 17% of acetonitrile. The retention times are 12.1, 17.1, and 24.7 min for benzylpenicillin, hexanoylpenicillin, and heptanoylpenicillin, respectively. The mobile phase used for the separation of octanoylpenicillin is 0.025 M KH_2PO_4–methanol (50 : 50). The retention time for this penicillin is 9.6 min. In these later conditions, the retention time for heptanoylpenicillin is 4.3 min, whereas hexanoylpenicillin is not retained.

3.4.2. Separation of PhAc and Derivatives by Thin-Layer Chromatography

In some cases, the identification of some PhAc catabolic intermediates requires their separation by thin-layer chromatography (TLC) (*see* **Subheading 3.2.1.4.**). To do so, aliquots are taken from the culture broths at different times and centrifuged at 12,000g for 10 min (at 4°C) to remove the cells.

In this chromatographic method, as a stationary phase we used Silica-Gel plates and diethylether, water containing ammonium sulfate (28% w/w), and methanol (50 : 25 : 5 vol) as a mobile phase. The different aromatic compounds are visualized with iodine vapors *(44)*. PhAc hydroxyderivatives appear as yellow spots, whereas PhAc and other nonhydroxylated derivatives are not stained.

Under the above conditions, the Rfs for 4-OH-PhAc, 3-OH-PhAc, PhAc, 2-OH-PhAc, 3,4-di-OH-PhAc, and 2,5-di-OH-PhAc (homogentisic acid) are 0.72, 0.68, 0.67, 0.60, 0.55, and 0.30, respectively. The Rf of PHAc, which is not stained with iodine, is established by using labelled [1-14C]-PhAc.

4. Notes

1. This technique differs from the one used for the collection of mutants by conjugational transposition because (1) different *E. coli* DH5α transformants harboring the desired fragments to disrupt each target gene are used as donor strains and (2) *E. coli* HB101(pRK600) is employed as a helper strain to facilitate the mobilization of the pK18:*mob* constructions from the donor strain (*E. coli* DH5α) to the receptor strain (*P. putida* U).

2. Most of these mutants are unstable under selective pressure. Under such conditions, some of them remove the pK18::*mob* to another place in the genome in order to restore the gene function and to maintain kanamycin resistance. The easy reversal of these mutants occurs because a second recombination event can restore the structure of the disrupted gene, therefore reacquiring the function that had been lost. Thus, whereas mutants affected in a nonbasic function are quite stable, those affected in a gene encoding a protein that is either required for growth or that plays an important function repair the mutagenized gene in less than 70 h.

3. It should be noted that 50% of the colonies selected after the second recombination event (able to grow in media containing sucrose) correspond to strains that have lost the plasmid (and therefore *sacB* and Gm resistance) but that have restored the gene disrupted by means of a single event of recombination similar to that reported in **Note 2**. Thus, about one-half of the bacteria selected by this procedure contain a faithful copy of the gene to be deleted. These strains can be readily identified because they have restored the ability to grow in chemically defined media containing, as the sole carbon source, the aromatic compound that is catabolized through the pathway one wished to mutate.

4. Although the overexpression of most proteins belonging to the phenylacetyl-CoA catabolon were approached successfully, some of them were either amplified to a lower extent (e.g., PhaZ, which has polyhydroxyalkanoate depolymerase activity; *see* **Fig. 8B**) or were poorly expressed (PhaF and PhaI, two phasins involved in the regulation of gene expression and in the maintenance of the structure of the PHA granule). The reason for this effect remains unknown, although in the latter cases, it could be argued that the overexpression of these proteins in the recombinant strain could affect the expression of other genes (PaaF) or the binding to lipid structures in a nonspecific fashion, causing cellular damage (PaaI).

5. Analysis of phenylacetyl-CoA ligase in the cell-free extracts of the bacteria transformed with the seven plasmids, each containing of the recombinant ligases, revealed that four of them (those mutated in Cys61, Cys161, Cys256, and Cys430) are unable to activate PhAc to phenylacetyl-CoA and, therefore, seem to be required for maintaining catalytic activity, whereas the other three (Cys52, Cys385, and Cys419) do not participate either in the catalysis or in the maintenance of the appropriate functional enzymatic structure.

6. Taking into account that bacterial viability might be strongly affected after the shocking procedure, this parameter should be determined in all of the experiments. From the bacterial suspension used to evaluate PhAc uptake after osmotic treatment,

serial dilutions of cells should be carried out and different aliquots must be seeded on Trypticase Soy Agar plates.

7. If *P. putida* is not able to catabolize a certain compound, its inducer capacity is studied by growing the bacteria in MM plus glucose (5 m*M*) instead of PhAc. Once the glucose had been exhausted, the molecule to be tested (1 m*M*) was added to the culture.

8. When purified preparations are required, the supernatant must be precipitated with ammonium sulfate and the complete procedure reported by us *(45)* should be followed.

9. The spectrophotometric and the spectrophotofluorimetric assays should only be used with a very pure preparation of enzyme in order to avoid the pitfalls caused by the presence of nonspecific ATPases. In this case, the colorimetric assay (*see* **Subheading 3.2.1.2.**) requires longer incubation times because of the low quantities of enzyme present in the purified preparations. For this reason, a more sensitive assay, such as the spectrophotofluorimetric, could be advantageous.

10. Using the plasmid pK*paa*, the catalytic activity of the functional complex involved in the transformation of PhAc to 2-OH-phenylacetyl-CoA can be studied. This reaction requires two enzymatic systems: the phenylacetyl-CoA ligase (PaaF), which converts PhAc to phenylacetyl-CoA, and the whole hydroxylation complex (PaaGHIJK), which catalyzes its transformation into 2-OH-phenylacetyl-CoA *(4,46)*.

11. When *E. coli* W14 is cultured in M9 containing glycerol and PhAc as carbon sources, the bacterium accumulated 2-OH-PhAc in the medium, indicating that, as we have previously shown in *P. putida* U *(4)*, the ligase–hydroxylation complex catalyzes the transformation of PhAc into 2-OH-PhAc (via 2-OH-PhAc-CoA and probably by means of an epoxide intermediate) (*see* **Fig. 2**). Surprisingly, when the same bacterium is cultured in MM + glycerol + 3-OH-PhAc, it accumulates a compound in the medium that—when the pH of the medium rises past 6.5—stains the growth medium a dark red color. This compound seems to be similar to the product accumulated by a *P. putida* U mutant blocked in the gene encoding homogentisate dioxygenase. However, HPLC analysis of the culture broths of *E. coli* W14*paaFGHIJK* cultured in minimal medium + glycerol + 3-OH-PhAc revealed that homogentisic acid (HA) is not accumulated (at least at a concentration detectable by HPLC) and that the colored product was not present in those cultures in which 3-OH-PhAc had been omitted or replaced by L-Tyr. These results suggest that the dark red product is a 3-OH-PhAc derivative that—when oxidized—originates, as in the case of homogentisic acid, a quinoid derivative of a hydroxylated compound other than HA. This compound has been identified as 2,3-OH-PhAc, a non-natural intermediate that involves the activation of 3-OH-PhAc to 2,3-OH-PhAc-CoA, which is later released to the broth as free 2,3-OH-PhAc. It is interesting to note that whereas the hydroxylation of PhAc-CoA can occur in any *ortho* position, the hydroxylation of 3-OH-PhAc-CoA only occurs in specific position (2), but not in the equivalent one (2′). If this had been the case, 2,5-OH-PhAc-CoA (homogentisyl-CoA) would have been generated, a compound that once hydrolysed would release homogentisic acid; this product could be metabolized through

the tyrosine catabolic pathway. Because the biosynthesis of many natural occurring phenols very often involves epoxide intermediates, the specific hydroxylation of 3-OH-PhAc-CoA at position 2 might be explained in terms of an epoxidation by electrophilic attack (hydroperoxides, superoxide anion, oxaziridine intermediates, or O_2 coordinated to a metal ion) on the aromatic system. Such an attack could lead either to an epoxide (arene oxide) or directly to an α-hydroxy carbonium ion which then develops the 2,3-dihydroxy functionality with concomitant aromatization. Arene oxides have been prepared and it has been shown that they can be converted (presumably via carbonium ions) to end products in which the NIH shift *(47)* has occurred *(48,49)*. When other genetic constructions (containing separately the gene encoding ligase—*paaF*—or the five genes encoding the hydroxylation complex—*paaGHIJK*—) are expressed in the same bacteria, they are unable to generate 2-OH-PhAc (from PhAc) or the colored compound (from 3-OH-PhAc), suggesting that both ligase and the hydroxylation complex are required for the synthesis of these molecules. Moreover, the synthesis of 2,3-OH-PhAc is avoided when PhAc was supplemented to minimal media containing 3-OH-PhAc and glycerol, showing that in the presence of PhAc, the ligase and the hydroxylation complex do not generate it, probably as a consequence of their higher affinity for the natural substrate (PhAc).

12. When acyl-CoA synthetase from *P. putida* is used, the synthesis yield are about threefold higher. Furthermore, many other structural analogs can be synthesized.

13. To analyze other aromatic compounds related to the different catabolic pathways belonging to the phenylacetyl-CoA catabolon, other HPLC conditions were used. These protocols are available in the literature *(4,12,28,29,50)*.

Acknowledgment

We greatly acknowledge support from the Comisión Interministerial de Ciencia y Tecnología (CICYT), Madrid, España, grants BMC2000-0125-C04-01 and BIO2003-05309-C04-01.

References

1. Timmis, K. N. and Pieper, D. H. (1999) Bacteria designed for bioremediation. *Trends Biotechnol.* **17,** 201–204.

2. Wackett, L. P. and Bruce, N. C. (2000) Environmental biotechnology. Towards sustainability. *Curr. Opin. Biotechnol.* **11,** 229–231.

3. Luengo, J. M., García, J. L., and Olivera, E. R. (2001) The phenylacetyl-CoA catabolon: a complex catabolic unit with broad biotechnological applications. *Mol. Microbiol.* **39,** 1434–1442.

4. Olivera, E. R., Miñambres, B., García, B., et al. (1998) Molecular characterization of the phenylacetic acid catabolic pathway in *Pseudomonas putida* U: the phenylacetyl-CoA catabolon. *Proc. Natl. Acad. Sci. USA* **95,** 6419–6424.

5. Schleissner, C., Olivera, E. R., Fernandez-Valverde, M., and Luengo, J. M. (1994) Aerobic catabolism of phenylacetic acid in *Pseudomonas putida* U: biochemical

characterization of a specific phenylacetic acid transport system and formal demonstration that phenylacetyl-coenzyme A is a catabolic intermediate. *J. Bacteriol.* **176,** 7667–7676.

6. Martínez-Blanco, H., Reglero, A., Rodríguez-Aparicio, L. B., and Luengo, J. M. (1990) Purification and biochemical characterization of phenylacetyl-CoA ligase from *Pseudomonas putida.* A specific enzyme for the catabolism of phenylacetic acid. *J. Biol. Chem.* **265,** 7084–7090.

7. Miñambres, B., Martinez-Blanco, H., Olivera, E. R., et al. (1996) Molecular cloning and expression in different microbes of the DNA encoding *Pseudomonas putida* U phenylacetyl-CoA ligase. Use of this gene to improve the rate of benzylpenicillin biosynthesis in *Penicillium chrysogenum. J. Biol. Chem.* **271,** 33,531–33,538.

8. Ismail, W., Mohamed, M. E., Wanner, B. L., et al. (2003) Functional genomic by RMN spectroscopy. Phenylacetate catabolism in *Escherichia coli. Eur. J. Biochem.* **270,** 3047–3054.

9. García, B., Olivera, E. R., Miñambres, B., et al. (1999) Novel biodegradable aromatic plastics from a bacterial source. Genetic and biochemical studies on a route of the phenylacetyl-CoA catabolon. *J. Biol. Chem.* **274,** 29,228–29,241.

10. Abraham, G. A., Gallardo, A., San Roman, J., et al. (2001) Microbial synthesis of poly(beta-hydroxyalkanoates) bearing phenyl groups from *Pseudomonas putida:* chemical structure and characterization. *Biomacromolecules.* **2,** 562–567.

11. Olivera, E. R., Carnicero, D., Jodra, R., et al. (2001) Genetically engineered Pseudomonas: a factory of new bioplastics with broad applications. *Environ. Microbiol.* **3,** 612–618.

12. Olivera, E. R., Carnicero, D., García, B., et al. (2001) Two different pathways are involved in the β-oxidation of *n*-alkanoic and *n*-phenylalkanoic acids in *Pseudomonas putida* U: genetic studies and biotechnological applications. *Mol. Microbiol.* **39,** 863–874.

13. Qandt, J. and Hynes, M. F. (1993) Versatile suicide vectors which allow direct selection for gene replacement in Gram-negative bacteria. *Gene* **127,** 15–21.

14. Schäfer, A., Tsuch, A., Jäger, W., Kalinowski, J., Tierbach, G., and Pühler, A. (1994) Small mobilizable multi-purpose cloning vectors derived from the *Escherichia coli* plasmids pK18 and pK19: selection of defined deletions in the chromosome of *Corynebacterium glutamicum. Gene* **145,** 69–73.

15. Lipmann, F. and Tuttle, L. C. (1945) A specific micromethod for the determination of acyl phosphates. *J. Biol. Chem.* **159,** 21–28.

16. Selvaraj, G. and Iyer, V. N. (1983) Suicide plasmid vehicles for insertion mutagenesis in *Rhizobium meliloti* and related bacteria. *J. Bacteriol.* **156,** 1292–1300.

17. Herrero, M., de Lorenzo, V., and Timmis, K. N. (1990) Transposon vectors containing non-antibiotic selection markers for cloning and stable chromosomal insertion of foreing DNA in gram-negative bacteria. *J. Bacteriol.* **172,** 6557–6567.

18. Donnenberg, M. S. and Kaper, J. B. (1991) Construction of an *eae* deletion mutant of enteropathogenic *Escherichia coli* by using a positive-selection suicide vector. *Infect. Immun.* **59,** 4310–4317.

19. Bujard, H., Gentz, R., Lanzer, M., et al. (1987) A T5 promotor based transcription-translation system for the analysis of proteins in vivo and in vitro. *Methods Enzymol.* **155**, 416–433.
20. Fernández-Valverde, M., Reglero, A., Martínez-Blanco, H., and Luengo, J. M. (1993) Purification of *Pseudomonas putida* acyl coenzyme A ligase active with a range of aliphatic and aromatic substrates. *Appl. Environ. Microbiol.* **59**, 1149–1154.
21. Luengo, J. M., García, B., Sandoval, A., Naharro, G., and Olivera, E. R. (2003) Bioplastics from microorganisms. *Curr. Opin. Microbiol.* **6**, 251–260.
22. Rodríguez-Aparicio, L. B., Reglero, A., and Luengo, J. M. (1987) Uptake of *N*-acetylneuraminic acid by *Escherichia coli* K-235. Biochemical characterization of the transport system. *Biochem. J.* **246**, 287–294.
23. Brunner, R. and Rohr, M. (1975) Phenacyl:Coenzyme A ligase. *Methods Enzymol.* **43**, 476–481.
24. Bradford, M. M. (1976) A rapid and sensitive method for the quantification of microgram quantities of protein utilizing the principle of protein-dye binding. *Anal. Biochem.* **72**, 248–254.
25. Rodríguez-Aparicio, L. B., Reglero, A., Martínez-Blanco, H., and Luengo, J. M. (1991) Fluorometric determination of phenylacetyl-CoA ligase from *Pseudomonas putida:* a very sensitive assay for a newly described enzyme. *Biochim. Biophys. Acta* **1073**, 431–433.
26. O'Connor, K., Dobson, A. D. W., and Hartmans, S. (1997) Indigo formation by microorganisms expressing styrene monooxygenase activity. *Appl. Environ. Microbiol.* **63**, 4287–4291.
27. Fujioka, M., Morino, Y., and Wada, H. (1970) Metabolism of phenylalanine *(Achromobacter eurydice). Methods Enzymol.* **17a**, 585–596.
28. de Roo, G., Ren, Q., Witholt, B., and Kessler, B. (2000) Development of an improved in vitro activity assay for medium chain length PHA polymerases based on CoenzymeA release measurements. *J. Microbiol. Methods* **41**, 1–8.
29. Kraak, M. N., Smits, T. H. M., Kessler, B., and Witholt, B. (1997) Polymerase C1 levels and poly(R-3-hydroxyalkanoate) synthesis in wild-type and recombinant *Pseudomonas* strains. *J. Bacteriol.* **179**, 4985–4991.
30. Ellman, G. L. (1959) Tissue sulfhydryl groups. *Arch. Biochem. Biophys.* **82**, 70–77.
31. Martínez-Blanco, H., Reglero, A., Martín-Villacorta, J., and Luengo, J. M. (1990) Design of an enzymatic hybrid system: a useful strategy for the biosynthesis of benzylpenicillin *in vitro. FEMS Microbiol. Lett.* **72**, 113–116.
32. Luengo, J. M. (1995) Enzymatic synthesis of hydrophobic penicillins. *J. Antibiot.* **48**, 1195–1212.
33. Luengo, J. M., Alemany, M. T., Salto, F., Ramos, F., López-Nieto, M. J., and Martín, J. F. (1986) Direct enzymatic synthesis of penicillin G using cyclases of *Penicillium chrysogenum* and *Acremonium chrysogenum. Bio/Technology* **4**, 44–47.
34. Luengo, J. M. and Moreno, M. A. (1987) Separation by high performance liquid chromatography of penicillins with C4 to C10 aliphatic side chains. *Anal. Biochem.* **164**, 559–562.

35. Luengo, J. M. (1998) Enzymatic synthesis of penicillins, in *Amino Acids, Peptides, Porphyrins, and Alkaloids. Comprehensive Natural Products Chemistry* (Barton, D., Nakanishi, K., and Meth-Cohn, O., eds.), Pergamon, Amsterdam, Vol. 4, pp. 239–274.

36. Martínez-Blanco, H., Reglero, A., Fernández-Valverde, M., et al. (1992) Isolation and characterization of the acetyl-CoA synthetase from *Penicillium chrysogenum:* involvement of this enzyme in the biosynthesis of penicillins. *J. Biol. Chem.* **267,** 5474–5481.

37. Ferrero, M. A., Reglero, A., Martín-Villacorta, J., and Luengo, J. M. (1990) Biosynthesis of penicillins G, V and K from glutathione S-derivatives. *J. Antibiot.* **43,** 684–691.

38. Sánchez, F., Lozano, M., Rubio, V., and Peñalva, M. A. (1987) Transformation in *Penicillium chrysogenum. Gene* **51,** 97–102.

39. Díez, B., Gutierrez, S., Barredo, J. L., von Solingen, P., van der Voort, L. H. M., and Martín, J. F. (1990) The cluster of penicillin biosynthetic genes. Identification and characterization of the *pcb*AB gene encoding the alpha-aminoadipyl-cysteinyl-valine synthetase and linkage to the *pcb*C and *pen*DE genes. *J. Biol. Chem.* **265,** 16,358–16,365.

40. Lageveen, R. G., Huisman, G. W., Preusting, H., Ketelaar, P., Eggink, G., and Witholt, B. (1998) Formation of polyesters by *Pseudomonas oleovorans.* Effect of substrates on formation and composition of poly-*(R)*-3-hydroxyalkanoates and poly-*(R)*-3-hydroxyalkenoates. *Appl. Environ. Microbiol.* **54,** 2924–2932.

41. Beck, L. R., Cowsar, D. R., Lewis, D. H., et al. (1979) A new long-acting injectable microcapsule system for the administration of progesterone. *Fertil. Steril.* **31,** 545–551.

42. Menei, P., Croué, A., Daniel, V., Pouplard-Barthelaix, A., and Benoit, J. P. (1994) Fate and biocompatibility of three types of microspheres implanted into the brain. *J. Biomed. Mater. Res.* **28,** 1079–1085.

43. Gallardo, A., Eguiburu, J. L., Fernández Berridi, M. J., and San Román, J. (1998) Preparation and *in vitro* release studies of ibuprofen loaded films and microspheres made from graft copolymers of poly(L-lacticacid) on acrylic backbones. *J. Control Rel.* **55,** 171–179.

44. Sugumaran, M. and Vaidyanathan, C. S. (1979) Microsomal hydroxylation of phenylacetate by *Aspergillus niger. FEMS Microbiol. Lett.* **5,** 427–430.

45. Luengo, J. M. (1985) Precipitation of phenyl and phenoxy penicillin from solutions using ammonium sulphate. *Anal. Biochem.* **149,** 466–470.

46. Ferrández, A., Miñambres, B., García, B., et al. (1998) Catabolism of phenylacetic acid in *Escherichia coli:* characterization of a new aerobic hybrid pathway. *J. Biol. Chem.* **273,** 25,974–25,986.

47. Guroff, G., Daly, J. W., Jerina, D. M., Renson, J., Witkop, B., and Udenfriend, S. (1967) Hydroxylation-induced migration: the NIH shift. Recent experiments reveal an unexpected and general result of enzymatic hydroxylation of aromatic compounds. *Science* **157,** 1524–1530.

48. Boyd, D. R., Daly, J. W., and Jerina, D. M. (1972) Rearrangement of [1-^{2}H] and [2-^{2}H]-naphtalene 1,2-oxides to 1-naphtol. Mechanism of NIH shift. *Biochemistry* **11,** 1961–1966.

49. Kasperek, G. J., Bruice, T. C., Yagi, H., Kaubisch, N., and Jerina, D. M. (1972) Solvolytic chemistry of 1,4-dimethylbenzene oxide. A new and novel mechanism for the NIH shift. *J. Am. Chem. Soc.* **94,** 7876–7882.

50. Luengo, J. M., Iriso, J. L., and López-Nieto, M. J. (1986) Direct enzymatic synthesis of natural penicillins using phenylacetyl-CoA : 6-APA phenylacetyl transferase of *Penicillium chrysogenum:* minimal and maximal side chain length requirements. *J. Antibiot.* **39,** 1754–1759.

6

Recombinant Microorganisms for the Biosynthesis of Glycosylated Antitumor Compounds

Carmen Méndez and José A. Salas

Summary

Actinomycetes produce many bioactive compounds with clinical, veterinary, or agricultural applications. Many of these natural products contain sugars attached to the corresponding aglycons, which usually participate in the molecular recognition of the cellular target. The glycosylation pattern of these complex metabolites can be altered by chemical synthesis or synthetic modification of intermediates usually produced via fermentation and by the use of genetically engineered recombinant microorganisms and biosynthetic genes for in vivo experiments through combinatorial biosynthesis. Here, we describe procedures to generate novel glycosylated compounds derived from the antitumor compound elloramycin. This procedure involves biotransformation of different recombinant strains of *Streptomyces albus* harboring the elloramycin ElmGT glycosyltransferase and able to synthesize different dTDP-activated deoxysugars whose biosynthesis is directed by different plasmids.

Key Words: Glycosylation; glycosyltransferase; streptomycetes; antitumor.

1. Introduction

Actinomycetes are Gram-positive bacteria of great importance because of their capability of synthesizing a number of products with useful biological activities. These bioactive natural products include compounds with clinical application as antibiotics (tetracyclines, macrolides), antitumor (daunorubicin, doxorubicin), or immunosupresive agents (rapamycin) and also compounds with veterinary or agricultural applications, such as growth promoters (tylosin), insecticides (spinosyn), herbicides (bialaphos), and antiparasites (avermectins) agents. Many of these natural products contain sugars attached to the corresponding aglycons. These sugars contribute to the structural biodiversity and usually participate in the molecular recognition of the cellular target. The presence of these sugars is, therefore, important for the biological activity of the compounds and, in many cases,

From: *Methods in Biotechnology, Vol. 18: Microbial Processes and Products*
Edited by: J. L. Barredo © Humana Press Inc., Totowa, NJ

essential. These sugars are linked to the aglycon as monosaccharides, disaccharides, or oligosaccharides of variable sugar length through *O*- (more frequently), *C*-, or *N*-glycosylation (*see* **Fig. 1**). The sugars are transferred to the aglycon by glycosyltransferases, which are generally sugar-, aglycon- and site-specific.

There are a number of routes for altering the glycosylation pattern of these complex metabolites. The first is total chemical synthesis or synthetic modification of intermediates usually produced via fermentation. The disadvantage of this method is the enormous structural complexity of many of these natural products. The second is the use of microorganisms and biosynthetic genes for in vivo experiments through combinatorial biosynthesis. This strategy relies upon the generation of novel recombinant strains endowed with the capability of synthesizing new sugars and the use of "sugar flexible glycosyltransferases." This new technology, named combinatorial biosynthesis, manipulates genes in natural-product biosynthesis pathways as a way of producing natural-product analogs and thus resulting in the formation of novel derivatives with potential bioactivity. In the last years, great progress has been made in the identification and characterization of genes involved in the biosynthesis of the sugar moieties and in the isolation of genes coding for glycosyltransferases from many different antibiotic-producing microorganisms (*1*).

Elloramycin (*see* **Fig. 2**) is an anthracycline-like antitumor drug produced by *Streptomyces olivaceus* Tü2353. It is active against Gram-positive bacteria and also exhibits antitumor activity. Structurally, elloramycin belongs to the large and important family of the aromatic polyketides. Its aglycon closely resembles tetracenomycin C, but it has an additional *C-12a-O*-methyl group and, in contrast to tetracenomycin C, is glycosylated with a permethylated L-rhamnose residue at the C-8 hydroxyl group (*2*). This L-rhamnose moiety is transferred to the elloramycin aglycon by the ElmGT glycosyltransferase (*3*) and the sugar is further permethylated by the action of three methyltransferases (*4*). ElmGT has been shown to have a certain degree of "flexibility for the sugar" being able to transfer different sugars (*3*). Here, we describe procedures to generate novel glycosylated antitumor compounds derived from elloramycin by taking advantage of the flexibility of ElmGT for the transfer of different deoxysugars. This procedure involves biotransformation of different recombinant *Streptomyces albus* strains harboring the ElmGT glycosyltransferase and able to synthesize different dTDP-activated deoxysugars.

2. Materials

1. Forward primer (FP): 5′-ATCG<u>TCTAGA</u>GGCTGCGGGAAGCGACCTGACATG-3′, containing a restriction site for *Xba*I (underlined).
2. Reverse primer (RP): 5′-ATCG<u>AAGCTT</u>GTCGCAGCGGGCCGTGAGGTCA-3′, containing a restriction site for *Hind*III (underlined).

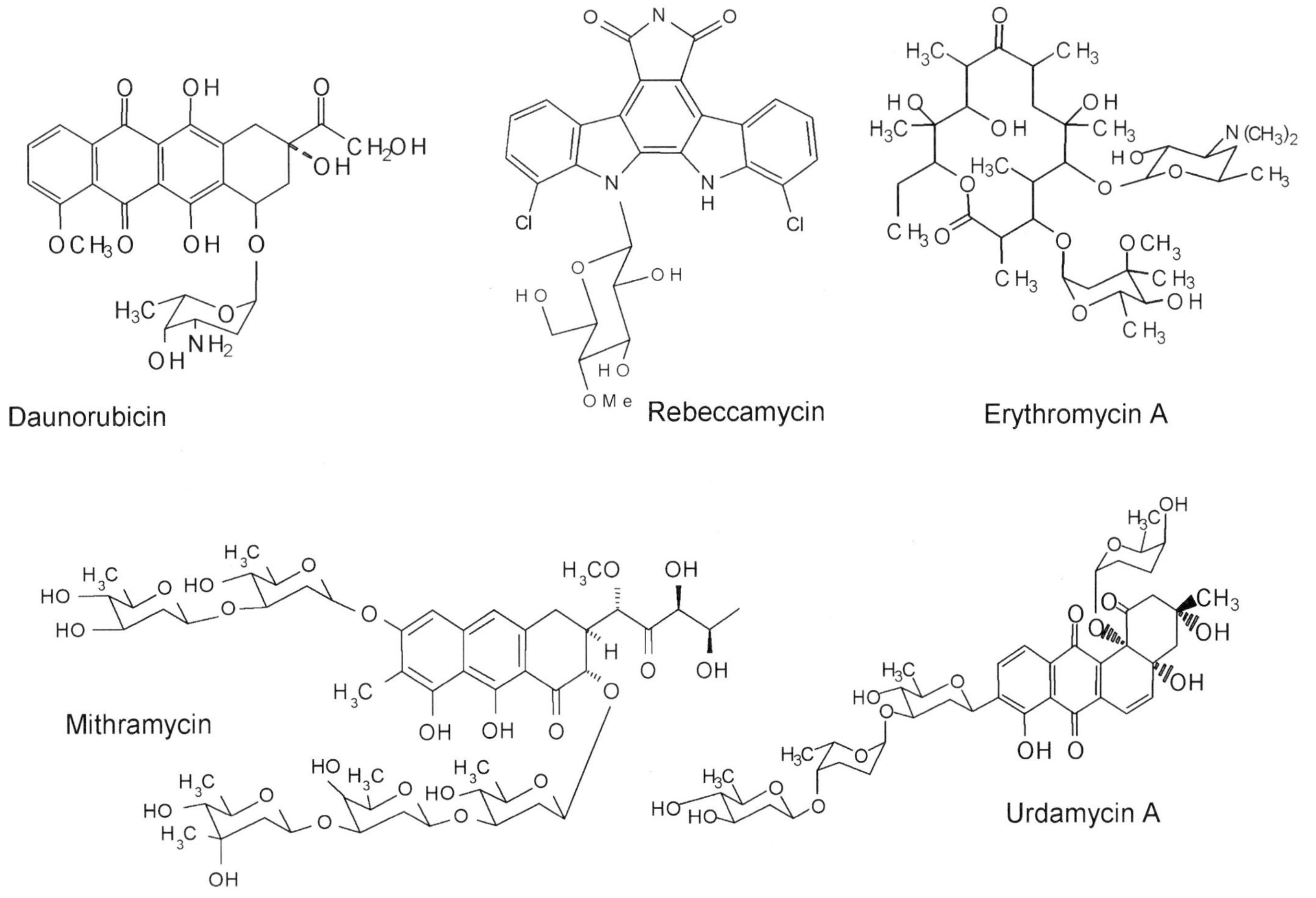

Fig. 1. Chemical structures of different glycosylated compounds.

133

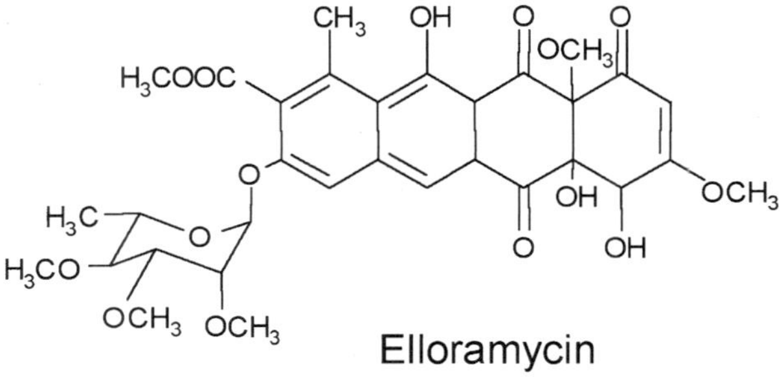

Elloramycin

Fig. 2. Chemical structure of elloramycin.

3. 7U primer: 5′-AGAGAAGCTTACTAGTCCCCCCCGAGAGGCAGCGGCC-CATG-3′, containing restriction sites for *Hind*III and *Spe*I (underlined).
4. 7L primer: 5′-AGTCTAGAGCTAGCTCATGCTGCTCCTCGCCGGGTCG-GTGGG-3′, containing restriction sites for *Xba*I and *Nhe*I (underlined).
5. 14U primer: 5′-AGAGAAGCTTGCTAGCCCGGACCACGCGAAGGAC-CTTTCACATG-3′, containing restriction sites for *Hind*III and *Nhe*I (underlined).
6. 14L primer: 5′-AGTCTAGATTAATTAAGCTAGTTGTCGTTCCAGAACG-GCTCCCGG-3′, containing restriction sites for *Xba*I and *Pac*I (underlined).
7. 1α primer: 5′-ACGTAAGCTTCCTAGGGCGCCGTCCTGGATCACAATG-3′, containing restriction sites for *Hind*III and *Avr*II (underlined).
8. 1β primer: 5′-ACGTGTACAGTTAACCTTCGGGGTGCTCAGCTCAGG-3′, containing restriction sites for *Xba*I and *Hpa*I (underlined).
9. 15α primer: 5′-ACGTAAGCTTGTTAACCCGAAGGGAACCCCATGCCC-3′, containing restriction sites for *Hind*III and *Hpa*I (underlined).
10. 15β primer: 5′-ACGTCTAGAACTAGTGCATCAGCACCAGCGCACCCG-3′, containing restriction sites for *Xba*I and *Spe*I (underlined).
11. RU primer: 5′-AGAGAAGCTTACTAGTCCCATCACAGAGATCAGGACGACG-CATG-3′, containing restriction sites for *Hind*III and *Spe*I (underlined).
12. RL primer: 5′-AGTCTAGAGCTAGCTCAGATACGGACGGCGGAGGTGAAGT-CA-3′, containing restriction sites for *Xba*I and *Nhe*I (underlined).
13. Z3U primer: 5′-AGAGAAGCTTACTAGTACAAGGCGGCCGCGCTGTCGGTG-3′, containing restriction sites for *Hind*III and *Spe*I (underlined).
14. Z3L primer: 5′-AGTCTAGAGCTAGCTCACTTCCTTTCGTCAACCGGCGCG-GTG-3′, containing restriction sites for *Xba*I and *Nhe*I (underlined).
15. QU primer: 5′-AGTCTAGACCGGTTGACGAAAGGAAGTGAC-3′, containing a restriction site for *Xba*I (underlined).
16. QL primer: 5′-AGTCTAGACTAGCCACGTGCGGCGACGA-3′, containing a restriction site for *Xba*I (underlined).
17. DMSO (dimethyl sulfoxide).
18. dNTPs (deoxynucleotides).

19. Vent DNA polymerase and ThermoPol buffer (10X) (New England Biolabs, Beverly, MA).
20. *Escherichia coli* vectors pIJ2925, pUK21, pUC18, pLITMUS29 (New England Biolabs).
21. *E. coli* strain DH10B (Invitrogen, Carlsbad, CA)
22. Thermocycler.
23. Restriction enzymes, T4 DNA ligase.
24. Agarose and gel electrophoresis equipment.
25. GFX polymerase chain reaction (PCR) DNA and Gel Band Purification Kit (Amersham Biosciences, Bucks, UK).
26. GFX PCR DNA and Gel Band Purification Kit (Amersham Biosciences, Bucks, UK).
27. TE buffer: 10 m*M* Tris-HCl, pH 8, and 1 m*M* EDTA.
28. *Streptomyces* vectors: pIAGO *(5)*, pKC796 *(6)*, pEM4 *(7)*.
29. *Streptomyces* strains: *S. albus* GB16 *(8)*, *S. fradiae* Tü2717, *S. lividans* GB16F4 *(3)*.
30. Cosmid DNAs: cos16F4 *(2)*, cosAB61 *(5)*.
31. X-gal (5-bromo-4-chloro-3-indolyl-β-D-thio-galactopyranoside).
32. IPTG (isopropyl-β-D-thio-galactopyranoside).
33. Antibiotics: ampicillin, apramycin, tobramycin, thiostrepton, kanamycin.
34. TSB (Trypticase soya broth): 30 g/L soy trypticaseine, pH 7.2.
35. R5: 103 g/L sucrose, 0.25 g/L K_2SO_4, 10.12 g/L $MgCl_2 \cdot 6H_2O$, 10 g/L glucose, 0.1 g/L Difco casaminoacids, 2 mL/L trace element solution, 5 g/L Difco yeast extract, 5.73 g/L TES buffer, 0.05 g/L KH_2PO_4, 20 m*M* $CaCl_2 \cdot 2H_2O$, 0.3 g/L proline, 0.007 *N* NaOH, and 22 g/L agar.
36. R5A: 103 g/L sucrose, 0.25 g/L K_2SO_4, 10.12 g/L $MgCl_2 \cdot 6H_2O$, 10 g/L glucose, 0.1 g/L Difco casaminoacids, 2 mL/L trace element solution, 5 g/L Difco yeast extract, 21 g/L MOPS, and 25 g/L agar. Adjust to pH 6.8 with KOH
37. Nutrient agar: nutrient broth with 20 g/L agar.
38. Bennett: 10 g/L glucose, 1 g/L peptone, 1 g/L yeast extract, 2 g/L tryptone, and 20 g/L agar.
39. 2X TY: 16 g/L tryptone, 5 g/L NaCl, 10 g/L yeast extract, and 22 g/L agar.
40. DNA sequencing equipment.
41. Refrigerated centrifuge. Beckman model J-21 B, with JA-14 rotor (Beckman Coulter, Inc., Fullerton, CA).
42. Supor membranes (Pall Corp., Ann Arbor, MI).
43. Extraction cartridges Sep-Pak Vac C18; 10 g (Waters, Milford, MA).
44. High-performance liquid chromatography (HPLC) equipment with photodiode array detector and Millenium software (Waters, Milford, MA).
45. μBondapak C_{18} column (Waters, Milford, MA).
46. μBondapak C_{18} radial compression cartridge (PrePak Cartridge, 25 × 100 mm; Waters, Milford, MA).

3. Methods

3.1. General Methods

DNA manipulations are performed by standard DNA recombinant methods *(9)* and are not described in detail. *E. coli* DH10B transformants are plated onto

2X TY agar plates containing the appropriate antibiotic (100 µg/mL ampicillin, 50 µg/mL kanamycin or 20 µg/mL tobramycin, final concentration), 50 m*M* IPTG, and 40 µg/mL X-gal, when required.

Genetic manipulation of *Streptomyces* strains are performed by standard procedures *(10)* and are not described in detail. *Streptomyces* transformants are plated on R5 medium and incubated at 30°C for 18–24 h. Selection of transformants is performed by adding 3 mL of Soft Nutrient Agar (SNA) containing the appropriate antibiotic (25 µg/mL apramycin or 50 µg/mL thiostrepton, final concentration).

3.2. Construction of a Recombinant Strain of Streptomyces albus GB16 Expressing the elmGT Glycosyltransferase

3.2.1. Amplification of elmGT

To amplify the *elmGT* gene, two oligoprimers (FP and RP) were designed corresponding to the 5′ end of the gene (including the ATG start codon) and a region downstream of the gene. The primers also include nucleotide sequences for two restriction enzymes at their 5′ ends; in this way, two restriction sites will flank the amplified DNA fragment, thus facilitating subcloning the DNA fragment into different vectors.

For amplifying *elmGT,* DNA of cos16F4 *(2)* is used as template. This is a cosmid isolated from a DNA library of *Streptomyces olivaceus* Tü2353 (elloramycin producer) that contains most of the elloramycin biosynthetic gene cluster, including the *elmGT* glycosyltransferase gene *(3).* The amplification reaction is carried out in a 50-µL total volume, using 0.5-mL Eppendorf tubes. Add to the tube in the following order:

1. 1 µL (100 ng/µL) Template DNA, 1 µL of FP and 1 µL of RP (each one at 30 pmol/µL), 5 µL of ThermoPol buffer (10X), 5 µL of dNTPs (each one at 2 m*M*), 1 µL of Vent DNA Polymerase (2 U/µL), 5 µL of DMSO, and 30 µL of distilled water.
2. Carry out the polymerization reaction in a thermocycler using the following conditions: an initial denaturation step of 3 min at 98°C; 30 cycles of 30 s at 95°C, 60 s at 68°C, and 90 s at 72°C; an extra extension step of 5 min at 72°C.
3. Once the polymerization reaction has finished, keep samples at 4°C until used

3.2.2. Subcloning of the elmGT Gene in an Integrative Streptomyces Vector

In order to express the *elmGT* glycosyltransferase gene in *S. albus,* the gene is first subcloned downstream of a constitutive promoter *(ermEp)* and then, together with the promoter, into an integrative *Streptomyces* vector. From the polymerization reaction product:

1. Take a sample of DNA and run an agarose gel electrophoresis. Cut out the DNA fragment of approx 1.3 kb.

2. Purify the DNA through a GFX column, using the GFX PCR DNA and Gel Band Purification Kit. Resuspend the DNA in 50 μL of TE buffer.
3. Digest the DNA fragment with *Xba*I plus *Hind*III and ligate to pIAGO, digested with the same restriction enzymes. pIAGO *(5)* is a bifunctional plasmid *(E. coli–Streptomyces)*, derived from the multicopy vector pWHM3 *(11)*, that contains the erythromycin-resistance promoter *(ermEp)* cloned into the polylinker and flanked by several unique restriction sites that can be used as cloning sites. These sites are located within the *lacZ* gene, which allows blue-white selection on plates containing X-gal plus IPTG. Selection for the antibiotic marker in this vector is with ampicillin in *E. coli* and with thiostrepton in *Streptomyces*. By subcloning the amplified fragment into pIAGO, the *elmGT* gene is under the control of the *ermE* gene promoter *(ermEp)*.
4. Transform *E. coli* cells and select transformants with 100 μg/mL ampicillin.
5. Purify and confirm the right construct, named pGB15.
6. Digest pGB15 with *Eco*RI and *Hind*III and subclone the 1.5-kb fragment (containing the *elmGT* gene together with the *ermEp*) into pIJ2925, digested with the same restriction enzymes. pIJ2925 *(12)* is a pUC18 derivative that contains a modified version of the polylinker without affecting the blue-white selection characteristic of this vector. This polylinker has the particularity of containing two flanking *Bgl*II sites, which are frequently used for subcloning in *Streptomyces* vectors.
7. From this construct, isolate the 1.5-kb *Bgl*II fragment, containing both the *ermEp* and *elmGT* gene, and subclone into the unique *Bam*HI site of vector pKC796 (*see* **Note 1**) to generate pGB16. pKC796 *(6)* is a bifunctional plasmid *(E. coli–Streptomyces)* that contains the apramycin-resistance cassette (*see* **Note 2**) and the *attP* attachment site from phage ΦC31. This attachment site allows integration of the vector into the *Streptomyces* chromosome through site-specific recombination *(6)*.

3.2.3. Transformation of S. albus Protoplasts

Transform *S. albus* protoplasts with DNA from pGB16. Protoplasts are obtained using conventional methods for *Streptomyces (10)* with some modifications.

1. Inoculate 250-mL Erlenmeyer flasks containing 25 mL of TSB medium, with 50 μL of a dense spore suspension of *S. albus*.
2. Incubate for 24 h at 30°C and 1.14*g*.
3. Use 1 mL of this culture to inoculate 250-mL Erlenmeyer flasks containing 25 mL of TSB with 0.75% glycine.
4. Incubate the culture for 36 h and process the mycelium to generate protoplasts, using established protocols *(10)*.
5. Plate transformed protoplasts and select transformants with 25 μg/mL apramycin.
6. Pick up transformant colonies and plate out on Bennett agar medium *(10)* containing 25 μg/mL apramycin.
7. Confirm that pGB16 has been integrated in the *S. albus* chromosome by Southern analysis *(10)*. The resultant strain is designated as *S. albus* GB16

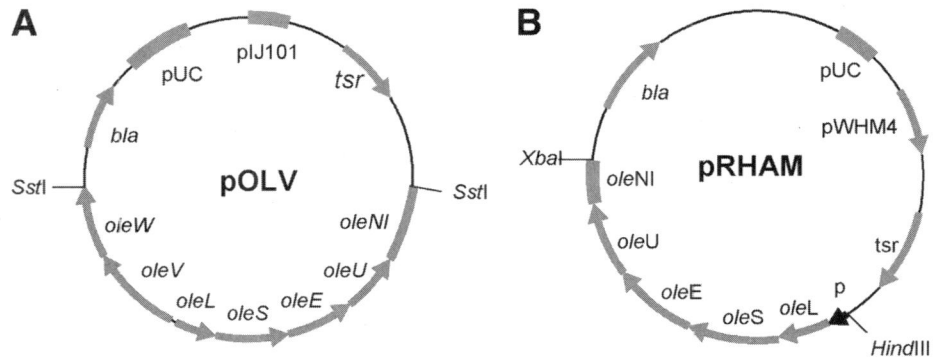

Fig. 3. Maps of pOLV (**A**) and pRHAM (**B**).

3.3. Construction of Plasmids Directing the Biosynthesis of dTDP-Activated Sugars

Generation of plasmids able to direct the biosynthesis of dTDP-activated sugars follows two different approaches: (1) subcloning of a DNA region containing sugar biosynthetic genes from the chromosome of the oleandomycin producer *Streptomyces antibioticus* (this is the case of pOLV and pRHAM) and (2) construction of a "plug and play" system by PCR amplification of genes involved in sugar biosynthesis, flanked by unique restriction sites. pLN2 represents such a case.

3.3.1. Construction of pOLV

pOLV is a plasmid that contains all of the genes required for the biosynthesis of dTDP-L-olivose from the oleandomycin gene cluster in *S. antibioticus* (*5*). Oleandomycin is a 14-membered macrolide that contains 2 deoxysugars attached to the aglycone: D-desosamine and L-oleandrose. It has been established that L-oleandrose is synthesized *in situ,* once L-olivose is attached to the aglycon, through specific 3-*O*-methylation of this sugar residue by the OleY methyltransferase (*13*). cosAB61 is a cosmid containing oleandomycin genes involved in sugar biosynthesis (*5*). Within the DNA insert in this cosmid, all of the genes required for the biosynthesis of dTDP-L-olivose (*oleW* 3-ketoreductase, *oleV* 2,3-dehydratase, *oleL* 3,5-epimerase, *oleS* dTDP-glucose synthase, *oleE* 4,6-dehydratase, and *oleU* 4-ketoreductase genes) are located in a *Sst*I fragment. For constructing pOLV (*see* **Fig. 3A**):

1. Digest cosAB61 DNA with *Sst*I.
2. Run an agarose gel electrophoresis and purify the 7.6-kb *Sst*I fragment from the gel. To purify the fragment, cut out an agarose slice containing this frag-

ment and purify the fragment using the GFX PCR DNA and Gel Band Purification Kit.

3. Resuspend the DNA in a final volume of 10 µL.
4. Ligate the purified fragment to pIAGO previously digested with *Sst*I. The *Sst*I site is located upstream of the *ermEp*.
5. Transform *E. coli* cells and select transformants with 100 µg/mL ampicillin.
6. Purify and confirm the right construct, named pOLV (*see* **Note 3**).

3.3.2. Construction of pRHAM

pRHAM is a plasmid that contains four genes (*oleL, oleS, oleE,* and *oleU*) from the oleandomycin gene cluster, required for the biosynthesis of dTDP-L-rhamnose *(8)*. These genes are in a 4.1-kb *Sph*I–*Xmn*I fragment in cosAB61. To construct this plasmid (*see* **Fig. 3B**):

1. Digest cosAB61 DNA with *Sph*I and *Xmn*I.
2. Run an agarose gel electrophoresis and purify the 4.1-kb *Sph*I–*Xmn*I fragment from the gel.
3. Resuspend the DNA in a final volume of 10 µL.
4. Ligate the purified fragment to pUK21 vector *(14)*, previously digested with *Sph*I and *Sma*I (*see* **Note 4**).
5. Transform *E. coli* cells and select transformants with 50 µg/mL kanamycin.
6. Purify and confirm the right construct, named pLR2347 (*see* **Note 5**). Generation of pLR2347 has the objective of introducing appropriate restriction sites at both sites of the DNA fragment to facilitate further subcloning.
7. Digest pLR2347 with *Spe*I and *Xba*I (both enzymes cut in the polylinker, at both sites of the fragment) and purify the insert (4.1-kb fragment).
8. Ligate the *Spe*I–*Xba*I fragment to vector pEM4 previously digested with *Xba*I. pEM4 *(7)* is a bifunctional plasmid *(E. coli–Streptomyces)* derived from the multi-copy vector pWHM4 *(11)* and containing the erythromycin-resistance promoter (*ermEp**) cloned into the polylinker and flanked by several unique restriction sites that can be used as cloning sites. These cloning sites are located within the *lacZ* gene, which allows blue-white selection on plates containing 40 µg/mL X-gal plus 50 m*M* IPTG. Selection of clones containing this vector is done with 100 µg/mL ampicillin in *E. coli* and with 50 µg/mL thioestrepton in *Streptomyces*. By subcloning the fragment in pEM4, transcription of the genes will be under the control of the *ermEp**.
9. Transform *E. coli* cells and select transformants with ampicillin.
10. Purify and confirm the right construct, named pRHAM (*see* **Note 6**)

3.3.3. Construction of pLN2

This is a plasmid that contains all of the genes involved in the biosynthesis of L-oleandrose from the oleandomycin biosynthetic pathway in *S. antibioti-cus (15)*. This plasmid directs the biosynthesis of dTDP-L-olivose. To construct this plasmid, three genes (*oleL, oleS,* and *oleE*) are subcloned as a single

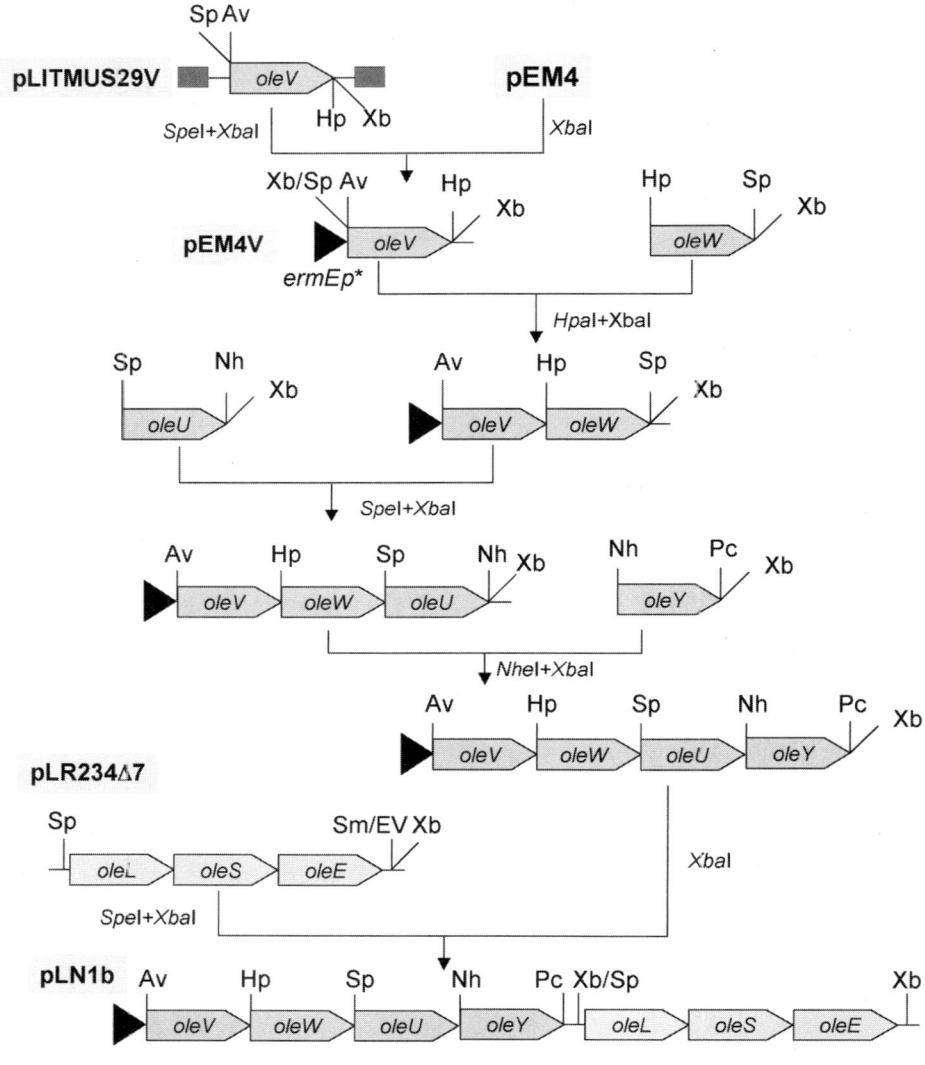

Fig. 4. Strategy for constructing plasmid pLN2: generation of the intermediate pLN1b.

DNA fragment. They code for enzymes catalyzing the first three common steps in the biosynthesis of all 6-deoxysugars. The rest of the genes are incorporated into the construct as PCR-amplified DNA fragments flanked by unique restriction sites not frequently found in *Streptomyces* DNA. To construct pLN2, first pLN1b has to be generated (*see* **Fig. 4**):

1. Amplify by PCR the *oleU, oleV, oleW,* and *oleY,* using cosAB61 DNA as the template, and primers 7U and 7L *(oleU)*, 1α and 1β *(oleV)*, 15α and 15β *(oleW)*, and 14U and 14L *(oleY)*. These primers are designed to introduce two restriction sites at both sides of each gene. Two of these sites are common for all the genes (*Hind*III and *Xba*I), and two other are specific for each gene.

2. Digest all PCR products (except *oleV*) with *Hind*III and *Xba*I. In the case of *oleV,* digest with *Avr*II and *Xba*I.

3. Run an agarose gel electrophoresis and purify the different fragments from the gel.

4. Ligate each of the fragments (except *oleV*) to pUC18 DNA, previously digested with *Hind*III and *Xba*I. In the case of *oleV,* ligate the fragment to pLITMUS29 digested with *Avr*II and *Xba*I.

5. Transform *E. coli* cells and select transformants with 100 μg/mL ampicillin.

6. Purify and confirm the different DNA constructs. The different pUC18 constructs (pUC18W, pUC18U, pUC18Y) and pLITMUS29V (containing the *oleV* gene) will be used as source of individual genes (*see* **Note 7**).

7. Digest pLITMUS29V with *Spe*I and *Xba*I. *Spe*I cuts in the polylinker, outside the *Avr*II site. Purify the 1.5-kb *Spe*I–*Xba*I fragment.

8. Ligate the purified DNA fragment to pEM4 previously digested with *Xba*I (*see* **Note 8**).

9. Transform *E. coli* cells and select transformants with ampicillin.

10. Purify and confirm the DNA construct, which is named pEM4V (*see* **Note 9**).

11. Sequentially add all of the genes to pEM4V. To do this, digest the constructs that successively are generated with the two restriction enzymes that cut at the 3′ ends of the last incorporated gene and insert the new DNA fragments flanked by the same restriction sites. For example, *oleW* is the first gene that is added to pEM4V. To clone this gene, pUC18W is digested with *Hpa*I and *Xba*I and the DNA fragment containing *oleW* is ligated to pEM4V, previously digested with the same restriction enzymes. Confirmation of the right inserts in every cloning step is achieved by digesting the successive constructs with the same enzymes used for subcloning the specific gene. By sequentially adding all of these genes, construct pEM4VWUY is obtained, which contains genes *oleV, oleW, oleU,* and *oleY* downstream of the *ermEp**.

12. In parallel and in order to clone *oleL, oleS,* and *oleE* genes, digest plasmid pLR2347 (*see* **Subheading 3.3.2.**) with *Sma*I and *Eco*RV and then, after 10 times dilution, religate the DNA (*see* **Note 10**). This has the objective of deleting a DNA fragment that contains *oleU*. The *Sma*I site cuts within *oleU*, and *Eco*RV does this in the polylinker. Transformed cells are selected with 50 μg/mL kanamycin. Purify and confirm the DNA construct pLR234Δ7 (*see* **Note 11**).

13. Digest pLR234Δ7 with *Spe*I and *Xba*I and also with *Ssp*I, to obtain the *Spe*I–*Xba*I fragment of 2.9 kb that contains *oleL, oleS,* and *oleE* genes (*see* **Note 12**).

14. Ligate the 2.9-kb *Spe*I–*Xba*I fragment from pLR234Δ7 to pEM4VWUY, previously digested with *Xba*I.

15. Transform *E. coli* cells and select transformants with ampicillin.

16. Purify and confirm the DNA construct pLN1b (*see* **Note 13**).

17. From this construct, isolate the *Hind*III (there is a site upstream of *ermEp**)–*Xba*I fragment, containing both the *ermEp** and all of the "sugar" genes, and subclone into pWHM3 *(11)*, previously digested with the same restriction enzymes, to generate pLN2.

pLN2 can be easily modified to create other plasmids able to direct the biosynthesis of different deoxysugars. These modifications include exchanging, adding, or deleting genes.

3.3.4. Exchanging Genes: Generation of pLNR

oleU codes for a 4-ketoreductase that can be exchanged by equivalent genes. The exchange of *oleU* by *urdR* generates pLNR, which directs the biosynthesis of dTDP-D-olivose. This is done as follows (*see* **Fig. 5**):

1. Amplify *urdR,* using as the template, total DNA from *Streptomyces fradiae* Tü2717 (urdamycin producer) and primers RU and RL.
2. Digest the amplified DNA fragment with *Spe*I and *Nhe*I.
3. Digest pLN2 with *Spe*I and *Nhe*I to liberate the *oleU* gene. This generates two DNA fragments. Isolate from an agarose gel the larger fragment that contains the vector.
4. Ligate this purified fragment with the *urdR*-amplified fragment.
5. Transform *E. coli* cells and select transformants with 100 µg/mL ampicillin.
6. Purify and confirm the DNA construct pLNR (*see* **Note 14**).

3.3.5. Deleting Genes: Generation of pLNΔ

Deletion of *oleV* plus *oleW* will generate plasmid pLNΔ, which directs the biosynthesis of dTDP-L-rhamnose. This is done as follows:

1. Digest pLN2 with *Avr*II and *Spe*I, in order to delete *oleV* and *oleW*. This digestion will generate two fragments.
2. Isolate the larger DNA fragment from an agarose gel, and after 10 times dilution, religate the DNA (*see* **Note 15**).
3. Transform *E. coli* cells and select transformants with 100 µg/mL ampicillin.
4. Purify and confirm the DNA construct pLNΔ (*see* **Note 16**).

3.3.6. Exchanging and Inserting Genes: Generation of pLNRHO

Exchange of *oleU* by the 4-ketoreductase *urdZ3,* and addition of the 3,4-dehydratase *urdQ,* will generate plasmid pLNRHO which directs the biosynthesis of dTDP-L-rhodinose. This is done as follows:

1. First, pLNZ3 has to be generated. Amplify *urdZ3* and *urdQ,* using, as the template, total DNA from *Streptomyces fradiae* Tü2717 (urdamycin producer) and primers Z3U and Z3L, and QU and QL, respectively.
2. Digest the amplified *urdZ3* gene with *Spe*I and *Nhe*I.
3. Digest pLN2 with *Spe*I and *Nhe*I to liberate the *oleU* gene. This generates two DNA fragments. Isolate the larger fragment from an agarose gel.

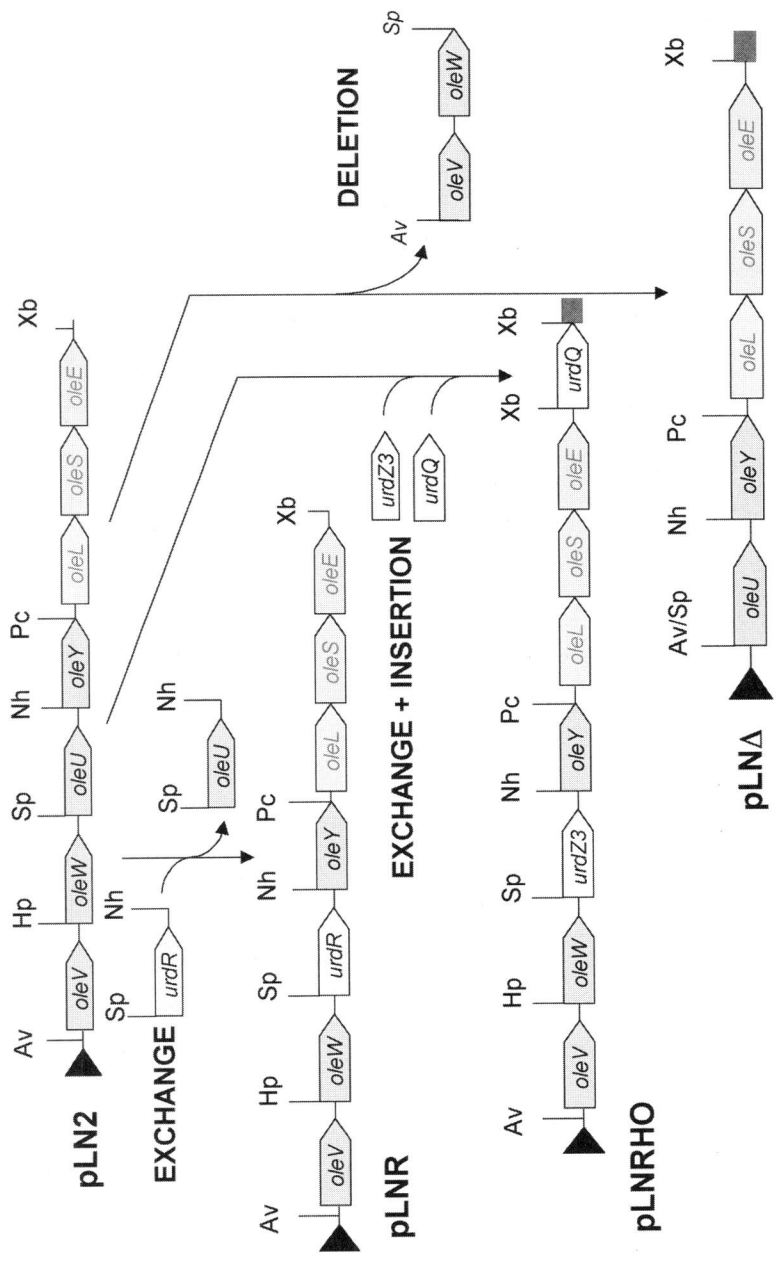

Fig. 5. Construction of pLN2 derivatives.

4. Ligate this fragment with the amplified *urdZ3* gene.
5. Transform *E. coli* cells and select transformants with 100 μg/mL ampicillin.
6. Purify and confirm the DNA construct, named pLNZ3 (*see* **Note 17**).
7. Digest pLNZ3 with *Xba*I and ligate with the *urdQ* amplicon, previously digested with *Xba*I.
8. Transform *E. coli* cells and select transformants with 100 μg/mL ampicillin.
9. Purify and confirm the DNA construct pLNRHO (*see* **Note 18**).

3.4. Generation of S. albus GB16 Recombinant Strains Containing Different Plasmids Directing Deoxysugar Biosynthesis

Once the "sugar plasmids" have been constructed, they are introduced in the *S. albus* GB16, which express the *elmGT* glycosyltransferase.

1. Grow *S. albus* GB16 in TSB with 0.75% glycine, to generate protoplasts.
2. Independently, transform *S. albus* GB16 protoplasts with the different "sugar plasmids" (i.e., pOLV, pRHAM, pLN2, pLNR, pLNΔ and pLNRHO), generated in **Subheading 3.3.**
3. Plate transformed protoplasts on R5 and select transformants with 50 μg/mL thiostrepton.
4. Incubate plates at 30°C for 5–6 d.
5. Confirm and select one recombinant strain from each transformation for further experiments. These strains are named *S. albus* GB16-OLV, *S. albus* GB16-RHAM, *S. albus* GB16-LN2, *S. albus* GB16-LNR, *S. albus* GB16-LNΔ and *S. albus* GB16-LNRHO.

3.5. Generation of Glycosylated Derivatives of 8-Demethyl-Tetracenomycin C

The different *S. albus* GB16-derived strains generated in **Subheading 3.4.** are used as hosts for biotransformation experiments to generate glycosylated derivatives of 8-demethyl-tetracenomycin C (8-DMTC), through the combined action of the ElmGT glycosyltransferase and the activated deoxysugars synthesized by the different "sugar plasmids."

3.5.1. Preparation of 8-Demethyl-Tetracenomycin C

The aglycon 8-DMTC is purified from *S. lividans* GB16F4 (*3*). This is a *S. lividans* derivative harboring cos16F4 (*2*). This cosmid contains most of the elloramycin biosynthetic gene cluster and is able to direct the biosynthesis of 8-DMTC (*16*). Preparation of 8-DMTC is as follows:

1. Inoculate 250-mL Erlenmeyer flasks containing 25 mL of TSB and 2.5 μg/mL of apramycin with spores of *S. lividans* GB16F4.
2. Incubate for 24 h at 30°C and 1.18*g*.
3. Use this culture as preinoculum to inoculate (at 2.5% v/v) eight 2-L Erlenmeyer flasks containing 400 mL of R5A medium (*17*) with 2.5 μg/mL of apramycin.

4. Incubate for 5 d at 30°C and 250 rpm.
5. Centrifuge cultures in a JA-14 rotor using 250-mL bottles at 12500*g* for 30 min at 4°C.
6. Filter the supernatants through Supor membranes, 0.2 μm pore size.
7. Apply the filtered solution to a solid-phase extraction cartridge Supelclean LC18, 10 g.
8. Wash the cartridge with 0.1% trifluoroacetic acid (TFA) in water.
9. Elute with a mixture of acetonitrile and 0.1% TFA in water, using a linear gradient from 0% to 100% acetonitrile in 60 min at 10 mL/min.
10. Take samples every 5 min.
11. Analyze samples by HPLC, using a μBondapak C$_{18}$ column and a linear gradient from 10% to 100% acetonitrile in 0.1% TFA in water for 30 min at 1.5 mL/min. Detection and spectral characterization of peaks are performed with a photodiode array detector and Millenium software. 8-DMTC has maximum absorption at 290 nm and elutes at 18 min under these chromatographic conditions.
12. Dry down samples containing 8-DMTC under vacuum.
13. Dissolve extracts in a small volume of a mixture of DMSO : acetonitrile (1 : 1).
14. Further purify 8-DMTC by reverse-phase preparative HPLC in a μBondapak C$_{18}$ radial compression cartridge (PrePak Cartridge, 25 × 100 mm).
15. Elute with a mixture of acetonitrile and 0.1% TFA in water (30 : 70). Approximately 600 mg of pure compound can be obtained.

3.5.2. Biotransformation of 8-DMTC by S. albus GB16 Recombinant Strains

The different *S. albus* GB16 recombinant strains generated in **Subheading 3.4.** are independently used for biotransformation experiments, following the same culture conditions, as follows:

1. Inoculate 50-mL Erlenmeyer flasks containing 5 mL of TSB and 5 μg/mL of thiostrepton, with spores of the appropriate *S. albus* GB16 strain derivative.
2. Incubate for 24 h at 30°C and 250 rpm.
3. Use 100 μL of this culture to inoculate 50-mL Erlenmeyer flasks containing 5 mL of R5A liquid medium.
4. After 24 h of incubation, add 100 μg/mL (final concentration) of 8-DMTC to the culture.
5. Further, incubate at 30°C and 250 rpm taking 900 μL samples at 24-h intervals during 3 d.
6. Mix samples with 300 μL ethyl acetate in an Eppendorf tube.
7. Centrifuge for 1 min to separate the aqueous and organic phases.
8. Take the organic phase and evaporate under vacuum.
9. Resuspend the residue in a small volume of methanol (i.e., 200 μL).
10. Analyze 50-μL samples by HPLC and identify glycosylated derivatives of 8-DMTC by looking for new peaks with the same absorption spectra as 8-DMTC but with different retention times.

4. Notes

1. *Bam*HI and *Bgl*II generate compatible cohesive ends.
2. The apramycin-resistance cassette confers resistance to apramycin, both in *E. coli* and *Streptomyces,* and to tobramycin in *E. coli.*
3. Digestion of pOLV must liberate two fragments, one corresponding to the vector (7.5 kb) and another one to the amplified DNA fragment (7.6 kb).
4. *Xmn*I and *Sma*I generate blunt ends.
5. Digestion of pLR2347 must liberate two fragments, one corresponding to the vector (3.1 kb) and another one to the insert DNA (4.1 kb).
6. The 4.1 kb *Spe*I–*Xba*I fragment can be inserted in both orientations into pEM4. The right orientation generates pRHAM and can be confirmed by digesting the DNA with *Hind*III plus *Xba*I, which generates two fragments, one corresponding to insert and *ermEp** (4.3 kb) and the other corresponding to the rest of the vector (6.7 kb).
7. All PCR products are sequenced to identify potential mutations introduced during PCR reaction.
8. *Spe*I and *Xba*I generate compatible cohesive ends.
9. The 1.5-kb *Spe*I–*Xba*I fragment can be inserted in the *Xba*I site of pEM4 in both orientations. In the right orientation, the gene will be under the control of the *ermEp**. The right construct pEM4V can be confirmed by digesting the DNA with *Hind*III (it cuts upstream of the *ermEp**) and *Xba*I (it cuts downstream of *oleV*), which must generate a 1.7-kb fragment (*oleV* plus *ermEp**) and a 6.7-kb fragment, corresponding to the rest of the vector.
10. *Eco*RV and *Sma*I generate blunt ends.
11. Confirmation of the right construct pLR234Δ7 is achieved by digesting this DNA with *Spe*I, which must originate two fragments of 2.9 kb, and by the lack of digestion with *Eco*RV or *Sma*I.
12. Digestion with *Spe*I and *Xba*I will generate two fragments of the same size, but because *Ssp*I cuts twice in the vector pUK21 but not in the insert, this digestion will allow one to distinguish between the insert and the vector.
13. Confirmation of the right construct pLN1b can be achieved by digesting the DNA with *Pac*I and *Xba*I. If the fragment containing *oleL–oleE* has been inserted in the right orientation, this digestion will originate two fragments of 2.9 and 12.1 kb. If the fragment has been cloned in the wrong orientation, this digestion will originate only one fragment.
14. Confirmation of the right construct pLNR is done by two approaches. In the first one, the possibility of amplifying DNA using specific oligonucleotides for *urdR* and *oleU* is tested. Positive amplification with *urdR*-specific oligoprimers and negative amplification with *oleU*-specific oligoprimers will be a proof to identify the right construct. In the second one, the orientation of the *urdR* gene is confirmed. The *urdR* gene can be inserted in both orientations, because *Spe*I and *Nhe*I generate compatible cohesive ends. If *urdR* has been cloned in the right orientation, digestion of the DNA with *Spe*I and *Nhe*I will result in the liberation of the *urdR* gene, otherwise no digestion of DNA should occur.
15. *Avr*II and *Spe*I generate compatible cohesive ends.

16. Confirmation of pLNΔ can be done by digesting the DNA with *Avr*II and *Spe*I. These enzymes should not be able to digest this DNA.
17. Confirmation of pLNZ3 can be done by testing the possibility of amplifying DNA using specific oligonucleotides for *urdZ3* and *oleU*. Positive amplification with *urdZ3*-specific oligoprimers and negative amplification with *oleU*-specific oligo-primers will be a proof to confirm the exchanging of genes. To confirm the right orientation of *urdZ3*, digest this construct with *Spe*I and *Nhe*I, which will result in the liberation of the *urdZ3* gene.
18. Confirmation of pLNRHO can be done by digesting the DNA with *Xba*I (it should liberate a fragment corresponding to the *urdQ* gene). The right orientation of the fragment is confirmed by sequencing one end of the gene using the direct primer for pUC.

Acknowledgments

Research in the authors laboratory was supported by grants of the Plan Nacional en Biotecnología (BIO2000-0274, to J. A. S.), Programa Nacional de Promoción General del Conocimiento (BMC2002-03599, to C. M.), and Plan Regional de Investigación del Principado de Asturias (GE-MED01-05). We wish to thank all the people who have worked in our laboratory for helpful suggestions and discussions, especially those contributing to research on the subject of this article: G. Blanco, A. F. Braña, E. Patallo, L. Rodríguez, and I. Aguirrezabalaga. We also wish to thank Juergen Rohr (University of Kentucky, USA) for his contribution to research in this topic.

References

1. Méndez, C. and Salas, J. A. (2001) Altering the glycosylation pattern of bioactive compounds. *Trends Biotechnol.* **19,** 449–456.
2. Decker, H., Rohr, J., Motamedi, H., Zahner, H., and Hutchinson, C. R. (1995) Identification of *Streptomyces olivaceus* Tu 2353 genes involved in the production of the polyketide elloramycin. *Gene* **166,** 121–126.
3. Blanco, G., Patallo, E. P., Braña, A. F., et al. (2001) Identification of a sugar flexible glycosyltransferase from *Streptomyces olivaceus,* the producer of the antitumor polyketide elloramycin. *Chem. Biol.* **8,** 253–263.
4. Patallo, E. P., Blanco, G., Fischer, C., et al. (2001) Deoxysugar methylation during biosynthesis of the antitumor polyketide elloramycin by *Streptomyces olivaceus:* characterization of three methyltransferase genes. *J. Biol. Chem.* **276,** 18,765–18,774.
5. Aguirrezabalaga, I., Olano, C., Allende, N., et al. (2000) Identification and expression of genes involved in biosynthesis of L-oleandrose and its intermediate L-olivose in the oleandomycin producer *Streptomyces antibioticus. Antimicrob. Agents Chemother.* **44,** 1266–1275.
6. Kuhstoss, S., Richardson, M. A., and Rao, R. N. (1991) Plasmid cloning vectors that integrate site-specifically in *Streptomyces* spp. *Gene* **97,** 143–146.

7. Quiros, L. M., Aguirrezabalaga, I., Olano, C., Méndez, C., and Salas, J. A. (1998) Two glycosyltransferases and a glycosidase are involved in oleandomycin modification during its biosynthesis by *Streptomyces antibioticus*. *Mol. Microbiol.* **28,** 1177–1185.

8. Rodríguez, L., Oelkers, C., Aguirrezabalaga, I., et al. (2000) Generation of hybrid elloramycin analogs by combinatorial biosynthesis using genes from anthracycline-type and macrolide biosynthetic pathways. *J. Mol. Microbiol. Biotechnol.* **2,** 271–276.

9. Sambrook, J., Fritsch, E. F., and Maniatis, T. (1989) *Molecular Cloning: A Laboratory Manual,* 2nd ed., Cold Spring Harbor Laboratory Press, Cold Spring Harbor, NY.

10. Kieser, T., Bibb, M. J., Buttner, M. J., Chater, K. F., and Hopwood, D. A. (2000) *Practical* Streptomyces *genetics.* The John Innes Foundation, Norwich, UK.

11. Vara, J., Lewandowska-Skarbek, M., Wang, Y. G., Donadio, S., and Hutchinson, C. R. (1989) Cloning of genes governing the deoxysugar portion of the erythromycin biosynthesis pathway in *Saccharopolyspora erythraea (Streptomyces erythreus)*. *J. Bacteriol.* **171,** 5872–5881.

12. Janssen, G. R. and Bibb, M. J. (1993) Derivatives of pUC18 that have *Bgl*II sites flanking a modified multiple cloning site and that retain the ability to identify recombinant clones by visual screening of *Escherichia coli* colonies. *Gene* **124,** 133–134.

13. Rodríguez, L., Rodríguez, D., Olano, C., Braña, A. F., Méndez, C., and Salas, J. A. (2001) Functional analysis of OleY L-oleandrosyl 3-*O*-methyltransferase of the oleandomycin biosynthetic pathway in *Streptomyces antibioticus*. *J. Bacteriol.* **183,** 5358–5363.

14. Vieira, J. and Messing, J. (1991) New pUC-derived cloning vectors with different selectable markers and DNA replication origins. *Gene* **100,** 189–194.

15. Rodríguez, L., Aguirrezabalaga, I., Allende, N., Braña, A. F., Méndez, C., and Salas, J. A. (2002) Engineering deoxysugar biosynthetic pathways from antibiotic-producing microorganisms. A tool to produce novel glycosylated bioactive compounds. *Chem. Biol.* **9,** 721–729.

16. Wohlert, S.- E., Blanco, G., Lombó, F., et al. (1998) Novel hybrid tetracenomycins through combinatorial biosynthesis using a glycosyltransferase encoded by the *elm*-genes in cosmid 16F4 which shows a broad sugar substrate specificity. *J. Am. Chem. Soc.* **41,** 10,596–10,601.

17. Fernandez, E., Weissbach, U., Sanchez Reillo, C., et al. (1998) Identification of two genes from *Streptomyces argillaceus* encoding glycosyltransferases involved in transfer of a disaccharide during biosynthesis of the antitumor drug mithramycin. *J. Bacteriol.* **180,** 4929–4937.

7

Assay Methods for Detection and Quantification of Antimicrobial Metabolites Produced by *Streptomyces clavuligerus*

Paloma Liras and Juan F. Martín

Summary

Streptomyces clavuligerus is used as model to illustrate the determination and quantification of antimicrobial metabolites with different biological activities produced by cultures of this strain. *S. clavuligerus* produces an array of compounds with different structures. It produces cephamycin C, a β-lactam antibiotic acting on cell wall biosynthesis, the β-lactamase inhibitor clavulanic acid, which contains a clavam nucleus, as well as other clavams structures with antibacterial or antifungal activities. Also, *S. clavuligerus* has been described as a producer of the compounds with pyrrothin structure holomycin and holothin, which might act as antitumor agents. The production of each of these compounds depends on the culture medium and the culture conditions used and, occasionally, is related to specific strains or mutants of *S. clavuligerus*. Although a final characterization of the different compounds depends on fine chemical analytical methods, the detection and quantification of these compounds by high-performance liquid chromatography or by microbiological assays based in their different properties is possible. In this chapter, we provide some of the protocols routinely used for this purpose.

Key Words: *Streptomyces clavuligerus;* clavulanic acid; cephamycin C; clavams; holomycin.

1. Introduction

Microbial secondary metabolites are small molecules, with a great variety of very complex chemical structures, which are formed by the microorganisms after the growth phase is complete. These substances are dispensable (i.e., nonproducer mutants show growth patterns similar to those of the producer wild-type strains) (*1*). The role of secondary metabolites in the producer microorganisms is not completely clear, but they probably provide ecological advantages to the pro-

From: *Methods in Biotechnology, Vol. 18: Microbial Processes and Products*
Edited by: J. L. Barredo © Humana Press Inc., Totowa, NJ

ducer strain in its natural habitat (2), although this has been proven only in a few particular cases.

Secondary metabolites include substances with different biological activities, many of them of great interest to mankind. Some secondary metabolites have antibiotic activity, inhibiting the growth of other microorganisms, which makes them of clinical interest. Others act as enzyme inhibitors or as antitumoral agents or might be used as phytohormones enhancing the growth of plants.

Actinomycetes (Gram-positive filamentous bacteria) and filamentous fungi are the main producers of secondary metabolites in the microbial world. The soil is their natural habitat and both groups share a mycelial morphology, forming colonies composed of substrate mycelium, frequently embebbed in the soil, and aerial mycelium that proliferates above the soil surface. Actinomycetes and fungi produce characteristic spores, borne in the tips of the hyphae, for dispersion of the strain; they act also as resistance forms, which allow them to cope with adverse environmental conditions. However, in spite of these similarities, the actinomycetes are bacteria with prokaryotic structure, whereas the filamentous fungi are eukaryotic organisms.

Very frequently, a single microbial strain produces more than one secondary metabolite, and some microorganisms produce a wealth of compounds with different structures and biological activities. The compounds produced by one particular strain might have similar chemical structures, belonging to a family of structurally similar compounds, or they might have completely different structures. microorganisms contain clusters of genes for the formation of each secondary metabolite produced by the cell. Each cluster contains genes encoding structural biosynthetic enzymes but also genes encoding regulatory proteins for the control of the metabolite formation, genes for exporting the metabolite, as well as genes required to make the producer strain resistant to the metabolite in those cases in which the secondary metabolite produced has a lethal biological activity (3). Frequently, these gene clusters extend from 30 to 100 kb in the genome, and if the strain produces several compounds, a large amount of the genome in the organism is devoted to the formation of these secondary metabolites (4,5), which indicates that they might be important for survival of the producer strain in their natural habitat.

In this chapter, we present Streptomyces clavuligerus as a model microorganism to illustrate the detection and quantification of different antibiotic activities. This strain produces clavulanic acid, cephamycin C, alanylclavam, and other antifungal clavams, as well as holomycin, holothin, and tunicamycin. Identification of each of these secondary metabolites relies on advanced organic chemistry and biochemistry techniques such as high-performance liquid chromatography (HPLC), nuclear magnetic resonance (NMR), liquid chromatography–mass spectrometry (LC-MS), and so forth. However, routine quantification of the different secondary metabolites in a given culture is frequently based on the determination of their biological activities (i.e., antimicrobial activities, enzyme inhibitory action, antifungal activity, etc.). HPLC separation and quantification by comparison with pure standards is also a routine technique for those components that lack easily measurable biological activities (e.g., antitumor compounds).

2. Materials

1. *Streptomyces clavuligerus* ATCC 27064 (ATCC, Manassas, VA).
2. *S. clavuligerus cyp::aph* (**6**).
3. *Klebsiella pneumoniae* ATCC 29665 (ATCC).
4. *Escherichia coli* Ess22-35 (**7**).
5. *Bacillus* UL-1 (CM-ULE, León, Spain).
6. *Bacillus* sp. ATCC 27860 (ATCC).
7. *Enterobacter cloacae* CECT960 (CECT, Valencia, Spain).
8. TSB medium: 30 g/L soy trypticaseine, pH 7.2.
9. TSA medium: 30 g/L soy trypticaseine and 20 g/L agar, pH 7.2.
10. SA medium: 10 g/L starch, 2 g/L L-asparagine, 0.6 g/L $MgSO_4 \cdot 7H_2O$, 4.4 g/L K_2HPO_4, 21 g/L MOPS, and 1 mL/L trace elements, pH 6.8.
11. Trace elements solution: 1 g/L $FeSO_4 \cdot 7H_2O$, 1 g/L $ZnSO_4$. $7H_2O$, 1 g/L $MnCl_2 \cdot 4H_2O$, and 1.3 g/L $CaCl_2 \cdot 3H_2O$.
12. ME medium: 5 g/L glucose, 0.5 g/L yeast extract, 0.5 g/L beef extract, 1 g/L NZ Amine, 21 g/L MOPS, and 20 g/L agar, pH 7.0.
13. Davis Mingioli modified medium: 8 g/L D-glucose, 0.5 g/L sodium citrate $\cdot 3H_2O$, 7 g/L K_2HPO_4, 3 g/L KH_2PO_4, 0.1 g/L $MgSO_4 \cdot 7H_2O$, and 1 g/L $(NH_4)_2SO_4$, pH 7.0.
14. GSPG medium: 15 g/L glycerol, 20 g/L sucrose, 2.5 g/L proline, 1.5 g/L glutamic acid, 5 g/L NaCl, 2 g/L K_2HPO_4, 0.4 g/L $CaCl_2$, 0.1 g/L $MnCl_2 \cdot 4H_2O$, 0.1 g/L $FeCl_3 \cdot 6H_2O$, 0.05 g/L $ZnCl_2$, 1 g/L $MgSO_4 \cdot 7H_2O$, pH 7.0.
15. Bacto-Difco penicillinase (Difco Laboratories, Detroit, MI).
16. Silica gel 60 plates (Merck KgaA, Darmstadt, Germany).
17. Tunicamycin A2, B, and C2 complexes (Sigma, St. Louis, MO).
18. Augmentin (SmithKlineBeecham, Surrey, UK).
19. Rotary shaker (e.g., Innova 4300) (New Brunswick Scientific, Edison, NJ).
20. Water bath.
21. Static incubator.
22. Centrifuge Eppendorf 5417C (Eppendorf, Hamburg, Germany).
23. Bench centrifuges.
24. Optical microscope.
25. Laminar flow cabinet.
26. HPLC (e.g. Beckman System Gold equipped with a detector Mod. 166; Beckman Instruments, San Ramón, CA).
27. Chromatography columns: μBondapak C_{18} column (300 × 4 mm) and Nucleosil C_{18} column (30 × 4 mm).

28. Imidazole reagent: dissolve 8.25 g imidazole in 24 mL of distilled water. Adjust pH to 6.8 with 5 M HCl and raise final volume to 40 mL.
29. Isocratic buffer phase: 0.1 M NaH$_2$PO$_4$ and 6% methanol, adjusted to pH 3.7 with glacial acetic acid.

3. Methods

Streptomyces clavuligerus has the ability to produce several secondary metabolites. Initially described as a cephamycin C-producing strain (*see* **Fig. 1A**) *(8)*; it also produces small amounts of the β-lactam compounds isopenicillin N and deacetylcephalosporin C, both intermediates in the cephamycin pathway. In a screening program searching for β-lactamase inhibitory compounds at Beecham Pharmaceutical, *S. clavuligerus* was found to produce a compound with β-lactamase inhibitory properties that was called clavulanic acid *(10)*. This compound, which has a low antibacterial activity, is a bicyclic β-lactam with a 3R,5R stereochemistry. The nucleus of clavulanic acid (*see* **Fig. 1B**) resembles that of the penicillins, but the five-membered (oxazolidine) ring has oxygen instead of sulfur and lacks the C-6 acylamino side chain of penicillins. Clavulanic acid inhibits most class A β-lactamases and the cloxacillin hydrolyzing enzymes of class D. It binds irreversibly to the serine hydroxyl group at the active center of β-lactamases, producing a stable acylated intermediate and resulting in the inactivation of the enzyme.

Additional compounds with clavam structure have been isolated from *S. clavuligerus* culture broths. As shown in **Fig. 1,** they include clavam-2-carboxylate, 2-formyloxymethylclavam, 2-hydroxymethylclavam, hydroxyethylclavam *(11)* and alanylclavam *(12)*. Only clavulanic acid among the clavam molecules shows β-lactamase inhibitory activity, which is related to its unique 3R,5R stereochemistry.

Clavams (i.e. clavam-2-carboxylate, 2-formyloxymethylclavam, 2-hydroxymethylclavam, and hydroxyethylclavam) are antifungal agents, but alanylclavam possess antibacterial properties. This antibacterial activity is the result of the noncompetitive inhibition of the bacterial homoserine *O*-succinyl transferase, an enzyme required for methionine biosynthesis. The clavulanic acid nonproducing mutant *S. clavuligerus dcl-8* produces, in addition, several *N*-acylderivatives of clavaminic acid *(9)* that might be of interest in understanding clavulanic acid biosynthesis. Mutants of *S. clavuligerus* are known to produce the non-β-lactam antibiotics holomycin, *n*-propionyl-holothin (*see* **Fig. 1C**), and tunicamycin (*see* **Fig. 1D**) *(13–15)*. Most of these molecules were detected initially in the HPLC chromatographic analysis of *S. clavuligerus* broths from cultures grown in different media and growth conditions. The compounds were isolated, purified, occasionally derivatized, and crystallized; then, their chemical structures were characterized by NMR, circular dichroism spectroscopy, and mass spectrometry.

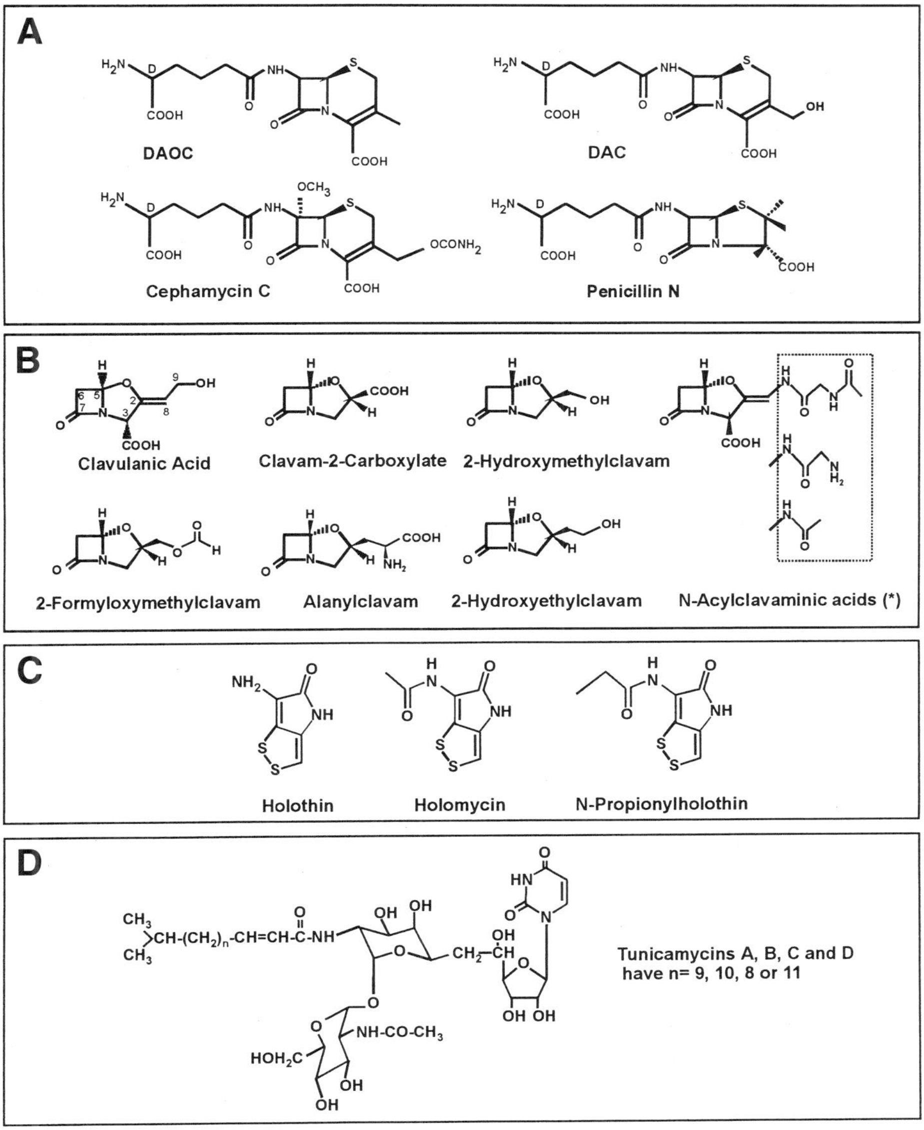

Fig. 1. Antibiotics produced by *S. clavuligerus:* (**A**) β-lactam antibiotics; (**B**) clavulanic acid and clavam type of antibiotics; (**C**) pyrrothin type of antibiotics; (**D**) tunicamycins. (*) *N*-Acylclavaminic acids were detected in the mutant *S. clavuligerus* dcl-8. (Reprinted with permission from **ref. 9**.)

An important question is how is possible to distinguish in simple experiments which compound(s) is produced at a determined time or medium in *S. clavuligerus* cultures. There are different simple procedures to detect or quantify each of these compounds, but the most general procedure for most of them is (1) to analyze them by HPLC chromatography, providing that the pure compound is available, and (2) the use of microbiological assays that relay in the biological activity of the compounds (i.e., in modified bioassays specific for the different types of activities).

3.1. Growth of Streptomyces clavuligerus

1. Seed plates of ME medium to obtain spores of *S. clavuligerus*.
2. Seed a TSB culture (100 mL) with 10^8 spores/mL and incubate at 28°C and 220 rpm in an orbital shaker up to an optical density (OD_{600}) from 4 to 7.
3. Mix aliquots (1 mL) of this culture with an equal volume of sterile 40% glycerol and keep frozen at –75°C.
4. Inoculate TSB medium (100 mL) in 500-mL triple baffled flasks (to obtain good aeration) with 1-mL aliquots of the frozen mycelium. The culture is incubated for about 24 h at 28°C and 220 rpm. This stage is called preinoculum and is used to get a rapid and good growth as well as reproducibility in the fermentations.
5. Inoculate final cultures of the desired media (100 mL) in 500-mL tripled baffled flasks with 5 mL of the preinoculum culture adjusted to an OD_{600} of 5.0. The cultures are incubated at 28°C and 220 rpm for up to 96 h.
6. Take aliquots (3 mL) at different times, centrifuge at 18,000g for 15 min, and use the supernatant to test the presence and amount of the different antibiotics. Usually, the cell pellet is used to determine the growth of the culture as dry weight or as total DNA content.

Steptomyces clavuligerus produces both cephamycin C and clavulanic acid in several media. However, specific media favor the formation of one or the other compound because of their composition.

3.2. Quantification of Cephamycin C by Bioassay

Cephamycin C is a β-lactam compound. In a manner similar to penicillins and cephalosporins, it acts on cell wall biosynthesis inhibiting the crosslinking of the peptidoglycan. However, cephamycin C, opposite to penicillin G or cephalosporin C, is only produced by bacteria. *S. clavuligerus* produces cephamycin C in several media; TSB is the choice medium to produce this antibiotic. Cephamycin C is usually quantified by a microbiological assay *(7)* by using a strain supersensitive to β-lactam antibiotics of the cephalosporin type called *E. coli* Ess22-35.

1. Inoculate 50 mL of TSB medium with a loop of cells from a TSA slant of *E. coli* Ess22-35 and incubate the culture at 30°C and 220 rpm for 16 h.
2. Adjust the OD_{600} of the culture to 1.0 with distilled water.

3. Keep tubes containing 40 mL of TSA medium with 1% agar (or 140 mL of TSA medium with 1% agar when using 23-cm square plates) at 45°C and add 0.5 mL of the *E. coli* suspension to each tube (1.75 mL to prepare square plates).
4. Homogenize by vortexing and pour into 9-cm-diameter Petri dishes.
5. Once the medium is solid, make 6-mm holes with a cork borer and apply 60 µL of *S. clavuligerus* broth, or samples of pure antibiotic.
6. Keep the plates at 4°C for 2 h to allow diffuse the antibiotic present in the samples.
7. Incubate the plates at 30°C for 12–18 h.

The bioassay measure all the penicillinase-resistant antibiotics present in the samples (i.e., cephamycin C, deacetoxycephalosporin C [DAOC], and deacetyl-cephalosporin C [DAC]). A standard plot of inhibition zones obtained at increasing cephamycin C concentration should be made in every assay. When cephamycin C is not available, cephalosporin C can be used as standard and the antibiotic is measured as "total cephalosporins" present in the sample. Penicillin N, DAOC, and DAC are intermediates of the cephamycin pathway *(16)* and are present in small amounts in the culture broth. However, the concentration of these compounds might be significant in specific mutants blocked in the genes *cefE*, *cmcH*, *cmcJ*, or *cmcI*. Penicillin N does not affect the bioassay against *E. coli* Ess32-35 and can be removed by treatment of the sample with a small amount of Bacto-Difco penicillinase or by the addition of penicillinase (cephalosporinase-free) from the supernatant of a *Bacillus subtilis* UL-1 culture. In those cases in which the presence of high concentrations of DAOC or DAC is suspected, the culture should be analyzed by HPLC. Holomycin also might interfere with the bioassay if the concentration is above 20 µg/mL (*see* **Subheading 3.7.**). Cephalosporins present in the samples are stable at low temperature and the samples of the fermentation can be kept for at least 6 d at –20°C before analysis.

3.3. Quantification of Cephalosporins by HPLC

To quantify cephalosporins by HPLC (*see* **Notes 1** and **2**), proteins in the samples of *S. clavuligerus* cultures are precipitated with an equal volume of methanol and then the proteins are eliminated by centrifugation at 18,000*g* for 5 min (*see* **Note 3**). The analysis of cephalosporins is made in the supernatant by using HPLC.

1. Inject samples of 100 µL in a µBondapak C_{18} column (300 × 4 mm).
2. Elute the antibiotics by using 200 m*M* NaH_2PO_4, pH 4.0, with a flow of 1.5 mL/min.
3. Detect cephalosporins at 260 nm.

Cephalosporin C, showing a retention time of 35 min under these conditions, is used as a control. The retention times for other cephalosporins are as follows: DAC, 3.5 min; DAOC, 10.3 min; cephamycin C, 4.7 min (*see* **Note 4**).

3.4. Quantification of Clavulanic Acid by Bioassay

Clavulanic acid is a β-lactamase inhibitor whose biochemistry and molecular genetics is now partially understood *(17)*. This compound can be quantified by bioassay or by HPLC (*see* **Subheadings 3.4.** and **3.5.**). The bioassay is based in the inhibition by clavulanic acid of the β-lactamase produced by *Klebsiella pneumoniae* ATCC29665. This assay is made in the presence of penicillin G, which is degraded by the β-lactamase activity of the test strain *(7)*. However, in the presence of clavulanic acid, the β-lactamase activity is inhibited and the penicillin G remains active. Therefore, in the plate area in which clavulanic acid diffuses, the active penicillin G produces an inhibition zone on the *K. pneumoniae* culture (*see* **Note 5**).

1. Grow *K. pneumoniae* in TSB medium at 30°C to an OD_{600} of 1.0. The culture is kept at 4°C and used for a maximum of 1 wk.
2. Maintain TSA medium at 45°C and add penicillin G (100 µg) and 3.3 mL of the *K. pneumoniae* culture to every 100 mL of the medium.
3. Pour 40 mL of inoculated medium into leveled 9-cm-diameter Petri plates (alternatively, 140 mL are poured in 23-cm square Petri dishes).
4. When the medium is solid, make 6-mm-diameter wells with a cork borer (6 wells in 9-cm-diameter plates and 16 wells in 23-cm square plates).
5. Add 60 µL of broth to every well (*see* **Note 6**). Clavulanic acid is allowed to diffuse for 2 h at 4°C.
6. Incubate the plates at 30°C for 12–14 h and measure the inhibition zones (maximum diameter = 32 mm).
7. Calculate the concentration of clavulanic acid as follows:

$$a = -1.711 \times 10^{-4} \, b^3 + 0.01547b^2 - 0.3369b + 1.5303 \qquad (1)$$

where *a* is the log of clavulanic acid concentration (in µg/mL) and *b* is the halo diameter (in mm). This equation is deduced from the standard plots of clavulanic acid (*see* **Note 7**).

3.5. Quantification of Clavulanic Acid and Clavams by HPLC

Clavulanic acid, as well as other β-lactam compounds, contains a clavam nucleus that reacts with imidazole giving stable compounds absorbing at 312 nm *(18)*.

1. One volume of the culture broth (0.4 mL) is mixed with 0.25 vol of imidazole reagent (*see* **Note 8**).
2. The reaction is kept at room temperature for 15 min to allow derivatization. The clavulanate–imidazole product is relatively stable and is kept at –20°C until analysis by HPLC.
3. Derivatized samples (50 or 100 µL) are injected in an HPLC system equipped with a Waters µBondapack C_{18} column (4.0 × 300 mm) as described by Foulstone and Reading *(19)*.

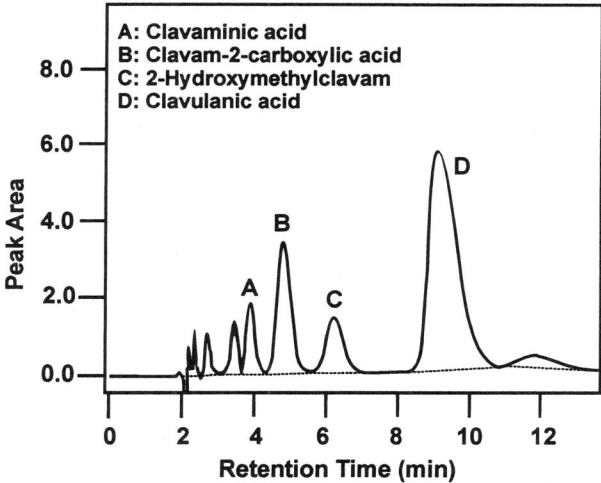

Fig. 2. HPLC analysis of TSB-modified medium culture supernatants of *S. clavuligerus*. Elution of (**A**) clavaminic acid (retention time = 4.0 min), (**B**) clavam-2-carboxylic acid (retention time = 5.1 min), (**C**) 2-hydroxymethilclavam (retention time = 6.1 min), and (**D**) clavulanic acid (retention time = 9.1 min) under the conditions is described in **Subheading 3.5.** (Reprinted with permission from **ref. 20.**)

4. The products are eluted with an isocratic buffer phase consisting in 0.1 M NaH$_2$PO$_4$ and 6% methanol, adjusted to pH 3.7 with glacial acetic acid, with a flow of 1.0 mL/min, and derivatized clavams are detected at 312 nm.

Under these conditions *(20,21)* and using samples of the authentic standards, clavulanic acid and clavams show the following retention times: clavulanic acid, 9.1 min; clavaminic acid, 4.0 min; clavam-2-carboxylate, 5.1 min; and 2-hidroxymethylclavam, 6.1 min (*see* **Fig. 2**). The quantification of the clavams can be given as equivalents of clavulanic acid from which pure standards are normally available (*see* **Note 9**). Clavulanic acid is freshly prepared at 50 µg/mL in 0.1 M phosphate buffer, pH 7.0, and adequately diluted in the same buffer. After derivatization the samples are stable and can be kept at –20°C until analyzed.

3.6. Quantification of Alanyl Clavam by Bioassay

Alanylclavam (*see* **Fig. 1**) has a clavam structure and shows antibacterial activity apparently because of their noncompetitive inhibition of homoserine-O-succinyl transferase, an enzyme required for methionine biosynthesis. The antibacterial activity is reversed by addition of D-methione, L-methionine,

L-cystathionine, L-homocysteine or *O*-acetyl-L-homoserine (but not by L-homoserine) to the bioassay *(12)*.

Production of alanylclavam is high in soy-bean-based media containing glycerol, maltose, or dextrins as carbon source, but this clavam is not produced when D-glucose, manitol, sucrose, or sorbose are present in the medium *(12)*. In GSPG medium, this antibiotic is formed late in the fermentation (at about 72 h). This medium is a choice medium to quantify this compound by bioassay because the components of the medium do not interfere in the bioassay. The bioassay is based on *Bacillus* sp. ATCC27860, a highly sensitive strain with an minimum inhibitory concentration (MIC) of 0.03 μg/mL to alanylclavam (*see* **Note 10**).

1. Seed the indicator organism, *Bacillus* sp. ATCC 27860, in TSB medium and incubate at 37°C for 3 d.
2. Then, the culture is subject to a temperature shock of 80°C for 10 min to eliminate vegetative cells.
3. Centrifuge the spores at 6000*g* for 10 min, wash twice with saline solution, quantify by dilution and plating, and keep at –20°C as a suspension in 20% glycerol.
4. Tubes containing 30 mL of Davis-Mingioli medium with 2 g/L glucose, kept at 60°C, are supplemented with 10^6 spores of *Bacillus* sp. ATCC 27860.
5. Pour the medium in 9-cm-diameter plates, allow to solidify, and make wells (6-mm in diameter) with a cork borer.
6. Apply supernatant broth of a *S. clavuligerus* culture (60 μL) into the wells and allow the broth to diffuse for 2 h at 4°C. As negative controls, supplement some samples with methionine (200 μg/mL) to confirm that the inhibition zones are the result of alanylclavam (*see* **Fig. 3**).
7. Incubate the plates at 37°C for 20–24 h and measure the inhibition zones in the range of 14–36 in mm diameter (*see* **Note 11**).
8. Calculate concentration of alanylclavam as follows:

$$a = 0.0557b - 1.6928 \tag{2}$$

where *a* is the log of the alanylclavam concentration (in μg/mL) and *b* is the diameter of the inhibition zone (in mm).

3.7. Determination of Holomycin Production

Holomycin is a pyrrothin type of antibiotic. These types of compound have been described to be potential antitumor agents. Holomycin formation by the wild-type *S. clavuligerus* is practically negligible. However, certain mutants altered in clavulanic acid biosynthesis such as *S. clavuligerus cyp::aph* produce 100- to 120-fold in relation to the wild type *(15)*.

1. Extract the culture broth with butanol (1 : 1).
2. Analyze and quantify the organic extract by HPLC with a Nucleosil C18 column (30 × 4 mm), eluting with an isocratic methanol : water (40 : 60) solvent mixture at a flow rate of 1 mL/min. The compound, detected at 360 nm, gives peaks that are concentration dependent.

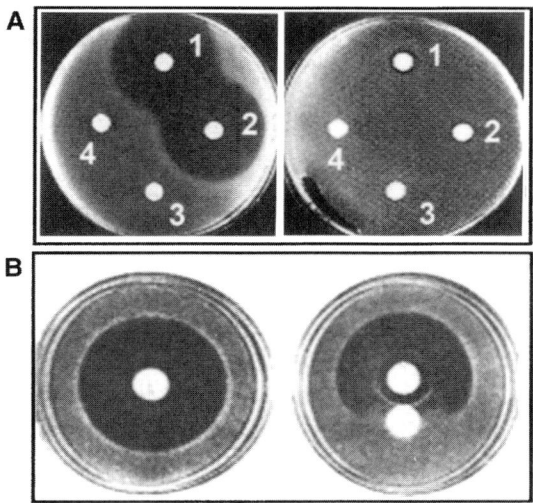

Fig. 3. Identification of alanylclavam activity by inhibition of *Bacillus* sp. ATCC 27860 and reversal by L-methionine. (**A**) Production of alanylclavam by *S. clavuligerus* in modified TSB medium *(1,2)* and lack of production in SA medium *(3,4)* as detected by the inhibition of *Bacillus* sp. ATCC 27860 (left side). The plate at the right side has been supplemented with 200 μg/mL methionine *(21)*. (**B**) Inhibition zone of *Bacillus* sp. ATCC 27860 (left) by a broth filtrate of *S. clavuligerus*. Reversal of the inhibition occurs when a disk embedded in a L-methionine solution is placed on the plate. (Reprinted with permission from **ref. *12*.**)

3. Calculate concentration of holomycin as follows:

$$a = 0.261b - 0.182 \tag{3}$$

where *a* corresponds to the concentration of holomycin (in μg/mL) and *b* is the area of the peak (*see* **Fig. 4**). The previously indicated control plot was obtained with pure holomycin obtained from cultures of *S. clavuligerus cyp::aph* *(14)*, because holomycin is not available as standard.

Holomycin, at concentrations above 20 μg/mL, might interfere in the *E. coli*-based bioassay of cephamycin C, giving an internal clear inhibition zone (as a result of full lysis of *E. coli*). In addition, the presence of holomycin is easily detected by the yellow-green color present in the culture broth, especially in colorless media such as GSPG medium. Additionally, this pyrrothin compound can be extracted from the broth with 1-butanol (1 : 1) and detected by thin-layer chromatography in silica gel 60 plates developed in chloroform–methanol (9 : 1), where it shows a relative mobility R_f of 0.45. The presence of holomycin can be

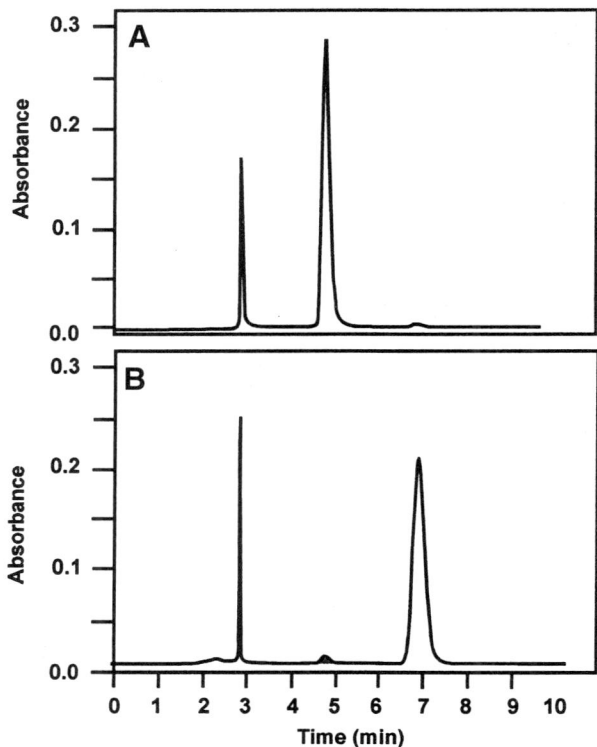

Fig. 4. HPLC chromatography of **(A)** holothin (retention time = 4.7) and **(B)** holomycin (retention time = 6.7) separated under the conditions described in **Subheading 3.7.** (Reprinted with permission from **ref. *15*.**)

detected by bioautography of the plates *(15)*. Under these extraction conditions, small amounts of tunicamycin (*see* **Fig. 1D**) can be detected as a bioactive compound with an R_f of 0.05 *(15)* and compared with pure samples of tunicamycin complexes. Small amounts of *N*-propionyl-holothin and holothin have been described to be present in *S. clavuligerus* cultures. Holothin can be separated by HPLC in the conditions indicated for holomycin (*see* **Fig. 4**); this compound has 100 times less biological activity than holomycin and should not interfere with holomycin determination in microbiological assays.

4. Notes

1. If a chromatographic peak in the broth is suspected to be a cephalosporin, a treatment of the sample with a cephalosporinase (e.g. the wide-spectrum cephalospori-

nase from *Enterobacter cloacae* CECT 960) for 30 min at 37°C should eliminate that peak from the chromatogram.

2. This is a general method to detect all the cephalosporin-type molecules. The literature presents methods for better quantification of particular cephalosporins.

3. The deproteinization treatment of the samples with an equal volume of methanol followed by centrifugation is essential to prolong the lifetime of the HPLC column.

4. The retention times might change slightly with the particular HPLC column used and even at different times in the life of the column, but the relative retention times of the different compounds remains quite constant.

5. Clavulanic acid in the samples should be assayed immediately after taking the samples from the culture because of its unstability, and every sample should be assayed in duplicate or triplicate in different plates. Stability increases in neutral buffered media.

6. When the clavulanic acid concentration is high, samples should be adequately diluted so that the inhibition zones remain below 32 mm.

7. A standard plot using freshly prepared clavulanic acid (1–10 µg/mL) should be made in every bioassay. As a source of clavulanic acid, a commercial preparation of Augmentin 500® (Beecham), containing clavulanic acid and amoxicillin, can be used.

8. The imidazole reagent is stable over a period of 2–3 mo.

9. Comparison of HPLC and microbiological assays for clavulanic acid gave consistent results when the samples are immediately tested. However the stability of the imidazole-derivatized clavulanic acid makes the HPLC assay more convenient.

10. Concentrations of alanylclavam above 5 µg/mL might produce square inhibition zones.

11. Frequently, an inner clear zone of inhibition in the assay (complete lysis of *Bacillus* sp.) might be related with the presence of abundant β-lactam antibiotic production (penicillin N, cephamycin C). There is certain variability in the inhibition zone produced by alanylclavam; therefore, it is convenient to make two or three assays of every sample and even different dilutions of the sample.

Acknowledgments

The authors acknowledge the support of grants BIO2000–0272 from the Spanish Ministry of Education, Culture, and Sports and from the Areces Foundation (Madrid).

References

1. Martín, J. F., Gutiérrez, S., and Aparicio, J. F. (2000) Secondary metabolites, in *Encyclopedia of Microbiology* (Lederberg J., ed.), Academic, San Diego, CA, Vol. 4, pp. 213–237.

2. Davies, J. (1990) What are antibiotics? Archaic functions for modern activities. *Mol. Microbiol.* **4,** 1227–1232.

3. Martín, J. F. and Liras, P. (1989) Organization and expression of genes involved in the biosynthesis of antibiotics and other secondary metabolites. *Annu. Rev. Microbiol.* **43,** 173–206.

4. Bentley, S. D., Chater, K. F., Cerdeño-Tárraga, et al. (2002) Complete genome sequence of the model actinomycete *Streptomyces coelicolor* A3(2). *Nature* **417,** 141–147.

5. Omura, S., Ikeda, H., Ishikawa, J., et al. (2001) Genome sequence of an industrial microorganism *Streptomyces avermitilis:* deducing the ability of producing secondary metabolites. *Proc. Natl. Acad. Sci. USA* **98,** 12,215–12,220.

6. Mellado, E., Lorenzana, L. M., Rodríguez-Sáiz, M., Díez, B., Liras, P., and Barredo, J. L. (2002) The clavulanic acid biosynthetic cluster of *Streptomyces clavuligerus:* genetic organization of the region upstream of the *car* gene. *Microbiology* **148,** 1427–1438.

7. Romero, J., Liras, P., and Martín, J. F. (1984) Dissociation of cephamycin and clavulanic acid biosynthesis in *Streptomyces clavuligerus. Appl. Microbiol. Biotechnol.* **20,** 318–325.

8. Higgens, C. E. and Kastner, R. E. (1971) *Streptomyces clavuligerus* sp. nov., a β-lactam antibiotic producer. *Int. J. System. Bacteriol.* **21,** 326–331.

9. Elson, W. E., Gillet, J., Nicholson, N. H., and Tyler, J. W. (1988) N-Acyl derivatives of clavaminic acid produced by a mutant of *Streptomyces clavuligerus. J. Soc. Chem. Chem. Commun.* 979–980.

10. Brown, A. G., Butterworth, D., Cole, M., et al. (1976) Naturally ocurring β-lactamase inhibitors with antibacterial activity. *J. Antibiot.* **29,** 668–669.

11. Brown, D., Evans, J. R., and Fletton, R. A. (1979) Structures of three novel β-lactams isolated from *Streptomyces clavuligerus. J. Chem. Soc. Chem. Commun.* 282–283.

12. Pruess, D. L. and Kellett, M. (1983) Ro-22-5417, a new clavam antibiotic from *Streptomyces clavuligerus.* 1. Discovery and biological activity. *J. Antibiot.* **36,** 208–212.

13. Okamura, K., Soga, K., Shimauchi, Y., Ishikura, T., and Lein. J. (1977) Holomycin and *N*-propionylholothin, antibiotics produced by a cephamycin C producer. *J. Antibiot.* **30,** 334–336.

14. Kenig, M. and Reading, C. (1979) Holomycin and an antibiotic (MM19290) related to tunicamycin, metabolites of *Streptomyces clavuligerus. J. Antibiot.* **32,** 549–554.

15. Fuente, A., Lorenzana, L. M., Martín, J. F., and Liras, P. (2002) Mutants of *Streptomyces clavuligerus* with disruptions in different genes for clavulanic acid biosynthesis produce large amounts of holomycin: possible cross-regulation of two unrelated secondary metabolic pathways. *J. Bacteriol.* **184,** 6559–6565.

16. Liras, P. (1999) Biosynthesis and molecular genetics of cephamycins. Cephamycins produced by actinomycetes. *Antonie van Leeuwenhoek* **75,** 109–124.

17. Liras, P. and Rodríguez-García, A. (2000) Clavulanic acid, a beta-lactamase inhibitor: biosynthesis and molecular genetics. *Appl. Microbiol. Biotechnol.* **54,** 467–475.

18. Bird, A. E., Bellis, J. M., and Gasson, B. C. (1982) Spectrophotometric assay of clavulanic acid by reaction with imidazole. *Analyst* **107,** 1241–1245.

19. Foulstone, M. and Reading, C. (1982) Assay of amoxicillin and clavulanic acid, the components of augmentin, in biological fluid with high-performance liquid chromatography. *Antimicrob. Agents Chemother.* **22,** 753–762.

20. Mosher, R. H., Paradkar, A. S., Anders, C., Barton, B., and Jensen, S. E. (1999) Genes specific for the biosynthesis of clavam metabolites antipodal to clavulanic acid are clustered with the gene for clavaminate synthase 1 in *Streptomyces clavuligerus. Antimicrob. Agents Chemother.* **43,** 1215–1224.
21. Paradkar, A. S. and Jensen, S. E. (1995) Functional analysis of the gene encoding the clavaminate synthase 2 isoenzyme involved in clavulanic acid biosynthesis in *Streptomyces clavuligerus. J. Bacteriol.* **177,** 1307–1314.

8

Purification of Plasmid DNA Vectors Produced in *Escherichia coli* for Gene Therapy and DNA Vaccination Applications

Maria Margarida Diogo, João António Queiroz, and Duarte Miguel F. Prazeres

Summary

A bench-scale protocol for the purification of plasmid DNA (pDNA) produced in *Escherichia coli* is described. The method is specifically designed to prepare pDNA vectors for gene therapy and DNA vaccination applications. The method comprises alkaline lysis, concentration with isopropanol, prepurification by $(NH_4)_2SO_4$ precipitation and purification by hydrophobic interaction chromatography (HIC), and desalting in gravity-operated 10-mL plastic columns. Analytical techniques used to control the performance of the method and to assess the quality of the pDNA vaccine are also described. Anion-exchange HPLC is used to determine pDNA concentration, whereas the presence of impurities such as RNA, proteins, *E. coli* genomic DNA, and endotoxins is determined by agarose gel electrophoresis, micro-bicinchoninic acid (BCA), real-time polymerase chain reaction and kinetic-quantitative chromogenic *lymulus amoebacyte lysate* (QCL LAL) assays, respectively. The method performs very well in terms of yield, purity and biological activity of the final pDNA. Furthermore, it is rapid, very easy to perform, and cost-efficient.

Key Words: DNA vaccines; endotoxins; gene therapy; hydrophobic interaction chromatography; plasmid DNA; purification.

1. Introduction
1.1. Plasmids for Gene Therapy

The preparation of plasmid DNA (pDNA) is a crucial laboratory task for applications such as cloning, transfection, and enzymatic modification *(1)*. pDNA can also be used as a nonviral vector to deliver therapeutic genes to human cells and tissues in gene therapy and DNA vaccination procedures *(2)*. pDNA for gene therapy experiments (e.g., cell transfection, animal studies) should be free from contaminants and host impurities in order to minimize pathogenic effects

From: *Methods in Biotechnology, Vol. 18: Microbial Processes and Products*
Edited by: J. L. Barredo © Humana Press Inc., Totowa, NJ

and adverse reactions *(3)*. Endotoxin or lipopolysaccharide (LPS) contamination is a major concern for in vivo gene therapy experiments with mammalian systems, because it could cause immunoresponsive reactions, induce a characteristic shock syndrome, and lead to low transfection efficiency *(3,4)*.

1.2. Existing Methods and Drawbacks

Although adequate for most molecular biology applications, many bench-scale methods fail to yield pDNA with an adequate purity for gene therapy experiments. The inability to reduce LPS to acceptable levels and the use of toxic compounds (e.g., phenol, chloroform, CsCl) and animal-derived enzymes (lysozyme, protease K, RNase), which often raises safety concerns with regulatory agencies, are common drawbacks of many pDNA protocols. Methods for the purification of pDNA comprehend a series of steps designed to release pDNA from the host *(Escherichia coli)* and remove impurities such as LPS, RNA, denatured pDNA, and genomic DNA (gDNA). Cells are usually disrupted by the alkaline lysis method *(5)* and variations thereof, which generate a clear lysate. This lysate can be concentrated and partially purified by extraction with organic solvents (e.g., phenol, chloroform, isoamylic alcohol) or by precipitation with isopropanol or ethanol *(1)*. This partially purified lysate is further purified by cleanup procedures that bind pDNA to solid matrices such as membranes and porous or nonporous beads packed in gravity flow or spin columns *(6)*. A fraction of the impurities is washed away readily if an adequate buffer is used. pDNA is then displaced and separated from co-bound impurities by selective elution. Most commercial kits use anion exchange matrices (silica based or polymeric) to bind the anionic pDNA under adequate salt and pH conditions *(6)*. Impurities are washed away with medium-salt buffers, the pDNA is eluted with high salt, precipitated with an appropriate agent and redissolved in a buffer. The use of anion exchange matrices has some limitations. LPS, RNA and gDNA often co-elute under the salt conditions used for pDNA release. The cleavage of RNA with RNase during lysis usually tackles RNA co-elution, but LPS and gDNA contamination is more difficult to deal with. The low capacity of many matrixes for pDNA binding is also an important limitation if large amounts of pDNA are needed. Many manufacturers also suggest discarding the matrices after each use due to limited time stability, a procedure that can be very costly.

1.3. Method Outline

The method described here for pDNA purification overcomes many of the above limitations. A key feature is the use of a hydrophobic interaction chromatography (HIC) matrix to bind impurities from cell lysates. Promoting the binding of impurities instead of pDNA is an innovation when compared to conventional wisdom *(6)*. The method was developed by scaling-down a large-scale process used for the preparative purification of pDNA *(7,8)*. Following alkaline lysis, pDNA is concentrated by precipitation with isopropanol. After resuspension, ammonium sulfate precipitation is used to reduce RNA and endotoxin levels, and to condition the solution prior to HIC *(7,8)*. HIC takes advantage of the more hydrophobic character of RNA, gDNA and LPS, when compared with double stranded pDNA *(7,8)*. pDNA is rapidly eluted with a high concentration of ammonium sulfate while bound impurities are removed afterwards by decreasing the ionic strength. Single-stranded RNA molecules bind to the matrix through the exposed bases, independently of their size. The single-stranded gDNA fragments that arise from lysis interact similarly with the HIC matrix. LPS, bind to the matrix via their hydrophobic lipidic moiety (lipid A). Other single-stranded nucleic acids with a high exposure of the bases, such as denatured pDNA and single-stranded oligonucleotides, also bind *(9)*. The gel filtration properties of the HIC matrix can be further used to perform a buffer exchange.

1.4. Support Methods

Analytical techniques used to control the performance of the method and to assess the quality of the pDNA are also described here. Anion-exchange HPLC is used to determine pDNA concentration, whereas the presence of impurities such as RNA, proteins, gDNA, and endotoxins is determined by agarose gel electrophoresis, micro-bicinchoninic acid (BCA), real-time polymerase chain reaction (PCR), and kinetic-QCL LAL assays, respectively.

2. Materials

1. Dimethyl sulfoxide (DMSO)-competent *E. coli* DH5α cells harboring the pDNA to be purified.
2. Luria–Bertani (LB) medium: 10 g/L tryptone, 5 g/L yeast extract, and 10 g/L NaCl.
3. Appropriate selective antibiotic.
4. Resuspension buffer: 50 m*M* glucose, 25 m*M* Tris-HCl, pH 8.0, and 10 m*M* ethylene diaminetetraacetic acid (EDTA).
5. Lysis buffer: 200 m*M* NaOH and 1% (w/v) sodium dodecyl sulfate (SDS).
6. Neutralization buffer: 3 *M* potassium acetate, pH 5.0.
7. 0.6 *M* NaOH, 2 mg/mL of sodium borohydride, and 1 *M* NaOH.
8. Sepharose CL-6B (Amersham Biosciences, Uppsala, Sweden).
9. Econo-Pac polypropylene disposable columns 20 mL (1.5 cm × 11 cm), including flow adapters (Bio-Rad, Hercules, CA).
10. Elution buffer 1 for HIC: 1.5 *M* ammonium sulfate and 10 m*M* Tris-HCl, pH 8.0.
11. Elution buffer 2 for HIC: 10 m*M* Tris-HCl, pH 8.0.
12. Phosphate-buffered saline (PBS) solution: 8 g/L NaCl, 1.44 g/L Na_2HPO_4, and 0.24 g/L KH_2PO_4, pH 7.4 *(1)*.
13. HPLC apparatus with UV detector at 260 nm.

14. POROS 20 PI weak anion-exchange resin (Perseptive Biosystem, Foster City, CA).
15. Elution buffer 1 for anion-exchange HPLC: 2 M NaCl and 10 mM Tris-HCl, pH 8.0.
16. Elution buffer 2 for anion-exchange HPLC: 10 mM Tris-HCl, pH 8.0.
17. RNase (DNase-free).
18. Pure pDNA standards (5–50 µg/mL).
19. Horizontal agarose gel electrophoresis equipment.
20. Agarose for electrophoresis.
21. TAE buffer: 40 mM Tris base, 20 mM acetic acid, pH 8.0, and 1 mM EDTA *(1)*.
22. Single-cutting restriction enzyme for the pDNA to be purified.
23. Molecular-weight marker λDNA/*Hind*III.
24. Ethidium bromide solution 10 mg/mL (*see* **Note 1**).
25. Sodium deoxycholate (DOC) 0.15% (w/v).
26. Trichloroacetic acid (TCA) 72% (w/v) (*see* **Note 2**).
27. SDS 5% (w/v) and 0.1 M NaOH.
28. BCA working reagents A and B (Pierce, Rockford, IL). Working solution: 50 parts of reagent A : 1 part of reagent B.
29. Bovine serum albumin (BSA) standards (10–100 µg/mL).
30. *E. coli* DH5α gDNA standard.
31. "Light Cycler-FastStart DNA master Syber Green I" mixture, glass capillaries, MgCl$_2$ solution, and LightCycler equipment for real-time PCR (Roche, Hercules, CA).
32. Forward primer: 5′-ACA CGG TCC AGA ACT CCT ACG-3′.
33. Reverse primer: 5′-GCC GGT GCT TCT TCT GCG GGT AAC GTC A-3′.
34. DMSO-competent *E. coli* DH5α cells.
35. Kinetic-QCL LAL assay kit and respective equipment (Biowhittaker, Walkersville, MD).

3. Methods

The following text describes the methods used to perform (1) the *E. coli* DH5α transformation and bacterial culture, (2) the isolation of pDNA, (3) the quantification of the pDNA, (4) the assessment of the pDNA purity, and (5) the assessment of pDNA transformation efficiency.

3.1. Transformation of E. coli DH5α Cells and Bacterial Culture

1. Transform *E. coli* DH5α cells with the target pDNA using standard molecular-biology protocols *(1)*.
2. Plate cells on a LB/agar plate *(1)* containing the appropriate selective antibiotic and incubate overnight at 37°C.
3. Pick a single colony and grow overnight at 37°C and 250 rpm in a 100-mL shake flask containing 25 mL of LB medium with the appropriate selective antibiotic.
4. Inoculate 200 mL of LB medium containing the appropriate selective antibiotic with the overnight culture and grow at 37°C and 250 rpm in a 500-mL shake flask.
5. Harvest the cells at late log phase and centrifuge the suspension at 14,300g during 15 min at 4°C. Discard the supernatant.

3.2. Isolation of pDNA

3.2.1. Alkaline Lysis

1. Resuspend the cell pellet by adding 8 mL of resuspension buffer.
2. Add 8 mL of lysis buffer to the cell suspension in order to disrupt cells and denature gDNA and proteins.
3. After careful homogenization, keep the mixture on ice for 10 min. The solution obtained after adding the lysis buffer must be handled very carefully in order to avoid breaking gDNA. The solution should not be vortex-mixed but gently homogenized by inverting the container four to six times.
4. Precipitate cellular debris, gDNA, proteins, and other impurities by adding 8 mL of prechilled neutralisation buffer.
5. After careful homogenization, keep the mixture on ice for 10 min in order to maximize the formation of insoluble material (flocs and precipitates).
6. Separate the solid material from the supernatant by centrifuging twice at 30,200g for 30 min at 4°C. Fine particulates can be removed by filtration through filter paper. Keep the supernatant with pDNA for further handling.

3.2.2. Concentration

1. Precipitate the pDNA contained in the 24 mL of supernatant by adding 0.7 vol of cold isopropanol (16.8 mL).
2. Centrifuge the solution at 39,200g for 30 min at 4°C. Discard the supernatant.
3. Wash the pellet by adding 2 mL of 70% ethanol (w/w). The purpose is to remove precipitated salts from alkaline lysis that make resuspension of the pellet in a low volume very difficult.
4. Recover the pellet by centrifugation at 39,200g for 10 min at 4°C and air-dry it for 5–10 min.

3.2.3. Clarification

1. Resuspend the washed pellet in 500 μL of 10 mM Tris-HCl, pH 8.0. In some cases, this operation could be difficult to perform.
2. Precipitate impurities (proteins, RNA, and LPS) in the solution by adding solid ammonium sulfate up to 2.5 M. After solubilization of the ammonium sulfate, the volume of the clarified solution increases.
3. Incubate the resulting solution on ice for 15 min.
4. Remove precipitate by centrifugation at 20,937g for 15 min at 4°C. Decant the supernatant (a pipet tip might destabilize the pellet) and keep it for further handling.

3.2.4. Preparation of the HIC Gel (see **Note 3**)

This protocol describes the preparation of 5 g of wet HIC gel.

1. Wash 5 g of suction-dried Sepharose CL-6B gel with water on a glass filter funnel.
2. Mix the gel with 5 mL of 1,4-butanediol-diglycidylether.

3. Add 5 mL of 0.6 M NaOH containing 2 mg/mL of sodium borohydride.
4. Mix the suspension by orbital shaking for 8 h at 25°C.
5. Stop the reaction by abundantly rinsing the gel on a glass filter funnel with water.
6. Keep the gel overnight at room temperature in 1 M NaOH and then wash it with water.

3.2.5. Purification of pDNA by HIC

1. Pack a 20-mL column with 10 mL of the HIC gel (*see* **Note 4**). Carefully place the porous glass flow adapter on top of the gel and position the column vertically.
2. Equilibrate the column by gravity flow with 20 mL of 1.5 M ammonium sulfate in 10 mM Tris-HCl, pH 8.0.
3. Load the pDNA-containing supernatant obtained after ammonium sulfate precipitation (approx 500 µL) onto the column (*see* **Note 5**).
4. Perform an isocratic elution by adding 5.5 mL of 1.5 M ammonium sulfate in 10 mM Tris-HCl, pH 8.0. This will elute the pDNA. Simultaneously collect the eluent at the column outlet as 2-mL fractions in individual tubes adequately labeled.
5. Add 14.5 mL of 10 mM Tris-HCl, pH 8.0, in order to elute bound and weakly retained species.
6. Measure the absorbance at 260 nm of the fractions collected at the column outlet and draw the corresponding chromatogram.
7. Wash the column with 10 mL 0.5 M NaOH and rinse abundantly with water prior to reuse.

A typical chromatogram obtained after loading 109 µg of pDNA is shown in **Fig. 1** (*see* **Note 6**). Analysis of the fractions showed that a high percentage of the injected pDNA is concentrated in a 2-mL fraction collected after discarding the first 3.5 mL of eluent. This fraction is kept for further desalting.

3.2.6. Desalting

1. Pack another 20-mL column with 10 mL of the HIC gel (*see* **Note 7**). Carefully place the porous glass flow adapter on top of the gel and position the column vertically.
2. Equilibrate the column by gravity flow with 20 mL of PBS buffer (*see* **Note 8**).
3. Load the pDNA pool (2 mL) obtained after HIC and allow it to enter the flow adapter placed on the top of the gel (*see* **Note 5**).
4. Perform an isocratic elution by adding 20 mL of PBS buffer (*see* **Note 8**). This will separate pDNA from ammonium sulfate. Collect the eluent at the column outlet as 2-mL fractions in individual tubes adequately labeled.
5. Measure the absorbance at 260 nm and the conductivity of the fractions and draw the corresponding chromatogram.
6. Rinse the column with buffer prior to reuse and wash periodically with 0.5 M NaOH.

The chromatogram obtained (*see* **Fig. 2**) shows that pDNA is completely separated from the salt. After discarding the first 3.5 mL of eluent, a high per-

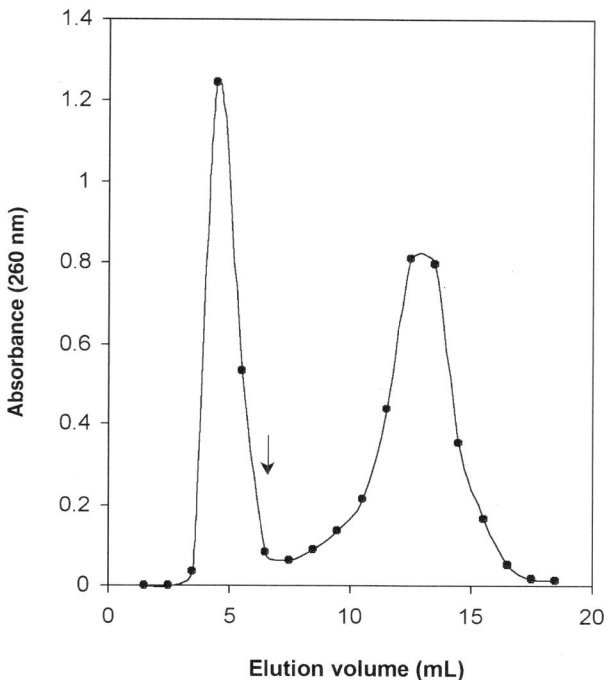

Fig. 1. Purification of 109 μg of pDNA by HIC. Impurities were eluted (↓) by reducing the concentration of ammonium sulfate in the eluent from 1.5 *M* to 0 *M*. The first peak corresponds to purified pDNA that was collected in a 2-mL fraction after discarding the first 3.5 mL of eluent.

centage of the desalted pDNA is recovered in a 3-mL fraction. Use pDNA or store at 4°C.

3.3. pDNA Quantification

If required, the concentration of pDNA in the solutions obtained after each step in the method can be accurately determined by HPLC, as follows:

1. Pack a 4.6 × 10-cm polyetheretherketone (PEEK) column with POROS 20 PI weak anion-exchange resin (*see* **Note 9**).
2. Connect the column to a HPLC system equipped with an UV detector to measure the absorbance of the exiting eluent at 260 nm.
3. Equilibrate the column with 0.7 *M* NaCl in 10 m*M* Tris-HCl, pH 8.0, at 1 mL/min.
4. Incubate the samples to be analyzed with 10 μL of RNase (DNase-free) during 30 min at 37°C prior to HPLC analysis.

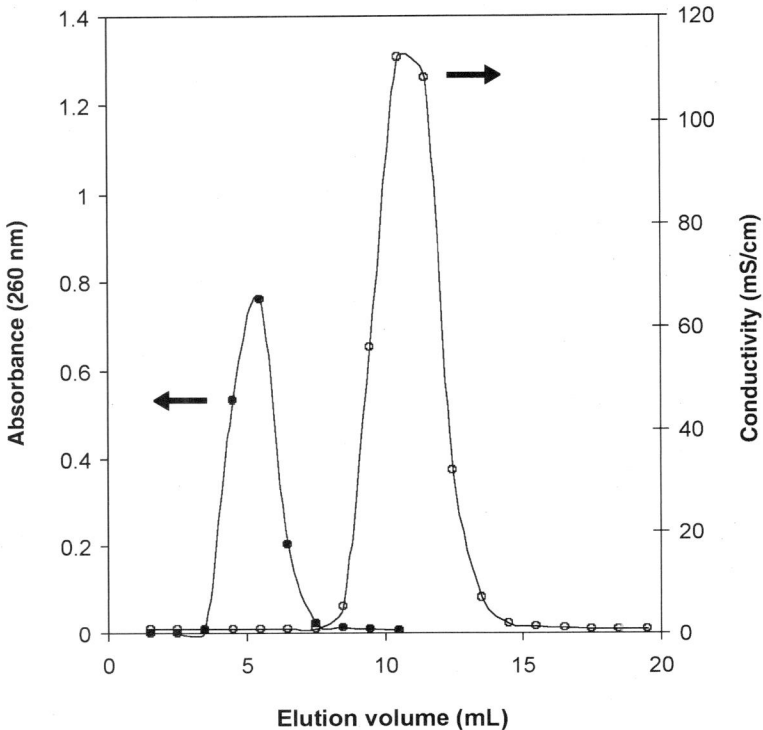

Fig. 2. Desalting of 76 μg of pDNA. The purified pDNA (●) is separated from 1.5 *M* ammonium sulfate (○) and collected in a 3-mL fraction after discarding the first 3.5 mL of eluent.

5. Inject the sample (100 μL) onto the column at a flow rate of 1 mL/min.
6. Keep the salt concentration at 0.7 *M* NaCl during the first 7.5 min. In the case of heavily contaminated samples, such as the clarified lysis solution, keep the salt concentration at 0.7 *M* NaCl for 12.5 min.
7. Increase the salt concentration linearly from 0.7 *M* to 2 *M* NaCl in 5 min.
8. Keep the salt concentration at 2 *M* NaCl during the next 2.5 min in order to elute all bound species.
9. Decrease the salt concentration to the initial value of 0.7 *M* NaCl and keep this salt concentration during the next 2.5 min in order to re-equilibrate the column for the next analysis.
10. Construct a calibration curve relating the known concentration of pure pDNA standards with the area of the corresponding peaks in the chromatogram.

The HPLC chromatograms of samples kept after performing each method step are shown in **Fig. 3.** The typical pDNA step yields are as follows: 70% for

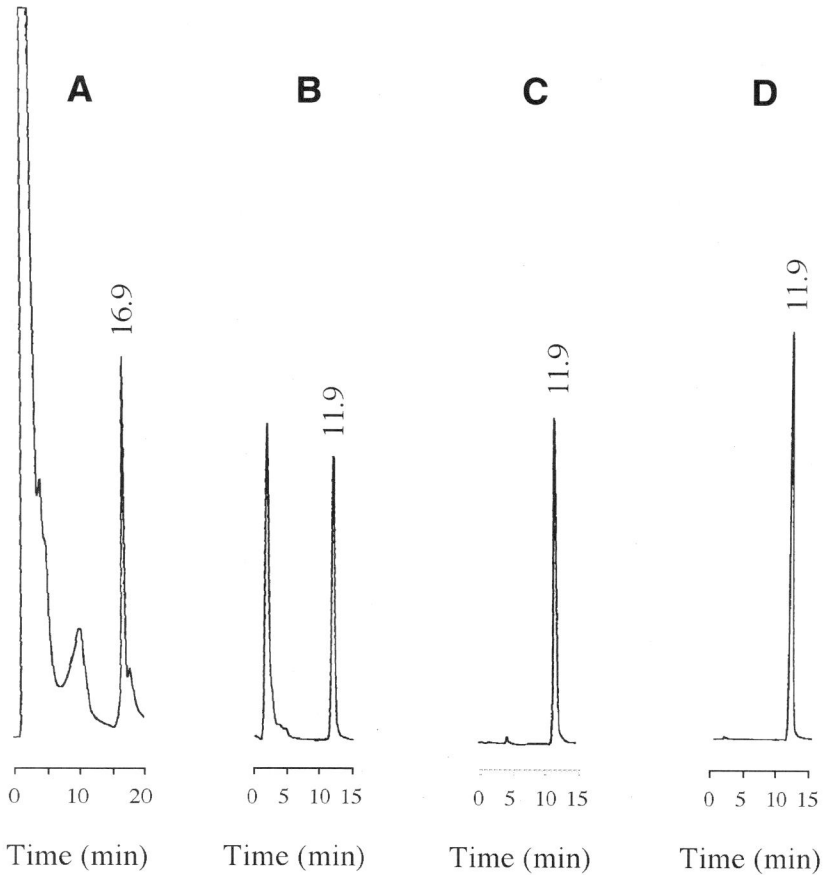

Fig. 3. Anion-exchange HPLC analysis of pDNA containing samples collected throughout the isolation process: **(A)** neutralized lysate, **(B)** pDNA solution prior to HIC, **(C)** pDNA pool after HIC, and **(D)** pDNA pool after desalting. The peaks at 1.66, 1.78, 4.14, 10.08, and 17.88 min correspond to impurities, whereas the sharp peaks at 16.9 **(A)**, 11.9 **(B, D)**, and 11.9 **(C)** min correspond to pDNA.

the precipitation with ammonium sulfate, 81% for HIC, and 95% for desalting. Overall pDNA yield of the method is 54%.

3.4. Assessment of pDNA Quality

A rigorous control of the quality of the final pDNA might be needed, especially if cell transfection and animal model studies are envisaged. The content of impurities such as RNA, host proteins, gDNA, and endotoxins can be determined as described next.

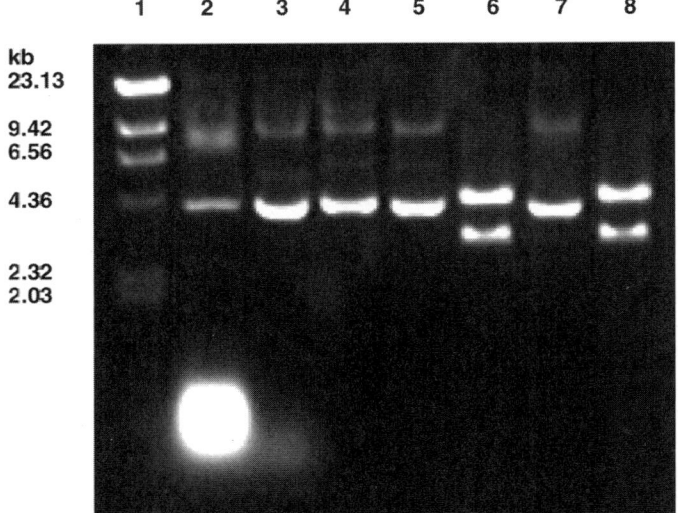

Fig. 4. Agarose gel electrophoresis analysis of samples kept after neutralization *(lane 2)*, clarification with ammonium sulfate *(lane 3)*, HIC *(lane 4)*, and desalting *(lane 5)*. *Lane 6* shows pure pDNA digested with a single-cutting restriction enzyme. pDNA obtained with a standard commercial kit is also shown, undigested *(lane 7)* and digested with a restriction enzyme *(lane 8)*. *Lane 1:* molecular-weight markers.

3.4.1. HPLC Purity

After performing a HPLC analysis (as described in **Subheading 3.3.**), the HPLC purity can be defined as the percentage of the plasmid peak area in the overall chromatogram (*see* **Fig. 3**). Although only the components that absorb at 260 nm are accounted for, this constitutes a very good indicator of purity. The HPLC purity of samples kept after each method step were as follow: 5.3% after isopropanol precipitation, 40% after precipitation with ammonium sulfate, and 100% after HIC and desalting.

3.4.2. RNA and pDNA Variants

The presence of RNA and nonsupercoiled pDNA isoforms can be easily determined by horizontal agarose gel electrophoresis.

1. Cast a 1% agarose gel with an adequate number of wells and place it in the horizontal electrophoresis unit.
2. Add loading buffer to samples and apply an adequate amount to the gel wells.
3. Run for 20 h in TAE buffer at 1.5 V/cm and stain the gel for 30 min in 1 µg/mL ethidium bromide.

A typical gel (*see* **Fig. 4**) shows about 60% RNA contamination after precipitation with ammonium sulfate. After desalting, however, RNA is undetectable. In this case, the percentage of supercoiled pDNA in all samples is higher than 90%, as recommended by regulatory agencies. The identity of the purified pDNA can be confirmed by performing restriction analysis and comparing the obtained pattern with the restriction map.

3.4.3. Proteins

Protein impurities can be determined by the micro-BCA assay, modified to include a previous precipitation of proteins with TCA. The purpose of this alteration is to eliminate interference by substances used during purification (e.g., ammonium sulfate). A calibration curve is made with BSA standards from 10 to 100 μg/mL.

1. Add 800 μL of distilled H_2O and 100 μL of DOC to 100 μL of sample in a microcentrifuge tube (1.5 mL).
2. Incubate the mixture for 10 min at room temperature.
3. Add 100 μL of 72% TCA (w/v) and vortex until complete homogenization.
4. Centrifuge the mixture at 3000*g* for 15 min at room temperature.
5. Aspire off the supernatant using a needle attached to a vacuum flask.
6. Solubilize the pellets in 50 μL of SDS 5% (w/v) with 0.1 *N* NaOH.
7. Add 200 μL of BCA working solution and vortex the mixture.
8. Incubate the mixture at 60°C for 30 min.
9. Read the absorbance of 200 μL of the mixture at 595 nm in a microplate reader.

A typical protein content of 0.032 μg/μg pDNA is measured after ammonium sulfate precipitation, but no protein contamination is detected on the final pDNA product.

3.4.4. gDNA

Contamination by gDNA can be assessed by amplifying a 181-bp fragment of the 16S rRNA gene from the *E. coli* DH5α gDNA, using real-time PCR in a "Light Cycler" *(10)* (*see* **Note 10**).

1. Prepare a reaction mixture (20 μL) containing 0.75 μ*M* of each primer (forward and reverse), 3 m*M* $MgCl_2$, and 2 μL of the "Light Cycler-FastStart DNA master Syber Green I" inside a glass capillary.
2. Program the following PCR cycle: denaturation for 10 min at 95°C (only before the first cycle), denaturation for 10 s at 95°C, annealing for 5 s at 65°C, and extension for 7 s at 72°C.
3. Amplify gDNA standards and samples with 35 cycles. Samples are usually diluted between 1000-fold and 8000-fold.
4. Follow the change in fluorescence during the annealing step.

Standard curves were constructed by a serial dilution of *E. coli* DH5α gDNA (1.4 pg to 140 ng). The gDNA content on the final pDNA product was determined to be 1.9 ng/µg pDNA. This value is well below the maximum of 10 ng of gDNA/µg pDNA recommended for therapeutic pDNA.

3.4.5. Endotoxins

Endotoxin contamination can be assessed with the kinetic-QCL LAL assay kit. The detection level of the method is 0.005 EU/mL. The endotoxin content in the final pDNA was determined to be 0.000241 EU/µg pDNA (*see* **Note 11**). This level of contamination is excellent because therapeutic pDNA should possess less than 0.1 EU/µg pDNA.

3.5. Assessment of pDNA Transformation Efficiency

The *E. coli* DH5α DMSO-competent cells were transformed with the purified pDNA using a standard protocol *(1)*. A control study was performed using a pDNA purified with a commercial kit.

Calculate the transformation efficiency by platting 100 µL of 1 : 100 and 1 : 1000 dilutions of the transformed cells suspension on LB plates containing the appropriate selective antibiotic.

The transformation efficiency obtained with the purified and control pDNA is typically similar and around 1.4×10^6 colony-forming units (cfu)/µg pDNA. This confirms the adequacy of the method to prepare biologically active pDNA.

4. Notes

1. Ethidium bromide is mutagenic.
2. Trichloroacetic acid is light sensitive.
3. Similar results are obtained with the commercial HIC resin Phenyl-Sepharose (Amersham Biosciences). In this case, go directly to **Subheading 3.2.5.**
4. The method will work as well with larger-diameter columns packed up to the same height (approx 6 cm), as long as the packed gel is uniform.
5. Each buffer added to the columns should be allowed to enter the flow adapter (a porous glass disk) placed on the top of the gel before the next buffer is added. If this advice is not observed, mixing of buffers will occur and a poor separation will be obtained. As soon as the liquid has entered the adapter, the next buffer can be added. This should be performed immediately; otherwise, the gel on the top of the column could dry.
6. Up to 500 µg of pDNA can be loaded onto this column without compromising the purity of the final pDNA. Higher pDNA amounts can be eventually purified in one HIC run, but it is advisable to check the final purity, as the gel capacity for impurities might have been exceeded. The sample volume, however, should not be increased. In this case columns, with larger diameters can be used (*see* **Note 4**).

7. Any chromatographic resin suitable for desalting can be used to separate ammonium sulfate from pDNA.

8. Because this is the last step in the method, a buffer that is suitable for the intended application of the pDNA can be used instead of PBS, as long as it is compatible with the desalting resin used.

9. The HPLC analysis can be performed using a different chromatographic resin. An HIC–HPLC technique was recently described *(11)* using Phenyl-Sepharose Source (Amersham Biosciences). The method is very rapid, does not require the digestion of the samples with RNase, is capable of handling heavily contaminated solutions, and yields very reproducible results.

10. The gDNA contamination in the final pDNA product can also be determined by the Southern slot blot assay *(7,8,10)* if a real-time PCR apparatus is not available. In fact, our real-time PCR results were confirmed by this method. However, the Southern slot blot method is highly time-consuming and presents a lower sensitivity; this is an important disadvantage in the case of very pure samples.

11. No special precautions are needed with buffers and equipment to routinely obtain such low levels of endotoxin contamination. Major endotoxin clearance is obtained with the ammonium sulfate step, with HIC taking care of residual contamination.

Acknowledgments

This work was partially supported by the Portuguese Ministry of Science & Technology (POCTI/BIO/43620/2000 and PhD grant BD/21241/99 to M. M. Diogo). The authors would like to thank Sofia Martins and Gabriel Monteiro for performing gDNA analysis.

References

1. Sambrook, J., Fritsch, E. F., and Maniatis, T. (1989) *Molecular Cloning: A Laboratory Manual,* 2nd ed., Cold Spring Harbor Laboratory Press, Cold Spring Harbor, NY.

2. Ledley, F. D. (1995) Nonviral gene therapy: the promise of genes as pharmaceutical products. *Hum. Gene Ther.* **6,** 1129–1144.

3. Wicks, I. P., Howell, M. L., Hancock, T., Kohsaka, H., Olee, T., and Carson, D. A. (1995) Bacterial lipopolysaccharides copurifies with plasmid DNA: implications for animal models and human gene therapy. *Hum. Gene Ther.* **6,** 317–323.

4. Butash, K. A., Natarajan, P., Young, A., and Fox, D. K. (2000) Reexamination of the effect of endotoxin on cell proliferation and transfection efficiency. *BioTechniques* **29,** 610–619.

5. Birnboim, H. C. and Doly, J. (1979) A rapid alkaline extraction procedure for screening recombinant plasmid DNA. *Nucleic Acids Res.* **7,** 1513–1523.

6. DeFrancesco, L. (1997) Mini Preps are here to stay. *Scientist* **11,** 22–25.

7. Diogo, M. M., Queiroz, J. A., Monteiro, G. A., et al. (2000) Purification of a cystic fibrosis plasmid vector for gene therapy using hydrophobic interaction chromatography. *Biotechnol. Bioeng.* **68,** 576–583.

8. Diogo, M. M., Ribeiro, S., Queiroz, J. A., et al. (2001) Production, purification and analysis of an experimental DNA vaccine against rabies. *J. Gene Med.* **3,** 577–584.

9. Diogo, M. M., Queiroz, J. A., Monteiro, G. A., and Prazeres, D. M. F. (2000) Separation and analysis of plasmid denatured forms using hydrophobic interaction chromatography. *Anal. Biochem.* **275,** 122–124.

10. Martins, S. A. M., Prazeres, D. M. F., Cabral, J. M. S., and Monteiro, G. A. (2003) Comparison of real-time polymerase chain reaction and hybridization assays for the detection of *Escherichia coli* genomic DNA in process samples and pharmaceutical-grade plasmid products. *Anal. Biochem.* **322,** 127–129.

11. Diogo, M. M., Queiroz, J. A., and Prazeres, D. M. F. (2003) Assessment of purity and quantification of plasmid DNA in process solutions using high performance hydrophobic interaction chromatography. *J. Chromatogr. A* **998,** 109–117.

9

Genome Breeding of an Amino Acid-Producing *Corynebacterium glutamicum* Mutant

Masato Ikeda, Junko Ohnishi, and Satoshi Mitsuhashi

Summary

The classical strain breeding method based on random mutation and selection cannot avoid introducing detrimental or unnecessary mutations into the genome. A methodology that overcomes the limitations of the classical method is "genome breeding." In this approach, biotechnologically useful mutations identified through the genome analysis of classical mutants are systematically introduced into the wild-type genome in a pinpointed manner, thus allowing creation of a defined mutant that carries only useful mutations. This methodology was applied to generate an efficient L-lysine-producing mutant of *Corynebacterium glutamicum*. Introduction of the Val-59→Ala mutation in the homoserine dehydrogenase gene and the Thr-311→Ile mutation in the aspartokinase gene into the wild-type strain by allelic replacement resulted in accumulation of 8 and 55 g/L of L-lysine, respectively. The two mutations were then reconstituted on the wild-type genome, which led to a synergistic effect on production and accumulation of 75 g/L of L-lysine in a relatively short period of cultivation. The procedure and the impact of this methodology are described.

Key Words: Genome breeding; *Corynebacterium glutamicum;* amino acid production; L-lysine; homoserine dehydrogenase; aspartokinase.

1. Introduction

Industrial fermentation strains of useful metabolites such as amino acids and nucleotides have generally been constructed by repeating random mutation and selection *(1)*. This classical method has greatly contributed to the progress of the fermentation industry, but has serious disadvantages. In this approach, chemical mutagens (e.g., nitrosoguanidine) or ultraviolet (UV) are used to induce mutants that exhibit improved production. Consequently, uncharacterized secondary mutations are introduced into the genome and accumulated in one background at every cycle of mutagenesis. Because of these undesirable

From: *Methods in Biotechnology, Vol. 18: Microbial Processes and Products*
Edited by: J. L. Barredo © Humana Press Inc., Totowa, NJ

mutations, production strains are generally inferior to corresponding wild-type strains in biotechnologically important properties such as growth and sugar consumption, which hampers the establishment of highly productive processes. In addition, the production mechanism remains a black box and efficient improvement has become difficult.

Currently, we have entered the genomic era. The availability of genomic data in industrial organisms has allowed polymerase chain reaction (PCR)-based cloning and sequencing of any desired genes of production strains derived from the organisms. By comparing the two sequences from a classically derived producer and its parental wild type, it has become possible to decipher the results obtained from mutation-selection and define the genetic basis for high-level production. These studies have now allowed us to take a novel approach employing genomic information to reconstruct an existing production strain *(2–4)*. This genome-based strain breeding, which we call "genome breeding," creates a "minimal mutation strain" through characterization and reconstitution of a mutation set essential for high-level production.

The subject here involves the genome breeding of L-lysine-producing *Corynebacterium glutamicum*. Our previous analysis of a classically derived L-lysine producer of *C. glutamicum* has already identified two specific mutations, the Val-59→Ala mutation in the homoserine dehydrogenase gene *(hom)* and the Thr-311→Ile mutation in the aspartokinase gene *(lysC)*, as key mutations for L-lysine production *(3)*. Following the findings, this chapter demonstrates generation of an efficient L-lysine producer by reconstitution of those key mutations in the wild-type genome.

2. Materials

1. *Escherichia coli* vector pHSG299 (Takara Bio, Otsu, Japan).
2. Levansucrase gene *sacB* of *Bacillus subtilis* *(5)*.
3. *Corynebacterium glutamicum* wild-type strain ATCC 13032 (ATCC, Manassas, USA).
4. Oligonucleotide primer *hom*-a: 5′-ATGGTCATGGTGAAGGCCTG-3′.
5. Oligonucleotide primer *hom*-b: 5′-GATATCAGAA_G_CAGCAATGCC-3′. Underlined G represents the intended mutation.
6. Oligonucleotide primer *hom*-c: 5′-GGCATTGCTG_C_TTCTGATATC-3′. Underlined C represents the intended mutation.
7. Oligonucleotide primer *hom*-d: 5′-GGTGTGGAAAGCGATGGATG-3′.
8. Oligonucleotide primer *lysC*-a: 5′-GCGATGTCACCACGTTGGGT-3′.
9. Oligonucleotide primer *lysC*-b: 5′-GCAGGTGAAG_A_TGATGTCGGT-3′. Underlined A represents the intended mutation.
10. Oligonucleotide primer *lysC*-c: 5′-ACCGACATCA_T_CTTCACCTGC-3′. Underlined T represents the intended mutation.
11. Oligonucleotide primer *lysC*-d: 5′-ACCTCGATTTCCGTGCCACG-3′.

12. Oligonucleotide primer *lysC311*-F: 5'-GACGGCACCACCGACATCA<u>T</u>-3'. Underlined T represents the single-base mutation to discriminate mutants from wild-type sequences.

13. Oligonucleotide primer *lysC311*-R: 5'-AGACCAGCGGCATCGTGAAGTGGCT-3'.

14. Oligonucleotide primer *hom59*-F: 5'-AGGTTCGTGGCATTGCTG<u>C</u>-3'. Underlined C represents the single-base mutation to discriminate mutants from wild-type sequences.

15. Oligonucleotide primer *hom59*-R: 5'-CGGTTGGCGCGCCACCTGCAC-CGTTTCCGT-3'.

16. Restriction enzymes, T4 DNA ligase, and *Taq* polymerase.

17. QIAquick PCR purification kit (Qiagen, Valencia, CA).

18. DNA Blunting Kit (Takara Bio).

19. Agarose gel electrophoresis, PCR, and electroporation equipments.

20. Gene pulser and pulse controller.

21. Kanamycin.

22. BY medium: 0.3% NaCl, 0.5% yeast extract, 0.7% meat extract, and 1% peptone.

23. Flask production medium: 100 g/L glucose, 10 g/L corn steep liquor, 45 g/L $(NH_4)_2SO_4$, 2.5 g/L urea, 0.5 g/L KH_2PO_4, 0.5 g/L $MgSO_4·7H_2O$, 10 mg/L thiamine-HCl, 0.3 mg/L D-biotin, and 30 g/L $CaCO_3$. Adjust to pH 7.0.

24. LSS1 medium: 50 g/L sucrose, 40 g/L corn steep liquor, 8.3 g/L $(NH_4)_2SO_4$, 1 g/L urea, 2 g/L KH_2PO_4, 0.83 g/L $MgSO_4·7H_2O$, 10 mg/L $FeSO_4·7H_2O$, 1 mg/L $CuSO_4·5H_2O$, 10 mg/L $ZnSO_4·7H_2O$, 10 mg/L β-alanine, 5 mg/L nicotinic acid, 1.5 mg/L thiamine-HCl, 0.5 mg/L D-biotin, and 30 g/L $CaCO_3$. Adjust to pH 7.2.

25. LPG-1 medium: 50 g/L glucose, 20 g/L corn steep liquor, 25 g/L NH_4Cl, 1 g/L urea, 2.5 g/L KH_2PO_4, 0.75 g/L $MgSO_4·7H_2O$, 50 mg/L $FeSO_4·7H_2O$, 13 mg/L $MnSO_4·5H_2O$, 50 mg/L $CaCl_2·2H_2O$, 6.3 mg/L $CuSO_4·5H_2O$, 1.3 mg/L $ZnSO_4·7H_2O$, 5 mg/L $NiCl_2·6H_2O$, 1.3 mg/L $CoCl_2·6H_2O$, 1.3 mg/L $(NH_4)_6Mo_7O_{24}·4H_2O$, 23 mg/L β-alanine, 14 mg/L nicotinic acid, 7 mg/L thiamine-HCl, and 0.5 mg/L D-biotin. Adjust to pH 7.0.

26. Jar fermentors.

27. U-1080 Auto Sipper Photometer (Hitachi, Tokyo, Japan).

28. Determinar GL-E (Kyowa Medex, Tokyo, Japan).

29. HPLC (Shimadzu, Kyoto, Japan).

3. Methods

The following methods described outline (1) the construction of the recombinant plasmids for the replacement of the wild-type chromosomal gene sequences with the mutated gene sequences, (2) the generation of the defined mutants through the allelic replacement, and (3) the L-lysine production by the mutants.

3.1. Plasmid Construction

The construction of the recombinant plasmids containing the mutated *hom* and *lysC* sequences is described in **Subheadings 3.1.1.–3.1.3.** This includes (1)

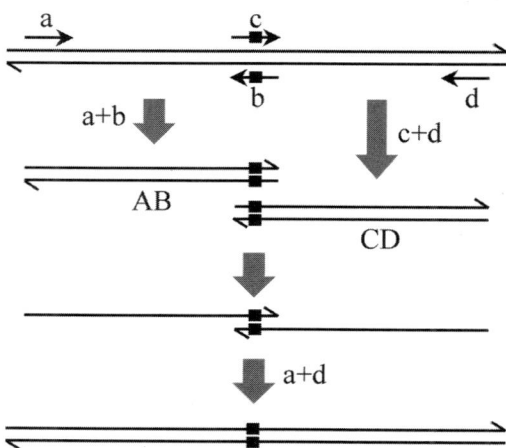

Fig. 1. PCR-mediated site-directed mutagenesis via overlap extension. The double-stranded DNA and primers are shown by lines with arrows indicating the 5′-to-3′ orientation. The site of the desired mutation is indicated by the black square.

the construction of the vector for gene replacement in *C. glutamicum*, (2) the construction of the recombinant plasmid for the replacement of the wild-type chromosomal *hom* sequence with the mutated sequence, and (3) the construction of the recombinant plasmid for the replacement of the wild-type chromosomal *lysC* sequence with the mutated sequence. DNA manipulations were performed according to standard procedures (*6*) and are not described here in detail because of space limitations. The DNA sequences of *hom* and *lysC* have been submitted to GenBank and have been assigned accession numbers AX123536 and AX120365, respectively (*see* **Note 1**).

3.1.1. Gene Replacement Vector

Plasmid pESB30 is a vector for the introduction of intended mutations into the *C. glutamicum* genome by two-step homologous recombination. It can be constructed by ligating a 2.6 kb *Pst*I DNA fragment containing *sacB* (*5*), the levansucrase gene of *Bacillus subtilis,* with *Pst*I-digested pHSG299 (*7*), an *E. coli* vector carrying the kanamycin gene. It carries only a replication origin of the *E. coli* vector pHSG299 and, thus, is nonreplicative in *C. glutamicum* (*see* **Note 2**).

3.1.2. Plasmid With Mutated hom Sequence

The generation of a DNA fragment containing the mutated *hom* sequence can be performed by the PCR-mediated site-directed mutagenesis via overlap extension, as shown in **Fig. 1.**

1. Amplify two segments of the *hom* sequence independently by PCR.
2. For each PCR reaction, use one flanking primer that hybridizes at one end of the target gene sequence (primer a or d in **Fig. 1**) and one internal primer that hybridizes at the site of the mutation and contains the desired mutation (a T to C exchange at position 176 in *hom,* leading to an amino acid replacement of Val-59 by Ala, designated *hom59*) in the center of the primer (primer b or c in **Fig. 1**). The following PCR primer sequences can be used for this purpose: *hom*-a, *hom*-b, *hom*-c, and *hom*-d. Because the internal primers *hom*-b and *hom*-c are complementary, the two fragments AB and CD generated in the first PCR have an overlap that can be fused by denaturing and annealing them in a subsequent primer extension reaction (*see* **Fig. 1**).
3. Add the outside primers *hom*-a and *hom*-d to the second PCR reaction and amplify the fused mutant product to generate the DNA fragment containing the mutated *hom* sequence.
4. Purify the resulting fragment using a QIAquick PCR purification kit and clone into pESB30 using the TA cloning method *(6)*.
5. The T vector can be made by digesting pESB30 with *Bam*HI, followed by blunting with DNA Blunting Kit and adding T to the blunt ends using *Taq* polymerase and dTTP. The recombinant plasmid thus obtained is designated pChom59 (*see* **Note 3**).

3.1.3. Plasmid With Mutated lysC Sequence

The generation of a DNA fragment containing the mutated *lysC* sequence can be also performed according to the scheme shown in **Fig. 1.**

1. Amplify two segments of the *lysC* sequence independently by PCR.
2. For each PCR reaction, use one flanking primer that hybridizes at one end of the target gene sequence (primers a or d in **Fig. 1**) and one internal primer that hybridizes at the site of the mutation and contains the desired mutation (a C to T exchange at position 932 in *lysC,* leading to replacement of Thr-311 by Ile, designated *lysC311*) in the center of the primer (primer b or c in **Fig. 1**).
3. The following PCR primer sequences can be used for this purpose: *lysC*-a, *lysC*-b, *lysC*-c, and *lysC*-d.
4. Add the outside primers *lysC*-a and *lysC*-d to the second PCR reaction and amplify the fused mutant product to generate the DNA fragment containing the mutated *lysC* sequence.
5. Purify the resulting fragment using a QIAquick PCR purification kit and clone into pESB30 using the TA cloning method as described in **Subheading 3.1.2.** The recombinant plasmid thus obtained is designated pClysC311.

3.2. Mutant Generation

The *hom59* and *lysC311* mutations on pChom59 and pClysC311, respectively, can be introduced into *C. glutamicum* ATCC 13032 through two recombination events. **Figure 2** schematically illustrates the recombination event and demonstrates how it leads to introduction of each mutation into the corresponding gene on the genome.

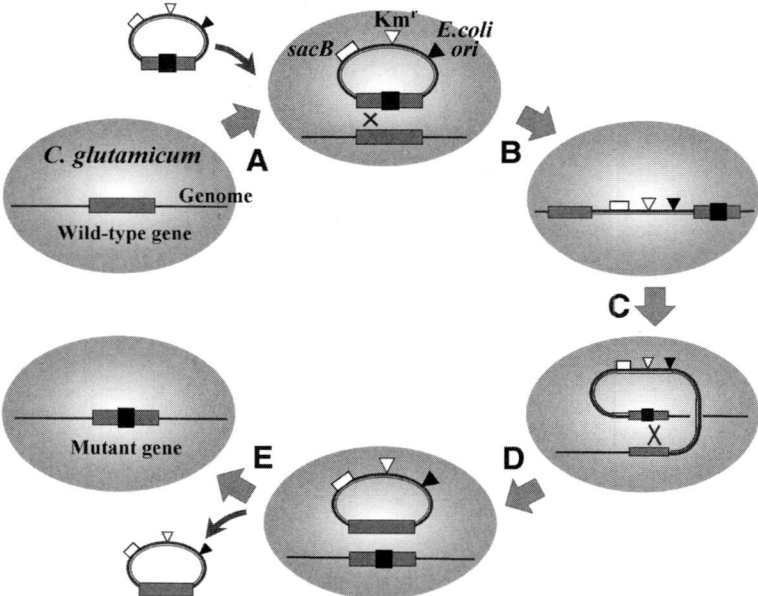

Fig. 2. Schematic diagram of the homologous recombination event leading to the allelic replacement. The site of the desired mutation is indicated by the black square.

3.2.1. Single Mutant

1. Transform *C. glutamicum* ATCC 13032 with plasmid pChom59 or pClysC311 by electroporation *(8)*, using the gene pulser and pulse controller.
2. Plate the cells on a BY plate containing kanamycin (20 µg/mL) and incubate for 1 d at 30°C. The plasmids cannot replicate in *C. glutamicum*. Transformation of each plasmid and subsequent selection for the plasmid marker yield transformants that have integrated plasmid DNA into the genome via a single-crossover homologous recombination (*see* **Fig. 2A,B**).
3. Select single colonies and grow in BY medium without kanamycin for 1 d to allow a second recombination event to take place (*see* **Fig. 2C,D**).
4. Spread the appropriate dilution (10^4–10^5 cells) of the culture on a BY plate containing 10% sucrose and incubate for 1 d at 30°C. Usually, about 10^2 colonies can be obtained in a plate. This selection involves the *sacB*-positive selection system *(9)*. Because the expression of integrated plasmid-borne *sacB* in the presence of sucrose is lethal to *C. glutamicum*, only cells where *sacB* is deleted as a consequence of the second homologous recombination can grow on the selective plate. In this recombination, the allelic replacement arises when the wild-type *hom* or *lysC* sequence is deleted from the genome, together with *sacB* (*see* **Fig. 2E**).

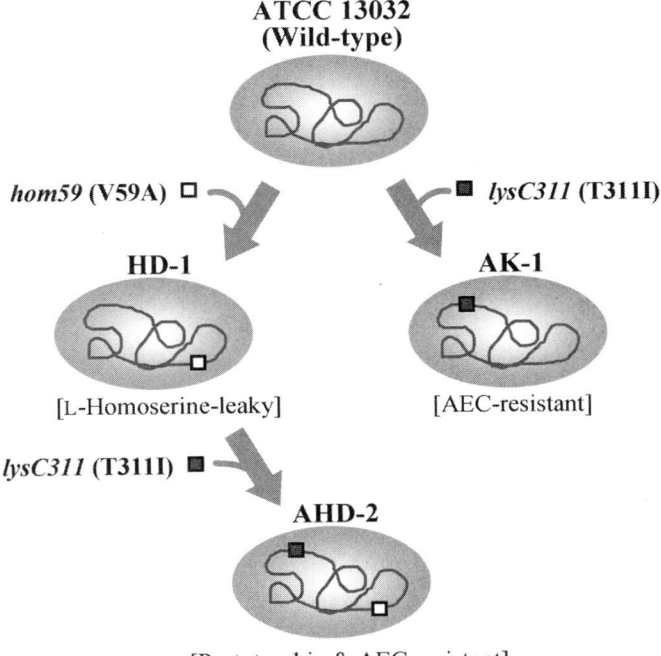

Fig. 3. Schematic diagram of the creation of the defined mutants HD-1, AK-1, and AHD-2. The sites of the *hom59* and *lysC311* mutations are indicated by the open and gray squares, respectively. AEC: *S*-(2-aminoethyl)-L-cysteine.

5. Select 10 sucrose-resistant colonies for each transformant randomly and examine for the presence of the *hom59* or *lysC311* mutation by allele-specific PCR *(10)* to isolate the positive recombinant carrying the mutation.
6. For this PCR analysis, the following site-specific primers can be used: *lysC311*-F, *lysC311*-R, *hom59*-F, and *hom59*-R.
7. The presence of the single-base mutations can be confirmed by DNA sequencing. Strains carrying the *hom59* and *lysC311* mutations on the wild-type background are designated HD-1 and AK-1, respectively (*see* **Fig. 3** and **Note 4**).

3.2.2. Double Mutant

Introduction of the *lysC311* mutation into single mutant HD-1 can be carried out with the use of plasmid pClysC311 in the same way as described in **Subheading 3.2.1.** to construct the *hom–lysC* double mutant AHD-2 (*see* **Fig. 3**). For allele-specific PCR to examine the presence of the *lysC311* mutation, the same site-specific primers shown in **Subheading 3.2.1.** can be used (*see* **Note 5**).

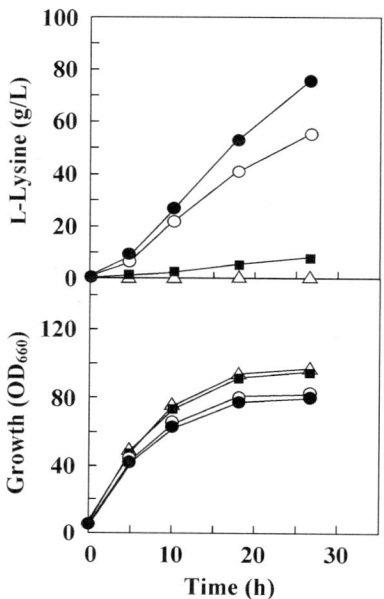

Fig. 4. L-Lysine fermentation by strains HD-1 (■), AK-1 (○), and AHD-2 (●) in fed-batch fermentor cultivation. For comparison, the profiles of wild-type strain ATCC 13032 (△) are shown as controls.

3.3. L-Lysine Production

L-Lysine production by strains HD-1, AK-1, and AHD-2 can be performed using flasks or 5-L jar fermentors, with wild-type ATCC 13032 as a control (*see* **Fig. 4**).

3.3.1. Flask Fermentation

1. Grow strains HD-1, AK-1, and AHD-2 on BY plates at 30°C for 1 d.
2. Inoculate cells grown on a BY plate into a large test tube containing 6 mL BY medium supplemented with 2% glucose and cultivate for 12 h at 30°C.
3. Transfer a 3.0-mL amount of the seed culture into a 300-mL flask containing 30 mL flask production medium.
4. Incubate on a rotary shaker at 34°C and 210 rpm for 72 h.

3.3.2. Jar Fermentor Cultivation

1. Grow strains HD-1, AK-1, and AHD-2 on BY plates at 30°C for 1 d.
2. Inoculate the cells into 250 mL of LSS1 medium in a 2-L flask and cultivate on a rotary shaker at 30°C and 210 rpm to the early stationary phase (usually grown for 12 h).

3. Transfer the seed broth into a 5-L jar fermentor containing 1400 mL of LPG-1 medium and cultivate with an agitation speed of 800 rpm, aeration at 2 L/min, and 34°C. The pH should be maintained at 7.3 with NH_4OH.

4. After the sugar initially added is consumed, feed a solution containing 50% (w/v) glucose, 4.5% (w/v) NH_4Cl, and 0.5 mg/L of D-biotin continuously until the total amount of sugar in the medium reaches 25%. The feeding rate of the solution should be controlled to maintain the glucose concentration in the medium at a low concentration (below 0.5%).

3.3.3. Analysis

In the fed-batch fermentor cultivation, cell growth can be monitored by measuring the optical density (OD_{660}) of the culture broth with the U-1080 Auto Sipper Photometer. Glucose concentration can be determined using Determinar GL-E. L-Lysine titer can be determined as L-lysine HCl by HPLC after derivatization with *o*-phthalaldehyde *(11)*. As indicated in **Fig. 4**, strains HD-1 and AK-1 grow and consume glucose as fast as the wild-type strain, thus leading to completion of fermentations within only 30 h. In the case of strain AK-1, the final growth level is somewhat limited, as can be seen from the OD_{660}. In these fermentations, strains HD-1 and AK-1 accumulate about 8 and about 55 g/L of L-lysine, respectively. Furthermore, strain AHD-2 shows almost the same growth profile as strain AK-1 and produces much more L-lysine (about 75 g/L) than strain AK-1. Such high-productive L-lysine production has illustrated the potential of the genome breeding approach to innovate traditional biotechnological processes (*see* **Note 6**).

4. Notes

1. The complete genome sequence of the representative wild-type strain of *C. glutamicum* ATCC 13032 has been assigned the accession number of BA000036 in DDBJ/GenBank/EMBL. The sequence is available from the webpage www.gib.genes.nig.ac.jp/single/index.php?spid=Cglu_ATCC13032.

2. In the coryneform bacteria, various systems have been developed to force the homologous recombination between plasmid-borne and chromosomally located gene copies *(12–16)*. Although this chapter describes the system using a nonreplicating plasmid that carries a mutated allele and an antibiotic-resistance marker, other systems using, for example, a plasmid with temperature-sensitive replication *(15)* or two incompatible plasmids *(16)*, are also effective in the generation of an intended mutant through the gene replacement.

3. If a mutant strain that carries a desired mutation is available, it is unnecessary to perform the site-directed mutagenesis via overlap extension. Only PCR from the genomic DNA of the mutant strain can generate a DNA fragment containing the mutation. In this case, PCR primer sequences are designed based on the nucleotide sequences of gene regions flanking the desired mutation. The cloning of the PCR product into pESB30 can be carried out in the same way as described earlier using the TA cloning method.

4. As shown in **Fig. 3,** the *hom* single mutant HD-1 exhibits the phenotype of the partial requirement for L-homoserine. On the other hand, the *lysC* single mutant AK-1 exhibits the phenotype of resistance to an L-lysine structural analog, *S*-(2-aminoethyl)-L-cysteine (AEC). These phenotypes can be used for the selection of the desired mutant, instead of the allele-specific PCR or DNA sequencing.

5. It should be noted that the *hom–lysC* double mutant AHD-2 no longer requires L-homoserine for maximum growth (*see* **Fig. 3**). The suppression of the *hom59*-induced phenotype of the partial requirement of L-homoserine by the coexistence with the *lysC311* mutation would occur because the mutated homoserine dehydrogenase of strain AHD-2 might proceed better with the reaction when the substrate L-aspartate-β-semialdehyde was oversupplied through AEC-resistant aspartokinase activity.

6. One of the potential advantages of the genome breeding strain for the use in biotechnological processes is its very high production rate. In addition to the biotechnological aspect, it is also worth noting that the genome breeding makes it possible to rationalize the production mechanism through characterization of the genetic basis for high-level production.

Acknowledgments

The authors would like to acknowledge all their co-workers, especially M. Hayashi, S. Ando-Hayashi, T. Nakano, K. Tanaka, M. Maeda, T. Abe, Y. Yonetani, H. Mizoguchi, S. Nakagawa, N. Hirao, K. Ochiai, and H. Yokoi.

References

1. Ikeda, M. (2003) Amino acid production processes, in *Microbial Production of L-Amino Acids* (Faurie, R. and Thommel, J., eds.), *Advances in Biochemical Engineering and Biotechnology* Vol. 79. Springer-Verlag, Berlin, pp. 1–35.
2. Ikeda, M. and Nakagawa, S. (2003) The *Corynebacterium glutamicum* genome: features and impacts on biotechnological processes. *Appl. Microbiol. Biotechnol.* **62,** 99–109.
3. Ohnishi, J., Mitsuhashi, S., Hayashi, M., et al. (2002) A novel methodology employing *Corynebacterium glutamicum* genome information to generate a new L-lysine-producing mutant. *Appl. Microbiol. Biotechnol.* **58,** 217–223.
4. Ohnishi, J., Hayashi, M., Mitsuhashi, S., and Ikeda, M. (2003) Efficient 40°C fermentation of L-lysine by a new *Corynebacterium glutamicum* mutant developed by genome breeding. *Appl. Microbiol. Biotechnol.* **62,** 69–75.
5. Schweizer, H. P. (1992) Allelic exchange in *Pseudomonas aeruginosa* using novel ColE1-type vectors and a family of cassettes containing a portable *oriT* and the counter-selectable *Bacillus subtilis sacB* marker. *Mol. Microbiol.* **6,** 1195–1204.
6. Sambrook, J. and Russell, D. W. (2001) *Molecular Cloning: A Laboratory Manual,* 3rd ed., Cold Spring Harbor Laboratory Press, Cold Spring Harbor, NY.
7. Takeshita, S., Sato, M., Toba, M., Masahashi, W., and Hashimoto-Gotoh, T. (1987) High-copy-number and low-copy-number plasmid vectors for *lacZ* α-complementation and chloramphenicol- or kanamycin-resistance selection. *Gene* **61,** 63–74.

8. Rest, M. E. van der, Lange, C., and Molenaar, D. (1999) A heat shock following electroporation of *Corynebacterium glutamicum* with xenogenein plasmid DNA. *Appl. Microbiol. Biotechnol.* **52,** 541–545.

9. Jäger, W., Schäfer, A., Pühler, A., Labes, G., and Wohlleben, W. (1992) Expression of the *Bacillus subtilis sacB* gene leads to sucrose sensitivity in the Gram-positive bacterium *Corynebacterium glutamicum* but not in *Streptomyces lividans. J. Bacteriol.* **174,** 5462–5465.

10. Kwok, S., Chang, S.-Y., Sinskey, A. J., and Wang, A. (1995) Design and use of mismatched and degenerate primers, in *PCR Primer: A Laboratory Manual* (Dieffenbach, C. W. and Dveksler, G. S., eds.), Cold Spring Harbor Laboratory Press, Cold Spring Harbor, NY, pp. 143–155.

11. Hill, D. W., Walters, F. H., Wilson, T. D., and Stuart, J. D. (1979) High performance liquid chromatographic determination of amino acids in the picomole range. *Anal. Chem.* **51,** 1338–1341.

12. Schwarzer, A. and Pühler, A. (1991) Manipulation of *Corynebacterium glutamicum* by gene disruption and replacement. *Bio/Technology* **9,** 84–87.

13. Reyes, O., Guyonvarch, A., Bonamy, C., Salti, V., David, F., and Leblon, G. (1991) "Integron"-bearing vectors: a method suitable for stable chromosomal integration in highly restrictive Corynebacteria. *Gene* **107,** 61–68.

14. Vertès, A. A., Hatakeyama, K., Inui, M., Kobayashi, M., Kurusu, Y., and Yukawa, H. (1993) Replacement recombination in Coryneform bacteria: high efficiency integration requirement for non-methylated plasmid DNA. *Biosci. Biotechnol. Biochem.* **57,** 2036–2038.

15. Kimura, E., Abe, C., Kawahara, Y., Nakamatsu, T., and Tokuda, H. (1997) A *dtsR* gene-disrupted mutant of *Brevibacterium lactofermentum* requires fatty acids for growth and efficiently produces L-glutamate in the presence of an excess of biotin. *Biochem. Biophys. Res. Commun.* **234,** 157–161.

16. Ikeda, M. and Katsumata, R. (1998) A novel system with positive selection for the chromosomal integration of replicative plasmid DNA in *Corynebacterium glutamicum. Microbiology* **144,** 1863–1868.

10

Metabolic Activity Profiling by [13]C Tracer Experiments and Mass Spectrometry in *Corynebacterium glutamicum*

Christoph Wittmann and Elmar Heinzle

Summary

This chapter introduces ^{13}C tracer techniques combined with labeling measurement by mass spectrometry (MS) for the quantification of metabolic pathway fluxes. As an example, the relative contribution of (1) *de novo* synthesis from glucose and (2) synthesis from building blocks contained in complex substrates to the formation of proteinogenic amino acids in the industrial amino acid producer *Corynebacterium glutamicum* is investigated.

Key Words: Mass spectrometry, *Corynebacterium glutamicum;* metabolic flux; ^{13}C tracer; complex substrate.

1. Introduction

For effective application and optimization of biotechnological processes, a detailed understanding of metabolic functioning and regulation of the used organism is crucial *(1,2)*. Among the most important approaches in metabolic physiology are different techniques aimed at the quantitative investigation of biological networks. In this regard, ^{13}C tracer experiments have recently received great interest in biotechnology and medicine to identify metabolic pathways and to quantify their contribution to the overall activity of the cell *(3–5)*.

This chapter introduces ^{13}C tracer techniques by a simple and interesting example on the quantification of pathways contributing to protein synthesis. Hereby, the influence of complex substrates is investigated. Complex substrates are widely applied in biotechnological processes, but little is known about the exact impact of their constituents on the process. This lack of knowledge and the lot-to-lot variation of complex substrates currently requires extensive screening of many substrate mixtures to identify the few that support the desired fermentation performance *(6)*. The studied organism, *Corynebacterium*

From: *Methods in Biotechnology, Vol. 18: Microbial Processes and Products*
Edited by: J. L. Barredo © Humana Press Inc., Totowa, NJ

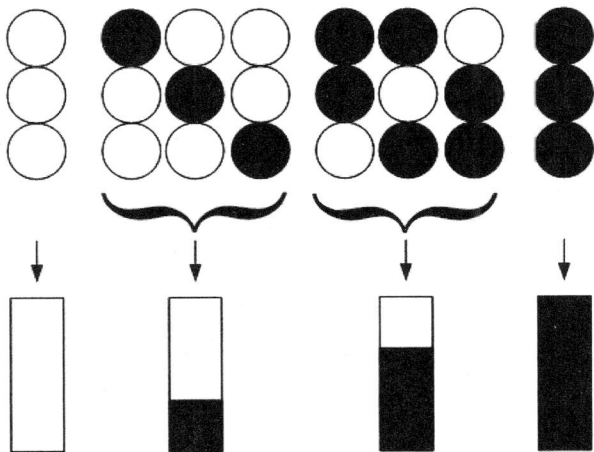

Fig. 1. Pools of *positional isotopomers* and *mass isotopomers* exemplified for a molecule with three carbon atoms, either ^{12}C (white) or ^{13}C (black).

glutamicum, is widely applied for the production of amino acids such as glutamate or lysine and is, therefore, of high industrial relevance.

The following definitions for the description of the ^{13}C labeling state of a compound for carbon are used in the present work. A *positional isotopomer* has an exactly determined labeling pattern, described by a specific number of ^{13}C atoms in specific positions of the molecule. Generally, 2^n isotopomers are possible for a compound with n carbon atoms. The eight different positional isotopomers of a compound with three carbon atoms are shown in **Fig. 1.** A *mass isotopomer X_m+i* is specified by the number i of ^{13}C atoms in the molecule, but not by their positions. A compound with n carbons has $n+1$ different mass isotopomers, ranging from the nonlabeled U-^{12}C mass isotopomer (X_m) to the fully labeled U-^{13}C mass isotopomer (X_m+n). Thus, four different mass isotopomers, comprising nonlabeled, single-labeled, double-labeled, and triple-labeled form, can occur for the C_3 compound in the given example (*see* **Fig. 1**). The nonlabeled and fully labeled mass fractions contain one positional isotopomer, whereas three positional isotopomers belong to the $m+1$ and to the $m+2$ fraction, respectively. The sum of molar fractions of all mass isotopomers is equal to 1. The most important mass isotopomers in the current work are nonlabeled and fully labeled, respectively.

Generally, an organism has different possibilities of supplying the building blocks for biomass synthesis. It can either (1) derive them from building blocks contained in the medium or (2) carry out a *de novo* synthesis from the major

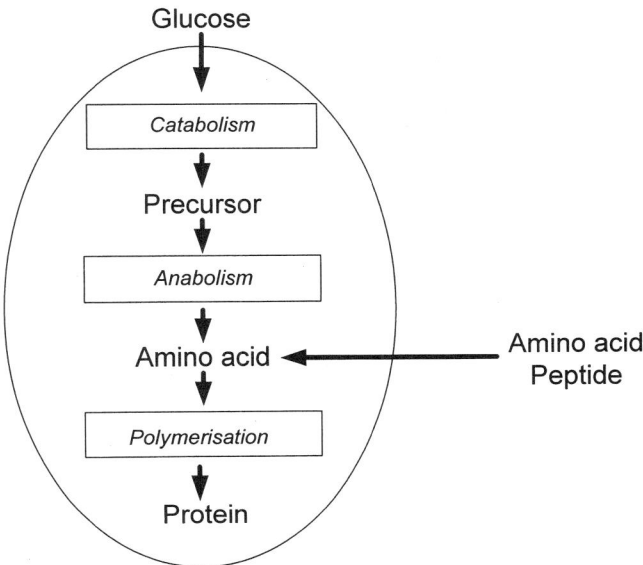

Fig. 2. Alternative synthesis of cell protein by (1) *de novo* formation from glucose or (2) uptake of building blocks from complex substrates.

carbon source. This is exemplified in **Fig. 2** for growth on a medium with glucose as the main carbon and energy source, and yeast extract as complex substrate. Amino acids required for cell protein formation can be directly derived from peptides or amino acids contained, for example, in yeast extract. Alternatively, they can be formed via the different biosynthetic routes of catabolism, anabolism, and polymerization.

Knowledge of the contribution of the different routes can hardly be achieved by sole stoichiometric measurements. [13]C tracer experiments using different ^{13}C enrichments of the different substrate pools, however, provide an elegant and straightforward method for this purpose. The strategy is shown in **Fig. 3**. In the examined case, fully labeled glucose is applied as the major carbon source together with nonlabeled yeast extract. The ^{13}C-labeling pattern of the protein contained amino acids formed during cultivation of the organism depends on their origin. Each amino acid molecule directly supplied from the complex substrate is nonlabeled (*see* **Fig. 3B**), wheras each amino acid molecule synthesized from glucose is fully labeled (*see* **Fig. 3A**). In the case where both routes are active, a mixture of fully labeled and nonlabeled molecules will result (*see* **Fig. 3C**) The ratio between nonlabeled (*m*) and fully labeled mass isotopomers

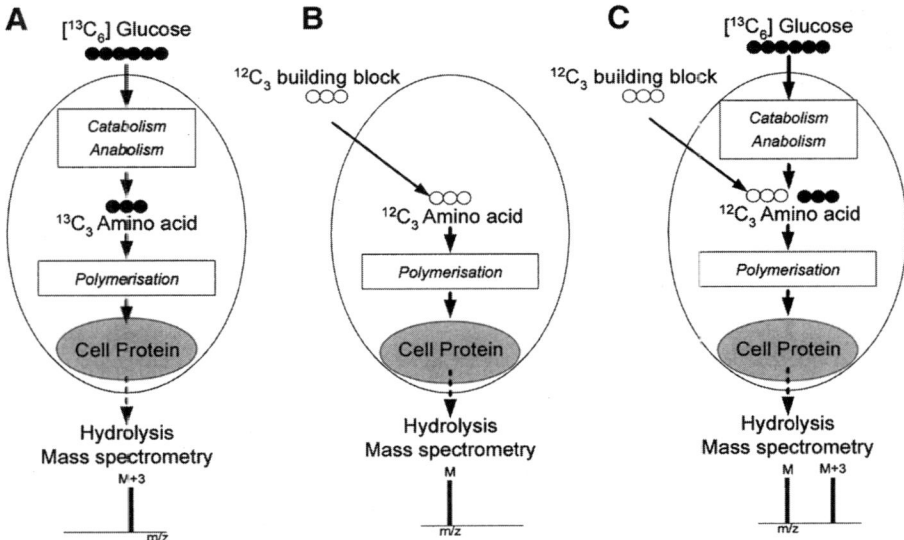

Fig. 3. Experimental strategy for quantification of the relative contribution of (1) *de novo* formation from glucose or (2) uptake of building blocks from complex substrates to the synthesis of proteinogenic amino acids.

($m+x$) of amino acids contained in the protein therefore provides information about their synthesis, whereby mass spectrometry is very useful for measurement of the ^{13}C enrichment:

$$\frac{r_{de\ novo\ synthesis}}{r_{synthesis\ from\ building\ blocks}} = \frac{A_{m+x}}{A_m} \tag{1}$$

where A_m denotes the abundance of the nonlabeled mass isotopomer and $A_{m}+x$ denotes the abundance of the fully labeled mass isotopomer of the analyte in the mass spectrum. The above situation is based on some assumptions (*see* **Note 1**), but allows a straightforward data processing and evaluation and should be applicable to many cases.

2. Materials

1. *Corynebacterium glutamicum* strain ATCC 13032.
2. Cultivation equipment (autoclave, shaker, shake flask, laminar hood).
3. 99% $[^{13}C_6]$ glucose.
4. Sterile NaCl (0.9%) solution.
5. Photometer.

6. Preculture agar plates: 5 g/L glucose, 5 g/L yeast extract, 10 g/L tryptone, 5 g/L NaCl, and 12 g/L agar.
7. Preculture medium: 5 g/L glucose, 5 g/L yeast extract, 10 g/L tryptone, and 5 g/L NaCl.
8. Solution M 100X: 200 mg/L $FeCl_3\cdot6H_2O$, 20 mg/L $CuCl_2\cdot2H_2O$, 200 mg/L $MnSO_4\cdot H_2O$, 50 mg/L $ZnSO_4\cdot7H_2O$, 20 mg/L $Na_2B_4O_7\cdot10H_2O$, and 10 mg/L $(NH_4)_6Mo_7O_{24}\cdot H_2O$.
9. Solution A: 3 mg 3,4-dihydroxybenzoate, 5.5 mg $CaCl_2$, 20 mg $MgSO_4\cdot7H_2O$, 2 mg $FeSO_4\cdot7H_2O$, 100 mg NaCl, 1 mL 100X solution *M,* and 60 mL deionized water. Adjust to pH 5.0 with HCl.
10 Solution B: 800 mg K_2HPO_4, 100 mg KH_2PO_4, 50 μg biotin, 100 μg thiamine·HCl, and 10 mL deionized water.
11. Solution C: 500 mg $(NH_4)_2SO_4$ and 10 mL deionized water.
12. Solution D: 50 mg 99% $[^{13}C_6]$ glucose and 0.5 mL deionized water.
13. Solution E: 50 mg yeast extract and 0.5 mL deionized water.
14. 6 *M* HCl.
15. 6 *M* NaOH.
16. Incubator for 105°C.
17. Centrifugal filter (0.22 μm).
18. – 70°C Freezer.
19. Freeze-drying equipment.
20. Dimethyl formamide (DMF) with 0.1% pyridine.
21. *N*-Methyl-*N-t*-butyl-dimethylsilyltrifluoroacetamide (MBDSTFA).
22. Quadrupole gas chromatography/mass spectrometry (GC/MS) with electron impact ionization.
23. HP5MS column (5% phenylmethyl-siloxan-diphenylpolysiloxan, 30 m × 250 μm).

3. Methods

This chapter describes the whole protocol for metabolic activity profiling of *C. glutamicum* involving (1) cultivation of the organism on ^{13}C tracer medium, (2) sampling and analysis of ^{13}C labeling patterns from protein hydrolysates, and (3) data processing and evaluation.

3.1. Cultivation of C. glutamicum on ^{13}C Tracer Medium

3.1.1. Preculture

1. Grow cells of *C. glutamicum* on preculture agar plates for 24 h at 30°C.
2. Pick one colony and transfer into a 50-mL baffled shake flask containing 5 mL preculture medium.
3. Incubate for 24 h at 30°C and 250 rpm on a rotary shaker (shaking diameter = 2.5 cm).
4. Harvest cells by centrifugation (5 min, 9300*g,* 4°C).
5. Wash with 10 mL sterile 0.9% NaCl solution. This washing step is required to remove residues from the preculture medium.

6. Resuspended the cells in 2 mL sterile 0.9% NaCl solution.
7. Measure the cell concentration of the cell suspension by the optical density at 660 nm (OD_{660}) in order to estimate the exact volume of the cell suspension required to inoculate the tracer experiments with a starting OD_{660} of 0.1.

3.1.2. Main Culture and Cell Harvest

The tracer experiment is carried out in a 50-mL baffled shake flask containing 5 mL medium.

1. Autoclave separately solutions A, B, C, D, and E at 120°C for 15 min.
2. After cooling down to room temperature, combine them to obtain the final medium with (1) the complex carbon source and (2) the tracer substrate [$^{13}C_6$] glucose at a final concentration of 10 g/L (see **Note 2**). For this purpose, sterile solution A (3 mL), solution B (0.5 mL), solution C (0.5 mL), solution D (0.5 mL), and solution E (0.5 mL) are mixed.
3. Inoculate the cells into the flask with an initial OD_{660} of about 0.1 (see **Note 3**).
4. Incubate the flask at 250 rpm and 30°C on a rotary shaker (2.5 cm shaking diameter) for 24 h.
5. At the end of the cultivation, harvest the cells by centrifugation (5 min, 9300g, 4°C).

3.2. Protein Hydrolysis

The preparation of protein hydrolysates involves (1) washing of the harvested cells, (2) hydrolysis, and (3) lyophilization.

1. Carefully wash the harvested cells to completely remove any complex medium constituents: centrifuge (5 min, 9300g, 4°C) and resuspend the cell pellet in 10 mL deionized water. The washing step is important because the later GC/MS labeling analysis of ^{13}C-enriched proteinogenic amino acids must not be interfered with by the overlay with nonenriched amino acids present in the culture medium.
2. Wash again under same conditions.
3. Measure the OD_{660} to determine cell concentration. According to the result, an appropriate volume containing about 1 mg cell dry mass (CDM) is transferred into an Eppendorf tube (see **Note 4**).
4. Harvest the washed biomass pellet by centrifugation (5 min, 9300g, 4°C) and mix with 50 µL 6 N HCl on a vortex.
5. Incubate the mixture at 105°C for 24 h for hydrolysis of the cell protein.
6. Leave the hydrolysate to cool down to room temperature and neutralize by the addition of 6 N NaOH. This is required to avoid sublimation of HCl during the freeze-drying procedure, which would be collected at the steel-made condenser of the freeze-dryer and would lead to its corrosion.
7. Centrifuge the mixture (5 min, 9300g, 4°C) through a centrifugal filter device (0.22 µm) to remove insoluble components.

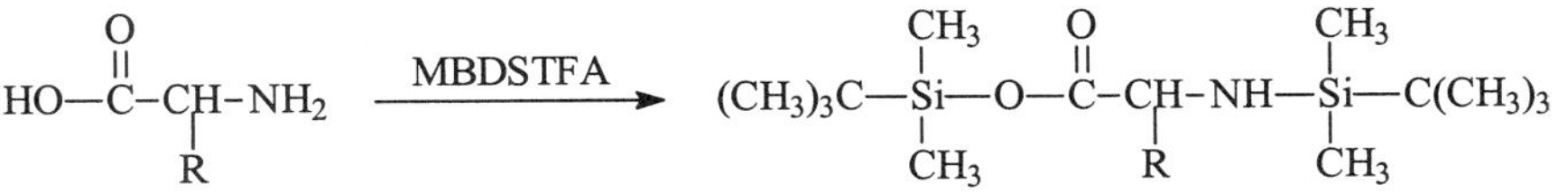

$$
\underset{\overset{\displaystyle R}{|}}{\text{HO}-\overset{\overset{\displaystyle O}{\|}}{\text{C}}-\text{CH}-\text{NH}_2} \quad \xrightarrow{\text{MBDSTFA}} \quad (\text{CH}_3)_3\text{C}-\underset{\overset{|}{\text{CH}_3}}{\overset{\overset{|}{\text{CH}_3}}{\text{Si}}}-\text{O}-\underset{\overset{|}{R}}{\overset{\overset{\|}{O}}{\text{C}}}-\text{CH}-\text{NH}-\underset{\overset{|}{\text{CH}_3}}{\overset{\overset{|}{\text{CH}_3}}{\text{Si}}}-\text{C}(\text{CH}_3)_3
$$

Fig. 4. Reaction of an amino acid with MBDSTFA to the corresponding TBDMS derivate. Amino acid side chains containing a functional group (e.g., –OH, –SH, –COOH, –CONH$_2$, –NHR) are additionally converted into TBDMS-derivatized residues.

8. Transfer the clear filtrate into a 1-mL glass vial and freeze at –70°C (*see* **Note 5**).
9. Freeze-dry the frozen filtrate.

3.3. GC/MS Analysis

For GC/MS analysis, the amino acids are converted into *t*-butyl-trimethylsilyl (TBDMS) derivates. The scheme of the derivatization reaction of an amino acid with MBDSTFA (*N*-methyl-*N*-*t*-butyl-dimethylsilyltrifluoroacetamide) is shown in **Fig. 4.** TBDMS derivatization of amino acids is especially useful in metabolic network analysis because [M-57] fragments containing the entire carbon skeleton of the analyte can be observed by GC/MS with high signal intensities *(7)*.

3.3.1. Derivatization

1. Add 50 μL DMF containing 0.1% pyridine and 50 μL MBDSTFA on the top of the solid sample.
2. Incubate the mixture at 80°C for 60 min (*see* **Note 6**).
3. Transfer the derivatized sample into a 100-μL GC-micro insert.

3.3.2. GC/MS Analysis

The labeling analysis is carried out with 0.1-μL sample injection on a GC with a HP5MS capillary column (30 m × 250 μm), electron impact ionization at 70 eV, and quadrupole detector. The approach is based on a protocol for the ^{13}C-labeling measurement of TBDMS-derivatized amino acids in cell extracts *(8)*. In contrast to previous work, the temperature gradient was modified to reduce the analysis time. The temperature gradient is 120°C for 1 min, 10°C/min up to 310°C, and 310°C for 1 min. Further operation temperatures are 320°C (inlet), 320°C (interface), and 325°C (quadrupole). The carrier gas flow is helium at 0.7 mL/min. Measurement is carried out in scan mode (m/z = 150–650).

A total ion current (TIC) spectrum of a protein hydrolysate originating from a cultivation of *C. glutamicum* on a medium with [13C$_6$] glucose and nonenriched yeast extract is shown in **Fig. 5.** Complete separation of all expected derivatized proteinogenic amino acids can be done within 20 min (*see* **Note 7**).

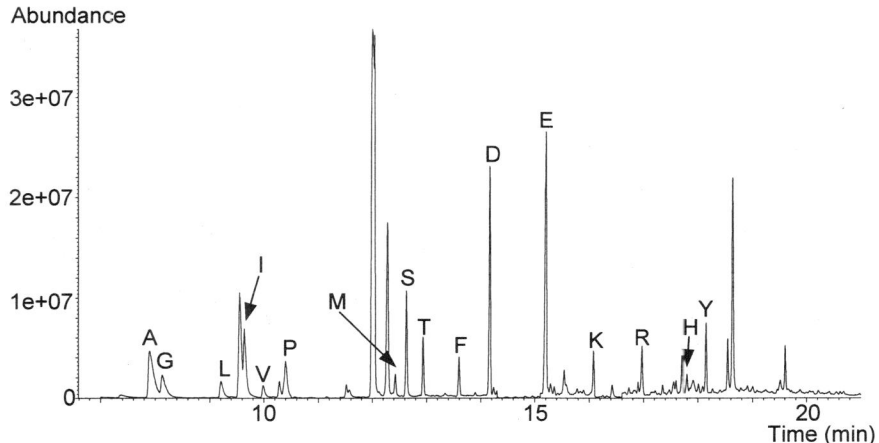

Fig. 5. The GC/MS spectrum of a protein hydrolysate from *C. glutamicum* ATCC 13032 grown on a complex medium with 10 g/L [^{13}C$_6$]glucose and 10 g/L yeast extract. The proteinogenic amino acids were converted into TBDMS derivates prior to the measurement. The GC/MS measurement was carried out in scan mode (m/z = 150–650).

In order to quantify the pathway fluxes toward proteinogenic amino acids, each of the amino acid derivates has to be analyzed for its ^{13}C-labeling pattern. Hereby, the monoisotopic signal and the fully ^{13}C-enriched signal have to be integrated. To illustrate the data processing, the mass spectrum of TBDMS$_2$–alanine is depicted in **Fig. 6.** As typically observed in GC/MS, different fragment ions of the analyte are present. The major ion clusters are found at m/z = 260–263, m/z = 232–234, and m/z = 158–160. The intact derivatized molecule has a mass-to-charge (m/z) ratio of 317. The corresponding fragment ions originate from (1) the release of a *t*-butyl group from the derivatization residue (for the [M-57] fragment at 260–263), (2) the release of a *t*-butyl group from the derivatization residue plus a CO group including carbon C$_1$ of alanine (for the [M-85] fragment at 232–234), and (3) the release of a COO–TBDMS group including the carbon C$_1$ of alanine (for the [M-159] fragment at 158–160).

3.4. Estimation of Carbon Fluxes Toward the Formation of Proteinogenic Amino Acids

In the performed ^{13}C tracer experiment, the signal abundances at m/z = 260 and 263 represent the nonenriched [^{12}C$_3$] and the fully enriched [^{13}C$_3$] derivatized alanine, m and $m+3$, respectively. The ratio of these two signals is used to estimate the relative flux from (1) the nonenriched complex substrate and (2) the fully enriched [^{13}C$_6$] glucose toward alanine.

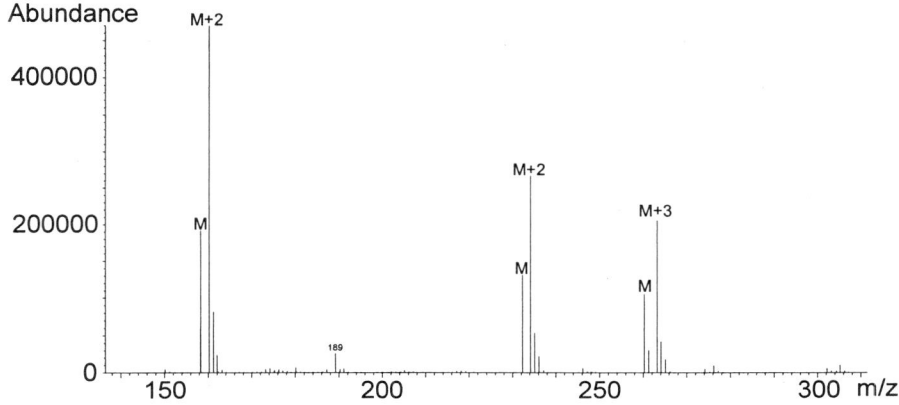

Fig. 6. Mass spectrum of TBDMS-derivatized alanine from a protein hydrolysate of *C. glutamicum* ATCC 13032 grown on a complex medium with 10 g/L [$^{13}C_6$]glucose and 10 g/L yeast extract.

Alternatively, also the *m* and *m*+2 signals of fragment ions comprising the carbon C_2C_3 of alanine can be used for this purpose.

Table 1 lists suitable fragment ions of each proteinogenic amino acid to be used for labeling analysis. Additionally, the amino acid carbon atoms present in the considered fragment and the resulting *m/z* values of the monoisotopic *(m)* and the fully $^{13}C_3$ enriched mass isotopomer *(m+x)* are given (*see* **Note 8**).

In the given example, ^{13}C-labeling analysis of amino acids from the cell protein of *C. glutamicum* provides a deep insight into the metabolism of this strain (*see* **Table 1**). The relative abundance of the nonenriched and the fully ^{13}C-enriched mass isotopomer allows a precise estimation of the relative flux from (1) the complex substrate and (2) glucose toward the different amino acids in the cell protein. Usually, the relative error for the quantification of relative mass isotopomer fractions by GC/MS is about 1.0%. Therefore, rather accurate data can be obtained by this approach. For most of the amino acids, the [M-57] fragment ion containing the entire amino acid carbon skeleton can be considered. The only exceptions are leucine and isoleucine, for which the [M-57] fragment is not specific because of the potential release of the C_4H_9 group from the derivatization residue and the amino acid residue itself, respectively. The [M-57] fragment ion of proline is interfered by isobaric overlay *(8)*. Here, [M-159] fragment ions are used for calculation.

It can be seen that the relative contribution of *de novo* synthesis and uptake from the medium varied tremendously for the different amino acids. Whereas some amino acids such as alanine, aspartate, or glutamate are predominantly

Table 1
GC/MS Analysis of Nonlabeled *(m)* and Fully ^{13}C-labeled *(m+x)* Proteinogenic Amino Acids of *C. glutamicum* Grown on a Medium Containing [^{13}C$_6$]glucose and Nonlabeled Yeast Extract

Compound	Considered fragment	m [m/z]	$m+x$ [m/z]	Analyte carbons	Mass isotopomer fraction (%) m	$m+x$
Ala-(TBDMS)$_2$	[M-57]	260	263	$C_1C_2C_3$	33.8	66.2
Gly-(TBDMS)$_2$	[M-57]	246	248	C_1C_2	49.9	50.1
Val-(TBDMS)$_2$	[M-57]	288	293	$C_1C_2C_3C_4C_5$	51.3	48.7
Leu-(TBDMS)$_2$	[M-159]	200	205	$C_2C_3C_4C_5C_6$	75.0	25.0
Ile-(TBDMS)$_2$	[M-159]	200	205	$C_2C_3C_4C_5C_6$	86.9	13.1
Pro-(TBDMS)$_2$	[M-159]	184	188	$C_2C_3C_4C_5$	57.0	43.0
Met-(TBDMS)$_2$	[M-57]	320	325	$C_1C_2C_3C_4C_5$	78.2	21.8
Ser-(TBDMS)$_3$	[M-57]	390	393	$C_1C_2C_3$	29.1	70.9
Thr-(TBDMS)$_3$	[M-57]	404	408	$C_1C_2C_3C_4$	66.1	33.7
Phe-(TBDMS)$_2$	[M-57]	336	345	$C_1C_2C_3C_4C_5C_6C_7C_8C_9$	98.6	1.4
Asp-(TBDMS)$_3$	[M-57]	418	422	$C_1C_2C_3C_4$	29.1	70.9
Glu-(TBDMS)$_3$	[M-57]	432	437	$C_1C_2C_3C_4C_5$	11.8	88.2
Lys-(TBDMS)$_3$	[M-57]	431	437	$C_1C_2C_3C_4C_5C_6$	71.0	29.0
Arg-(TBDMS)$_3$[a]	[M-57]	442	448	$C_1C_2C_3C_4C_5C_6$	60.5	39.5
His-(TBDMS)$_2$	[M-57]	440	446	$C_1C_2C_3C_4C_5C_6$	65.3	34.7
Tyr-(TBDMS)$_3$	[M-57]	466	475	$C_1C_2C_3C_4C_5C_6C_7C_8C_9$	54.0	46.0

[a] The intact molecule has a total m/z of 499, because an NH_3 group is released from arginine during the derivatization.

synthesized from glucose, other amino acids such as leucine, isoleucine, methionine, or phenylalanine are mainly derived from the complex carbon source. It can be concluded that *C. glutamicum* would especially benefit from the additional supply of amino acids that originate to a large extent from the complex substrate. In contrast, a lower effect is expected for those amino acids that are mainly synthesized from glucose despite their presence in the medium.

To summarize, the presented approach is a simple and straightforward method for detailed investigations of amino acid metabolism during growth of microorganisms. In addition to the case study presented here, ^{13}C tracer experiments can be applied in various cases and offer a multitude of possibilities to investigate cellular metabolism in detail. The protocol described in the present work can be easily modified for other organisms. By appropriate experimental design (e.g., choice of the tracer substrate, medium, or growth conditions), it is applicable to study various aspects of metabolic networks. A large number of tracer compounds with stable isotopes is commercially available. Because of the very small amount of sample biomass (1 mg CDM), analysis as described here can be carried out at low costs. In many cases, such as the given example, valuable data about certain metabolic properties can be obtained with very simple data processing.

In addition to experiments, as in the example given in the present work, more sophisticated appoaches aim at the quantification of intracellular reactions of large biological networks (^{13}C metabolic flux analysis). The quantification of metabolic fluxes in complete biological networks follows the same principle as shown here, where ^{13}C tracer experiments are designed and applied in cultivations of the investigated organism. Also, in these cases, the information on fluxes within the examined network is contained in the ^{13}C-labeling patterns of, for example, secreted products, intracellular metabolites, or biomass constituents. However, because of the complexity of the regarded networks, the accurate quantification of metabolic fluxes relies on comprehensive approaches involving stoichiometric and isotopomer balancing. The reader interested in experimental and computational tools developed and applied for such studies is referred to recent review artices (*6,9,10*).

4. Notes

1. The current example is simplified in that way that only the direct pathways from the two substrates are regarded. This allows a straightforward and simple data processing and evaluation. In a more complete view, one has to consider that nonlabeled compounds from the complex substrate and ^{13}C-labeled compounds from glucose all lead into the pathways of the central metabolism, where the rearrangement of carbon skeleton in the different reactions leads to a combination of labeled and nonlabeled carbon in metabolites. Because of this, also partly ^{13}C-labeled amino acids can be observed. To fully account for this and include also other mass isotopomers in the analysis, data processing is more complex. First, each mass isotopomer distribution has to be corrected for natural isotopes (e.g., Si, O, N, H, S in the measured ion and of C in the derivatization residue) to obtain the labeling pattern of the carbon skeleton of the analyte itself. Correction for isotopes can be carried out by matrix calculus applying isotope correction matrices and mathematical software (*11*). From the corrected mass isotopomer fractions (*y*) of an amino acid with *x* carbon atoms, the summed fractional labeling (SFL) can be calculated:

$$SFL = \frac{\sum_{i=1}^{x} i \cdot y_{m+i}}{x} \tag{2}$$

The SFL describes the total ^{13}C enrichment of a molecule. It can range from 1 (100% ^{13}C) to 0 (100% ^{12}C). To estimate the relative contribution of the two carbon sources, glucose, and complex substrate, the SFL of each amino acid is calculated and compared to the value of the two substrates.

2. Commonly, citrate is used as a complexing agent for bivalent metal ions in cultivations of *C. glutamicum*. As previously shown, citrate is utilized by *C. glutamicum* as a carbon and energy source *(12)*. In the present work, citrate would contribute to the labeling pattern of the cell protein and interfere with the studied substrates. Because of this, 3,4-dihydroxybenzoate, required only in very low amounts, is used instead of citrate as the complexing agent in the present work.

3. The inoculum size should be low, so that the influence of nonlabeled biomass present at the beginning of the cultivation can be neglected in the later GC/MS measurements. For the case that the final cell concentration is about 10 OD_{660} units, the inoculum accounts only for about 1% of it.

4. The amount of cell dry mass (CDM) in the sample can be estimated by a correlation between optical density and cell dry mass. For *C. glutamicum* ATCC 13032 a correlation factor of 0.29 g CDM per OD_{660} unit has been previously measured *(13)*. This value can be used, whereby it should be noted that it can vary to some extent depending on the photometer used for optical density measurement.

5. The sample should be frozen, placing the glass vials into a vertical position to ensure that all the amino acids from the cell protein are at the bottom of the vial and later come in contact with the derivatization reagent. The reagent applied later for derivatization of the amino acids prior to their GC/MS analysis is unstable in the presence of water, so complete dryness of the sample after the lyophilization is required.

6. The use of a water bath for the incubation is not recommended in order to avoid contact of the reaction mixture with water, which would cause hydrolysis of the reagent.

7. The samples should be measured within 1 d after their derivatization. Derivates of lysine and histidine are especially unstable. It should be noted that, usually, no signals are to be expected for glutamine (converts to glutamate during hydrolysis), asparagine (converts to aspartate during hydrolysis), tryptophane (decomposes during hydrolysis), and cysteine (decomposes during hydrolysis).

8. It is necessary to take into account the possibility of ambiguous fragments (e.g., $m/z = 274$ of TBDMS–leucine, which can be produced through cleavage of a *t*-butyl rest from the derivatization residue or from the side chain of leucine.

References

1. Bailey, J. E. (1991) Toward a science of metabolic engineering. *Science* **252,** 1668–1675.
2. Stephanopoulos, G. and Kelleher, J. (2001) Biochemistry. How to make a superior cell. *Science* **292,** 2024–2025.
3. de Meer, K., Roef, M. J., Kulik, W., and Jakobs, C. (1999) In vivo research with stable isotopes in biochemistry, nutrition and clinical medicine: an overview. *Isotopes Environ. Health Stud.* **35,** 19–37.
4. Christensen, B. and Nielsen, J. (2000) Metabolic network analysis. A powerful tool in metabolic engineering. *Adv. Biochem. Eng. Biotechnol.* **66,** 209–231.

5. Wittmann, C. (2002) Metabolic flux analysis using mass spectrometry. *Adv. Biochem. Eng. Biotechnol.* **74,** 39–64.

6. Zhang, J., Reddy, J., Buckland, B., and Greasham, R. (2003) Toward consistent and productive complex media for industrial fermentations: studies on yeast extract for a recombinant yeast fermentation process. *Biotechnol. Bioeng.* **82,** 640–652.

7. Kitson, F. G., Larsen, B., and McEven, C. N. (eds.) (1996) *Gas Chromatography and Mass Spectrometry: A Practical Guide,* Academic, San Diego, CA.

8. Wittmann, C., Hans, M., and Heinzle, E. (2002) In vivo analysis of intracellular amino acid labelings by GC/MS. *Anal. Biochem.* **307,** 379–382.

9. Kelleher, J. K. (2001) Flux estimation using isotopic tracers: common ground for metabolic physiology and metabolic engineering. *Metab. Eng.* **2,** 100–110.

10. Wiechert, W. (2002) An introduction to ^{13}C metabolic flux analysis. *Genet. Eng. (NY).* **24,** 215–238.

11. Wittmann, C. and Heinzle, E. (1999) Mass spectrometry for metabolic flux analysis. *Biotechnol. Bioeng.* **62,** 739–750.

12. Kiefer, P., Heinzle, E., and Wittmann, C. (2002) Influence of glucose, fructose and sucrose as carbon sources on kinetics and stoichiometry of lysine production by *Corynebacterium glutamicum. J. Ind. Microbiol. Biotechnol.* **28,** 338–343.

13. Wittmann, C. and Heinzle, E. (2002) Genealogy profiling through strain improvement using metabolic network analysis—metabolic flux genealogy of several generations of lysine producing corynebacteria. *Appl. Environ. Microbiol.* **68,** 5843–5859.

11

Protein and Vitamin Production in *Bacillus megaterium*

Heiko Barg, Marco Malten, Martina Jahn, and Dieter Jahn

Summary

The Gram-positive bacterium *Bacillus megaterium* is frequently employed for the industrial production of exoenzymes including amylases, proteinases, and penicillin acylases, for instance. Here, we describe the utilization of the excellent protein secretion capacity of this bacterium for the production and export of homologous and heterologous recombinant proteins. Necessary genetic tools like various expression vectors, protoplast transformation protocols, and genomic integration strategies are outlined. Practical examples for the directed development of *B. megaterium* to a vitamin B_{12} production host are also described.

Key Words: *Bacillus megaterium;* gene expression; protein production; protein export; vitamin production.

1. Introduction

The Gram-positive bacterium *Bacillus megaterium* fascinates microbiologists ever since it was first described over 100 yr ago. Biologically, it is an interesting organism because of its unusual physiology, its capacity to sporulate, its production of useful enzymes and products, and the wide range of inhabited ecological niches. The large size of *B. megaterium* cells and spores, responsible for the name, makes the bacterium especially amenable to morphological studies. Even if it is taxonomically placed into the *Bacillus subtilis* group of *Bacilli*, its relation to *B. subtilis* is of clearly distant nature, which is documented by multiple genetic, biochemical, and physiological differences. However, similar to *B. subtilis, B. megaterium* carries multiple advantages for a broad industrial use in the production of proteins and vitamins. Its impressive protein secretion capacity makes it a suitable host for the production of exoenzymes like various amylases, penicillin amidase, or steroid hydrolases *(1)*. For the production of heterologous recombinant proteins, necessary molecular biology tools like an efficient protoplast transformation system, stable free-replicating plasmids, and

From: *Methods in Biotechnology, Vol. 18: Microbial Processes and Products*
Edited by: J. L. Barredo © Humana Press Inc., Totowa, NJ

directed genomic gene integration systems were developed *(2)*. Species imminent characteristics like its natural proteinase deficiency makes *B. megaterium* an especially suitable and commercially effective tool for biotechnological applications. In this chapter, we describe methods using *B. megaterium* for the production and secretion of homologous and heterologous proteins.

2. Materials

1. Basic molecular biology materials described elsewhere *(3)*.
2. Luria–Bertani (LB) medium: 10 g/L bacto-tryptone, 5 g/L bacto-yeast extract, 5 g/L NaCl, and 20 g/L agar. Adjust to pH 7.0 with 5 *N* NaOH. Autoclave at 121°C for 15 min.
3. pHBintE expression and chromosomal integration plasmid (Dr. D. Jahn, Technical University of Braunschweig, Braunschweig, Germany).
4. Antibiotic Medium No. 3 (Difco Laboratories, Detroit, MI).
5. SMMP buffer: 17.5 g/L Antibiotic Medium No. 3, 500 mM saccharose, 20 mM Na–maleinat, pH 6.5, and 20 mM MgCl$_2$.
6. PEG-P solution: 40% (w/v) polyethylene glycol 6000, 500 mM saccharose, 20 mM Na–maleinat, pH 6.5, and 20 mM MgCl$_2$.
7. CR5 top agar: 50% CR5 solution A and 50% CR5 solution B.
8. CR5 solution A: 300 mM saccharose, 31.1 mM 3-morpholino-propanesulfonic acid (MOPS), pH 7.3, 15 mM NaOH, 1.3 mM K$_2$SO$_4$, 45.3 mM MgCl$_2$·6 H$_2$O, 313 µM KH$_2$PO$_4$, 13.8 mM CaCl$_2$, 52.1 mM L-proline, and 50.5 mM D-glucose.
9. CR5 solution B: 4.0% (w/v) agar, 0.2% (w/v) casaminoacids, and 10% (w/v) yeast extract.
10. Lysis buffer: 20 mM ethylenediaminetetraacetate disodium salt (EDTA), 100 mM Na$_3$PO$_4$, and 5 mg/mL lysozyme. Adjust to pH 6.5 with H$_3$PO$_4$.
11. Signal prediction algorithm: SignalP V2.0 (www.cbs.dtu.dk/services/SignalP-2.0/) *(4)*.
12. *Leuconostoc mesenteroides* NRRL B512-F (DSMZ, Braunschweig, Germany).
13. *Escherichia coli* strain DH10B *(3)*.
14. *Bacillus megaterium* strain MS941 *(5)*.
15. *B. megaterium* strain DSM509 (DSMZ).
16. Restriction enzymes, T4 polynucleotide kinase, T4 ligase, antibiotics.
17. Oligonucleotide A: 5′-CGA GCT CCC GGG TAA GAA TAT AAG GAG GTG AAA ATT TAT GCC ATT TAC AG-3′.
18. Oligonucleotide B: 5′-AAG ATC TGC ATG CGA AAG CTT ATG CTG ACA CAG CAT TTC C-3′.
19. Oligonucleotide C: 5′-CAA AGG GGG AAA TGt acA ATG GTC AAA CT AGT TCG-3′. Exchanged bases are shown in lowercase letters.
20. Oligonucleotide D: 5′-GAA CTA GTT TGG ACC ATT gta CAT TTC CCC CTT TG-3′. Exchanged bases are shown in lowercase letters.
21. Oligonucleotide E: 5′-GTA CAA TGA AAA AAG TAC TTA TGG CAT TCA T-3′.
22. Oligonucleotide F: 5′-TTA CTT TTT TCA TGA ATA CCG TAA GTA TAT AAC-3′.
23. Oligonucleotide G: 5′-TAT TTG TTT ATC GCT GAT TCT ATC TGT TTT AG-3′.

24. Oligonucleotide H: 5′-AAA TAG CGA CTA AGA TAG ACA AAA TCG GCG-3′.
25. Oligonucleotide I: 5′-CCG CTC CGC CGT CTG GCG CAG GCG CCG CAT T-3′.
26. Oligonucleotide J: 5′-AGG CGG CAG ACC GCG TCC GCG GCG TAA GC-3′.
27. Phosphate-buffered growth medium, pH 6.8: 2 g/L $(NH_4)_2SO_4$, 0.3 g/L $MgSO_4$, 0.5 g/L yeast extract, 40 mg/L $MnCl_2 \cdot 2H_2O$, 53 mg/L $CaCl_2 \cdot 2H_2O$, 2.5 mg/L $FeSO_4 \cdot 7H_2O$, 2.5 mg/L $(NH_4)_6Mo_7O_{24} \cdot H_2O$, 2.5 mg/L $CoCl_2 \cdot 6H_2O$, 35.2 g/L KH_2PO_4, 72.6 g/L $Na_2HPO_4 \cdot 2H_2O$, and 30 g/L D-glucose. Prepare phosphate buffer as well as glucose separately as 10X stock solutions and add after sterilization of the media.
28. Trace elements 1000X stock solution: 40 g/L $MnCl_2 \cdot 2H_2O$, 53 g/L $CaCl_2 \cdot 2H_2O$, 2.5 g/L $FeSO_4 \cdot 7H_2O$, 2.5 g/L $(NH_4)_6Mo_7O_{24} \cdot H_2O$, and 2.5 g/L $CoCl_2 \cdot 6H_2O$.
29. 25% (w/v) Xylose stock solution.
30. Sodium dodecyl sulfate (SDS) sample buffer without mercaptoethanol *(3)*.
31. 6% SDS polyacrylamide gels *(3)*.
32. Activity stain washing buffer: 20 mM sodium acetate, pH 5.2–5.4, 50 µg/L calcium chloride, and 0.1% (v/v) Triton X-100. Add 100 g/L (approx 300 mM) sucrose as substrate for the development of dextran overnight.
33. 50% (v/v) Ethanol solution in water.
34. Protoplast buffer (PB): 20 mM potassium phosphate buffer, pH 7.5, 15 mM $MgCl_2$, and 20% sucrose. Mix 100 mM solutions of K_2HPO_4 and KH_2PO_4 to final pH 7.5, add $MgCl_2$ and sucrose, and dilute five times to obtain 20 mM solution.
35. PB buffer with lysozyme: PB buffer and 1 mg/mL lysozyme.
36. PB buffer with trypsin: PB buffer and 1 mg/mL trypsin.
37. Triton stock: 20% (w/v) Triton X-100.
38. TCA: 20% (w/v) trichloroacetic acid.

3. Methods

The following methods cover (1) the construction and use of two different xylose-inducible expression vectors, (2) the transformation of *B. megaterium* with these plasmids, (3) the induction of protein production, (4) the chromosomal integration of the plasmid, and (5) the excretion of recombinantly produced proteins into the medium. DNA manipulations were performed according to standard procedures *(3)* and are not described here in detail because of space limitations.

3.1. Expression Plasmids

For the production of recombinant proteins in *B. megaterium,* we constructed the xylose-inducible expression plasmids pMM1520 and pHBintE (*see* **Fig. 1**). Both are based on an expression system initially developed by Rygus and Hillen *(2)*. Transcription of cloned target genes in *B. megaterium* cells is controlled by the *xylA* promoter. It is induced by the addition of 0.5% of xylose to the growth medium (*see* **Note 1**). Expression of genes inserted into the multiple cloning site generally leads to fusion proteins carrying the first three amino acids of the *xylA*-

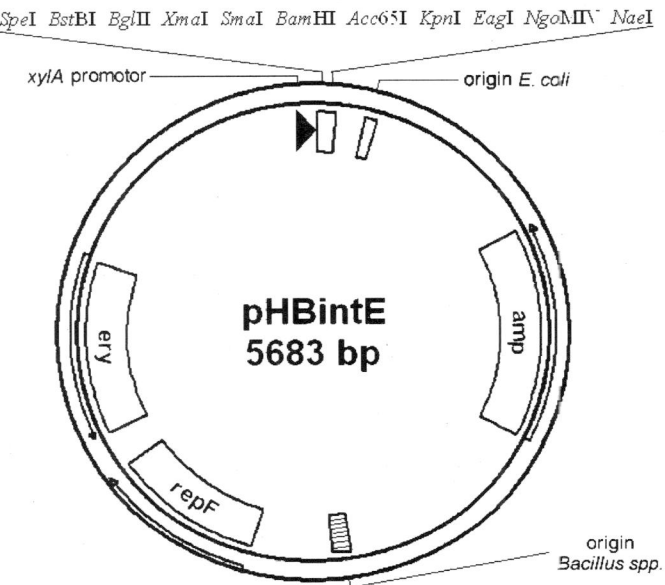

Fig. 1. Schematic structure of the pHBintE expression and chromosomal integration plasmid. pMM1520 differs only in the nontemperature-sensitive origin.

encoded protein (*see* **Note 2**). Both expression plasmids contain two origins of replication. One origin is required for replication in *Bacillus* spp., and a second one, originating from pBR322, is used for replication in *E. coli.* The difference between pMM1520 and pHBintE is the origin for replication in *Bacillus* spp. The Bacillus-specific origin in pHBintE is derived from pE194ts and is temperature sensitive (ts). This temperature sensitivity allows integration of pHBintE into the *B. megaterium* chromosome, described in **Subheading 3.3.** The origin of pMM1520 allows stable free replication of the plasmid at various temperature conditions. Plasmids pMM1520 and pHBintE contain ampicillin-resistance genes for plasmid selection in *E. coli.* An erythromycin (pHBIntE) or a tetracycline (pMM1520) resistance gene allows selection for transformants and maintenance of both plasmids in *B. megaterium* (*see* **Note 3**). DNA manipulations and cloning in *E. coli* were performed by standard procedures *(3)*.

3.2. Transformation of B. megaterium

Prior to the transformation of *B. megaterium* cells, protoplasts isolation is necessary to remove the solid cell wall of this bacterium.

3.2.1. Protoplasts Generation

Protoplasts are very sensitive to physical stress and must be handled very carefully.

1. Inoculate 50 mL of LB medium in a 100 mL flask with 1 mL of a *B. megaterium* overnight culture and incubate at 37°C under aerobic conditions (*see* **Note 4**).
2. Harvest the cells by centrifugation (4500g, 15 min) when the cell density has reached an optical density (OD$_{578}$) of 1.
3. Resuspend in 5 mL of fresh SMMP buffer.
4. Add 50 µL of 1 mg/mL lysozyme in SMMP buffer for cell wall removal.
5. Incubate at 37°C for 60 min (*see* **Note 5**).
6. Harvest the cells carefully by centrifugation (1300g, 10 min).
7. Resuspend in 5 mL of fresh SMMP buffer. Avoid vigorous shaking because protoplasts are very sensitive to physical stress.
8. Repeat **steps 6** and **7** one more time.
9. After resuspension of the cell pellet, the protoplasts are either directly used for transformation or frozen for later use. Protoplast samples are storable for up to 3 mo at –80°C in SMMP buffer containing 15% (v/v) glycerol. However, the transformation rate is optimal with freshly prepared protoplasts.

3.2.2. Transformation With a Free-Replicating Plasmid

1. Air-dry about 1 µg of pHBintE comprising the cloned target gene.
2. Add 500 µL of protoplast suspension prepared as outlined under **Subheading 3.2.1.**
3. Add 1.5 mL of PEG-P solution and incubate for 2 min at room temperature. This procedure makes the cell membrane permeable to DNA.
4. Add 5 mL of SMMP buffer and mix very gently.
5. Harvest the cells carefully by centrifugation (1300g, 10 min) at room temperature.
6. Decant the supernatant immediately after centrifugation and resuspend the pellet carefully in 500 µL of SMMP. The pellet might not be visible, so do not mix vigorously!
7. Incubate the cells for 90 min at 37°C and gentle shaking.
8. While shaking, prepare 2.5-mL aliquots of CR5 top agar. Combine the sterile CR5 solutions A and B (melt prior to use!) and keep the complete CR5 top agar liquid at 42°C.
9. Add aliquots of 50 µL, 100 µL, and the rest of the transformed *B. megaterium* cells to the liquid 2.5-mL aliquots of CR5 top agar.
10. Mix gently by rotating the vessel.
11. Pour the top agar containing the suspended cells on prewarmed LB plates (37°C) containing 20.8 µM tetracycline (pMM1520) or 6.8 µM erythromycin (pHBIntE).
12. Incubate for 24 h at 37°C.
13. Select single colonies and grow overnight at 37°C on LB plates containing 20.8 µM tetracycline or 6.8 µM erythromycin.

3.3. Protein Production and Extraction

After successful transformation of *B. megaterium* with a free-replicating expression plasmid like pMM1520, the next step is the induction of gene expression with xylose to initiate protein production.

1. Grow *B. megaterium* cells transformed with the expression plasmid of choice overnight at 30°C under aerobic conditions in LB medium supplemented with 20.8 μM tetracycline.
2. Samples of the overnight culture can be frozen at -80°C in LB medium containing 25% (v/v) glycerol.
3. Inoculate individual amounts of LB medium containing 6.8 μM erythromycin with aliquots (1% of the culture volume) of the aerobically grown overnight culture and allow the cells to grow at 30°C under aerobic conditions to an OD_{578} of 1 (corresponds to a cell density of approx 10^9 cells/mL).
4. Induce gene expression from the *xylA* promoter by the addition of 33 mM xylose (*see* **Note 1**).
5. The maximum of protein production is reached 3–5 h after induction of gene expression.
6. Harvest the cells by centrifugation (4500g, 15 min).
7. Resuspend in 1–2 vol of lysis buffer.
8. Incubate for 30 min at 37°C.
9. Examine the degree of lysis under the microscope.
10. Destroy the protoplasts by vigorous shaking.
11. The lysate is ready for protein extraction or analysis by sodium dodecyl sulfate–polyacrylamide gel electrophoresis (SDS-PAGE) using standard molecular biology methods *(3)*.

3.4. Chromosomal Integration

Stable maintenance of recombinant genes on a free-replicable plasmid requires selection via expensive antibiotics. Especially in industrial production processes, this cost-intensive procedure is not desired. Alternatively, stable complementation of auxotrophic phenotypes by plasmid-borne open reading frames serves the same purpose. However, this method requires a selectable marker (i.e., appropriate mutants and complementing cloned genes). Moreover, conventional expression plasmids are limited in their efficiency to carry and simultaneously express multiple genes of interest. Therefore, stable integration of the complete expression construct into the *B. megaterium* chromosome is the method of choice to circumvent outlined problems. Additionally, the simple integration of the inducible *xylA* promoter upstream the chromosomal copy of *B. megaterium* genes or operons of interest allows the high-level production of homologous proteins. More detailed information about the integration procedure is provided in **Subheading 3.4.1.**

3.4.1. General Integration Instruction

Integration of plasmids via recombination into the *B. megaterium* chromosome is a very rare process, even if the bacterium is successfully transformed. It is necessary to use a screening method for this integration event to distinguish strains with integrated plasmids from strains with free-replicable expression plasmids. For this purpose, we use the temperature-sensitive origin of replication of the expression plasmid pHBintE. This origin of replication admits reliable replication only below temperatures of 37°C. No plasmid replication is possible above the permissive temperature of 42°C, because a component of the plasmid-encoded replication system is heat sensitive. The chromosomal integration is based on a single recombination event. It requires crossing over between a DNA region of the chromosome and an identical DNA sequence of the plasmid. In order to obtain efficient crossing-over rates, the identical DNA of the plasmids requires a minimal length of a least 300 basepairs. For successful chromosomal integration of a DNA of interest use the following steps.

1. Grow *B. megaterium* cells transformed with the expression plasmid pHBintE carrying the gene of interest overnight at 30°C under aerobic conditions in LB medium with 6.8 μM erythromycin (*see* **Note 6**).
2. Inoculate 150 mL of LB medium containing 6.8 μM erythromycin with 1 mL of the overnight culture and allow growing at 30°C under aerobic conditions for about 12 h.
3. Inoculate in fresh LB medium analogous to **step 2** and cultivate for 6 h more.
4. Shift the temperature from 30°C to 42°C and cultivate an additional 6 h.
5. Inoculate fresh medium again from the culture of **step 4** and cultivate at 42°C for another 12 h. Repeat this step four times.
6. Samples of the culture taken at several time-points of incubation can be stored at –80°C in LB medium containing 25% (v/v) glycerol.

3.4.2. Integration of the xylA Promoter Downstream of the B. megaterium Operon hemAXCDBL Coding for Proteins Involved in Tetrapyrrole Synthesis

The pathways for the biosynthesis of the tetrapyrroles heme and vitamin B_{12} share the initial part for the formation of uroporphyrinogen III from glutamyl-tRNA. Involved genes of *B. megaterium* are encoded by the *hemAXCDBL* operon. Elevated expression of these genes overcomes the initial limiting steps of tetrapyrrole biosynthesis and, thus, increases vitamin B_{12} production by a factor of 30. In order to achieve significantly increased *hemAXCDBL* transcription, the xylose-inducible *xylA* promoter was integrated into the *B. megaterium* chromosome upstream of the target operon. For this purpose, recombination of a plasmid encoded *hemA* copy under the control of the *xylA* promoter with its chromosomal counterpart was designed. Therefore, the first gene of the operon, *hemA* (1341 bp), was placed under the control of the xylose-inducible promoter

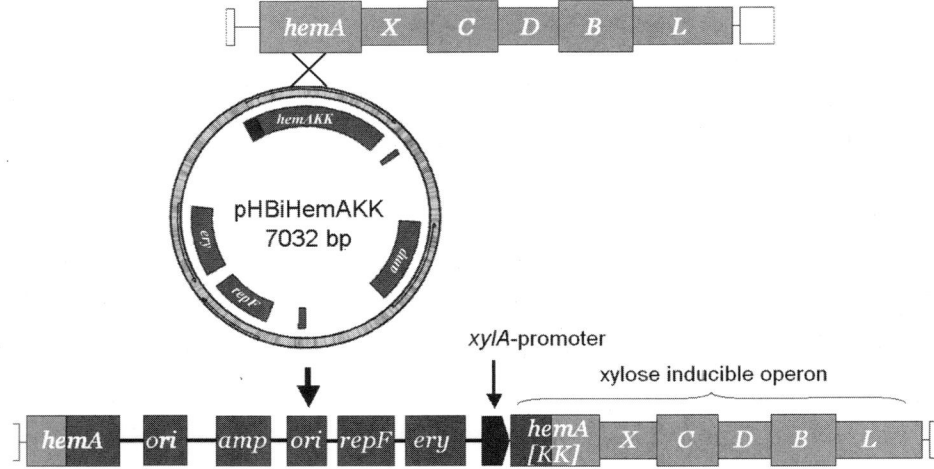

Fig. 2. Principle of the single-crossing-over integration of pHBiHemAKK into the *B. megaterium* chromosome. This chromosomal integration of the *xylA* promoter upstream of *B. megaterium hemA* allows the xylose-induced transcription of the whole chromosomal *hemAXCDBL* operon.

of pHBintE. Using standard polymerase chain reaction (PCR) and cloning techniques *(3)*, the *hemA* gene was isolated and ligated into *Spe*I–*Kpn*I restriction sites of the multiple cloning site of pHBintE (*see* **Fig. 1**). Second, this *hemA* gene-containing vector, called pHBiHemAKK, was transformed into *B. megaterium* DSM509 (*see* **Subheading 3.2.**). The freshly transformed strain allowed expression of *hemA* by xylose induction (*see* **Subheading 3.3.**) of the free-replicable pHBiHemAKK when the strain was cultured at 30°C. Third, to allow expression of the complete *hemAXCDBL* operon, chromosomal integration of the plasmid pHBiHemAKK was necessary. For this reason, transformants were treated as outlined in **Subheading 3.4.1.** leading to a site-specific chromosomal integration via single-crossing-over recombination into the *B. megaterium* chromosome (*see* **Fig. 2**). The integration process was controlled by PCR and Southern blot hybridization (data not shown). Protein formation was checked by SDS-PAGE of the xylose-induced strain.

3.5. Protein Secretion

Microorganisms secrete enzymes, so-called exoenzymes, into their habitats to attack and utilize highly polymeric nutrients. In *B. megaterium*-type strain DSM319, several proteins are secreted. One of these is the only extracellular protease NprM. Significantly increased stability of the secreted proteome was achieved by the elimination of this protease by Wittchen and Meinhardt *(5)*, creating the strain MS941. This strain provides a suitable host for the production and secretion of heterologous proteins. Here, we describe the recombinant production and secretion of a glucosyltransferase, the dextransucrase DsrS from *Leuconostoc mesenteroides* NRRL B512-F. This approach also illustrates that the strain is applicable for the successful formation and secretion of high-molecular-weight mass proteins, up to a M_r = 200,000.

Five secretion pathways are known in *Bacilli*, of which two are of main interest for the transport of heterologous proteins (*see* **refs.** *6* and *7* for reviews). The best characterized is the SEC-dependent protein transport. The SEC system recognizes a certain signal peptide of the protein. The protein is completely unfolded and transported through a canal in the cytoplasmic membrane into the surrounding growth medium. A signal peptidase at the surface of the cell removes the signal peptide by cleavage at its recognition sequence, releasing the mature protein into the medium. The following methods describe the (1) cloning of the dextransucrase gene with its native signal peptide into the expression plasmid pMM1520, (2) the construction of the ubiquitously applicable export vector pMM1533, (3) an SDS-PAGE-based assay for the analysis of the glucosyltransferase activity, and (4) methods to determine the efficiency of the secretion process by tracking the glucosyltransferase on its way out of the cell.

3.5.1. Construction of a Vector for the Export of Proteins

The construction of a *DsrS* vector for the production and export of a glucosyltransferase includes (1) the evaluation of the signal peptide of *dsrS* using bioinformatics (*see* **Note 7**), (2) the optimization of the ribosomal binding site, and (3) the final cloning of the *dsrS* gene into pMM1520.

3.5.1.1. EVALUATION OF THE DSRS LEADER SEQUENCE USING THE SIGNALP ALGORITHM

Signal peptides are N-terminal amino acid extensions of a protein that mediate its export out of the bacterial cell. Examples of signal peptides from *B. subtilis* and *B. megaterium* proteins are shown in **Table 1.** During the secretion process, the signal peptide is cleaved off by a signal peptidase releasing the mature protein into the broth. Secreted proteins with signal peptides from 24 to 65 amino acids were found in *B. subtilis* *(8)*. The signal peptides in *Bacilli* consist of three characteristic domains: (I) The positively charged N-terminal domain contains usually two to three basic residues such as arginine (R) or lysine (K). (II) The hydrophobic core region structurally organizes itself into α-helices via interaction with the membrane. A glycine or proline residue usually disrupts the structure of the hydrophobic helix. (III) The more hydrophilic

Table 1
Signal Peptides of Secretory Proteins of *B. subtilis* and *B. megaterium*

Protein	No.[a]	aa[b]	Signal peptide	
B. subtilis				
α-Amylase	NP_388186	660	MFAKRFKTSLLPLFAGFLLLFHLVLAGPAAASA	ETA
Bacillopeptidase F	NP_389413	1433	MRKKTKNRLISSVLSTVVISSLLFPGAAG	ASS
Chitosanase	NP_390566	277	MKISMQKADFWKKAAISLLVFTMFFTLMMSETVFA	AGL
Extracellular serine Protease	NP_391719	645	MKNMSCKLVVSVTLFFSFLTIGPLAHA	QNS
Levansucrase	NP_391325	473	MNIKKFAKQATVLTFTTALLAGGATQAFA	KET
Levanase	NP_390581	677	MKKRLIQVMIMFTLLTMAFSADA	ADS
Endo-β-1,4-glucanase	JN0111	499	MKRSISIFTCLLITLTMGGMLASPASA	AGT
B. megaterium				
α-Amylase	CAA30247	520	MKGKKWTALALTLPLAASLSTGVDA	ETV
α-Amylase	AAK0059	533	MRGSLRIKYANVFFRGDGVQMFKRITTVGLSVVMFLPSIYGGSKAYA	DTV
β-Amylase	CAB61483	543	MKQLCKKGLAFVLMFIFVNAFILSPINGAAA	VDG
Neutral protease	CAA52964	562	MKKKKQALKVLLSVGILSSSFAFAHTSSA	APN
Penicillin amidase	CAA85774	802	MKTKWLISVIILFVFIFPQNLVFA	GED
Estcrasc	CAD23620	210	MKKVLMAFIICLSLJLSVLAAPPSGAKA	ESV

Note: Domains and cleavage sites of signalpeptidase were predicted by the SignalP-HMM-algorithm *(4)*. n-terminal basic domain; hydrophobic core region, c-terminal domain with cleavage site (depicted as blank) of signal peptidase type I.

[a] Accession number of Entrez database of NCBI.
[b] Amino acid count of protein with signal peptide.

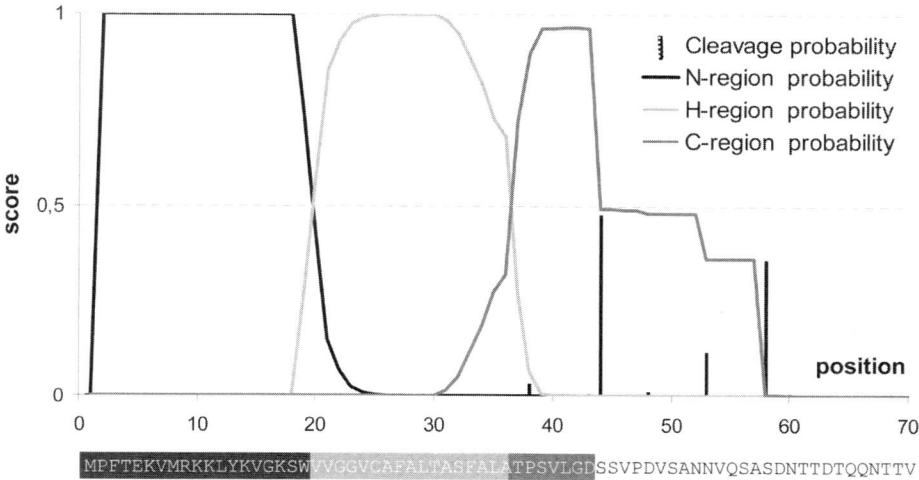

Fig. 3. SignalP V2.0 analysis by a hidden Markov model of the three characteristic domains of dextransucrase DsrS (AAA53749.1) and the predicted most probable cleavage site at G43.

C-terminal domain contains the cleavage site for a type I signal peptidase *(6,7)*. Using the algorithm of SignalP V2.0, it was verified that the dextransucrase DsrS contains a signal peptide that is predicted to be functional in Gram-positive *Bacilli* (*see* **Fig. 3**).

3.5.1.2. OPTIMIZATION OF THE RIBOSOME-BINDING SITE

Retaining the original amino acid sequence of the signal peptide of the *L. mesenteroides* dextransucrase DsrS, a transcriptional fusion using the expression vector pMM1520 placing the *dsrS* gene under control of the *xylA* promoter was constructed. Moreover, the open reading frame of *xylA* was terminated by a stop codon. The *dsrS* gene contained its own *L. mesenteroides*-specific ribosome-binding site (RBS). In *Bacilli,* according to experiments with *B. subtilis* *(9)*, the RBS sequence plays a more pronounced role compared to Gram-negative bacteria. Optimal RBSs have from 7 to 10 basepairings between the mRNA and the 3′ end of the 16S rRNA. The start codon AUG has the highest translation efficiency, followed by UUG and GUG. The spacer region between RBS and the start codon, with an optimal length from 7 to 9 basepairs, should not include cytosines or guanines. The formation of secondary structures in this region is detrimental to the binding of the ribosome. According to these prerequisites, the RBS of *dsrS* from *L. mesenteroides* was analyzed and systematically optimized.

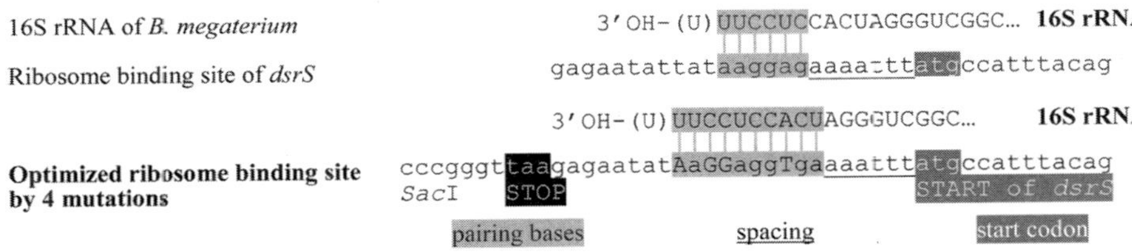

16S rRNA of *B. megaterium*   3′ OH- (U) UUCCUC CACUAGGGUCGGC... **16S rRNA**

Ribosome binding site of *dsrS*   gagaatattat aaggag aaaa·ttatg ccatttacag

                               3′ OH- (U) UUCCUCCACU AGGGUCGGC... **16S rRNA**

Optimized ribosome binding site   cccggt taa gagaatat AaGGaggTga aaatttatg ccatttacag
by 4 mutations   SacI   STOP   START of *dsrS*

	pairing bases	spacing	start codon
Original ribosome binding site of *dsrS*	6	7	ATG
Optimized ribosome binding site of *dsrS*	10	6	ATG
Optimal ribosome binding site according to Vellanoweth [9]	7–10	7–9	ATG

Fig. 4. Strategy for optimization of ribosome binding site of *L. mesenteroides dsrS* for expression in *B. megaterium*. Four more basepairings of the *dsrS* mRNA and the 16S rRNA were introduced via primer of a PCR amplificating *dsrS*. The optimal spacing between the RBS and the start codon was retained.

The gene encoding 16S rRNA from *B. megaterium* QM B1551 was recently sequenced (*9*). The 3′ end of the deduced 16S rRNA sequence is completely conserved between *B. megaterium* and *E. coli* except for the adenine forming the 3′ end of the 16S rRNA of *E. coli*. This base is absent or substituted by an uracil residue in *B. megaterium* 16S rRNA. Four bases enhancing the pairing of the *dsrS* mRNA RBS with the 16S rRNA were introduced using a PCR-based strategy (the approach is depicted in **Fig. 4**).

3.5.1.3. Cloning of the dsrS Gene Into pMM1520

The dextransucrase DsrS is a glucosyltransferase originated from *L. mesenteroides* NRRL-B512F. Its gene was sequenced and cloned in *E. coli* (*10*).

1. PCR reactions are set up in a thermocycler with a heated lid as follows: 20 pmol of oligonucleotide primers A and B, 50 ng chromosomal DNA. 1.25 mM MgCl$_2$, 0.2 mM dNTPs, and 5 µL *Taq* polymerase (5 U) in a total volume of 20 µL. The 5′ ends of the primers include restriction sites for *Sac*I and *Sph*I, respectively. *Taq* polymerase is added after an initial denaturing step of 95°C for 8 min.
2. PCR reactions are programmed as follows: initial denaturing step, 8 min 95°C, 30 cycles, denaturing, 1 min 95°C, annealing, 1 min 59.2°C, elongation 1 min 72°C; final elongation step 8 min 72°C.
3. Separate the PCR product from primer by agarose gel electrophoresis (*3*).
4. Extract the DNA from the gel by standard procedures (*3*).
5. The integration of the *Sac*I and *Sph*I restriction sites in the oligonucleotide primers enables a cleavage of the PCR product with these restriction endonucleases and the

cloning of the product into the multiple cloning site of the equivalently cleaved, purified, and dephosphorylated vector pMM1520 *(3)*.

6. Transform ligation mixture into *E. coli* DH10B cells by electroporation and select colonies on LB–ampicillin (100 μg/mL) plates.
7. Perform miniplasmid preparation to obtain plasmid DNA *(3)*.
8. Perform a restriction endonuclease-digest analysis of the obtained plasmids. The expected sizes of the vector and of the insert are 7.4 and 4.5 kb, respectively.
9. Determine the sequence of the insert.
10. Transform the new expression plasmid pMM1520dsrS into *B. megaterium* MS941 as outlined under **Subheading 3.2.**

3.5.2. Construction of a Vector Encoding the LipA Signal Peptide From B. megaterium

Here, we describe a method for the construction of an expression plasmid encoding a signal peptide from *B. megaterium* protein. The new plasmid (pMM1533) enables the secretion of heterologous proteins by cloning their genes in frame into the multiple cloning site downstream of the signal peptide-encoding DNA. The signal peptide of choice was the recently discovered extracellular esterase LipA from *B. megaterium* *(11)* (*see* **Note 8**). This whole system is xylose inducible and is based on the pMM1520 plasmid.

The required cloning strategy was chosen as described below. First, the new restriction site 5′-TGTACA-3′ of the restriction enzyme *Bsp*1407I was incorporated prior to the translational start codon into pMM1520. The primers C and D were used in a Quick Change reaction. The resulting new shuttle vector was designated pMM1522. This vector was cleaved with *Bst*BI and *Bsp*1407I, dephosphorylated, and purified. The signal peptide of LipA contains 28 amino acids with a corresponding DNA sequence of 84 nucleotides. Three pairs of oligonucleotides of a length of approx 30 basepairs for the *de novo* construction of the 84-nucleotide DNA sequence were designed as such that after insertion into pMM1522, synthesis of the signal peptide sequence was controlled by the *xylA* promoter.

1. Design appropriate sequences of oligonucleotides (*see* **Fig. 5**).
2. Phosphorylate 200 pmol of each oligonucleotide E to J by 10 U of T4 polynucleotide kinase in 20 μL for 1 h at 37°C, followed by heat inactivation at 65°C for 20 min *(3)*.
3. Combine the reaction mixtures of the pairs of oligonucleotides, denature for 3 min at 95°C, and hybridize at the appropriate temperature for 1 min and at 55°C for another minute.
4. Ligate the now double-stranded oligonucleotides with 400 U of T4 ligase (10 min at 25°C and 12 h at 16°C).
5. Ligate the resulting DNA fragment into the vector pMM1522 with a 100X molar excess of the oligonucleotides.

218

A Original DNA sequence and signal peptide of extracellular esterase LipA

```
5' gtgaaaaagtacttatggcattcattattgtttatcgctgattcatcgttttagccgtccgcgtctggcgcaaagct 3'
   M  K  K  V  L  M  A  F  I  I  C  L  S  L  I  L  S  V  L  A  A  P  P  S  G  A  K  A ...
```

B Its construction with 6 oligo nucleotides and addition of a new *KasI* restritcion site directly at the signalpeptidase cleavage site AXA

```
Bsp1407I                                                  KasI      BstBI
TGTACA                                                    GGCGCC    TTGCAA

5' GTACAAtgaaaaagtacttatggcattcattattgtttatcgctgattcatcgttttagccgtccgcgtctggcgcaGGCGCCGCATT 3'
3'     TTacttttcatgaataccgtaagtaataacaaatagcgactaagatagacaaatcggcaggcggcagaccggtCCGCGGCGTAAGC 5'
        M  K  V  L  M  A  F  I  I  C  L  S  L  I  L  S  V  L  A  A  P  P  S  G  A  G  A  A  F
```

Fig. 5. (**A**) The original DNA sequence and the resulting amino acids of the signal peptide of the extracellular esterase of *B. megaterium* are shown. The cleavage site AXA of the signal peptidase typeI is underlined. (**B**) Six oligonucleotides were phosphorylated, ligated, and cloned into the depicted restriction sites *Bsp1407I* and *BstBI* of pMM1522, placing the signal peptide under the control of the promoter of *xylA*. At the variable amino acid X in the signal peptidase cleavage site AXA, LipA contained a lysine. This was changed into a glycine, creating a new restriction enzyme site of *KasI*, *SfoI*, and *NarI* for direct cloning of target genes into the cleavage site.

6. Transform ligation reactions into *E. coli* cells.
7. Select transformants on ampicillin containing LB plates.
8. Check mini preparations of plasmid DNA of the clones for the new restriction site *Kas*I (*see* **Note 9**).
9. Sequence plasmids of at least 10 positive clones.

3.5.3. Detection of Secreted Proteins

3.5.3.1. ACTIVITY STAIN OF GLUCOSYLTRANSFERASES

The staining procedure relies on the formation of insoluble dextran from sucrose by partially renatured dextransucrase after SDS-PAGE separation. A visible white precipitate is formed in the SDS gel.

1. Prepare preculture of *B. megaterium* MS941 containing the plasmid pMM1520dsrS in 50 mL LB and shake gently at 37°C (*see* **Note 10**).
2. After 16 h, raise agitation to vigorous shaking for 30 min, transferring the culture in the exponential growth phase.
3. Inoculate 100 mL of phosphate-buffered medium supplemented with yeast extract with 1 mL of this preculture and grow again with vigorous agitation at 37°C (*see* **Note 11**).
4. Induce the DsrS production at OD_{578} of 0.3 with the addition of 0.5% (w/v) xylose.
5. Take samples at 3, 6, 9, and 12 h after induction.
6. Harvest the cells by centrifugation (4500g, 10 min).
7. Add 10 µL of SDS sample buffer without mercaptoethanol to 90 µL of the supernatant.
8. Load 6% SDS-PAGE with 15 µL of the samples and perform gel electrophoresis.
9. Wash the gel three times in activity stain washing buffer at 4°C to dilute the SDS. This results in refolding of the enzyme in the gel matrix.
10. Incubate the gel at 37°C in this buffer supplemented with 100 g/L sucrose for 16 h. Dextransucrase catalyzes the transfer of the glucose residue from sucrose to form dextran, which is an α 1–6 linked chain of glucose residues. White bands in the transparent gel show the formation of dextran.
11. Deposit gel in 50% ethanol to visualise traces of dextran by shrinkage of the gel matrix and precipitation of the dextran polymer.

In the case of the *L. mesenteroides*-derived dextransucrase (DsrS) expressed in *B. megaterium* MS941, a secretion maximum was found after 3 h. No obvious protein degradation was observed (*see* **Fig. 6** and **Note 12**).

3.5.3.2. LOCALIZATION OF EXPORTED RECOMBINANT PROTEIN

Generally, the location of a secreted protein is experimentally determined via a protease sensitivity protection assay. This test probes the accessibility of the protein of interest to protease degradation. During the assay, exported protein is cleaved by trypsin treatment, whereas intracellular, membrane-, or cell-wall-localized proteins are at least partially protected from the protease attack. To

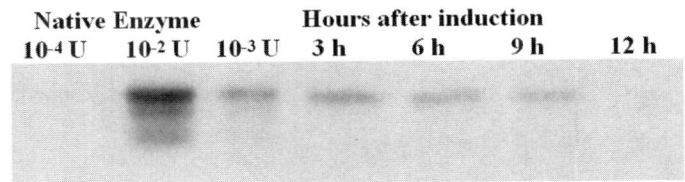

Fig. 6. Secretion of dextransucrase DsrS by *B. megaterium* MS941. Thirteen microliters of cell-free culture broth of *B. megaterium* MS941 pMM1520dsrS were obtained at indicated time-points after induction of DsrS production with 0.5% xylose. Dextran production of partially retained DsrS in the gel is used to stain active DsrS following SDS-PAGE (activity stain).

test for potential cell wall localization, *B. megaterium* protoplasts, generated by lysozyme treatment, can be additionally treated with trypsin. However, protoplast integrity has to be secured. Appropriate additional controls are indicated.

1. Take a sample containing 6×10^9 cells of MS941 after induction (*see* **Subheading 3.5.2.2., steps 1–5**).
2. Harvest the cells by centrifugation (4500*g*, 15 min, 4°C).
3. Resuspend the cell pellet in 1 mL of protoplast buffer (PB) plus lysozyme.
4. Examine the degree of protoplast formation under the microscope. Spherically appearing cells are protoplasted.
5. Incubate until 80–90% of the cells appear spheric-shaped.
6. Handle protoplast very carefully in order not to destroy them (*see* **Note 13**).
7. Divide into three aliquots A, B, and C of 300 µL each: (A) Add 450 µL of PB (negative control without protease); (B) add 83 µL of PB and 367 µL of PB plus trypsin; (C) add 83 µL of triton stock and 367 µL of PB plus trypsin.
8. Incubate 30 min at 37°C.
9. Divide into aliquots of 375 µL and mix immediately with 375 µL of TCA.
10. Keep 2 h on ice to precipitate proteins.
11. Centrifuge 15 min at 10,000*g* and 4°C.
12. Wash twice with 300 µL acetone by centrifugation (10,000*g*, 5 min, 4°C) (*see* **Note 14**).
13. Add 10 µL of 8 *M* urea to resuspend precipitated proteins.
14. Add 10 µL of SDS sample buffer.
15. Heat 5 min at 95°C.
16. Analyze one-half of A, B, and C preparations by 10% SDS-PAGE, followed by staining with Coomassie brilliant blue, and the other half by 6% SDS-PAGE, followed by activity stain (*see* **Subheading 3.5.2.2., steps 8–11**).

The results of the protease protection assay are shown in **Fig. 7.** Samples of *B. megaterium* MS941 containing pMM1520dsrS taken 1 h 30 min after induc-

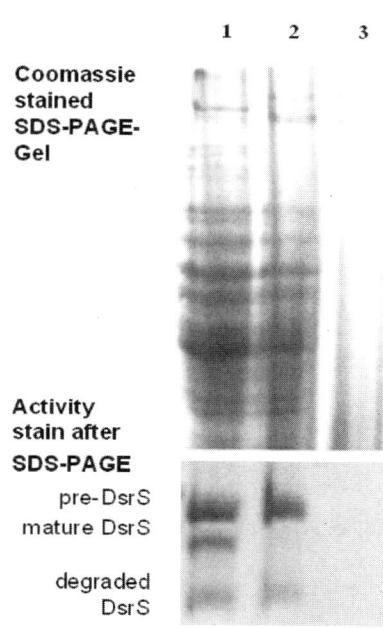

Fig. 7. Localization of dextransucrase DsrS during the production and export in *B. megaterium*. Cells of *B. megaterium* MS941 carrying pMM1520dsrS were harvested 1.5 h after induction of DsrS production with 0.5% xylose and protoplasted. To equivalent portions of the protoplasted solution either pure protoplast buffer, protoplast buffer containing lysozyme, or protoplast buffer containing lysozyme and 2% Triton X-100 were added. The proteins were precipitated by 10% TCA and analyzed by SDS-PAGE and activity staining. *Lane 1:* protoplasts; *lane 2:* protoplasts + trypsin; *lane 3:* protoplasts + trypsin + 2% Triton X-100. The Coomassie-stained SDS-PAGE shows that the protoplasts are intact because *lane 2* compared to *lane 1* shows no significant degradation by trypsin. The activity stain shows that the mature DsrS is located outside of the cytoplasmic membrane because it is degraded by trypsin (*lane 2* compared to *lane 1*). Trypsin is attacking pre-DsrS and degraded DsrS only in broken protoplasts (*lane 3*), demonstrating a localization in the cytoplasm.

tion of the *dsrS* gene expression by the addition of xylose were analyzed by SDS-PAGE. The obtained results indicated that the protoplasts were intact. The mature DsrS is located outside the cytoplasmic membrane because it is degraded by trypsin, whereas pre-DsrS stayed in the cell.

4. Notes

1. The induction time for the expression of the *xylA* promoter should be optimized for each different protein expression. Using the system described here with various

recombinant proteins, the optimal expression was achieved at cell densities of OD_{578} 0.3–1.5. The xylose concentration routinely varied from 0.5% to 1.5% (w/v).

2. Using pMM1520 as a *xylA* promoter-dependent expression vector, it is necessary to use the first restriction site of the multiple cloning site *Spe*I to obtain a fusion protein with the first three amino acids of the XylA protein. In this case, the RBS of the *xylA* promoter sequence is applied. Alternatively, it is also possible to use pMM1520 as an expression vector for genes carrying their own ribosome binding site.

3. The pMM1520 expression system has been successfully used for the production of recombinant proteins from *B. megaterium, B. subtilis, Clostridium cochlearium,* and *E. coli.*

4. We successfully used different *B. megaterium* strains (DSM319, DSM509, MS941, WH320) for protoplast transformation and protein production.

5. The time for cell wall removal by lysozyme treatment could differ for a variety of reasons. It depends on different cell amounts treated and the specific activity of the lysozyme used. The best time to stop lysozyme treatment is when about 60–80% of the cells are protoplasted. The progress in cell wall removal was routinely followed using the microscope: intact *B. megaterium* cells show a rod-shaped morphology, whereas the protoplasts are coccilike.

6. As described in **Subheading 3.4.,** integration or stabilization of the plasmid pMM1520 is controlled by the temperature of the transformed *B. megaterium* cultures. Hence, it is possible to use the same plasmid as a free-replicable expression vector if the transformed strain is always cultured at 30°C. Temperature conditions of 40–42°C lead to an integration of the plasmid into the *B. megaterium* chromosome via single-crossing-over recombination.

7. Prediction of the signal peptide cleavage site should be verified by the determination of the N-terminal amino acid sequence of the purified exported recombinant protein.

8. Always check signal peptide and the complete fusion proteins with SignalP.

9. Not all individual clones showed the new restriction site *Kas*I. Only 1 out of 10 of the sequenced "positive" clones resulted in the desired sequence. Although this method is very quick, it seems to be extremely prone to infidelities in the synthesis of the oligonucleotides.

10. Transformation of the extracellular protease-deficient *B. megaterium* strain MS941 *(5)* resulted in a significantly enhanced stability of the secreted enzymes. The best secretion of DsrS was obtained by cultivation of the cells in a phosphate-buffered medium supplemented with yeast extract.

11. In order to reduce the lag phase of the *B. megaterium* main culture, it is very important to inoculate from a fast aerobically growing preculture (usually OD_{578} from 3 to 4). The strain MS941 is a derivative of the type strain DSM319, which does not sporulate readily under the employed growth conditions. However, the preculture should not enter the stationary phase.

12. The described SDS-gel-based activity stain assay is also applicable to analysis of intracellularly produced glucosyltransferases. The potential accumulation of pre-

protein could be the bottleneck in the secretion process. This could be the result of aggregation of preprotein in the cell, as also seen for preDsrS.

13. The formation and integrity of the protoplasts is an extremely critical process. Do not use detergent-washed glassware. Detergents like Triton X-100 destroy the protoplasts. Some titration of the amount of lysozyme might be necessary either to enhance protoplast formation or to decrease it. The formation should not take longer than 30 min. An additional buffer exchange before **step 6** to lysozyme free PB by centrifugation (1260*g*, 10 min, room temperature) and resuspension could result in more stable protoplasts. Pipet slowly and do not introduce air bubbles. Mix gently by rolling the tubes.

14. After TCA precipitation, it is essential to completely remove residual TCA. Acidic samples will not run adequately during SDS-PAGE. The acetone used for washing is supplemented with bromophenol blue to check for remaining acidity (yellow). One microliter of 5 *M* NaOH can be added to restore neutral pH.

References

1. Vary, P. (1994) Prime time for *Bacillus megaterium. Microbiology* **140,** 1101–1013.
2. Rygus, T. and Hillen, W. (1991) Inducible high-level expression of heterologous genes in *Bacillus megaterium* using the regulatory elements of the xylose-utilization operon. *Appl. Microbiol. Biotechnol.* **35,** 594–599.
3. Sambrook, J., Fritsch, E. F., and Maniatis, T. (1989) *Molecular Cloning: A Laboratory Manual,* 2nd ed., Cold Spring Harbor Laboratory Press, Cold Spring Harbor, NY.
4. Nielsen, H. (1997) Identification of prokaryotic and eukaryotic signal peptides and prediction of their cleavage sites. *Protein Eng.* **10,** 1–6.
5. Wittchen, K. D. and Meinhardt, F. (1995) Inactivation of the major extracellular protease from *Bacillus megaterium* DSM319 by gene replacement. *Appl. Microbiol. Biotechnol.* **42,** 871–877.
6. van Wely, K. H. (2001) Translocation of proteins across the cell envelope of Gram-positive bacteria. *FEMS Microbiol. Rev.* **25,** 437–454.
7. Tjalsma, H. (2000) Signal peptide-dependent protein transport in *Bacillus subtilis:* a genome-based survey of the secretome. *Microbiol. Mol. Biol. Rev.* **64,** 515–547.
8. Antelmann, H. (2001) A proteomic view on genome-based signal peptide predictions. *Genome Res.* **11(9),** 1484–1502.
9. Vellanoweth, R. L. and Rabinowitz, J. C. (1992) The influence of ribosome-binding-site elements of translational efficiency in *B. subtilis* and *E. coli* in vivo. *Mol. Microbiol.* **6,** 1105–1114.
10. Monchois, V. (1997) Characterization of *Leuconostoc mesenteroides* NRRL B-512F dextransucrase (DSRS) and identification of amino-acid residues playing a key role in enzyme activity. *Appl. Microbiol. Biotechnol.* **48,** 465–472.
11. Ruiz, C. (2002) Analysis of *Bacillus megaterium* lipolytic system and cloning of LipA, a novel subfamily I.4 bacterial lipase. *FEMS Microbiol. Lett.* **217,** 263–267.

12

Strategies for Large-Scale Production of Recombinant Proteins in Filamentous Fungi

Heidi Sisniega, José-Luis del Río, María-José Amaya, and Ignacio Faus

Summary

A pilot-plant process for the production of a recombinant sweet-tasting protein in the filamentous fungus *Aspergillus awamori* is described. Each step of the scale-up and downstream processes (concentration, diafiltration and ion-exchange chromatography) for the production and purification of the secreted recombinant protein is detailed.

Key Words: Scale-up; *Aspergillus awamori;* thaumatin; fermentation; cultivation; diafiltration.

1. Introduction

Filamentous fungi have been in use for a long time as hosts for the production of recombinant proteins *(1)*. Among the different fungal species, *Aspergillus* strains and, specifically, *Aspergillus awamori* strains are widely utilized in the biotechnological industry because of their superior ability to secrete large quantities of protein into the culture medium *(2)*. *A. awamori* has been used for the production of recombinant proteins such us glucoamylase *(3)*, xylanase *(4)*, chymosin *(5,6)*, propectinase *(7)*, human lactoferrin *(8)*, and human interleukin-6 (IL-6) *(9)*. In addition, *A. awamori* is accepted in the food industry as a safe microorganism. It is nontoxigenic and nonpathogenic. Products extracted from *A. awamori* have long been used for human food applications (Food and Drug Administration Generally Regarded as Safe [FDA GRAS] approved).

Historically, most work on the optimization of heterologous gene expression in *A. awamori* has focused on strain improvements to increase the productivity of a given protein. These improvements involve overproduction strategies such as using strong promoters, construction of synthetic genes, increasing gene dosage, controlling the proper folding of the protein within the endoplasmic reticulum (ER), getting a correct proteolytic cleavage of preproteins during

From: *Methods in Biotechnology, Vol. 18: Microbial Processes and Products*
Edited by: J. L. Barredo © Humana Press Inc., Totowa, NJ

secretion, or reducing the fungal endogenous protease activity. Once a strain is improved, production has to be performed by fermentation. Fungal fermentation is a complex process in which several problems can occur. Scale-up and optimization of this process requires optimizing of the fermenting medium, the initial condition of the inoculum, morphology of the fungus, and operation parameters such as temperature, pH, and dissolved oxygen *(10,11)*.

Thaumatin is an intensely sweet protein that has been approved for human consumption. It is currently used as a sweetener in human food, as excipient in pharmaceutical products, and as a feed additive *(12)*. This sweet protein is at present extracted from the arils of the fruit of the plant *Thaumatococcus danielli* Benth. However, the availability of the fruit from which it is extracted is uncertain, and its extraction is difficult, thus limiting its widespread use. There have been several attempts to produce thaumatin by the use of recombinant DNA technology in many different microorganisms *(13–15)*. The results have met with several rates of success, but most of them have been considered disheartening. Nevertheless, more recently, acceptable yields have been reported from recombinant *A. awamori* strains *(16)*.

The expression of recombinant thaumatin secreted by a high-production *A. awamori* strain will be used to illustrate the process for large-scale production and purification of a recombinant protein in filamentous fungi.

2. Materials

1. Fungal strain: *Aspergillus awamori* TGDTh-4 strain (*see* **Note 1**).
2. PW medium: 1.5% sucrose, 0.1% $NaNO_3$, 0.025% KH_2PO_4, 0.0275% $MgSO_4 \cdot 7H_2O$, 1.5% lactose, 0.25% bacto-peptone, 0.025% corn step solids, 0.2% NaCl, 2.61% KCl, and 0.05% salt solution, pH 6.75.
3. Salt solution for PW medium, 100X: 0.1% $CuSO_4 \cdot H_2O$, 0.3% $FeCl_3 \cdot 6H_2O$, 6% KH_2PO_4, and 1% $FeSO_4 \cdot 7H_2O$.
4. CM medium: 0.5% malt extract, 0.5% yeast extract, and 0.5% glucose.
5. MDFA medium: 6% sucrose, 2.7% glucose, 1.2% $(NH_4)_2SO_4$, 14.4% salt solution I, 0.8% salt solution II, and 0.05% antifoam.
6. Salt solution I for MDFA medium: 2% $Fe(NH_4)_2(SO_4)_2 \cdot 6H_2O$.
7. Salt solution II for MDFA medium: 10.4% K_2HPO_4, 10.2% KH_2PO_4, 1.15% $Na_2SO_4 \cdot 10H_2O$, 0.24% $MgSO_4 \cdot 7H_2O$, 0.02% $ZnSO_4 \cdot 7H_2O$, 0.02% $MnSO_4 \cdot H_2O$, 0.005% $CuSO_4 \cdot 5H_2O$, and 0.05% $CaCl_2 \cdot 2H_2O$.
8. Antifoam: MAZU DF 7960 (Mazer Chemicals, Porett Drive, IL).
9. Phosphate buffer: 25 mM sodium phosphate buffer, pH 7.0.
10. Tris-buffer saline (TBS): 50 mM Tris-HCl, pH 8.0, and 150 mM NaCl.
11. Blocking buffer: 10% dry milk, and 0.05% NP-40 in TBS.
12. Phosphate-buffered saline (PBS): 0.8% NaCl, 0.02% KCl, 0.115% Na_2HPO_4, and 0.02% KH_2PO_4.
13. PBS-T: 0.1% Tween-20 in PBS.
14. 5-Bromo-4-chloro-3-indolylphosphate toluidinium – nitroblue tetrazolium solution (BCIP-NBT) (Sigma, Barcelona, Spain).

15. Nytal, pore diameter 200 μm (Maissa, Barcelona, Spain).
16. Prefilter Nucleopore, cat. no. 211114, 47-mm diameter (Quimigranel, Barcelona, Spain).
17. Magnetic shaker.
18. Braun Biostat® B bioreactor (Braun Biotech S.A., Barcelona, Spain).
19. Braun Biostat UD bioreactor (Braun Biotech S.A.).
20. Proflux TM M12 Tangencial Filttration System (Amicon, Barcelona, Spain).
21. HiPrep 16/10 CM–Sepharose column (Amersham Biosciences, Barcelona, Spain).
22. Chromatography equipment.
23. Capillary electrophoresis equipment.
24. Mini vertical electrophoresis system.
25. Stainless-steel filter holder (Millipore, Barcelona, Spain).
26. Scanning Autoreader and Microplate Workstation (Biotek Ceres 900C, Barcelona, Spain).
27. Thaumatin (Sigma).

3. Methods

This section has been divided into five subsections, sequential steps required to carry out the cultivation and the purification process of thaumatin (*see* **Fig. 1**).

3.1. The Preinoculum

1. Seed the *A. awamori* TGDTh-4 strain at 3×10^5 spores/mL in three Petri dishes containing PW solid media and allow to sporulate at 30°C for 3 d (*see* **Note 2**).
2. Scrap the spores and place them in an 1-L cultivation bottle containing 100 mL of sterilized CM media.
3. Set the stirring at 400 rpm using a magnetic stirrer and allow growth for 30–36 h at 30°C (*see* **Note 3**).

3.2. The Inoculum: Biostat B

1. Load the Biostat B bioreactor with 2 L of MDFA growth media and autoclave.
2. Inoculate 200 mL of preinoculum under sterile conditions (*see* **Note 4**).
3. Perform the cultivation at the following conditions: Set temperature at 30°C, airflow rate at 1 vvm (air volume/medium volume per minute), and the percentage of dissolved oxygen at 30%. Maintain O_2 by adjusting the stirring conditions through a multiphase cascade control with an agitation rate between 34.1% and 53.9% of the total power input (*see* **Note 5**). The pH should never be allowed to fall below 6.2. Maintain it at this level by the steady addition of 7.5 *N* NaOH.
4. Allow the process to proceed for a total of 48 h.

3.3. The Cultivation Run: Biostat UD

1. Fill up the Biostat® UD bioreactor, which has been previously sterilized *in situ,* with 50 L of MDFA growth media (*see* **Note 6**)
2. Inoculate the 2.2 L of inoculum grown in the Biostat B bioreactor through a pump system. Make sure that a sterile technique is always used.

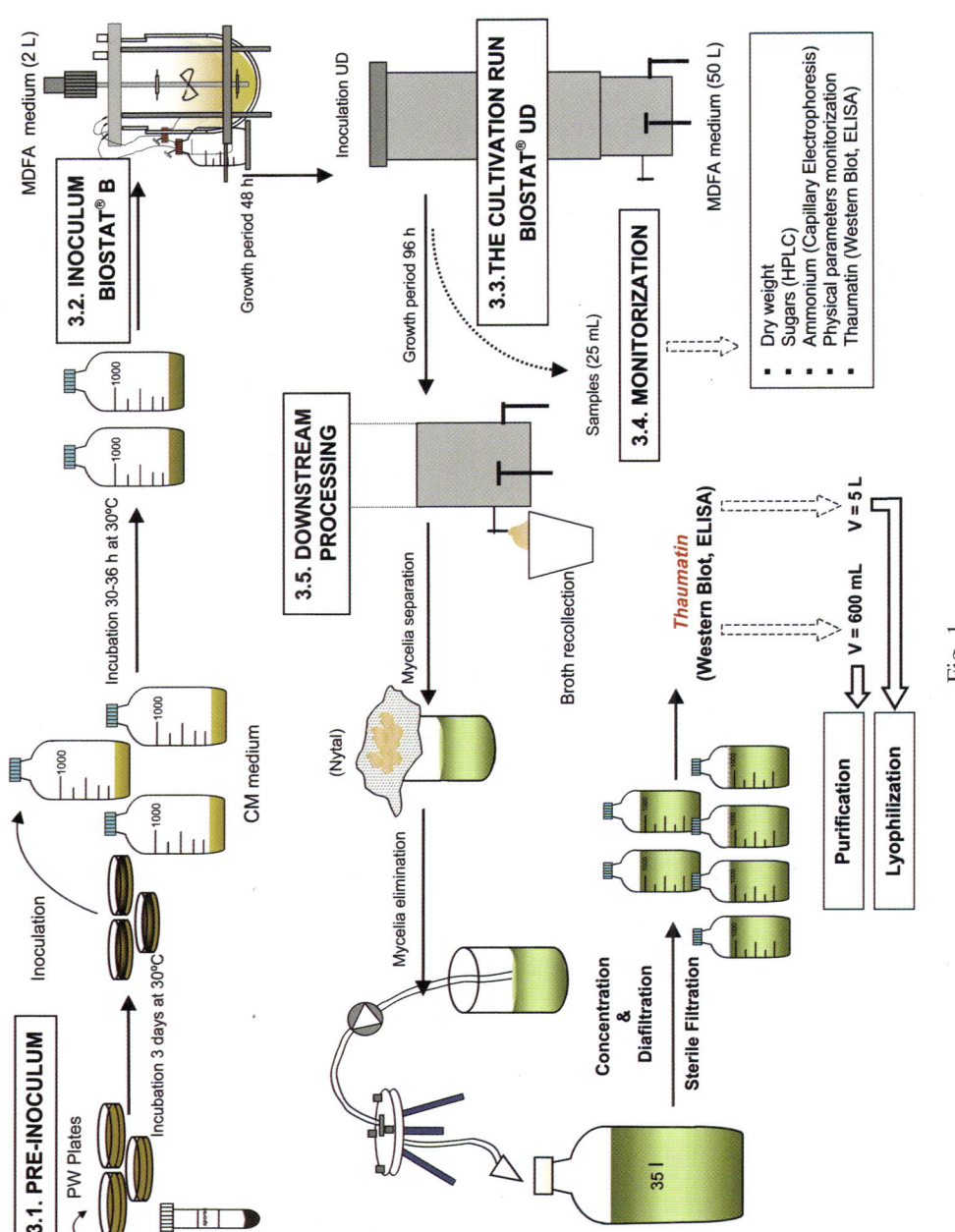

Fig. 1.

3. Perform the cultivation run under the same conditions as describe in **Subheading 3.2.**, except for the stirring range, which should be set between 16.2% and 30.2% of the total stirrer power input (*see* **Note 5**).
4. Proceed with fungal growth for 96 h. Thaumatin production reaches a maximum at about this time.
5. To monitor the cultivation, take samples of about 25 mL at 24-h intervals and process as detailed in **Subheading 3.4.**

3.4. Monitorization of the Cultivation

Temperature, pH, and dissolved oxygen are monitored automatically by the Biostat UD bioreactor software (Micro-MFCS) (*see* **Fig. 2**). The analytical methods used to determine dry weight, sugar and ammonium concentration, and thaumatin concentration are described as follows (*see* **Figs. 3** and **4**).

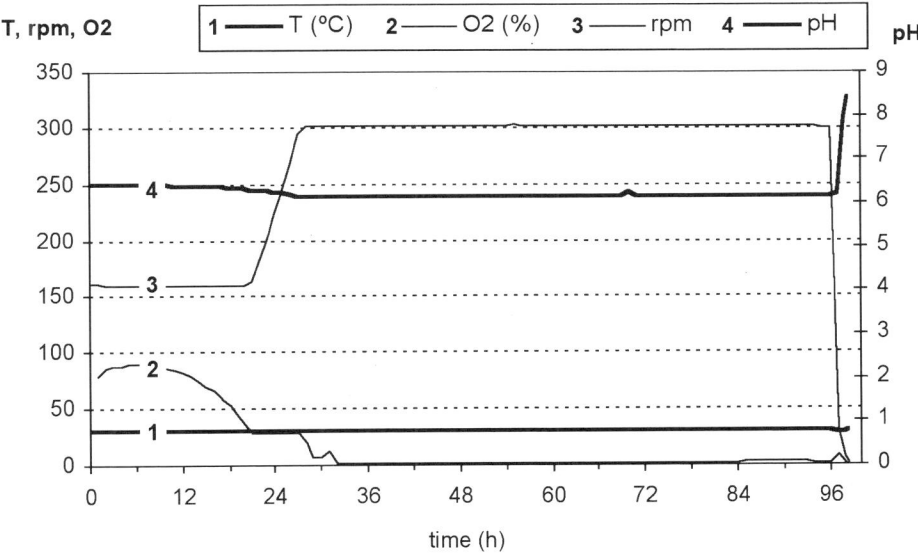

Fig. 2. Time course of those growth parameters that are automatically monitored in the cultivation run of *A. awamori* TGDTh-4 strain in the Biostat UD bioreactor.

Fig. 1. Flowchart of the scale-up cultivation process. The main steps are represented: preinoculum in the cultivation bottles, inoculum in the Biostat B bioreactor, and final growth in the Biostat UD bioreactor. The concentration, diafiltration, and downstream steps are also indicated, as well as the analytical methods used to characterize the fermentation process and subsequent thaumatin production.

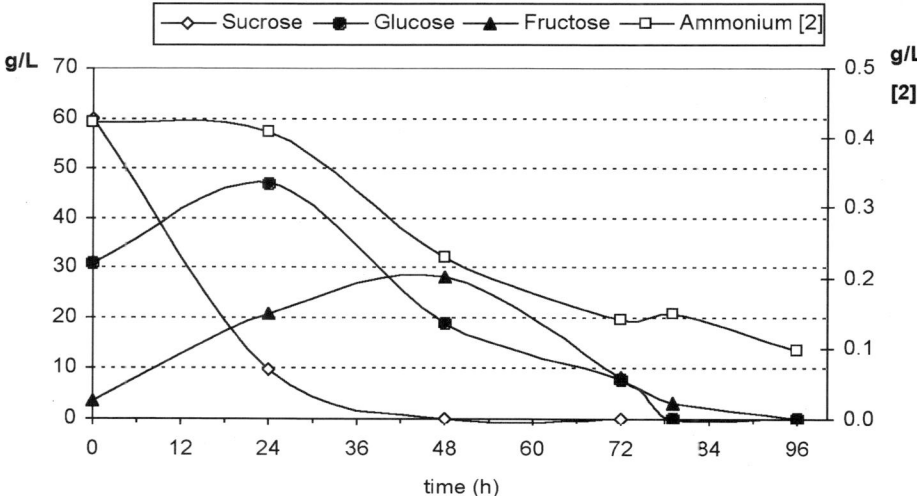

Fig. 3. Time course of sugar and ammonium consumption in the cultivation run of *A. awamori* TGDTh-4 strain in the Biostat UD bioreactor.

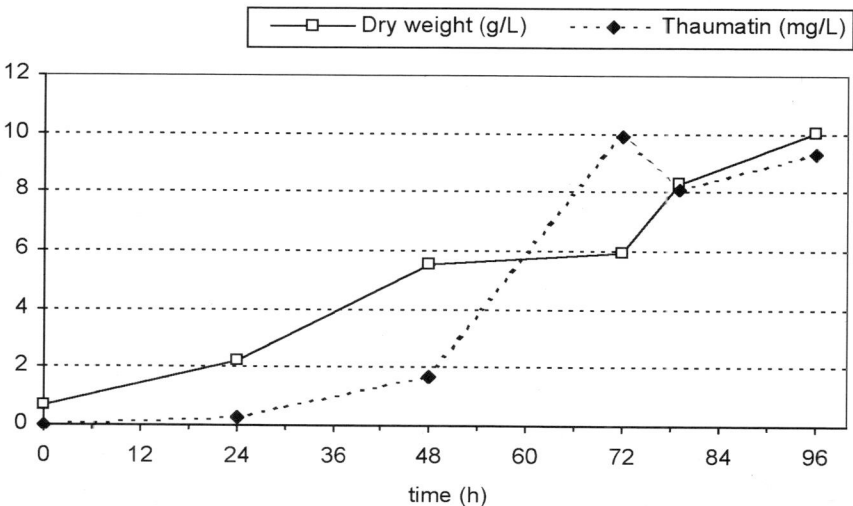

Fig. 4. Time course of thaumatin production and dry weight of mycelium in the cultivation run of *A. awamori* TGDTh-4 strain in the Biostat UD bioreactor.

3.4.1. Dry Weight

Dry weight is determined by passing 10 mL of sample through a prefilter Nucleopore that has been previously desiccated and tared. The biological material retained in the prefilter is washed with 40 mL of pure ethanol and 50 mL of distilled water. It is then incubated at 90°C and weight measurements are taken every 12 h until a constant weight is recorded. Divide the filtrate into aliquots. Freeze them until they are used for the analysis described in **Subheadings 3.4.2.** and **3.4.3.**

3.4.2. Sugar and Ammonium Concentration

Maltose, glucose, and fructose concentrations are determined by high-performance liquid chromatography (HPLC) with refraction index detector. Ammonium concentration is measured by capillary electrophoresis.

3.4.3. Thaumatin

3.4.3.1. IMMUNOBLOTTING

Perform immunoblot analysis with 35 µL of filtered supernatant sample. The samples can be loaded directly on a 14% sodium dodecyl sulfate (SDS)-polyacrylamide gel. After SDS–polyacrylamide gel electrophoresis (SDS-PAGE), transfer the proteins to a nitrocellulose membrane by using a semidry electroblotting system. Treat the membrane with antithaumatin polyclonal antibodies (*see* **Note 7**) diluted 1 : 5000 in blocking buffer for 2 h, wash three times (10 min for each wash) with TBS containing 0.05% NP-40, and incubate for 1 h with the commercial alkaline phosphatase conjugated secondary antibody diluted 1 : 5000 in the same buffer. After that, wash the membrane again and finally treat with a BCIP-NBT solution until the color is developed (*see* **Fig. 5**).

3.4.3.2. ELISA

Quantification of thaumatin in the culture medium is made by enzyme-linked immunoassay (ELISA). Coat the wells of plates with dilutions of supernatant samples from 1 : 10 to 1 : 1280 and incubate overnight at 4°C. The following day, wash the plates three times with PBS-T, block by incubating the wells with PBS-T containing 5% nonfat dry milk for 1 h, and wash again three times with PBS-T. Measure the thaumatin concentration by the addition of a 1 : 10,000 dilution of the antithaumatin polyclonal antibodies for 1 h, followed by a 1 : 5000 dilution of commercial alkaline phosphatase conjugate secondary antibody for 30 min. Quantify the antigen–antibody complex by measuring absorbance at 405 nm and 620 nm in a Scanning Autoreader and Microplate Workstation. Decreasing concentrations of commercial thaumatin diluted in PBS are used as a standard from 1 µg/mL to 1.6 ng/mL.

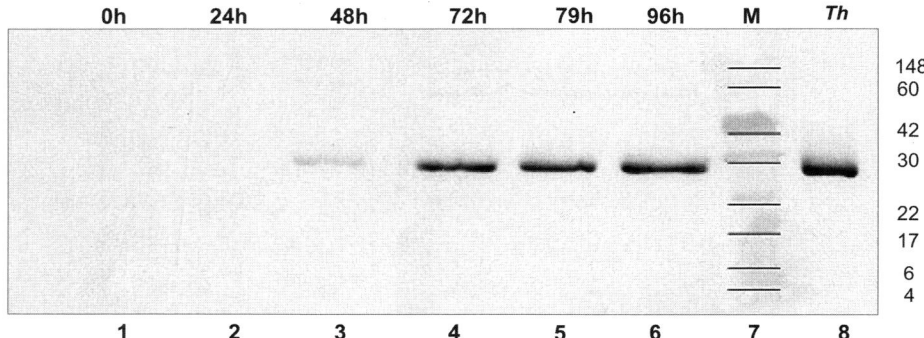

Fig. 5. Western blot of thaumatin produced at different times during the fermentation process. Samples were resolved in a 14% SDS-PAGE gel and transferred to a nitrocellulose membrane for Western blot as described in the text. The numbers to the right of the panel indicate the molecular weight of protein standards (M). *Lane 8* contains a sample of commercial thaumatin (200 ng).

3.5. Downstream Processing

3.5.1. Concentration and Diafiltration

At the end of the fermentation run:

1. Remove biomass by filtration through a Nylon cloth (Nytal).
2. Filter immediately, by pumping the precleared broth through two 142-mm-diameter prefilters using a stainless-steel filter holder (Millipore).
3. Concentrate and diafilter the cultivation broth using a Proflux™ M12 Tangential Filtration System. The system configuration consists of base unit, level switch, 2.5-L reservoir, cooling coil, inlet and outlet pressure transducers, secondary pump, and one spiral-wound membrane cartridge S1Y3 (molecular-weight cutoff = 3000 Daltons).

The steps taken to operate the system are as follows:

a. Calibration of pressure sensors.
b. Adjustments of alarm setpoints: low inlet pressure = 3.0 bars, high inlet pressure = 3.5 bars, differential pressure = 0.3 bars.
c. Washing of the system and the cartridges with deionized, distilled water.
d. Filling-up of the reservoir with process solution; the solution is kept at 8–10°C by recirculating cold water through the cooling coil using a refrigerated circulation system.
e. Setting of the level switch at the desired concentration volume (one-fourth to one-fifth of the initial volume).
f. Operation of the recirculation pump at 75%.
g. Adjustment of the back-pressure valve to obtain a 3.0-bar inlet. If necessary, reduce back-pressure during operation (*see* **Note 8**).

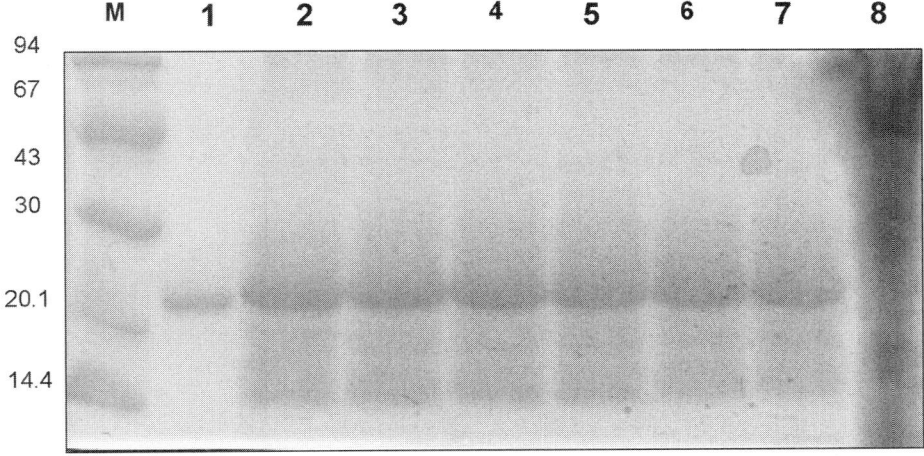

Fig. 6. Ion-exchange chromatography. Fractions collected from the CM–Sepharose column were resolved in a 14% SDS-PAGE gel and stained with Coomassie blue. The numbers to the left of the panel indicate the molecular weight of protein standards (M). *Lane 1* contains a sample of commercial thaumatin (4000 ng). *Lane 8* contains a sample of the concentrated and diafiltered fermentation broth before passage over the CM–Sepharose column. (Not all of the collected fractions are represented in the panel.)

4. After concentration, dialfilter against 5 vol of deionized water (*see* **Note 9**).
5. Drain the concentrated, diafiltered broth from the system.
6. Sterilize the protein solution by filtration.
7. Store at 4°C.

3.5.2. Ion-Exchange Chromatography

Recombinant thaumatin can be purified by cation-exchange chromatography using a carboxymethyl (CM)–Sepharose column.

1. Load the desired volume of diafiltered broth onto the column (23.5 × 1.6 cm), which has been previously equilibrated with 25 m*M* sodium phosphate buffer, pH 7.0, at a flux of 0.5 mL/min. We used 600 mL of the precleared, sterilized cultivation broth.
2. Wash the column with 25 m*M* sodium phosphate buffer, pH 7.0, until the absorbance at 280 nm returns to that of the buffer solution.
3. Elute the bound material with a linear gradient of 0–400 m*M* NaCl at 0.5 mL/min flow rate.
4. Collect samples of 2- to 5-mL intervals and identify fractions of interest by Coomassie blue staining on a SDS-PAGE gel (*see* **Fig. 6**). Do final verification of having the protein of interest by immunoblot (*see* **Fig. 7**).

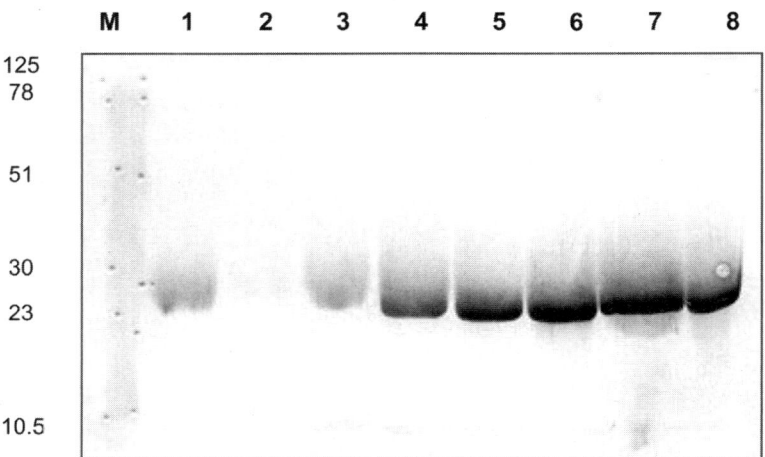

Fig. 7. Ion-exchange chromatography. Fractions collected from the CM–Sepharose column were resolved in a 14% SDS-PAGE gel and transferred to a nitrocellulose for Western blot. The numbers to the left of the panel indicate the molecular weight of protein standards (M). *Lane 1* contains a sample of commercial thaumatin (150 ng). (Not all of the collected fractions are represented in the panel.)

The purity of the samples was greater than 95% as determined by Coomassie blue staining (*see* **Note 10**).

3.5.3. Lyophilization

The lyophilization process was performed to test the stability of thaumatin, in order to assess its long-term storage properties. The lyophilization process consists of five steps:

1. Freezing of the sample. This should be done gradually for 4 h until a temperature of –40°C is reached.
2. Primary drying or sublimation. The sample is warmed by increasing the temperature very slowly to +20°C. The process takes about 40 h under vacuum conditions.
3. Secondary drying or desorption. The sample is maintained at +20°C for several hs under vacuum conditions.
4. Collect the dry solid and store at 4°C
5. Analyze protein by imunoblot analysis (*see* **Fig. 8**).

4. Notes

1. *Aspergillus awamori* TGDTh-4 strain is a derivative from an *A. awamori* Lpr66 mutant strain, which is deficient in aspergillopepsin A expression and has been

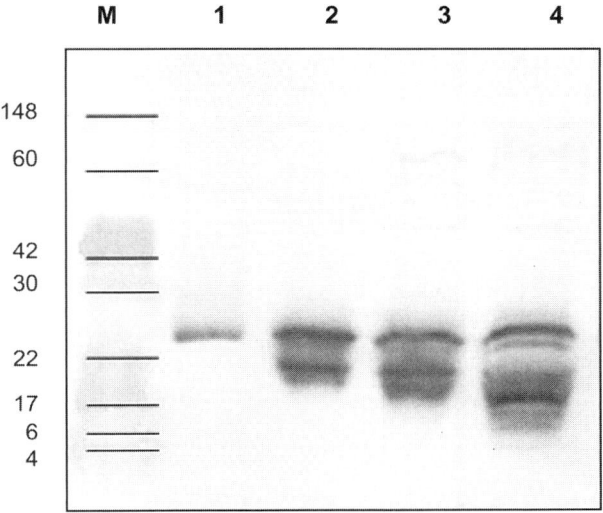

Fig. 8. Western blot of thaumatin from the lyophilized broth. A solution in water of 20 mg of lyophilized material per L was prepared. Three different volumes of 10, 20, and 40 µL were loaded (*lanes 2, 3,* and *4,* respectively) in a 14% SDS-PAGE gel . The numbers to the left of the panel indicate the molecular weight of protein standards (M). *Lane 1* contains a sample of commercial thaumatin (200 ng). It is important to note that in addition to the band corresponding to thaumatin, other bands of lower molecular weight could be detected, which indicates that some degradation of thaumatin during the lyophilization process occurred. Nevertheless, a solution made from the lyophilized material appeared to be as sweet as the nonlyophilized solution (palability tests made in our laboratory).

transformed with an expression plasmid containing a synthetic gene encoding thaumatin under the control of the glutamate dehydrogenase *(gdhA)* promoter *(16,17)*.

2. The plates should be completely covered by sporulated fungi. Sometimes, complete sporulation can take up to 1 wk.

3. It is very important that the inoculum has the proper morphology because this parameter can determine the amount of secreted protein. The bulk of protein secretion occurs at the hyphal tip and, therefore, factors that contribute to an increase in filamentous growth can have a very important effect on product yield. There are several parameters that influence growth, such as the viscosity of the media, dissolved oxygen tension in the bulk, shear forces, transition metal ions, and so forth. In this case, we have observed that growing the fungi in bottles and stirring the media with a magnetic stirrer at a high–moderate rate is the best way to get homogeneous and dispersed mycelia. It is better than using traditional flasks and shaking them in a rotatory shaker.

4. A suspension of spores can be also used for the inoculation of the Biostat B bioreactor. However, this method is not recommended because under these conditions, the mycelia can grow as pellets, which is not optimal for heterologous gene expression.

5. Mechanical forces as a result of stirrer power input and the dissolved oxygen tension in the bulk influence fungal morphology. In general, filamentous mycelia are the dominant morphology at a higher stirring rate *(18)*. During fermentation, the mycelia tend to stick to the oxygen probe. Therefore, it is advisable to increase the stirring rate for 2–3 min twice a day. Nevertheless, at the end of the fermentation run, the mycelia completely cover the probe and it is very difficult to remove the fungus completely. This does not affect the cultivation process, but it must be taken into account that from this moment, the dissolved oxygen concentration cannot be measured and the stirring is set automatically at a maximum percentage of the preset agitation power.

6. The MFDA medium for the Biostat UD bioreactor is sterilized separately from the bioreactor, as the ammonium evaporates when this medium is sterilized inside the bioreactor. The fungus cannot grow without ammonium.

7. Commercial thaumatin was purified by electroelution of the 22-kDa thaumatin band from a SDS–14% polyacrylamide gel. Antithaumatin antibodies were obtained by immunizing New Zealand rabbits by standard procedures *(19)*.

8. In this step, some quantities of protein can be lost if the protein is too highly concentrated or the molecular-weight cutoff of the cartridge is wrongly chosen. We suggest concentrating the protein no more than four times, as the protein can flocculate. A molecular-weight cutoff three times lower than the molecular-weight of the protein is also recommended.

9. The solution is diafiltered in order to remove salts and other low-molecular-weight contaminants. Other methods can be used, such as gel or ion-exchange chromatography.

10. The final goal of the purification scheme is to obtain a protein of sufficient purity, by means of an efficient and economically viable process. A well-chosen and optimized purification process should consist of no more than four steps to fulfill the aforementioned requirements (protein purity and economical process). Thaumatin purification is achieved in three steps: (1) concentration of the fermentation broth to reduce volume, (2) diafiltration to remove salts and trace impurities, and (3) ion-exchange chromatography to isolate the protein from the pretreated fermentation broth. Combination of other separation and chromatographic techniques could be chosen, bearing in mind they should be practical, effective, scalable, and economical.

Acknowledgments

We are indebted to Dr. Carlos Solá (Dean of the Universidad Autónoma de Barcelona, UAB) for allowing us the use of the bioreactors and to Dr. Pau Vila (Department of Chemical Engineering, UAB) for his technical assistance. We also thank Mrs. Mª Lluisa Vericat (Department of Toxicology of Grupo Uriach) for preparing the antithaumatin polyclonal antibodies and to Dr. Alberto

Fernández (Department of Pharmacology of Grupo Uriach) for his helpful technical assistance in the purification process.

References

1. Punt, P. J., van Biezen, N., Conesa, A., Albers, A., Mangus, J., and van den Hondel, C. A. M. J. J. (2002) Filamentous fungi as cell factories for heterologous protein production. *Trends Biotechnol.* **20,** 200–206.
2. Gouka, R. J., Punt, P. J., and van den Hondel, C. A. M. J. J. (1997) Efficient production of secreted proteins by *Aspergillus:* progress, limitations and prospects. *Appl. Microbiol. Biotechnol.* **47,** 1–11.
3. Verdoes, J. C., Punt, P. J., Schrickx, J. M. (1993) Glucoamylase overexpression in *Aspergillus niger:* molecular genetic analysis of strains containing multiple copies of the *gla* gene. *Trans. Res.* **2,** 84–92.
4. Smith, D. C. and Wood, T. M. (1991) Xylanase production by *Aspergillus awamori.* Development of a medium and optimization of the production of extracellular xylanase and β-xylosidase while maintaining low protease production. *Biotechnol. Bioeng.* **38,** 883–890.
5. Ward, M., Wilson, L., Kodama, K., Rey, R., and Berka, R. (1990) Improved production of chymosi in *Aspergillus* by expression of a glucoamylase–chymosin fusion protein. *Biotechnology* **8,** 435–440.
6. Dunn-Coleman, N., Bloebaum, P., Berka, R., et al. (1991) Commercial levels of chymosin production by *Aspergilus. Biotechnology* **9,** 976–982.
7. Hours, R. A., Katsuragi, T., and Sakai, T. (1994) Growth and protopectinase production of *Aspergillus awamori* in solid-state culture at different acidities. *J. Ferment. Bioeng.* **78,** 426–430.
8. Ward, P. P., Piddington, C. S., Cunningham, G. A., Zhou, X., Wyatt, R. D., and Connelly, O. M. (1995) A system for production of commercial quantities of human lactoferrin: a broad spectrum natural antibiotic. *Biotechnology* **13,** 498–502.
9. Broekhuijsen, M. P., Mattern, I. E., Contreras, R., Kinghorn, J. R., and van den Hondel, C. A. M. J. J. (1993) Secretion of heterologous proteins by *Aspergillus niger:* Production of active human interleukin-6 in a protease-deficient mutant by KEX2-like processing of a glucoamylase–hII 6 fusion protein. *J. Biotechnol.* **31,** 135–145.
10. Mackenzie, D. A., Gendron, L. C. G., Jeens, D. J., and Archer, D. B. (1994) Physiological optimization of secreted protein production by *Aspergillus niger. Enzyme Microb. Technol.* **16,** 276–280.
11. Archer, D. B., Mackenzie, D. A., and Ridout, M. J. (1995) Heterologous protein secretion by *Aspergillus niger* growing in submerged culture as dispersed or aggregated mycelia. *Appl. Microbiol. Biotechnol.* **44,** 157–160.
12. Faus, I. (2000) Recent developments in characterization and biotechnological production of sweet-tasting proteins. *Appl. Microbiol. Biotechnol.* **53,** 145–151.
13. Faus, I., Patiño, C., del Río, J. L., del Moral, C., Sisniega, H., and Rubio, V. (1996) Expression of a synthetic gene encoding the sweet-tasting protein thaumatin in *Escherichia coli. Biochem. Biophys. Res. Commun.* **229,** 121–127.

14. Faus, I., Patiño, C., del Río, J. L., et al. (1997) Expression of a synthetic gene encoding the sweet-tasting protein thaumatin in the filamentous fungus *Penicillium roquefortii. Biotechnol. Lett.* **19,** 1185–1191.

15. Faus, I., del Moral, C., Adroer, N., et al. (1998) Secretion of the sweet-protein thaumatin by recombinant strains of *Aspergillus niger va. awamori. Appl. Microbiol. Biotechnol.* **49,** 393–398.

16. Moralejo, F. J., Cardoza, R. E., Gutiérrez, S., and Martín, J. F. (2000) Thaumatin production in *Aspergillus awamori* by use of expression cassettes with strong fungal promoters and high gene dosage. *Appl. Environ. Microbiol.* **65,** 1168–1174.

17. Moralejo, F. J., Cardoza, R. E., Gutiérrez, S., Sisniega, H., Faus, I., and Martín, J. F. (2000) Overexpression and lack of degradation of thaumatin in an aspergillopepsin A-defective mutant of *Aspergillus awamori* containing an insertion in the *pep*A gene. *Appl. Micobiol. Biotechnol.* **54,** 772–777.

18. Cui, Y. Q., van der Lans, R. G., and Luyben, K. C. (1998) Effects of dissolved oxygen tension and mechanical forces on fungal morphology in submerged fermentation. *Biotechnol. Bioeng.* **57,** 409–419.

19. Dunbar, B. S. and Schoebel, E. D. (1990) Preparation of polyclonal antibodies. *Methods Enzymol.* **182,** 663–670.

13

Strain and Culture Conditions Improvement for β-Carotene Production With *Mucor*

Enrique A. Iturriaga, Tamás Papp, Jesper Breum, José Arnau, and Arturo P. Eslava

Summary

Carotenoids are chemical compounds that are in an increasing demand in the market because of their applications in the food, feed, cosmetics, and pharmaceutical industries. To date, most of the β-carotene is manufactured by chemical processes, but because it is used as feed and food additives, the interest is growing for β-carotene of biological origin. Several fungal species, particularly some included in the Mucorales, have been used to develop fermentation processes for the production of β-carotene. In this chapter, we describe an approach to obtain *Mucor circinelloides* strains that could be useful in the industrial production of carotenoids and that, for commercial interests, specifically avoids the use of molecular genetic engineering. The method relies on classical genetic techniques to isolate and characterize β-carotene overproducing mutants and to build up strains that better fit the industrial production. *M. circinelloides* is a dimorphic fungus that grows either as yeast cells or in a mycelial form. This feature can be used to further develop strains with a better industrial potential by isolating monomorphic (yeastlike) mutants or by controlling and modifying the morphology of the organism during batch cultivation.

Key Words: Batch culture; carotene; carotenoid production; dimorphism; food additives; fungi; isoprenoids; *Mucor;* mutagenesis; fungal morphology; strain improvement.

1. Introduction

Carotenoids comprise a group of naturally occurring pigments with a broad range of biological functions. They are derived from the general isoprenoid (terpenoid) pathway by tail-to-tail condensation of prenyl pyrophosphates. A conjugated double-bond system constitutes the most important part of the molecule and is responsible for their biological activity (*1,2*). A number of carotenoids are effective in preventing or controlling free-radical generation, providing protection against singlet oxygen and reactive oxygen species in photosynthetic

From: *Methods in Biotechnology, Vol. 18: Microbial Processes and Products*
Edited by: J. L. Barredo © Humana Press Inc., Totowa, NJ

organisms *(3)*. Other carotenoids have pro-vitamin A activity and are indeed essential components of the human and mammals diet, playing essential roles in nutrition, vision, and cellular differentiation *(4)*. In addition, there is increasing evidence that many carotenoids enhance the immune response *(5)* and inhibit proliferation of various types of cancer cell *(6)*. Many books and reviews have covered all aspects of the biochemistry and genetics of carotenoids in bacteria, fungi, and plants *(7–14)*. The genes involved in the pathway, isolated from many different sources, have also been used in biotechnological research and metabolic engineering of the production of carotenoids *(15–17)*.

A number of carotenoids are used in the animal and human alimentary industries: to intensify the color of the salmon or trout flesh and the egg yolk and as a human food colorant, because the carotenoid contents of foods change as a result of harvesting, handling, and processing methods *(18)*. Although today most of these added carotenoids are synthesized by chemical methods, there is an increasing interest for carotenoids of biological origin because of the public concern over the safety of artificial food colorants. This preference for carotenoids of natural origin has led to a search of natural sources of such compounds *(9,15,19,20)*.

Carotenoid-producing fungi often accumulate a single major carotenoid. Among the Mucorales, β-carotene is the main carotenoid accumulated in *Blakeslea trispora, Mucor circinelloides,* and *Phycomyces blakesleeanus,* and astaxanthin predominates in the basidiomycetous yeast *Xanthophyllomyces dendrorhous.* Many environmental factors influence the accumulation and composition of carotenoids in these fungi. Strain and culture conditions, certain chemicals *(21,22)*, or even the mating interaction in the Mucorales *(23)* affect the total amounts accumulated. Many fungi also increase their carotenoid content when grown under blue light *(24–27)*, although others such as *B. trispora* do not show this response.

The basic technology for the industrial production of carotenoids by fungi is already set up. Many fungi are currently used to produce a wide variety of compounds (different enzymes, antibiotics, amino acids, or recombinant proteins). Although today there is only one example, in Russia and Ukraine, of an industrial production of fungal β-carotene, the start-up in León (Spain) of a new production plant, Vitatene, is scheduled for 2004. The process was first developed in the United States *(28)*, and later improved in France *(29,30)* to get up to 30 mg of β-carotene/g dry weight or about 3 g/L. The fermentation process uses submerged cultures of *B. trispora* with aeration and agitation. Two of the disadvantages of this process are the need for joint fermentation of two strains of opposite sex and the use of chemical additives.

The possibility of using other fungi for the industrial production of β-carotene has been undertaken only at the laboratory level. The carotenoid accumulation in fungi is usually too low for profitable purposes, but different strategies can be used to increase the carotenoid content of these fungi. Strain improvement, optimization of the medium and culture conditions, and the search for chemical activators are some of the different approaches that have been used. In *P. blakesleeanus,* the use of certain mutations *(carS)* and some genetic tricks, such as the use of intersexual heterokaryons with balanced lethal mutations *(31)* or intersexual partial diploids *(32),* have allowed productions of up to 25 mg β-carotene/g dry weight. The potential use of these fungi also have some drawbacks: In *P. blakesleeanus,* there is a decrease in carotenoid accumulation in shaken cultures, so surface cultures must be employed; the growth temperature for *B. trispora* and *P. blakesleeanus* must never be above 25°C; the cell walls of *X. dendrorhous* are so thick that they cannot be used directly for animal feed and must be partially lysed *(33).*

In this context, several reasons make *M. circinelloides* a good candidate. (1) Different species of the genus *Mucor* are presently being used, with different technologies, for the industrial production of enzymes such as lipases, proteases, or rennin, among others. (2) It naturally produces β-carotene; the biosynthesis of carotenoids in this fungus is well known; great progress is being made about its regulation, and several structural and regulatory genes have been isolated, all of which can help in developing β-carotene-hyperproducing strains *(27,34–37).* (3) Strain improvement needs not be restricted to the accumulation of carotenoids, but it can be directed to develop strains more suitable for the industrial production: *M. circinelloides* is a dimorphic fungus *(38).* Depending on environmental conditions like the gas atmosphere or the presence of fermentable carbon sources, *M. circinelloides* can grow either aerobically as a coenocytic (i.e., without septa) mycelium or anaerobically as multipolar yeasts (*see* **Fig. 1A**). The morphological shift occurs rapidly upon the environmental change and is totally reversible (*see* **Fig. 1B**). The yeast morphology is usually preferred by the fermentation industry because it allows the growth in submerged conditions, the biomass production is higher, and the cells are more easily separated from the culture media. Moreover, these yeast cells grow perfectly well at about 30°C. One way to obtain this type of strain could be to isolate mutants carrying an alteration in a gene involved in filamentation (i.e., monomorphic) and that only grows in one of the two morphologies. Another possibility could be to develop a fermentation procedure in which a yeast-to-hyphal growth shift were induced during batch cultivation, because yeasts are optimal for high biomass production, and aerobiosis (hyphal growth) is an absolute requirement for carotenoid production. Thus, this *M. circinelloides* system should combine the advantages of yeast and filamentous fungi *(39).* (4) In addition, the industrial purification of β-carotene is

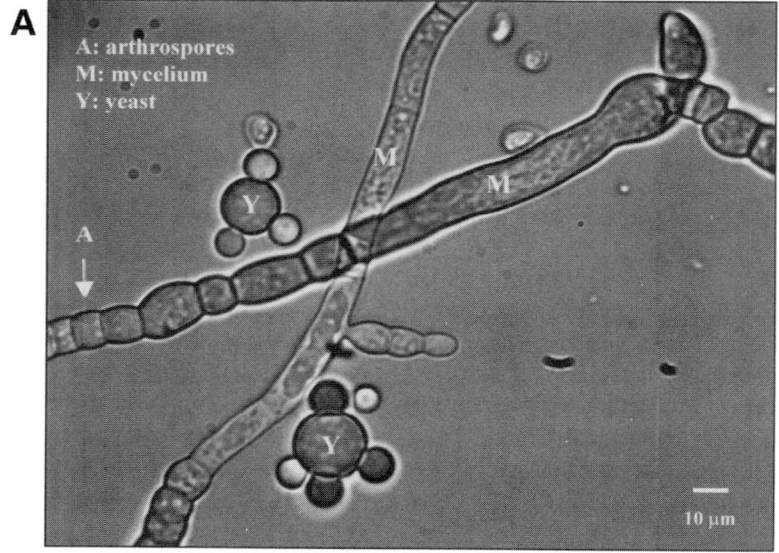

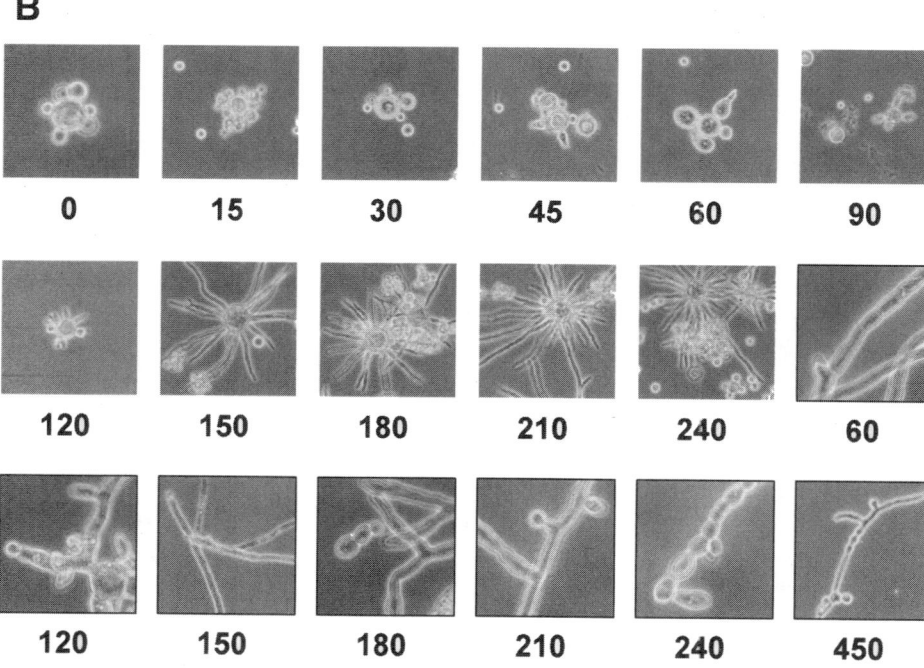

Fig. 1. Dimorphic growth of *M. circinelloides*. (**A**) An image showing some of the structures that *M. circinelloides* can form during submerged growth. (**B**) Reversibility of the dimorphic shift. The yeast-to-hyphae shift is faster; it starts at about 45 min of the environmental change and is fully completed at 240 min. The hyphae-to-yeasts (black-framed images) shift is slower, taking several hours to complete.

complicated and expensive and might not be necessary for many practical applications. In *Phycomyces* and other fungi, it has been shown that overproduction of β-carotene is accompanied by an increase in the total fatty acid content, which can reach up to the 50% dry weight *(40)*. Polyunsaturated fatty acids and γ-linolenic acid (which are essential to mammals) predominate in the composition of the total oils of both *Phycomyces* and *M. circinelloides (41)*. Oil dissolves and stabilizes β-carotene, so a carotene-rich oil saves in the extraction, purification, and conservation of β-carotene. Moreover, the carotene-rich, oil-rich mycelia might be used directly for animal feed. (5) *Mucor* mycelia are also used in microbial biotransformations of steroids *(42)*, and it has been shown that mycelial dead biomasses are effective in heavy-metal biosorption *(43)*. Of course, there are also some drawbacks. *M. circinelloides* could be classified as a crab-tree positive fungus (i.e., it produces ethanol during aerobic growth) *(39,44)*. The problem with aerobic ethanol production is that for *M. circinelloides,* relatively low ethanol concentrations (below 2%) are inhibitory and these levels are obtained during batch fermentation. Moreover, ethanol production is enhanced during anaerobic growth, yielding low biomass. In the next sections, we will illustrate the methods employed to optimize β-carotene production with *M. circinelloides.*

2. Materials

1. Strains: *M. circinelloides* f. *lusitanicus* (previously *M. racemosus*) wild-type strains CBS277.49 and ATCC1216b were obtained from the Centraalbureau voor Schimmelcultures and the American Type Culture Collection, respectively. Strain R7B (ATCC90608) is a leucine auxotroph derived from CBS277.49. Strains MCB1 to MCB 37 are β-carotene-hyperproducing mutants derived from wild-type strain ATCC1216b by NG mutagenesis. Monomorphic mutants MS55 to MS64 and YM1 to YM40 were isolated after spores mutagenesis of different *M. circinelloides* strains (*see* **Note 1**).
2. Media: The different media used in these procedures are described here. When solid medium is needed, agar is added at 20 g/L. Leucine, which is required by strain R7B and its derivatives, is added to the media at 200 mg/L.

    a. Vogels medium: 940 mL distilled water, 5 g/L casamino acids, 20 g/L glucose, 20 mL Vogels solution, 5 mL Vogels trace element solution, and 0.5 mg *d*-biotin. Vogels solution: 650 mL distilled water, 250 g KH_2PO_4, 125 g Na_3–citrate $\cdot$ $2H_2O$, 100 g NH_4NO_3, 10 g $MgSO_4 \cdot 7H_2O$, and 5 g $CaCl_2 \cdot 2H_2O$. Vogels trace element solution: 95 mL distilled water, 5 g $ZnSO_4 \cdot 7H_2O$, 5 g citric acid $\cdot$ H_2O, 1 g $Fe(NH_4)_2(SO_4) \cdot 6H_2O$, 0.25 g $CuSO_4 \cdot 5H_2O$, 0.05 g H_3BO_3, 0.05 g $MnSO_4 \cdot H_2O$, and 0.05 g $Na_2MoO_4 \cdot 2H_2O$.
    b. Yeast nitrogen base (YNB) NB medium: 1 L distilled water, 0.5 g YNB without amino acids, 10 g glucose, 1.5 g $(NH_4)_2SO_4$, and 1.5 g glutamic acid. Adjust pH to 4.5 or 3.0 for normal or colonial growth, respectively.

   c. YPG medium: 1 L distilled water, 20 g glucose, 10 g peptone, and 3 g yeast extract.

   d. Rice–YPG medium: to a sterile flask, add 29 g rice grains, 10 mL sterile distilled water, and 6 mL YPG liquid medium pH 4.5 and autoclave at 121°C for 20 min (longer times result in a too-dry substrate to allow growth).

3. Nitrosoguanidine or NG (*N*-methyl-*N'*-nitro-*N*-nitrosoguanidine; Sigma). Prepare 10 mL of a 1 mg/mL NG stock solution and dispense into aliquots of 1 mL. NG is light sensitive and tubes should be wrapped with aluminum foil. Store at –20°C (*see* **Note 2**).

4. Calcofluor white (CFW, Sigma), for the analysis of chitin deposition.

5. Duran glass filter no. 2 (Duran Schott, Germany). (Miracloth can also be used for the filter enrichment.)

6. High-performance liquid chromatography (HPLC) equipment (Waters) with on-line degasser and an on-line diode array detector. A NovaPack C18 4-μm column (3.9 mm × 150 mm) is used for carotenoid analyses. An Aminex HPX-87H column is used for separation and analyses of carbon sources.

7. A 1.5-L Braun bioreactor, equipped with automatic controllers for pH, temperature, gas flow rate, and agitation speed with rushton turbines. Sources of N_2 and CO_2 for anaerobic growth, and an automatic gas analyzer (Brüel & Kjær, Denmark) to determine the exhaust gases from the bioreactor.

3. Methods

The methods described outline (1) the mutagenesis procedure and the isolation of mutants of *M. circinelloides* with an increased carotenoid content, as well as the analysis of these carotenoids by HPLC, (2) the procedure for the isolation and characterization of monomorphic mutants, and (3) the best conditions for batch cultivation of wild-type and monomorphic mutants.

3.1. Isolation of β-Carotene-Hyperproducing Mutants

In order to isolate β-carotene-hyperproducing mutants of *M. circinelloides,* a series of mutagenesis experiments with NG were performed using the following protocol. The two wild-type strains were used, but, unfortunately, superyellow mutants were only isolated from strain ATCC1216b. Strains MCB1 to MCB 37 are β-carotene-hyperproducing mutants. Their carotenoid content was analyzed by HPLC, and further characterization of these mutants was carried out by analyzing their ability to synthesize β-carotene in light and darkness and their behavior in mating experiments with strains of opposite sex (*see* **Subheading 3.1.2.**).

3.1.1. Mutagenesis of M. circinelloides *Spores*

1. Incubate 10^6–10^8 spores of any of the *M. circinelloides* strains in 100 μg/mL NG, in an Eppendorf tube, in a final volume of 1 mL for 1 h and up to 6 h (*see* **step 6**) at room temperature with occasional mixing. Run a parallel control in which water

is added to the same amount of spores. Keep the tubes packed in aluminum foil to protect NG from light.

2. Wash the spores by centrifugation, discard the supernatant, and resuspend in 1 mL of sterile water (*see* **Note 2**). Vortex the tubes gently.
3. Repeat the washing step three times and resuspend in 1 mL of sterile water.
4. Make dilution series of both the control and the experiment tubes and plate onto YPG pH 3.0 plates to estimate the survival rate. Typically, a survival rate of 1–10% is acceptable.
5. Plate the remaining volume onto 10–20 YPG pH 4.5 plates to generate stocks of mutagenised spores that will be used for the screening (*see* **Note 3**).
6. Count colonies after 2–4 d incubation at 28°C. If the survival rate is higher than 10%, adjusting the NG concentration and/or the time of the NG treatment should be desirable.
7. After 7–8 d, collect the spores from the YPG plates and set up the spore stocks. From each stock, make dilutions to estimate spore concentrations (as in **step 4**) and inoculate 40–100 YPG pH 3.0 plates at a concentration of about 100–200 colonies per plate.
8. Incubate for 4–5 d under continuous illumination and then inspect the plates in order to isolate the colonies that show an increased yellow color. Select and transfer them to fresh YPG plates. Isolate spores and store them at –20°C. This will be the mutant spore stock.

3.1.2. Preliminary Characterization of β-Carotene-Hyperproducing Mutants

1. Plate about 10^3 spores from the individual isolates in parallel onto YPG pH 4.5 plates.
2. Incubate one set of plates under continuous illumination, and the other in the dark at 28°C. After 4–5 d, compare the two plates from each mutant to see if there is synthesis of β-carotene in darkness.
3. To do the mating experiments, plate spores or pieces of mycelium of the two strains to be tested in opposite parts of a YPG pH 4.5 plate and incubate in darkness at 22–25°C for 7–10 d. If the two strains react sexually, there will be an increase in the yellow color in the contact region, and special deep black structures (zygospores) will form.

All isolated mutants showed a higher accumulation of β-carotene than wild type. These assays showed that there are at least three classes among them: those that are able to produce carotenoids in darkness (wild type is not), those that accumulate more carotenoids only when grown under light conditions, and a mutant (MCB25) that accumulates carotenoids under light but does not show reaction in mating experiments (all but this one do).

3.1.3. Extraction and Analysis of Carotenoids by HPLC

3.1.3.1. PREPARATION OF MUCOR CIRCINELLOIDES MYCELIA

1. Plate about 10^3–10^4 spores per 10-cm Petri dish of YNB medium pH 4.5.
2. Incubate for 4 d at 22°C under the appropriate light illumination conditions.

3. Harvest the mycelium by removing it from the agar with tweezers and remove as much water (and agar) as possible by pressing it between two sheets of filter paper.
4. Freeze the mycelium with liquid nitrogen. Store at –80°C, wrapped in aluminum foil to avoid light degradation of carotenoids, or proceed directly to the next step.

3.1.3.2. Extraction of Carotenoids and Dry-Weight Determination

1. Grind the frozen mycelium with pestle and mortar with the continuous addition of liquid nitrogen
2. Wrap 100–200 mg of mycelium powder in a piece of a previously weighed slice of aluminum foil, dry at 100°C for 1 h, and weigh again to calculate the percentage of dry weight.
3. Use 70–140 mg of the mycelium powder per extraction (in a 1.5-mL Eppendorf tube). Store the remaining mycelium at –80°C
4. Add 500 µL of methanol and 500 µL of 10% diethyl ether in petroleum ether (boiling point = 50–70°C).
5. Vortex for 2–3 min and separate the two phases by centrifugation at maximum speed in a microcentrifuge for 4 min.
6. Collect the organic (upper) phase on a fresh tube and repeat the extraction with 500 µL of 10% diethyl ether in petroleum ether at least twice (or until the organic phase is colorless).
7. Pool the collected fractions and dry under nitrogen (preferably on ice), avoiding as much light as possible.
8. Store the dried samples at –80°C or proceed to the next step.

3.1.3.3. HPLC Analysis of Carotenoids (*see* **Note 4**)

1. Redissolve the samples in 200 µL of ethyl acetate, vortex, and centrifuge at maximum speed in a microcentrifuge for 2 min before injection.
2. Use 20 µL of each sample for the HPLC analysis, inject it into a NovaPack C18 4-µm column (3.9 mm × 150 mm), and monitor the elution of carotenoids between 200 and 550 nm. The elution conditions are as follows:

   a. Acetonitrile/2-propanol (85 : 15), 1.2 mL/min for 15 min.
   b. Gradually change to conditions in **step 2c** for 1 min.
   c. Acetonitrile/2-propanol (20 : 80), 1.5 mL/min for 6 min (*see* **Note 5**).
   d. Gradually change to conditions in **step 2a** for 1 min.
   e. Maintain these conditions for at least 7 min before injecting a new sample.

3. Under the previous settings, 15 min are enough to elute all of the different carotenoids present in *M. circinelloides* wild-type and mutant strains (*see* **Note 6**).

3.2. Isolation of Monomorphic Mutants of M. circinelloides

Monomorphic (yeastlike) mutants represent an alternative to wild-type (filamentous) *M. circinelloides* strains in biotechnological processes as diverse as carotenoid biosynthesis or the production of heterologous proteins. A method of choice for the isolation of monomorphic mutants is filter enrichment. The method is based on the fact that, in liquid culture, mutants able to grow as aerobic yeasts will remain smaller in size than wild-type hyphae. Yeast cells are 10–50 μm in diameter, whereas hyphae can grow into long filaments spanning up to several centimeters long (*see* **Fig. 1**). Unfortunately, the particular characteristics of this procedure only enable the isolation of mutants that grow as aerobic yeasts. After growth, hyphae are removed by filtration, which results in a preparation enriched in nonfilamentous cells, although arthrospores, ungerminated spores, and other slow-growing filaments can also be present (*see* **Note 7**). After the enrichment step, cells can be plated onto YPG pH 4.5 plates, and after a few days, the analysis of colonies should reveal individuals with abnormal or yeastlike morphology.

3.2.1. Filter Enrichment

1. Perform a mutagenesis experiment just as described in **Subheading 3.1.1., steps 1–3.**
2. Inoculate 0.1 mL of the mutagenized spores in 100 mL YPG pH 4.5 in a 250-mL baffled flask and grow aerobically overnight at 28°C at 180 rpm. Mycelial growth should be evident (*see* **Notes 8** and **9**).
3. Filter the cultures through Duran glass filter no. 2 (*see* **Note 10**) and wash with water or fresh YPG pH 4.5 medium.
4. Collect the cells in the eluate by passing it through a 0.5-μm Millipore filtration system (*see* **Note 11**).
5. Introduce the filter with sterile tweezers into a tube containing 5 mL water. Vortex briefly and remove the filter. If necessary, scrap the filter with the tweezers
6. Make dilution series and plate onto YPG pH 3.0 to estimate the number of viable cells and incubate for 2–3 d at 28°C. Store the remaining cells at 4°C until needed.
7. Once determined the number of viable spores, inoculate 20–40 YPG pH 4.5 or pH 3.0 plates at a concentration of about 50–200 colonies per plate.
8. Inspect the plates after 1–2 d for yeast colony morphology. A range of morphologies from yeast to mixed yeast–filamentous colonies and wild-type colonies can be expected (*see* **Fig. 2**).

The morphology exhibited by the mutants on YPG plates after 2 d will range from yeast colonies to others showing a mixed morphology, with a center of yeast cells from which filamentous growth occurs (*see* **Fig. 2**).

3.2.2. Preliminary Analyses of Monomorphic Mutants

3.2.2.1. GROWTH ON FERMENTABLE AND NONFERMENTABLE CARBON SOURCES

One way to distinguish the different mutants isolated by the above procedure is to check their ability to grow in fermentable and nonfermentable carbon sources. To do that, spores (or cells, because monomorphic mutants do not

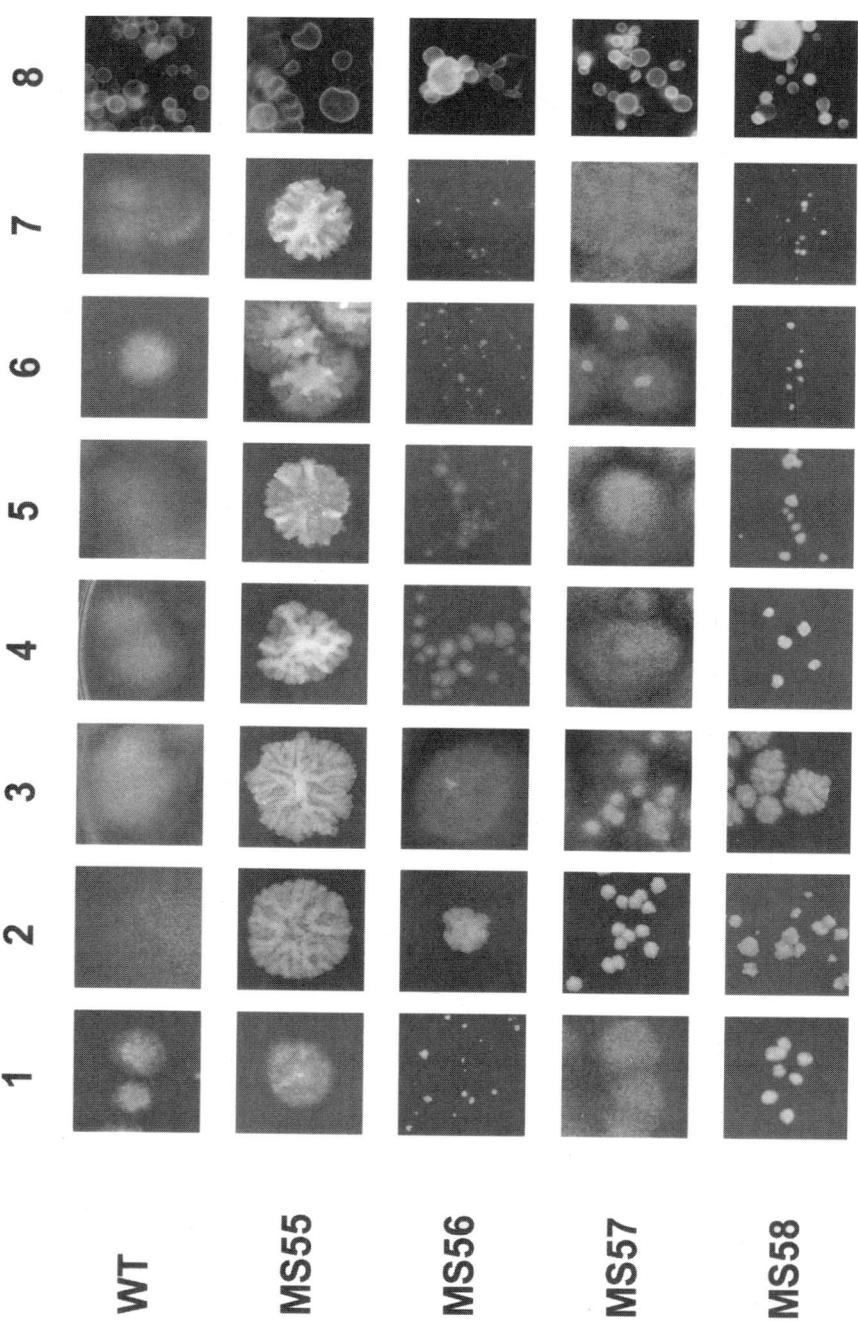

Fig. 2.

usually sporulate) are plated onto YNB or YPG media in which glucose has been substituted by different carbon sources, such as starch, maltose, or glycerol (in molar equivalents to glucose). The pH of the culture medium is also a potential environmental stimulus to induce the morphologic switch, so these carbon sources are tested also at pH 3.0 and pH 4.5 (*see* **Fig. 2**).

3.2.2.2. DEPOSITION OF CHITIN IN MONOMORPHIC MUTANTS

Calcofluor white (CFW) is used to analyze the deposition of chitin and glucans in cell walls and septa. Some of the mutations affecting the morphology of *M. circinelloides* would be expected to affect the structure and/or composition of the cell wall, and these would not necessarily be involved in filamentation. Thus, the isolated mutants are tested with CFW. An example is shown in **Fig. 2.** The steps are summarized as follows:

1. To a 1.5-mL Eppendorf tube coated with aluminum foil, add a diluted 100 µL sample of spores or cells.
2. Add 50 µL of CFW to a final concentration of 300 µg/mL.
3. Incubate at room temperature for 10 min with gentle shaking.
4. Wash twice with water and place a few drops of the sample on a microscope slide; cover with a microscope coverslip. Visualize on a fluorescence microscope with a filter block with a 330- to 380-nm excitation wavelength and a dichroic mirror with emission >420 nm.

3.3. Growth Culture Conditions of M. circinelloides Wild-Type and Monomorphic Mutants

An initial analysis of the *M. circinelloides* growth behavior, searching for pure yeasts or hyphal cells in the culture medium, was performed in five defined minimal media in shake flasks *(39)*. The overall procedures are outlined below.

3.3.1. Aerobic Fermentation of M. circinelloides Strains

Preliminary results showed that an inoculum of 5×10^4–10^6 spores/mL resulted in freely dispersed mycelial growth. Inoculating more spores resulted in entangling of hyphae and a very nonhomogeneous biomass in the reactor. With lower inoculum, a longer lag phase was typically observed, resulting also

---

Fig. 2. *M. circinelloides* monomorphic mutants. Mutants were grown on YNB pH 4.5 plates (1), YPG pH 3.0 (2), YPG pH 4.5 (3), YPM pH 3.0 (YPG in which glucose has been substituted with maltose) (4), YPM pH 4.5 (5), YPGly pH 3.0 (YPG in which glucose has been substituted with glycerol) (6), and YPGly pH 4.5 (7). Deposition of chitin in the monomorphic mutants and wild type (WT) (8).

in hyphal entangling. Therefore, batch fermentation should be started with a low pH. Sometimes, the hyphae can become very sticky, attaching to the metal structure of the fermentor. In our experience, it is possible to keep the mycelium freely dispersed by increasing the stirrer speed. However, care must be exercised not to increase the stirrer speed to a level at which the hyphae might break resulting in cell lysis. Vogels medium provided a high specific growth rate and biomass concentration. YPG medium was also used for fermentation experiments. The initial aim was to obtain either pure yeast or filamentous cultures and then to optimize the processes with each morphological form. In YPG and Vogels medium, purely filamentous cultures were obtained under aerobic cultivation. The protocol shown here typically yields, during aerobic cultivation, a maximum biomass of 5.2 g dry weight/L with a specific growth rate of 0.25 h^{-1}.

3.3.1.1. PREPARATION OF SPORES

1. Collect spores of the *M. circinelloides* strain after one vegetative cycle on YPG pH 4.5 plates. This should be the starting material for all the assay cultures.
2. Inoculate 10^7–10^8 spores onto a rice–YPG flask. This will increase the number of spores and the germination uniformity.
3. Incubate the inoculated flasks for 5–7 d at 28°C (this will result in a strong sporulation).
4. Use a 1% (v/v) Tween-80 solution to harvest the spores from the rice. Wash them several times by adding distilled water and centrifugation. Resuspend in an appropriate volume of distilled water; add glycerol to a final concentration of 20% and aliquot them in 1-mL tubes. Store at –80°C (*see* **Note 12**).

3.3.1.2. BATCH FERMENTATION: MEDIA AND CONDITIONS

1. Add sterilized Vogels medium and antifoam (SB2121; 0.1 ml/L) to the bioreactor. Set the temperature to 28°C and the pH to 3.5 at the beginning of the culture by the addition of 1 M NaOH or 1 M HCl. Set the agitation speed to 200 rpm at the beginning of the culture and a constant gas flow of 0.2 vvm (vessel volume per minute).
2. Inoculate 10^5 spores/mL.
3. Increase the pH to 5.0 as mycelium develops (between 5 and 30 h).
4. Increase the airflow to 1 vvm and the agitation up to 550 rpm as the level of biomass increases during fermentation.

3.3.1.3. GLUCOSE SOURCE AND ETHANOL ANALYSIS

The concentration of glucose and ethanol in culture filtrates from fermentations were measured by HPLC (*see* **Fig. 3**).

3.3.2. Characterization and Optimization of Growth Parameters of M. circinelloides *in Fermentor for Biotechnological Applications*

A protocol to induce the yeast-to-hyphal growth shift was also required because yeasts are optimal for high biomass production and hyphae are more

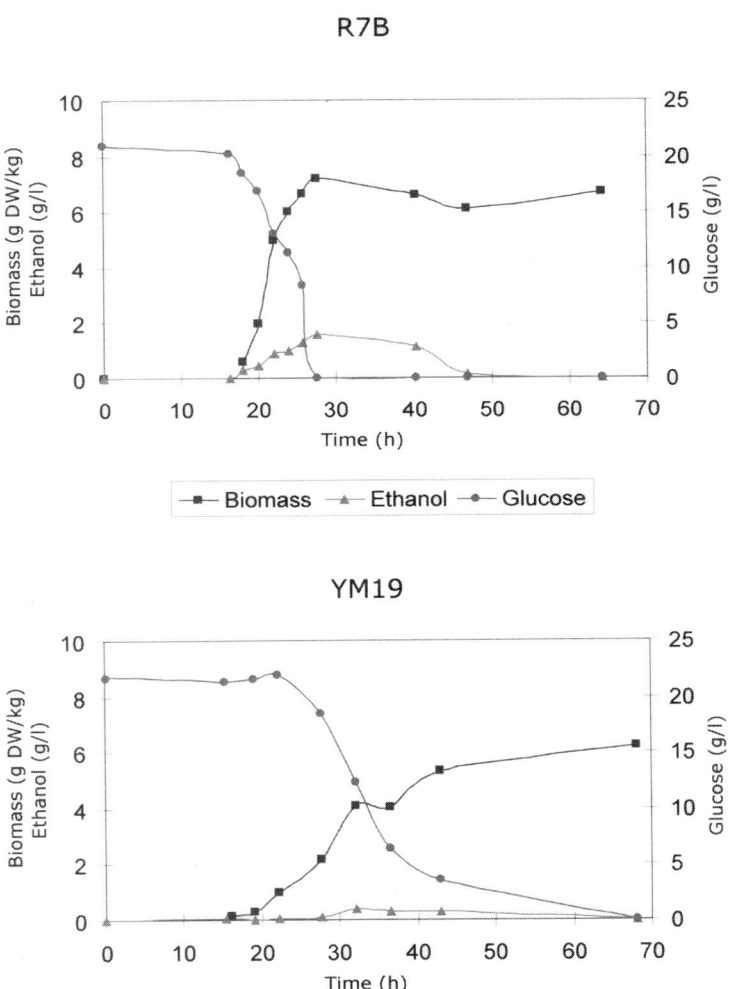

Fig. 3. Growth characteristics of wild-type (R7B) and a monomorphic mutant (YM19) during batch cultivation.

efficient in β-carotene accumulation and protein secretion. Having established conditions for the fermentation of *M. circinelloides* as yeasts and filamentous cells, fermentation was performed where the dimorphic shift was induced at a defined time-point in submerged batch cultivation. The organism was cultivated anaerobically in Vogels for the first 19 h, allowing for the growth of multipolar budding yeasts. The maximum specific growth rate was 0.19 h^{-1}.

Subsequently, the sparge gas was changed to air. A slight increase in specific growth rate was observed (maximum, 0.20 h^{-1}), and a maximum biomass concentration of around 5 g/L was obtained. The yield of biomass on glucose (Y_{SX}, c-mole basis) increased after the shift (0.15 compared to 0.24), whereas the yield of ethanol on glucose (Y_{SEtOH}, c-mole basis) decreased (0.43 compared to 0.52). The maximum ethanol concentration was 6.4 g/L. The values for the biomass yields on glucose were comparable to those obtained in experiments performed solely under anaerobic or aerobic conditions. The ethanol yield on glucose and the maximum ethanol concentration in the aerobic phase of the shift cultivation are higher than those obtained under solely aerobic conditions. This is likely to be the result of a degree of carryover from the anaerobic phase of the fermentation. The morphological shift was observed within a few hours of induction, as germ tubes appeared from mother yeast cells within 2 h of the shift to aerobic conditions. The germ tubes continued to extend and grow for the remainder of the process, resulting in a purely filamentous culture.

Batch fermentation is typically performed adding all media components to the fermentor prior to inoculation. This includes glucose at concentrations ranging from 10 to 100 g/L. Batch fermentation was carried out to address growth characteristics and to compare strains YM19 and R7B growing in Vogels supplemented with 20 g/L glucose (*see* **Fig. 3**). As shown, strain R7B reached a maximum biomass level within 24 h concomitantly with glucose depletion. This was accompanied by ethanol production (up to 2 g/L) and growth inhibition. Strain YM19 displayed a different growth profile. Glucose depletion occurred after 60 h. The final biomass levels were comparable to R7B and, more interestingly, reduced levels of ethanol were observed that remained below 1 g/L.

4. Notes

1. Genotypes of all source and derived strains are not relevant for this chapter. A table listing them will probably be tedious for the reader, but it can be obtained from the authors upon request.
2. Nitrosoguanidine is a dangerous carcinogenic chemical and handling should follow safety regulations as indicated by the manufacturer. All solutions, liquid waste, and glassware should be treated with 2% $Na_2S_2O_3 \cdot 10H_2O$ in a safety hood for 1 h to inactivate NG.
3. Some of the *M. circinelloides* strains (CBS277.49 and its derivatives) produce mostly multinucleate spores and only a small proportion of uninucleate spores. If one plates directly the mutagenized spores to perform the screening, it could happen that a spore contained mutant and wild-type nuclei. Because most mutations are recessive, the presence of the wild-type nucleus in the spore would mask the phenotype produced by the mutant nucleus for which we are searching. Plating the mutagenised spores in a rich medium and allowing a vegetative cycle to complete increases the chances to obtain uninucleate spores carrying the desired mutation.

4. These analyses were performed using the following Waters equipment: pump mod 600, controller mod 600, on-line degasser, and on-line diode array detector mod 996. Elution times can differ depending on the HPLC equipment, temperature, and so forth, and the procedure must be optimized in your lab.

5. The elution conditions specified are needed to remove all sterols of the sample before the next injection. Sterols are usually produced in higher concentrations than carotenoids and extracted with them, so they could mask their presence in the sample.

6. A previous run with known standards is essential. Some of them (i.e., lycopene) are commercially available, but others need to be prepared. The common way to do that is by transforming *Escherichia coli* with the plasmids containing the appropriate genes from other carotenogenic organisms and subsequent extraction of the carotenoid *(45)*.

7. Caution should taken to avoid the formation of arthrospores during liquid growth in the enrichment protocol, because arthrospores will pass through the filter and will germinate as wild-type filamentous cells. Basically, keeping the inoculum size of these cultures below 10^6 spores/mL reduces significantly the formation of arthrospores *(39)*.

8. Sometimes, hyphal growth is not so evident, because mutagenized spores germinate but grow weakly, so they need more time to form hyphae. Anyway, never grow cultures for more than 36 h.

9. There is an important issue with this enrichment procedure. If among the mutagenized spores there is a mutant one that germinates and grows as a yeast cell, it will give rise to hundreds of genetically identical cells during culture. All of them will be filtered and all of them will give rise to yeastlike colonies when plated. Thus, one can only trust one mutant per filter enrichment assay. This is the reason why the mutagenesis experiment is spliced into 10 enrichment experiments, just to isolate as many mutants as possible from a single experiment. However, sometimes, colonies arising from one of the enrichments show such a different morphology that two or more mutants could be isolated; even so, one should be cautious about them, because they could be the result of several mutational events.

10. Two sheets of Miracloth work equally well.

11. Centrifugation of the eluate can also be used, but the yield is usually lower.

12. Whenever possible, use the same batch of spores as the starting material for all subsequent batch experiments.

Acknowledgments

The authors wish to thank Professor Jens Nielsen (Biocentrum-DTU, Denmark), and Antonio Velayos (Neurogenetics Group, Wellcome Trust Centre for Human Genetics, Oxford) for their help in writing and critically reading the manuscript. This work was supported in part by grants IFD97-1476 (Spanish Ministerio de Educación y Cultura–Fondos FEDER), BIO2001-2040 (Spanish Ministerio de Ciencia y Tecnología), and SA067/01 (Junta de Castilla y León, Spain).

References

1. Britton, G., Liaaen-Jensen, S., and Pfander, H. P. (eds.) (1995) *Carotenoids, Vols I and II,* Birkhäuser, Basel.
2. Britton, G., Liaaen-Jensen, S., and Pfander, H. P. (eds.) (1998) *Carotenoids: Biosynthesis and Metabolism, Vol. III.* Birkhäuser, Basel.
3. Olson, J. A. (1993) Vitamin A and carotenoids as antioxidants in a physiological context. *J. Nutr. Sci. Vitaminol.* **39,** S57–S65.
4. Olson, J. A. (1993) Molecular actions of carotenoids, in *Carotenoids in Human Health* (Cantfield, L. M., Krinsky, N. I., and Olson, J. A., eds.), Annals of the New York Academy of Sciences. New York Academy of Sciences, New York, pp. 156–166.
5. Chew, B. P. (1993) Role of carotenoids in the immune response. *J. Dairy Sci.* **76,** 2804–2811.
6. Vainio, H. and Rautalahti, M. (1998) An international evaluation of the cancer preventive potential of carotenoids. *Cancer Epidemiol. Biomarkers Prev.* **7,** 725–728.
7. Goodwin, T. W. (ed.) (1980) *The Biochemistry of Carotenoids,* 2nd ed., Chapman & Hall, London.
8. Krinsky. N. I., Mathews-Roth, M. M., and Taylor, R. F. (eds.) (1989) *Carotenoids: Chemistry and Biology.* Plenum, New York.
9. Cerdá-Olmedo, E. (1989) Production of carotenoids with fungi, in *Biotechnology of Vitamin, Growth Factor and Pigment Production* (Vandamme, E. J., ed.), Elsevier Applied Science, London, pp. 27–42.
10. Sandmann, G. (1994) Carotenoid biosynthesis in microorganisms and plants. *Eur. J. Biochem.* **223,** 7–24.
11. Armstrong, G. A. (1994) Eubacteria show their true colors: genetics of carotenoid pigment biosynthesis from microbes to plants. *J. Bacteriol.* **176,** 4795–4802.
12. Britton, G. (1995) Structure and properties of carotenoids in relation to function. *FASEB J.* **9,** 1551–1558.
13. Armstrong, G. A. (1997) Genetics of eubacterial carotenoid biosynthesis: a colorful tale. *Annu. Rev. Microbiol.* **51,** 629–659.
14. Cunningham, F. X., Jr. and Gantt, E. (1998) Genes and enzymes of carotenoids biosynthesis in plants. *Annu. Rev. Plant Physiol. Plant Mol. Biol.* **49,** 557–583.
15. Sandmann, G. (2001) Carotenoid biosynthesis and biotechnological application. *Arch. Biochem. Biophys.* **385,** 4–12.
16. Burkhardt, P. K., Beyer, P., Wünn, J., et al. (1997) Transgenic rice *(Oryza sativa)* endosperm expressing daffodil *(Narcissus pseudonarcissus)* phytoene synthase accumulates phytoene, a key intermediate of provitamin A biosynthesis. *Plant J.* **11,** 1071–1078.
17. Misawa, N. and Shimada, H. (1997) Metabolic engineering for the production of carotenoids in non-carotenogenic bacteria and yeasts. *J. Biotechnol.* **59,** 169–181.
18. Giovannucci, E. (1999) Tomatoes, tomato-based products, lycopene, and cancer: review of the epidemiologic literature. *J. Natl. Cancer Inst.* **17,** 317–331.
19. Hirschberg, J. (1999) Production of high-value compounds: carotenoids and vitamin E. *Curr. Opin. Biotechnol.* **10,** 186–191.

20. Todd Lorenz, R. and Cysewski, G. R. (2000) Commercial potential for *Haematococcus* microalgae as a natural source of astaxanthin. *Trends. Biotechnol.* **18**, 160–167.

21. Eslava, A. P., Alvarez, M. I., and Cerdá-Olmedo, E. (1974) Regulation of carotene biosynthesis in *Phycomyces* by vitamin-A and beta-ionone. *Eur. J. Biochem.* **48**, 617–623.

22. Bejarano, E. R. and Cerdá-Olmedo, E. (1989) Inhibition of phytoene dehydrogenation and activation of carotenogenesis in *Phycomyces. Phytochemistry* **28**, 1623–1626.

23. Corrochano, L. M. and Cerdá-Olmedo, E. (1992) Sex, light and carotenes: the development of *Phycomyces. Trends Genet.* **8**, 268–274.

24. Cerdá-Olmedo, E. and Corrochano, L. M. (1996) Photoregulation of fungal gene expression, in *Light as an Energy Source and Information Carrier in Plant Physiology* (Jennings, R. C., Zucchelli, G., Ghetti, F., and Colombetti, G., eds.), Plenum, New York, pp. 285–292.

25. Linden, H., Ballario, P., and Macino, G. (1997) Blue light regulation in *Neurospora crassa. Fungal Genet. Biol.* **22**, 141–150.

26. Ruiz-Hidalgo, M. J., Benito, E. P., Sandmann, G., and Eslava, A. P. (1997) The phytoene dehydrogenase gene of *Phycomyces:* regulation of its expression by blue light and vitamin A. *Mol. Gen. Genet.* **253**, 734–744.

27. Velayos, A., Blasco, J. L., Alvarez, M. I., Iturriaga, E. A., and Eslava, A. P. (2000) Blue-light regulation of phytoene dehydrogenase *(carB)* gene expression in *Mucor circinelloides. Planta* **210**, 938–946.

28. Ciegler, A. (1965) Microbial carotenogenesis. *Adv. Appl. Microbiol.* **7**, 1–34.

29. Ninet, L. and Renaut, J. (1979) Carotenoids, in *Microbial Technology* 2nd ed. (Peppler, H. J. and Perlman, D., eds.), Academic, New York, pp. 529–544.

30. Filkenstein, M., Huang, C. C., Byng, G. S., Tsau, B. R., and Leach, J. (1993) Method for producing beta-carotene using a fungal mated culture. European Patent Office WO93/20183.

31. Murillo, F. J., Calderón, I. L., López-Díaz, I., and Cerdá-Olmedo, E. (1982) β-Carotene producing strains of the fungus *Phycomyces blakesleeanus.* US Patent Office 4,318,987.

32. Mehta, B. J. and Cerdá-Olmedo, E. (2001) Intersexual partial diploids of *Phycomyces. Genetics* **158**, 635–641.

33. Okagbue, R. N. and Lewis, M. J. (1985) Influence of mixed culture conditions on yeast-wall hydrolytic activity of *Bacillus circulans* WL-12 and on the extractability of astaxanthin from the yeast *Phaffia rhodozyma. J. Appl. Bacteriol.* **59**, 243–255.

34. Iturriaga, E. A., Velayos, A., and Eslava, A. P. (2000) The structure and function of the genes involved in the biosynthesis of carotenoids in the Mucorales. *Biotechnol. Bioprocess. Eng.* **5**, 263–274.

35. Velayos, A, Eslava, A. P., and Iturriaga, E. A. (2000) A bifunctional enzyme with lycopene cyclase and phytoene synthase activities is encoded by the *carRP* gene of *Mucor circinelloides. Eur. J. Biochem.* **267**, 5509–5519.

36. Velayos, A., Papp, T., Aguilar-Elena, R., et al. (2003) Expression of the *carG* gene, encoding geranylgeranyl pyrophosphate synthase, is up-regulated by blue light in *Mucor circinelloides. Curr. Genet.* **43,** 112–120.
37. Navarro, E., Ruiz-Pérez, V. L., and Torres-Martínez, S. (2000) Overexpression of the *crgA* gene abolishes light requirement for carotenoid biosynthesis in *Mucor circinelloides. Eur. J. Biochem.* **267,** 800–807.
38. Orlowski, M. (1994) Yeast/mycelial dimorphism, in *The Mycota. I. Growth, Differentiation and Sexuality* (Wessels, J. G. H. and Meinhardt, F., eds.), Springer-Verlag, Berlin, pp. 143–162.
39. McIntyre, M., Breum, J., Arnau, J., and Nielsen, J. (2002) Growth physiology and dimorphism of *Mucor circinelloides* during submerged batch cultivation. *Appl. Microbiol. Biotechnol.* **58,** 495–502.
40. Cerdá-Olmedo, E. and Avalos, J. (1994) Oleaginous fungi: carotene-rich oil from *Phycomyces. Prog. Lipid Res.* **33,** 185–192.
41. Stredanská, S. and Sajbidor, J. (1992) Oligounsaturated fatty acid production by selected strains of micromycetes. *Folia Microbiol. (Praha)* **37,** 357–359.
42. Madyastha, K. M. (1996) Novel microbial transformations of steroids. *Adv. Exp. Med. Biol.* **405,** 259–270.
43. Mogollón, L., Rodríguez, R., Larrota, W., Ramírez, N., and Torres, R. (1998) Biosorption of nickel using filamentous fungi. *Appl. Biochem. Biotechnol.* **70–72,** 593–601.
44. Deken, R. H. de (1966) The Crabtree effect: a regulatory system in yeast. *J. Gen. Microbiol.* **44,** 149–156.
45. Lee, P. C., Momen, A. Z. R., Mijts, B. N., and Schmidt-Dannert, C. (2003) Biosynthesis of structurally novel carotenoids in *E. coli. Chem. Biol.* **10,** 453–462.

14

Xanthophylls in Fungi

Metabolic Engineering of the Astaxanthin Biosynthetic Pathway in Xanthophyllomyces dendrorhous

Hans Visser, Gerhard Sandmann, and Jan C. Verdoes

Summary

Modification of *Xanthophyllomyces dendrorhous* toward a higher carotenoid synthesis will make the exploitation of *X. dendrorhous* as a natural source for astaxanthin more competitive. By recombinant DNA technology, the *X. dendrorhous* isopentenyl-pyrophosphate isomerase-encoding gene *(idi)* was inserted in a *X. dendrorhous* expression cassette that directs expression by the endogenous glyceraldehyde-3-phosphate dehydrogenase gene expression signals. *X. dendrorhous* transformation vector pPR2TN was used to deliver this expression cassette, by means of electroporation, into the ribosomal DNA of *X. dendrorhous* strain CBS 6938, which was confirmed by Southern analysis. High-performance liquid chromatography analysis of the carotenoids produced by *idi*-overexpressing transformant strains indicated a lower overall carotenoid formation when compared to the control strain. This finding is in contrast to that observed in bacterial systems, where enhanced expression of the *idi* gene increases the carotenoid content.

Key Words: Carotenoid; genetic engineering; HPLC; *Phaffia rhodozyma;* xanthophyll.

1. Introduction

Approximately 700 naturally occurring carotenoids are known. A few of them are of economical importance and applied as animal feed additive, food colorant, or nutritional supplement by the food and feed industry *(1)*. Carotenoids are terpenoid compounds that, in general, correspond to a 40-carbon-atom polyene chain *(2)*. A wide variety of natural carotenoids are produced by plant, algal, bacterial, and fungal species. In fungi, the central building block for terpenoid biosynthesis, isopentenyl pyrophosphate (IPP), is produced via the mevalonate pathway. Eight of these molecules are used to form phytoene, which is the first carotenoid in fungal carotenoid pathways leading to end products such as lycopene *(Ustilago violacea),* β-carotene *(Blakeslea trispora, Phycomyces blakesleeanus),* and the

From: *Methods in Biotechnology, Vol. 18: Microbial Processes and Products*
Edited by: J. L. Barredo © Humana Press Inc., Totowa, NJ

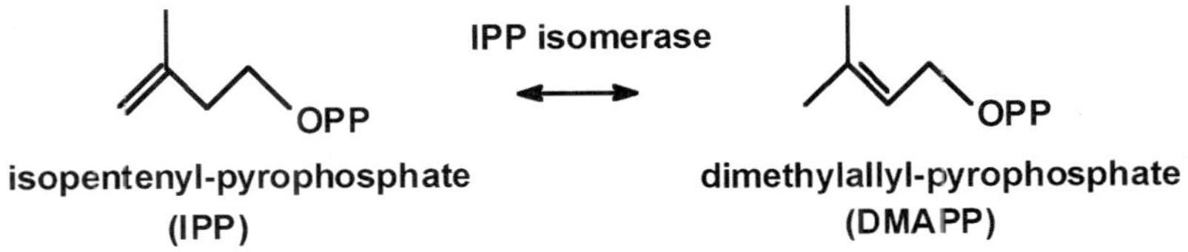

IPP isomerase

isopentenyl-pyrophosphate
(IPP)

dimethylallyl-pyrophosphate
(DMAPP)

Fig. 1. Scheme of the enzymatic reaction catalyzed by IPP-isomerase. *X. dendrorhous* IPP-isomerase is encoded by the *idi* gene.

oxygenated carotenoids or xanthophylls, canthaxanthin *(Cantharellus cinnabarinus)* and astaxanthin *(Xanthophyllomyces dendrorhous)* *(2,3)*.

The xanthophylls canthaxanthin and astaxanthin are produced synthetically on an industrial scale *(4)*. However, the current demand for natural xanthophylls favors the use of natural sources. Crustacea, the green alga *Haematococcus pluvialis,* and the yeast *X. dendrorhous* (formerly known as *Phaffia rhodozyma*) are regarded as the principal biological sources for astaxanthin *(1)*. Regarding the advantages and disadvantages of each of these sources, *X. dendrorhous* is quite favorable because of its rapid metabolism and production of high cell densities in fermentors. However, a disadvantage of *X. dendrorhous* is its rather low astaxanthin content. Consequently, this needs to be improved by either conventional mutagenesis or, alternatively, by genetic engineering.

Genetic engineering of isoprenoid pathways has shown that, in *Escherichia coli,* IPP-isomerase (*see* **Fig. 1**) could represent a key regulatory step in the production of isoprenoid compounds (i.e., overexpression of the corresponding *idi* gene increases the level of heterologous carotenoids synthesized by this bacterium) *(5,6)*. In this chapter, an example is given of metabolic (genetic) engineering of *X. dendrorhous*. To investigate whether overexpression of the *X. dendrorhous idi* gene in *X. dendrorhous* also results in an elevated carotenoid production as in *E. coli,* we describe the construction of such an *idi*-overexpressing *X. dendrorhous* strain. The applied methods describe the experimental route toward the engineered *X. dendrorhous* strain and include its analysis.

2. Materials

1. Plasmids pPR2TN and pPR15 (Fungal Genomics, Wageningen University, Netherlands).
2. Restriction enzymes *Cla*I and *Bam*HI.
3. Agarose gel electrophoresis equipment.
4. Phenol : cloroform : isoamylalcohol (25 : 24 : 1).
5. Ultrapure water.
6. *Xanthophyllomyces dendrorhous* strain CBS 6938 (Centraal Bureau voor Schimmelcultures, Utrecht, Netherlands).
7. Erlenmeyer flasks.

8. Oligonucleotide PgpdB: 5′-cccggatcctggtgggtgcatgtatgtac-3′.
9. Oligonucleotide idi1: 5′-gttgggcatggacatgatggtaagagtgttag-3′.
10. Oligonucleotide idi2: 5′-cttaccatcatgtccatgcccaacattgttcc-3′.
11. Oligonucleotide PT7B: 5′-cccggatccccctatagtgagtgctatta-3′.
12. YePD medium: 20 g/L bacto-peptone, 10 g/L yeast extract, and 20 g/L glucose.
13. Refrigerated incubator shaker Innova 4330 (New Brunswick Scientific, Nijmegen, Netherlands).
14. STM buffer: 270 mM sucrose, 10 mM Tris-HCl, pH 7.5, and 1 mM MgCl$_2$.
15. Electroporation cuvet, 0.2-cm electrode gap (Bio-Rad, Veenendaal, Netherlands).
16. Gene Pulser II and Pulse Controller Plus (Bio-Rad).
17. Spectrophotometer.
18. Potassium phosphate buffer: 50 mM, pH 7.0.
19. Dithiotreitol (DTT).
20. Geneticin (G418).
21. Freeze-dryer.
22. Separatory funnels.
23. High-performance liquid chromatography (HPLC) equipment.
24. C$_{30}$ column 3-μm (YMC, Wilmington, NC).
25. Sumipax OA-2000 column (Sumitomo Chemical Co., Osaka, Japan).
26. HyPurity C$_{18}$ 5-μm column (Hypersil, Kleinostheim, Germany).
27. Liquid nitrogen.
28. Alumina type-A5 (Sigma, Zwijndrecht, Netherlands).
29. DNA extraction buffer: 50 mM Tris-HCl, pH 7.4, 10 mM MgCl$_2$, 50 mM NaCl, and 1% (w/v) sodium dodecyl sulfate (SDS).
30. RNase A.
31. TE buffer: 10 mM Tris-HCl, pH 8.0, and 1 mM EDTA.
32. Nylon filter.
33. DIG nonradioactive nucleic acid labeling and detection system (Roche Diagnostics, Almere, Netherlands).
34. PCR DIG probe synthesis kit (Roche Diagnostics).
35. X-OMAT Kodak film (Sigma).
36. Film-developing equipment.

3. Methods

The methods describe the construction and analysis of *X. dendrorhous* strains in which the *X. dendrorhous idi* gene is overexpressed. A similar procedure can be used to overexpress any gene of interest in *X. dendrorhous*. The methods refer to the description of the *Xanthophyllomyces* transformation vector, as developed by others (*see* **Subheading 3.1.1.**), the construction of the *idi*-overexpression vector (*see* **Subheading 3.1.2.**), the transformation protocol of the wild *Xanthophyllomyces* strain with this vector (*see* **Subheading 3.2.**), the analysis of the carotenoid content and carotenoid composition of the transformed *Xanthophyllomyces* strain (*see* **Subheading 3.3.**), and the Southern analysis of the transformed *Xanthophyllomyces* strain (*see* **Subheading** 3.4.).

3.1. Transformation and idi Overexpression Vectors

3.1.1. Xanthophyllomyces Transformation Vector pPR2TN

Wery et al. *(7)* developed an efficient transformation system for *X. dendrorhous*. The transformation vector, pPR2TN, was constructed by using pUC18 as a backbone for plasmid propagation in *E. coli* and a marker gene cassette for selection of transformed *X. dendrorhous* strains. The latter was constructed by flanking the *nptII* gene from transposon Tn5, conferring resistance to kanamycin, by the promoter (P) and terminator (T) sequences of the *Xanthophyllomyces* glyceraldehyde-3-phosphate dehydrogenase gene *(gpd)* *(8)* by recombinant polymerase chain reaction (PCR) *(7)*. The *nptII* gene serves as a dominant selective marker and its expression results in resistance of the transformed *X. dendrorhous* strain to the antibiotic geneticin (G418).

This intermediate vector was supplemented with a large part of the ribosomal DNA sequences from *X. dendrorhous* to accomplish multiple copy integration of the vector into the chromosomal ribosomal DNA of *X. dendrorhous* after introduction of the plasmid DNA in the cell. However, in order to integrate, the transformation vector needs to be linearized in the ribosomal DNA first by a single-cutting endonuclease before transformation *(see* **Note 1**). In general, two endonuclease sites have been used for this goal: either *Cla*I (AT↓CGAT) or the rare cutter *Sfi*I (GGCCNNNN↓NGGCC).

A small multiple-cloning site was created (at the 3′ end of the Tgpd sequence) consisting of *Not*I, *Bam*HI, and *Hin*dIII, which can be used as cloning sites for DNA fragments of interest *(9)*. **Figure 2** shows the vector map of pPR2TN, and the transformation protocol is described in **Subheading 3.2.** *(see also* **Note 2**).

3.1.2. Construction of the idi Expression Vector

The isopentenyl-pyrophosphate (IPP) encoding gene, *idi,* from *X. dendrorhous* was isolated by Verdoes and van Ooyen *(6)*. The corresponding gene product is a key enzyme in the isoprenoid biosynthetic pathway, which catalyzes the interconversion of dimethylallyl-diphosphate and IPP. By recombinant PCR *(7),* the promoter region of the glyceraldehyde-3-phosphate dehydrogenase-encoding gene (P*gpd*) of *X. dendrorhous* was fused to the *idi* cDNA. A P*gpd* primer (PgpdB) whose target sequence is located approx 400 bp upstream of the GPD start codon and that contains a *Bam*HI endonuclease site for cloning purposes was used together with a downstream 3′-P*gpd*/5′-*idi* "fusion primer" (idi1), in a PCR reaction using plasmid pPRGDH6 *(8)* as a template. Additionally, the *idi* cDNA was amplified in a second PCR with an upstream 3′-P*gpd*/5′-*idi* "fusion primer" (idi2) and a downstream primer (PT7B), which is antisense to the T7 region of the *idi* cDNA vector (pBluescript) and supplemented with a *Bam*HI endonuclease site for cloning purposes, using plasmid pPRidi *(6)* as a template. In a third PCR, the P*gpd* and *idi* PCR products were fused and amplified using primers

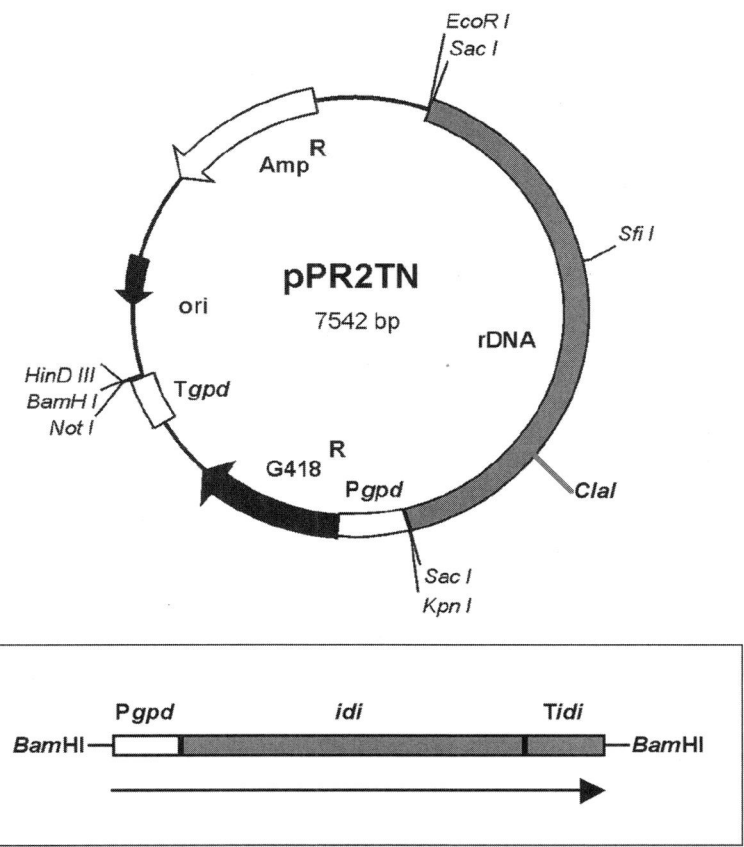

Fig. 2. Schematic representation of *X. dendrorhous* transformation vector pPR2TN. The *idi* expression cassette, which was inserted in the *Bam*HI site of pPR2TN to obtain pPR15, is indicated in the boxed area. For details, *see* **Subheadings 3.1.1. and 3.1.2**.

PgpdB and PT7B. The resulting PCR product is built up as follows: *Bam*HI–P*gpd*–idi–T*idi*–*Bam*HI, where (idi) represents the *idi* coding regions and T*idi* represents the *idi* transcription terminator sequence. This product was cloned in the pPR2TN vector. The corresponding plasmid was named pPR15 (*see* **Fig. 2**) and used to construct the *idi*-overexpressing *X. dendrorhous* strain.

3.2. Transformation of X. dendrorhous *CBS 6938 With pPR15*

The method of transformation of *X. dendrorhous* is subdivided by preparation of transforming DNA (*see* **Subheading 3.2.1.**), preparation of electrocompetent *X. dendrorhous* cells (*see* **Subheading 3.2.2.**), and transformation of *X. dendrorhous* by electroporation (*see* **Subheading 3.2.3.**).

3.2.1. Preparation of Transforming DNA

1. Digest 6 µg of pPR15 (and separately pPR2TN as a control) with 10 units of *Cla*I for 2 h at 37°C in order to linearize the vector.
2. Verify linearization by analyzing a portion of the digestion mixture corresponding to 0.1–0.5 µg pPR15 on a 1% agarose gel.
3. Purify the *Cla*I-linearized DNA by a phenol : chloroform : isoamylalcohol (25 : 24 : 1) extraction.
4. Concentrate the *Cla*I-linearized DNA in 10 µL of ultrapure water after ethanol precipitation (*see* **Note 3**).
5. This sample can be stored at –20°C or used the same day (store on ice).

3.2.2. Preparation of Electrocompetent Cells

1. Take a single *X. dendrorhous* CBS 6938 colony from a YePD agar plate, which had been incubating at 21°C for 48 h.
2. Transfer the colony to 30 mL of YePD medium in a 250 mL Erlenmeyer and incubate for 48 h at 21°C and 250 rpm (*see* **Note 4**).
3. Fill two 1-L Erlenmeyer flasks with 200 mL of YePD medium each and inoculate with precultured *X. dendrorhous* to an optical density (OD_{600}) of 0.01 and 0.02, respectively. Incubate at 21°C during the night (approx 18–20 h) at 250 rpm (*see* **Notes 4 and 5**).
4. In the morning, centrifuge the culture, which shows an OD_{600} of approx 1.2, for 5 min, 5500g, and at room temperature.
5. Resuspend the cell pellet in 25 mL of potassium phosphate buffer containing 25 m*M* DTT (freshly made; *see* **Note 3**), transfer to a 30-mL centrifuge tube, and incubate for 15 min at room temperature.
6. Centrifuge the cell suspension for 5 min at 4°C and 5500g (*see* **Note 6**).
7. Wash the cell pellet in 25 mL of ice-cold STM buffer and centrifuge for 5 min at 4°C and 5500g.
8. Repeat the washing step and resuspend the cells in 0.5 mL STM buffer (approx 3×10^9 cells/mL).
9. Store these electrocompetent cells on ice until use the same day.

3.2.3. Transformation of X. dendrorhous by Electroporation

1. On ice, mix 60 µL of electrocompetent *X. dendrorhous* cell suspension (*see* **Subheading 3.2.2.**) and the pPR15 DNA sample (*see* **Subheading 3.2.1.**) in a precooled Eppendorf tube. Follow the same procedure with the pPR2TN control sample throughout this method.
2. Transfer this mixture to an electroporation cuvet and tap the mixture to the bottom of the cuvet. Put it on ice.
3. For each transformation, prepare an Eppendorf tube filled with 0.5 mL of YePD medium and put it on ice.
4. Set the gene-pulser apparatus at 1000Ω, 800 V, and 25 µF.
5. Dry the out sides of the cuvet with a tissue and place it in the chamber.
6. Apply the pulse and immediately add 0.5 mL of ice-cold YePD medium.

7. Using a sterile Pasteur's pipet, transfer the cell suspension back to the Eppendorf tube and incubate for 2.5 h at 21°C.
8. Plate 100 μL of cell suspension onto YePD–agar plates containing 40 μg/mL G418 (five plates).
9. Incubate the plates at 21°C until colonies appear (48–96 h).
10. Refer to the pPR15 and pPR2TN transformants as CBS6938[pPR15] and CBS6938[pPR2TN], respectively.

3.3. Carotenoid Analysis of X. dendrorhous Transformants CBS6938[pPR2TN] and CBS6938[pPR15]

Here, the analysis of the carotenoid content and composition in transformed *X. dendrorhous* strains is described. The method that is used is an HPLC analysis, which is very sensitive and elucidates the composition and quantity of the carotenoids in the extract.

3.3.1. Cultivation of X. dendrorhous Transformants CBS6938[pPR15] and CBS6938[pPR2TN]

1. Take a transformant colony from a YePD–agar plate containing 40 μg/mL G418.
2. Transfer this colony to 35 mL of YePD medium containing 40 μg/mL G418 in a 250-mL Erlenmeyer.
3. Incubate the culture for 96 h at 21°C and 250 rpm (*see* **Note 4**).

3.3.2. Analysis of Astaxanthin and Other Carotenoids by HPLC

3.3.2.1. Relevant Physical Properties

Astaxanthin, the major carotenoid of *X. dendrorhous,* is accompanied by substantial amounts of the precursors 3-HO-canthaxanthin (also referred to as phoenicoxanthin) and 3-HO-echinenone as well as by small amounts of the 3,4-didehydrolycopene derivative 3-HO-4-keto-torulene. Also, traces of echinenone, torulene, and β-carotene can be detected *(10)*. The structures of the carotenoids from *X.dendrorhous,* their biosynthetic pathway, and the genes involved are outlined in **Fig. 3**.

Absorbance spectra of pigments are important criteria for their identification. The basic structure and some of the substituents determine the spectral properties of carotenoids *(11)*. For example, β-carotene possesses three distinct maxima at 425 nm (shoulder), 450 nm, and 475 nm (*see* **Fig. 4A**). The 3-HO groups have no effect on the spectrum. However, a 4-keto group in conjugation with the polyene chain like in echinenone is responsible for a bell-shaped spectrum shifting the single maximum to 460 nm. The effect of a second keto group at position 4′ is additive, as indicated by the absorbance maximum of canthaxanthin or astaxanthin at 470 nm (*see* **Fig. 4C**). Monocyclic 4-keto carotenoids have a less pronounced fine structure of the spectrum (i.e., the dip between the central maximum and the longer-wavelength maximum is much smaller) than

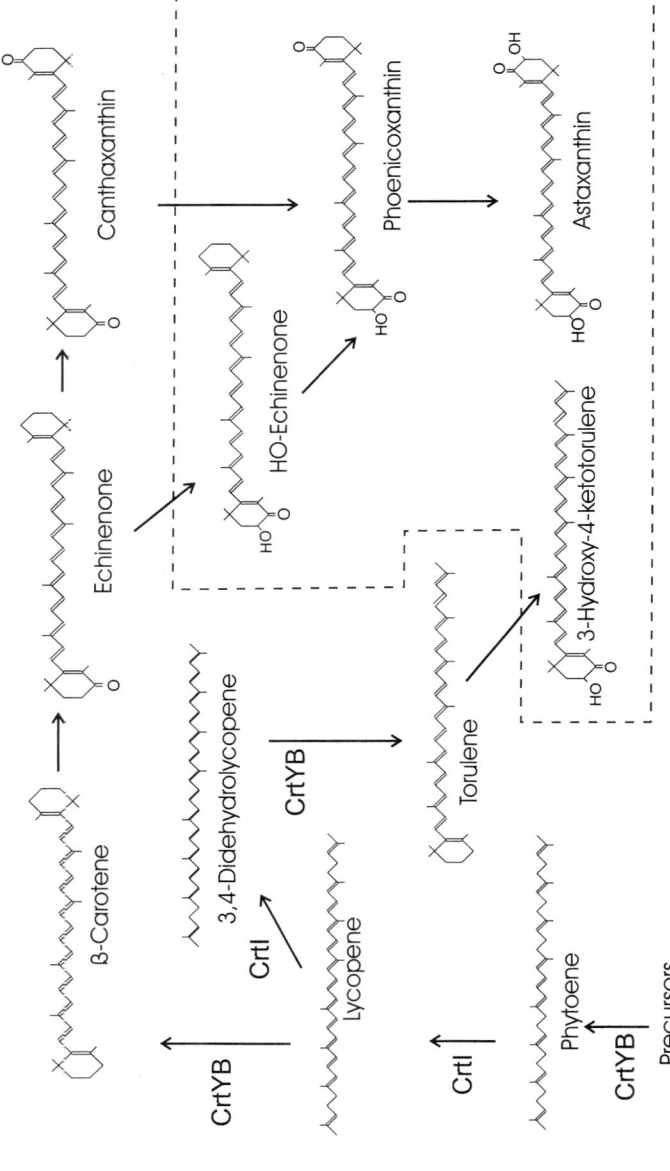

Fig. 3. Biosynthetic pathway of carotenoids in *X. dendrorhous*. The carotenoids in the box represent the major ones. Defined gene products catalyzing the individual reactions are indicated next to the arrows. The hydroxylation at position 3 and ketolation at position 4 can be catalyzed by a single P-450 enzyme.

264

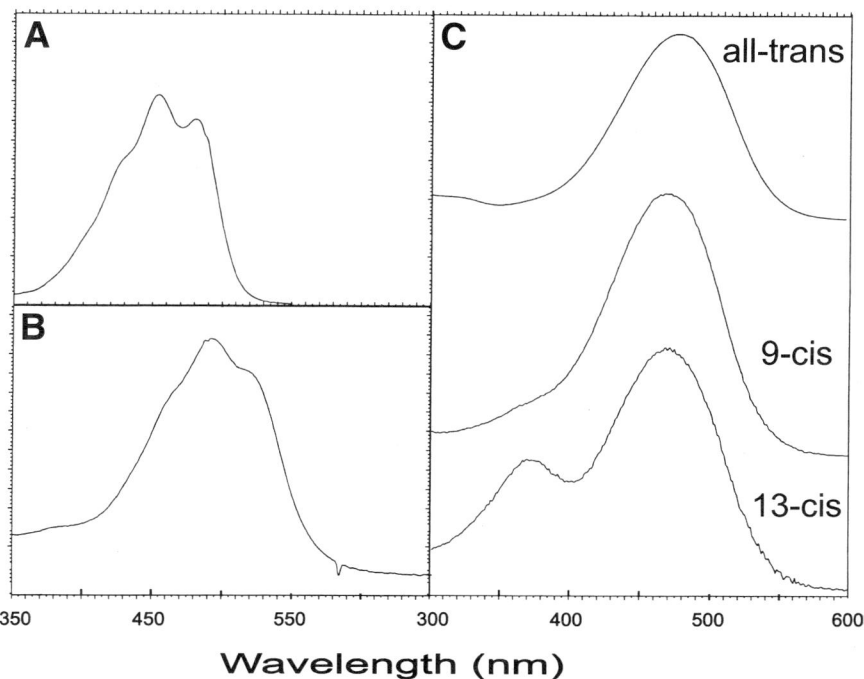

Fig. 4. Spectra of selected carotenoids: **(A)** β-carotene, **(B)** 3-HO-4-keto-torulene, and **(C)** astaxanthin isomers.

the corresponding derivative without the keto group but can be distinguished from bicyclic 4-keto carotenoids. The spectrum of monocyclic 3-HO-4-keto-torulene is shown (*see* **Fig. 4B**).

Most of the carotenoids in *X. dendrorhous* possess a 3-HO-4-keto-β-ionone end group. Increasing polarity decreases the mobility of a compound in absorption chromatography on stationary phases like silica or lowers the retention time in reversed-phase HPLC systems. However, the internal hydrogen-bonding of the α-ketol affects the overall polarity severely. Therefore, astaxanthin exhibits a higher mobility in thin-layer chromatography (TLC) on silica than 3-HO-cantaxanthin without the second HO group. During reversed-phase HPLC (partition chromatography), the hydrogen-bonding between the HO and the keto group has no effect on the expected order of elution (i.e., shorter retention time with increasing number of oxo groups).

Carotenoids exist in different isomeric forms. Cis/trans isomers occur at the double bonds of the polyene chain. Astaxanthin from *X. dendrorhous* is mainly all-trans, but also 9- and 13-*cis* isomers exist in a substantial portion. These cis forms have a slightly different optical spectrum. Compared to all-trans, the main absorbance maximum of 9-*cis* and 13-*cis* astaxanthin is shifted by 5 nm and a *cis* peak occurs in the latter at 370 nm (*see* **Fig. 4C**). The more central the *cis* bond, the more pronounced is this cis peak. In addition to these geometrical isomers, astaxanthin occurs in three optical isomeric forms with a 3S,3'S, 3R,3'S, and 3R,3'R chirality. In contrast to other 3-HO carotenoids and astaxanthin from other organisms, astaxanthin from *X. dendrorhous* is 3R,3'R *(12)*. Optical absorbance is not affected by the chirality.

3.3.2.2. EXTRACTION OF CAROTENOIDS FOR HPLC ANALYSIS

Polar organic solvents extract carotenoids from intact cells. However, acetone or methanol, which are normally used for many organisms, are not appropriate for *X. dendrorhous* because of the very rigid cell wall. Instead, dimethyl sulfoxide (DMSO) proved to extract carotenoid efficiently *(13)*. The boiling point of DMSO is rather high. Therefore, evaporation leads to destruction of carotenoids and is not advised. In order to remove the solvent for storage of the sample and for suspension in a small defined volume prior to separation of the carotenoids, the following procedure is recommended for carotenoid extraction from *X. dendrorhous* and was also applied to extract the carotenoids from CBS6938[pPR2TN] and CBS6938[pPR15] (*see* **Note 7**).

1. Suspend freeze-dried cell material (10 mg) in DMSO (3 mL) and heat it for 15 min at 60°C (*see* **Note 8**).
2. After centrifugation, the total carotenoid content can be determined from the supernatant by recording its OD_{470}.
3. Afterward, transfer the total extract into a separatory funnel, add 3 mL of diethylether, and keep the funnel on ice for 1–3 min.
4. Add 0.5 mL of water and remove the lower phase (*see* **Note 9**).
5. Add 5 mL of acetone, then 5 mL of 10% ether/petrol, and, finally, 8–10 mL of water to obtain phase separation.
6. Wash the collected upper phase with water (10 mL) and dry it in a stream of N_2.
7. In this state, the samples can be stored after sealing the vessel in a freezer. Upon demand, the extract can be resuspended in acetone and subjected to TLC or HPLC (*see* **Note 10**). Total carotenoid content can be calculated from the OD_{475} using the millimolar extinction coefficient of 125. This value determined for astaxanthin coincides with the coefficients for canthaxanthin and 3-HO-canthaxanthin and is close to the values of other 4-ketocarotenoids.

3.3.2.3. SEPARATION AND QUANTIFICATION BY HPLC

The most powerful HPLC support for separation of geometrical isomers is a C_{30}-bonded stationary phase *(14)*. Using a non-end-capped polymeric 3-μm C_{30} column, an isocratic system involving methanol/methyl-*tert*-butyl ether/water (81 : 15 : 4, v/v/v) at 11°C separates astaxanthin isomers from *X. dendrorhous*

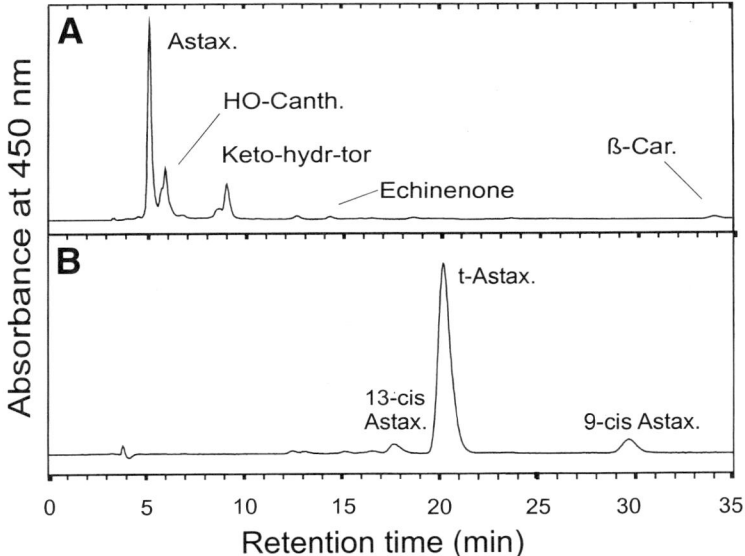

Fig. 5. HPLC separation of carotenoid from *X. dendrorhous:* (**A**) astaxanthin on a 3-µm C$_{30}$ column eluted with methanol/methyl-*tert*-butyl ether/water (81 : 15 : 4, by volume) at 11°C and (**B**) carotenoid extract on a HyPurity C$_{18}$ 5-µm column with acetonitrile/methanol/2-propanol (85 : 10 : 5, by volume) at 32°C.

(collected by TLC) into 9-cis, 13-cis, and all-trans (*see* **Fig. 5A**). Optical isomers separate on chiral stationary phases. In the case of astaxanthin, HPLC using a Sumipax OA-2000 column with hexane/dichloromethane/ethanol (48 : 15 : 0.8) as eluent resolves the three enantiomers (*15*).

In most cases, reversed-phase columns are used for HPLC separation of carotenoids. For routine measurement including the CBS6938[pPR15] and CBS6938[pPR2TN] transformants, the following HPLC system is recommended.

1. A HyPurity C$_{18}$ 5-µm column (*see* **Note 11**)
2. Acetonitrile/methanol/2-propanol (85 : 10 : 5, by volume) as the mobile phase for isocratic elution. The best separation was obtained at a temperature of 32°C.
3. A photodiode array detector for on-line registration of the spectra is favorable but not essential. Set detection wavelength to 470 nm (*see* **Note 12**).
4. Dissolve the sample in acetone (total concentration of about 0.1–0.3 µg/µL) and inject 20 µL.
5. The separation ends after about 40 min., when β-carotene has eluted as the least polar carotenoid.

Figure 5A shows an example of an HPLC separation of a carotenoid extract from *X. dendrorhous* at the above-mentioned conditions. The dominating

Table 1
Carotenoid Composition[a] of X. dendrorhous Transformants

Carotenoid	CBS6938 [pPR2TN] #1	CBS6938 [pPR2TN] #2	CBS6938 [pPR15] #1	CBS6938 [pPR15] #2
Lycopene	nd[b]	nd	nd	4.9
β-Carotene	9.0	7.4	3.3	6.1
Echinone	9.5	7.2	5.2	2.8
Hydroxyechinone	31	25	7.7	8.8
Phoenicoxanthin	57	67	31	34
Astaxanthin	73	88	66	67
Torulene	2.1	nd	nd	2.4
HDCO[c]	32	31	27	15
Total	214	226	140	141

[a] Measured as μg carotenoid/mg yeast dry weight.
[b] not detected.
[c] 3-Hydroxy-3′-4′-didehydro-β-ψ-caroten-4-one.

astaxanthin peak elutes as the first carotenoid with a retention time of 5.7 min. It is followed by 3-HO-canthaxanthin (6.3 min) and 3-HO-4-keto-torulene (9.9 min). Echinenone and β-carotene are minor carotenoids.

The carotenoids can be quantitated from the peak areas by integration of the peaks of the HPLC diagram. Calibration can be carried out by injecting different concentrations of astaxanthin or canthaxanthin (*see* **Note 13**), thus establishing a calibration curve. Both carotenoids are commercially available.

The carotenoid composition of four transformant strains of *X. dendrorhous* is given in **Table 1.** Contrary to what was expected (i.e., an increase in astaxanthin content), a decrease in astaxanthin content was observed for CBS6938[pPR15] when compared to the control strain CBS6938[pPR2TN]. Moreover, the total carotenoid production is reduced when the *X. dendrorhous idi* gene is overexpressed.

3.4. Southern Analysis of X. dendrorhous Transformants CBS6938[pPR2TN] and CBS6938[pPR15]

The integration of the transformation vectors can be analyzed by Southern blotting as described next.

3.4.1. Isolation of Chromosomal DNA from X. dendrorhous Transformants

1. Inoculate 20 mL of G418-supplemented (40 μg/mL) YePD medium with a single transformant colony and incubate at 21°C and 250 rpm (*see* **Note 4**) until the OD_{600} reaches 5.
2. Harvest cells by centrifugation (5 min, 5000*g*, 4°C) and wash with water.

3. Resuspend the cell pellet in 1 mL of water and transfer this to a mortar containing liquid nitrogen and 1 g of alumina type-A5.
4. Break the cells by grinding the frozen cell suspension (*see* **Note 14**).
5. Add the broken cells to a centrifuge tube containing 5 mL DNA extraction buffer.
6. The chromosomal DNA can be isolated from this suspension by repeated extractions with phenol : chloroform : isoamylalcohol (25 : 24 : 1).
7. Precipitate the chromosomal DNA by ethanol precipitation and dissolve it in 0.5 mL TE buffer containing 0.1 mg/mL RNase A.
8. Incubate this mixture for 5 min at 37°C.
9. Conduct a final phenol : chloroform : isoamylalcohol (25 : 24 : 1) extraction and ethanol precipitation.
10. Dissolve the chromosomal DNA in 0.5 mL TE.

3.4.2. Southern Analysis of X. dendrorhous *Transformants* CBS6938[pPR2TN] *and* CBS6938[pPR15]

1. Digest 5–10 µg of chromosomal DNA of the CBS6938[pPR15] and the CBS6938[pPR2TN] samples with 10 units of the endonuclease *Bam*HI for 4 h.
2. Separate the digested DNA samples on a 0.8% agarose gel.
3. Transfer DNA to a Nylon filter using standard recombinant DNA techniques *(16)*.

3.4.3. Probe Preparation, Hybridization, and Detection of Integrated Transformation Vectors

The DIG nonradioactive nucleic acid labeling and detection system was used in the Southern analysis following the instructions of the supplier (*see* **Note 15**). The *idi* cDNA served as a probe and was generated by using the PCR DIG probe synthesis kit. The probe's signal was trapped by exposure of an X-OMAT film to the blot for 1 h in a cassette. The signal was visualized by developing the film (*see* **Fig. 6**). The strong intensity of the 1.5 kb *idi*-fragment illustrates the multiple copy integration of pPR15.

4. Notes

1. Linearization of the transformation vector is essential, because integration in the genomic DNA takes place exactly at the corresponding endonuclease recognition sites. *Sfi*I is a rare cutter and often the pPR2TN-inserted DNA fragment does not contain a *Sfi*I site. Therefore, *Sfi*I is very useful for vector linearization. However, for optimal digestion of DNA, *Sfi*I needs two recognition sites *(17)*. Therefore, digestion of a pPR2TN transformation vectors with *Sfi*I often results in a mixture of predominantly cut and, to a lesser extend, uncut vectors. This should be kept in mind when determining the transformation efficiency (number of transformants/µg DNA).
2. The linearized pPR2TN vector integrates in the ribosomal DNA (rDNA) of *X. dendrorhous* in multiple copies *(18)*. When pPR2TN is used to deliver an expression cassette of a gene of interest (in this chapter, *idi*) into the rDNA, more than one cas-

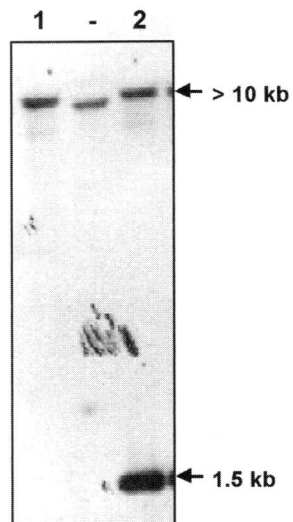

Fig. 6 Southern blot analysis of CBS6938[pPR2TN] and CBS6938[pPR15] transformants. Chromosomal DNA was digested with *Bam*HI. *Lane 1* contains
CBS6938[pPR2TN] DNA. The *idi* probe is hybridized to the endogenous *idi* gene,
which is on a large *Bam*HI DNA fragment (>10 kb). *Lane 2* contains CBS6938[pPR15]
DNA. In addition to the endogenous gene, the *idi* probe is also hybridized to the 1.5-kb
P*gpd–idi–*T*idi* expression cassette, which was released from the integrated pPR15 plasmid by *Bam*HI digestion.

     sette will be inserted. Consequently, this transformation setup will usually give rise
to the "overexpression" of the gene of interest.
3. Traces of, for example, phenol or ethanol will negatively influence the transformation efficiency. Therefore, the DNA sample should be as clean as possible. Use
   ultrapure water to dissolve the DNA. Additionally, use ultrapure water to make the
   solutions and buffers that are needed in the preparation of electrocompetent *X. dendrorhous* cells.
4. *Xanthophyllomyces dendrorhous* was cultivated in an Innova 4330 refrigerated
   incubator shaker. The shaking speed was 250 rpm. This shaking speed might not be
   equal to that of other incubators, which are set at 250 rpm. This does not pose a
   problem as long as the culture is well shaken (i.e., it should be well aerated).
   Cultures that have limited air (oxygen) supply do not produce astaxanthin.
   Generally, a shaking speed between 225 and 300 rpm is sufficient. If the culture
   does not turn pink, then try to increase the shaking speed or decrease the volume
   of YePD medium in the Erlenmeyer flask. The maximal amount of YePD medium
   should not exceed 10–20% (v/v) of the Erlenmeyer volume.
5. The doubling time of *X. dendrorhous* under these conditions is approx 2–3 h.

6. It is essential to keep everything on ice from this step onward because transformation efficiency drops when the temperature of the materials and samples is higher than that of ice water.

7. In order to avoid oxidation, carotenoid extracts should be prepared under dim light.

8. The α-ketol structure is susceptible to base-catalyzed oxidation of the 3-HO to a keto group by air. Therefore, alkaline conditions often used for saponification must be avoided. Otherwise, astaxanthin is rapidly oxidized to astacene. HO groups and keto groups (the latter with a smaller contribution) determine the polarity of carotenoids.

9. Make sure that the vial is cooled, otherwise the color might disappear upon addition of water.

10. When frozen in solution, the carotenoids will degrade gradually.

11. Most reversed-phase columns separate 3-hydroxy from 4-keto carotenoids very poorly. Apart from the column used here, the Vydac polymeric C_{18} column with methanol as eluent is also appropriate to separate the *X. dendrorhous* carotenoids.

12. The spectra from the elution peaks are very helpful for the identification of the carotenoids (*see* **Fig. 4**).

13. For quantitative calibration, carotenoid amounts starting from 1 μg and less should be used.

14. Previously, chromosomal DNA was isolated from *Xanthophyllomyces* using a method in which protoplasts were prepared after incubation with cell-wall-degrading enzymes (Novozym 234) for 3 h at 30°C. These protoplasts were subsequently lysed by addition of sodium dodecyl sulfate. Currently, we prefer to break *Xanthophyllomyces* cells by grinding them in liquid nitrogen with alumina type-A5 (a powder for disintegrating yeast cells by grinding or blending). This method is considerably faster and it yields high-quality chromosomal DNA.

15. Because of the hazardous properties and risks to human health and the environment, in our laboratory, radioactive labels (isotopes like ^{32}P) are being replaced by safe nonradioactive alternatives when possible. There are several kits for the nonradioactive labeling and detection of nucleic acid probes commercially available. We use the DIG system. A very detailed "DIG Application Manual" is available from the supplier.

Acknowledgment

This work was supported by grant QLK1-CT-2001-00780 from the European Commission.

References

1. Johnson, E. A. and An, G.-H. (1991) Astaxanthin from microbial sources. *Crit. Rev. Biotechnol.* **11,** 297–326.

2. Sandmann, G. and Misawa, N. (2002) Fungal carotenoids, in *The Mycota X Industrial Applications* (Osiewacz, H. D., ed.), Springer-Verlag, Berlin, pp. 247–262.

3. Cerdá-Olmedo, E. (1989) Production of carotenoids with fungi, in *Biotechnology of Vitamins, Pigments and Growth Factors* (Vandamme, E. J., ed.), Elsevier Applied Science, London, pp. 27–42.

4. Ernst, H. (2002) Recent advances in industrial carotenoid synthesis. *Pure Appl. Chem.* **74,** 2213–2226.

5. Kajiwara, S., Fraser, P. D., Kondo, K., and Misawa, N. (1997) Expression of an exogenous isopentenyl diphosphate isomerase gene enhances isoprenoid biosynthesis in *Escherichia coli. Biochem. J.* **324,** 421–426.

6. Verdoes, J. C. and van Ooyen, A. J. J. (1999) Isolation of the isopentenyl diphosphate isomerase encoding gene of *Phaffia rhodozyma;* improved carotenoid production in *Escherichia coli. Acta Bot Gallica* **146,** 43–53.

7. Wery, J., Verdoes, J. C., and van Ooyen, A. J. J. (1998) Efficient transformation of the astaxanthin-producing yeast *Phaffia rhodozyma. Biotechnol. Tech.* **12,** 399–405.

8. Verdoes, J. C., Wery, J., Boekhout, T., and van Ooyen, A. J. J. (1997) Molecular characterization of the glyceraldehyde-3-phosphate dehydrogenase gene of *Phaffia rhodozyma. Yeast* **13,** 1231–1242.

9. Verdoes, J. C., Krubasik, P., Sandmann, G., and van Ooyen, A. J. J. (1999) Isolation and functional characterisation of a novel type of carotenoid biosynthetic gene from *Xanthophyllomyces dendrorhous. Mol. Gen. Genet.* **262,** 453–461.

10. Andrews, A. G., Phaff, H. J., and Starr, M. (1976) Carotenoids of *Phaffia rhodozyma,* a red-pigmented yeast. *Phytochemistry* **15,** 1003–1007.

11. Britton, G. (1995) Structure and properties of carotenoids in relation to function. *FASEB J.* **9,** 1551–1558.

12. Andrews, A. G. and Starr, M. (1976) (3R,3′R)-Astaxanthin from the yeast *Phaffia rhodozyma. Phytochemistry* **15,** 1009–1011.

13. Sedmak, J. J., Weerasinghe, D. K., and Jolly, S. O. (1990) Extraction and quantitation of astaxanthin from *Phaffia rhodozyma. Biotechnol. Tech.* **4,** 107–112.

14. Sander, L. C., Sharpless, K. E., Craft, N. E., and Wise, S. A. (1994) Development of engineered stationary phases for the separation of carotenoid isomers. *Anal. Chem.* **66,** 1667–1674.

15. Maoka, T., Komori, T., and Matsuno, T. (1985) Direct diastereomic resolution of carotenoids I. 3-hydroxy-4-oxo-β-end group. *J. Chromatogr.* **318,** 122–124.

16. Sambrook, J., Fritsch, E. F., and Maniatis, T. (1989) *Molecular Cloning: A Laboratory Manual,* Cold Spring Harbor Laboratory Press, Cold Spring Harbor, NY.

17. Wentzell, L. M., Nobbs, T. J., and Halford, S. E. (1995) The *Sfi*I restriction endonuclease makes a four-stranded DNA break at two copies of its recognition sequence. *J. Mol. Biol.* **248,** 581–595.

18. Wery, J., Gutker, D., Renniers, A. C. H. M., Verdoes, J. C., and van Ooyen, A. J. J. (1997) High copy number integration into the ribosomal DNA of the yeast *Phaffia rhodozyma. Gene* **184,** 89–97.

15

Methodologies for the Analysis of Fungal Carotenoids

Paul D. Fraser and Peter M. Bramley

Summary

Carotenoids are high-value natural products utilized by several industrial sectors. Some filamentous fungi such as *Phycomyces blakesleeanus, Mucor circinelloides,* and *Blakeslea trispora* possess the ability to synthesise carotenoids. In this chapter, procedures for comprehensive analysis of fungal carotenoids are described. Protocols necessary for the growth of adequate biomass, carotenoid extraction, and analysis of carotenoids are provided. The techniques used include column chromatography, thin-layer chromatography, ultraviolet–visible spectrometry, and high-performance liquid chromatography with on-line photodiode array detection.

Key Words: Filamentous fungi; *Phycomyces blakesleeanus;* carotenoids; high-performance liquid chromatography; HPLC; photodiode array detection.

1. Introduction

Carotenoids are commercially important natural products used in the health, food, feed, pharmaceutical, and cosmetic industries *(1)*. The world market for carotenoids during 1999 was estimated at US$750–800 million and projections estimate this will increase to US$1 billion in 2005 *(2)*. At present, chemical synthesis is the method of choice used to produce carotenoids industrially. However, this approach does have several disadvantages such as contamination with reaction precursors or byproducts and the formation of racemic mixtures of nonbiological stereo isomers. In addition, chemical procedures are typically not environmentally friendly and often expensive. Therefore, opportunities exist for the production of carotenoids using biological sources. Carotenoid-forming fungi possess the potential to act as cell factories for the production of industrially useful carotenoids. Several filamentous fungi such as *Phycomyces blakesleeanus, Mucor circinelloides,* and *Blakeslea trispora* contain significant basal levels of carotenoids *(1)*. The application of strain improvement programs and/or recombinant DNA technologies to these organisms offer future potential

From: *Methods in Biotechnology, Vol. 18: Microbial Processes and Products*
Edited by: J. L. Barredo © Humana Press Inc., Totowa, NJ

for the development of these organisms as commercially viable, biological sources of carotenoids.

Carotenoids in fungi are synthesized via the mevalonate pathway *(3)*. The first committed step in the biosynthesis of fungal carotenoids occurs at the level of phytoene formation. Phytoene is a colorless C_{40} carotene formed from two molecules of the ubiquitous isoprenoid geranylgeranyl diphosphate (GGDP). Phytoene contains three conjugated double bonds. Further desaturation of the phytoene molecule by phytoene desaturase leads to lycopene via phytofluene, ζ-carotene, and neurosporene, which contain 5, 7, 9, and 11 conjugated double bonds, respectively. Cyclization of lycopene results in the formation of the bi-cyclic carotenoid β-carotene (provitamin A) formed via the monocyclic intermediate γ-carotene *(4)*.

The conjugated series of double bonds found in all carotenoid structures act as a chromophore, conveying characteristic red, yellow, and orange colors to these molecules. It also is responsible for the antioxidant activities of these compounds. The structural characteristics contribute to their unique ultraviolet–visible (UV/vis) spectral properties and chromatographic behavior, but also make them susceptible to oxidation from heat, light, and oxygen.

In order to evaluate carotenoid biosynthesis in fungi and mutant populations as well as assessing metabolite pools, it is important to have comprehensive analytical procedures both for qualitative and quantitative determination. In this chapter, we will illustrate how the chemical properties of carotenoids are exploited to devise efficient and accurate methods of analysis that are applicable not only to *P. blakesleeanus* but also to carotenoids present in other filamentous and nonfilamentous fungi.

2. Materials

1. *Phycomyces blakesleeanus* wild type and the mutants strains C115 *car*S42 *mad*-107(–), C9 *car*R21(–), and C5 *car*B (Professor E. Cerdá-Olmedo, Universidad de Sevilla, Sevilla, Spain).
2. Sutter's IV defined medium: add the following components to 800 mL of distilled water: 2 g L-asparagine, 20 g agar, 20 g D-glucose, and 20 mL (from a 50X concentrated stock solution) salts solution. When all the components are dissolved, the solution is brought to 1 L and pH 4.7.
3. Salts solution 50X: add the following components to 800 mL distilled water: 250 g KH_2PO_4, 25.0 g $MgSO_4 \cdot 7H_2O$, 5 mL trace element solution, 10 mL 14% (w/w) $CaCl_2$ solution, and 100 mg thiamine HCl. Adjust to a volume of 1 L.
4. Trace elements solution: 20 g/L citric acid, 15 g/L $Fe(NO_3)_3 \cdot 9H_2O$, 10 g/L $ZnSO_4 \cdot 7H_2O$, 3 g/L $MnSO_4 \cdot H_2O$, 0.3 g/L $CuSO_4 \cdot 5H_2O$, and 0.5 g/L $NaMoO_4 \cdot 2H_2O$.
5. Liquid medium: 25 g/L D-glucose, 0.5 g/L yeast extract, 0.5 g/L $MgSO_4$, 1.25 g/L L-asparagine, 1.5 g/L KH_2PO_4, 0.25 mg/L thiamine-HCl, and 1.25 g/L L-leucine. Adjust to pH 5.0 (*see* **Note 1**).

6. Autoclave for sterilization.
7. Laminar flow hood.
8. Petri dishes.
9. Universal bottles, 20 mL.
10. Inoculation loops.
11. Hockey stick.
12. Conical flasks, 2 L.
13. Diffuse light sources, 20 and 40 W/m^2.
14. Incubation facilities, 24°C.
15. Orbital shaker.
16. Freeze-dryer.
17. Freezer, –70°C.
18. Centrifuge for 1.5-mL and 15-mL tubes.
19. UV/vis spectrometer.
20. Nitrogen gas.
21. Water bath.
22. Oven, 120°C.
23. 1-mL Quartz cuvet.
24. Muslin for harvesting mycelia.
25. Sieve for breakage of mycelia, 355-μm aperture.
26. 15-mL Glass Pyrex tubes with screw caps.
27. Chromatography apparatus.
28. Thin-layer chromatography (TLC) tanks.
29. Brockmann grade III alumina chromatography column, 2.0 × 20 cm (NBS Biology Ltd., Hatfield, UK).
30. High-performance liquid chromatography (HPLC) system, e.g. Waters Alliance comprised of a 2690 separation modul, 996 PAD controlled by Millennium^{32} software (Waters, Milford, USA).
31. Nucleosil reversed-phase column C$_{18}$ octadocyl silane (ODS), 5 μm 150.0 × 4.6 mm with identical guard 10.0 × 4.0 mm (Jones Chromatography Ltd., Hengoed, UK).
32. YMC carotenoid column reversed-phase C$_{30}$, 5 μm, 150 × 4.6 mm with 10 × 4.0 mm guard of the same material (YMC, Wilmington, NC).
33. Heater/chillier column oven (Jones Chromatography Ltd.).
34. 50 mM Tris-HCl, pH 8.0.
35. Neutral alumina (NBS Biology Ltd).
36. Silica TLC plates, silica gel 60 F$_{254}$ (VWR International Ltd., Poole, UK).
37. Alumina TLC plates, aluminium oxide 60F$_{254}$ neutral.
38. Carotenoid standards (Craft Technologies Inc., Wilson, NC; Extrasynthese Corp, Genay, France; DHI, Hersholm, Denmark).

3. Methods

3.1. Growth and Maintenance of Phycomyces blakesleeanus

Carotenoid analysis is performed on *P. blakesleeanus* mycelia grown under defined culture conditions. Mycelia are produced from shaken liquid cultures.

These shaken cultures are initiated from heat-shocked spore suspensions. In order to obtain a suitable inoculum, *P. blakesleeanus* must first be propagated to sporulation on agar slopes containing Sutter's IV minimal medium (*5*). The following procedures are used to produce *P. blakesleeanus* spores and mycelia for carotenoid analysis.

1. Prepare agar plates, each containing about 20 mL of sterilized Sutter's IV defined media. Sterile the media by autoclaving at 121°C for 30 min and 15 lb/in^2. Cool the molten agar to about 60°C, pour into either Petri dishes or universal bottles (20 mL), and allow cooling and setting in a laminar flow hood.
2. Using aseptic techniques either (1) streak a small batch of hyphae from a mature culture over the surface of the agar with an inoculation loop, (2) cut small segments of mature culture with a sterile scarpel and place fungal side down onto fresh media, or (3) spread an aliquot of spore suspension over the surface of the agar with a hockey stick or a filter disk impregnated with spores placed on the agar. Incubate the cultures at 24°C in diffuse light (20 W/m^2). Mycelial growth proceeds, and after approx 2 wk, spore production is dense (*see* **Note 2**).
3. Collect spores from mature *P. blakesleeanus* cultures by placing sterile water (10 mL) onto the culture. Shake the solution to produce a spore suspension. This spore suspension (2 mL, approx 10^6 viable spores) can be used directly to inoculate liquid cultures. Alternatively, place 1.5 mL of spore suspension into a microcentrifuge tube and centrifuge at 8000*g* for 1 min. Remove the supernatant and resuspend the dark spore pellet in 100 μL of sterile 10% (v/v) glycerol. Inoculate liquid cultures with 100 μL of the concentrated spore suspension (*see* **Note 3**).
4. Carry out mycelia production in liquid culture using 2-L conical flasks as culture vessels, containing 500 mL of liquid media. Following inoculation, shake liquid cultures at 160 rpm on an orbital incubator at 25°C under diffuse light (48 W/m^2). The period of cultivation depends on the experimental objectives. For maximum pigment production, incubate cultures for 70 h (*see* **Note 4**).

3.2. Harvesting, Lyophilization, and Cell Breakage

1. Harvest mycelia from liquid shake cultures by filtration through two layers of prewashed muslin. Wash the mycelia with about 1 L of distilled water and squeeze to hand dryness. Freeze the mycelia immediately at –70°C in darkness for subsequent lyophilization.
2. Once frozen, quickly place mycelia onto the freeze-drier in vessels shielded from direct light. Allow lyophilization to proceed for 3 d, after which time the mycelia are biscuit dry and ready to use (*see* **Note 4**).
3. Prepare a fine-ground powder of the mycelia by passing the firm dry mycelia through a sieve of 355-μm aperture at room temperature. Collect the powder on aluminum foil and then extract for carotenoids.

3.3. Extraction of Carotenoids

Throughout the extraction procedure, take care to minimize exposure of the extract to heat, light, and oxygen.

1. Weigh accurately between 10 and 100 mg of fine mycelia powder into glass Pyrex tubes (15 mL) with screw caps. Add to the powder methanol (2 mL) and then chloroform (3 mL). Vortex the mixture vigorously (two times 30-s bursts), and then place the suspension on ice in the dark for 30 min (*see* **Notes 5** and **6**).
2. After cooling on ice, add 2 mL of 50 m*M* Tris-HCl, pH 8.0, and vortex the mixture for 3 s.
3. Centrifuge the mixture at 3500*g* for 5 min to create a clear partition, with an interface of protein and cellular debris.
4. Using a Pasteur pipet, pierce the protein interface between the aqueous and organic layers and remove the organic hypophase into a clean glass tube. Re-extract the aqueous hyperphase with an additional 3 mL of chloroform.
5. Discard the aqueous phase and residue. Pool the organic, colored phases containing carotenoids and then evaporate the solvent to complete dryness under a stream of nitrogen gas at room temperature or in a water bath at about 40°C.

3.4. Purification of Carotenoids Using Column Chromatography

Carotenoid mixtures can be separated from other components of crude mixtures by column chromatography (CC) on neutral alumina (*see* **Subheading 3.4.1.**) or by TLC (*see* **Subheading 3.4.2.**).

3.4.1. Column Chromatography

1. For the separation and purification of carotenes and carotenoids, use a Brockmann grade III alumina column. Prepare the column by preheating the alumina (10 g) in a small glass beaker at 120°C for 1 h in an oven. Cool the alumina in a desiccator and then add distilled water (0.6 mL). Thoroughly mix the water with the alumina. Add petroleum ether (boiling point, 40–60°C) onto the alumina until it just covers the surface. Mix well to create a slurry.
2. Prepare a glass chromatography column (1.5 cm in diameter) by adding a small quantity of nonabsorbent cotton wool to the base and a clip to the outlet. Add about 25 mL of petroleum ether 40–60°C to the column and then put in the alumina slurry. Allow the alumina to settle and then run off the excess petroleum until the meniscus is just above the top of the alumina.
3. Dissolve the dried carotenoid extract in petroleum ether (boiling point, 40–60°C) to achieve an optical density (OD_{450}) of about 0.1. Add the solution carefully to the top of the alumina column using a Pasteur pipet and remove the clip at the base. Allow the solution to enter the column and then carefully add more petroleum ether to the column without disturbing the alumina. Change the polarity of the solvent by adding increasing concentrations of diethyl ether (0.5–5% in 50-mL steps). The typical elution characteristics of carotenoids found in *P. blakesleeanus* mycelia are provided in **Table 1.**
4. Identify the carotenoids that are eluted by UV/vis spectrometry (*see* **Subheading 3.4.3.**).

Table 1
Chromatographic and Spectral Properties of Carotenoids Typically Found in Fungi

Carotenoid	TLC (R_f) System A	System B	CC (% DE)	HPLC Rt (min) System A	System B	System C	UV/vis*[a] (nm)	$A_{1cm}^{1\%}$	PDA spectrum (nm)
Phytoene	0.45	0.54	2.5–5.0	71.0	32.0	24.69	(276),286,297	1250	(275.8),285.6,297.3
ζ-Carotene	0.34	0.20	2.5–5.0	51.0	22.0	32.03	376,397,421	2555	379.6,400.1,425.4
Lycopene	0.24	0.10	2.5–5.0	33.0	15.0	40.69	444,470,502	3450	446.0,472.6,504.2
γ-Carotene	0.40	0.24	1.0–2.5	42.0	NR	36.3	437,462,494	3100	492.0,461.7,430.0
β-Carotene	0.42	0.27	1.0–2.5	58.0	26.0	28.22	(425),449,476	2592	(426),453.2,478.0

Abbr: TLC system A: Silica G stationary phase, 15% v/v toluene in petroleum ether (40–60°C) mobile phase; TLC system B: alumina stationary phase, 3% (v/v) toluene in petroleum ether (boiling point 40–60°C). CC: activated alumina Brockmann grade III; % DE: proportion of diethyl ether added to petroleum ether (boiling point 40–60°C); HPLC system A: acetonitrile, methanol, isopropanol (85 : 10 : 5), nucleosil reversed-phase C18 column (*6*); HPLC system B: ethyl acetate, isopropanol, water (35 : 70 : 5) (*7*); HPLC system C: reversed-phase C30 column, with mobile phases consisting of methanol (A), water/methanol (20 : 80) containing 0.25 ammonium acetate (B) and *tert*-methyl butyl ether (C). The gradient elution used with this column was 95% A, 5% B isocratically for 12 min, a step to 80% A, 5% B, 15% C at 12 min, followed by a linear gradient to 305 A, 5% B, 65% C by 30 min. A conditioning phase (30–60 min) was then used to return the column to the initial concentrations of A and B.

Note: $A_{1cm}^{1\%}$ is the absorbance of a 1% (v/v) solution in a low light path.

[a] Recorded in petroleum ether (boiling point, 40–60°C); $A_{1cm}^{1\%}$ for petroleum ether (boiling point, 40–60°C); NR = not recorded.

3.4.2. Thin-Layer Chromatography

Two TLC systems are used routinely to separate carotenoids found in *P. blakesleeanus* extracts. System A consists of a silica stationary phase and a mobile phase comprising 15% toluene in petroleum ether (boiling point, 40–60°C). System B utilizes an alumina stationary phase and mobile phase consisting of 3% toluene in petroleum ether (boiling point, 40–60°C). In both cases, the chromatography must be carried out in darkness. Alumina plates should be activated by heating at 120°C for about 1 h and then allowed to cool in a desiccator before use. Store activated plates in the desiccator.

1. Dissolve carotenoid extracts in acetone or chloroform and apply to either type of TLC plate at the origin. Do this quickly and evenly in a minimum volume of solvent. Allow the solvent to evaporate and then develop in the chromatographic solvent for about 1 h at room temperature in complete darkness.
2. Following development, remove the thin layer from the tank and allow the solvent to evaporate by gentle air-drying. The colored carotenoid bands can be visualized by eye.
3. Locate phytoene on the TLC by placing the plate in a chamber of iodine vapor. A brown zone will indicate the presence of phytoene.
4. Elute the carotenoids from TLC plates by scraping the stationary phase from the plate with a spatula in the region where the carotenoid is situated. Place the powder in a screw-capped glass tube (15 mL) and add 2 mL of diethyl ether. Vortex for 10 s.
5. Centrifuge the suspension at 3500g for 5 min and then transfer the carotenoid-containing supernatant removed into a fresh tube over ice and in the dark.
6. Identify carotenoids by comparison of their R_f values with those for authentic carotenoid standards, chromatographed under identical conditions.
7. Using a scanning spectrophotometer, obtain UV/vis spectra to identify carotenoids from their characteristic spectral properties (*see* **Subheading 3.4.3.**; examples are provided in **Table 1**).

3.4.3. Identification and Quantitation of Carotenoids by UV/Vis Spectrometry

1. Set the scanning spectrophotometer to a wavelength range between 250 and 600 nm. Use a quartz cuvet (1 mL), blank on air, and then petroleum ether.
2. Dissolve the carotenoid in petroleum ether (boiling point 40–60°C) and obtain its spectrum from 200 to 600 nm. Identify the carotenoid from this spectrum (*see* **Table 1**). The amount of carotenoid present in the sample is determined using the following equation:

$$X = (A \times Y)/(A^{1cm}/_{1\%} \times 100)$$

where X is the weight of the carotenoid (in g), Y is the total volume of sample solution (in mL), $A^{1cm}/_{1\%}$ = absorbance coefficient of a 1% solution in 1 cm light path. (*see* **Note 7**).

3.4.4. Separation, Quantitation, and On-Line Identification Using HPLC–Diode Array Detection

Numerous HPLC systems can be used to separate carotenoids typically found in fungi. Several have been summarized in **Table 1.** The use of a diode array detector allows spectra to be recorded simultaneously with elution of the carotenoids.

1. Dry the chloroform (*see* **Subheading 3.3.**), CC (*see* **Subheading 3.4.1.**), or TLC (*see* **Subheading 3.4.2.**) extracts under nitrogen until all the solvent is removed. To the residue, add 50 µL of ethyl acetate (HPLC grade) and ensure that the carotenoid is completely dissolved. Centrifuge the solution at 10,000g for 3 min to pellet any particles that will block the HPLC column. Remove the clear supernatant carefully, avoiding any pelleted debris. Add the ethyl acetate solution to an appropriate HPLC vial (typically 100 µL in volume) with an insert (*see* **Note 8**).
2. Carotenoids can be separated on the systems described in **Table 1.** Carry out continuous monitoring between 250 and 600 nm using the diode array detector. In all cases, use a flow rate of 1 mL/min and column pressure within 800–1200 lbs/in.^{2}. Set the column oven/chiller to 24° ± 1°C.
3. Identify carotenoids by (1) cochromatography and (2) comparison of spectral properties with authentic known standards.
4. Quantify the carotenoids present in an experimental sample by comparison of peak areas of each compound from the elution profile on HPLC (**Fig. 1**) using calibration curves. Prepare dose–response curves by injecting known amounts of authentic carotenoids onto the HPLC over a concentration range of 0.1–1.5 µg. A linear relationship between concentration and absorbance is observed over this range.

4. Notes

1. The trace element solution used in *P. blakesleeanus* media can be stored at 4°C for at least 10 yr, whereas the 50X concentrated salts solution can be kept at room temperature after adding 2 mL of chloroform as a preservative.
2. It is strongly advised to create master stocks of all fungal strains. *P. blakesleeanus* can be preserved at –70°C on solid medium for at least 10 yr. Working stocks on either plates or spore suspensions remain viable at 4°C for at least 1 yr. Avoid repeated subculturing from one stock, as this can lead to mutations. Instead, prepare a set of master stock cultures, store at –70°C, and use one at a time to prepare many working stock cultures, stored at 4°C.
3. For optimal growth, we find that heat shocking the spore solution (42°C for 1 h) and the combination of 2-L conical flasks containing 500 mL of media optimal. If it is necessary to increase the volume of growth medium, then maintain the vessel-to-media ratio.
4. The amount of mycelia produced from a typical 500-mL liquid culture is about 7 g dry weight. When placed on the freeze-dryer, it is best to use a container wrapped in foil to shield from direct sunlight. It is important that during the freeze-drying process none of the mycelia defrosts. To prevent any potential accidental thawing, it is best not to bulk mycelia batches but to maintain the tissue in discrete portions.

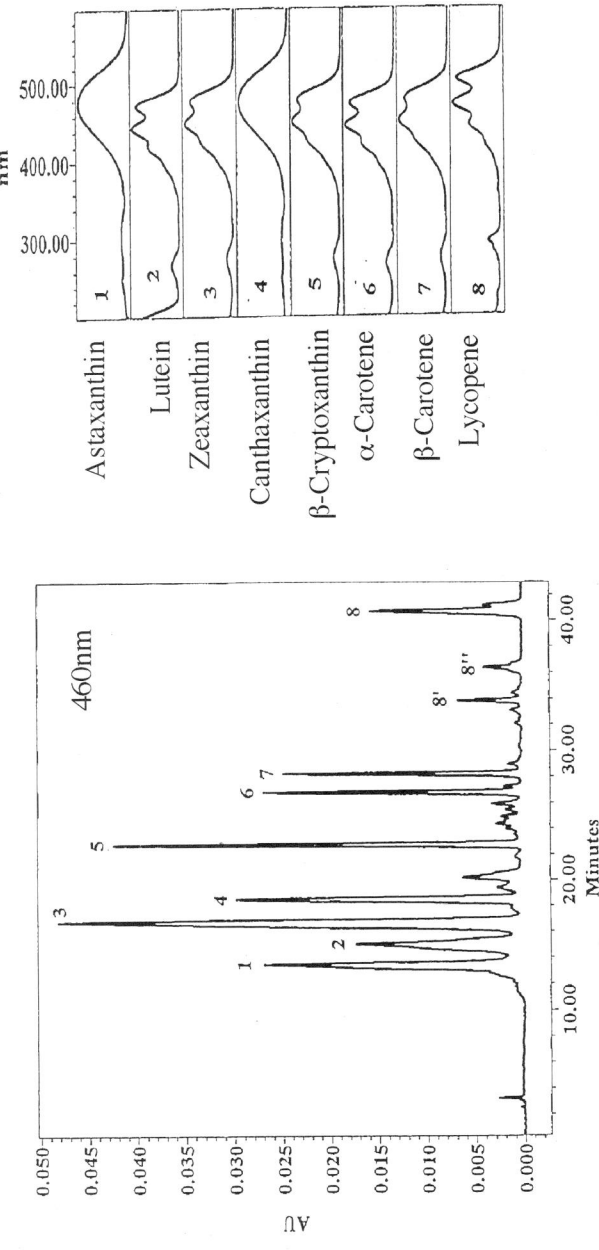

Fig. 1. Typical separation of carotenoid standards on a reversed-phase C$_{30}$ HPLC column. The procedure is described in **Subheading 3.4.4.** and reported in **ref. 8.**

Once freeze-dried completely (typically 3 d), the mycelia should be stored at
−20°C in an airtight container with a small quantity of silica gel desiccant.

5. Carotenoids are sensitive to light, heat, oxygen, acid, and, in some cases, alkali.
Once carotenoids have been extracted from biological tissue, it is important to
carry out manipulations in subdued light or darkness. Carotenoids are particularly
prone to degradation during chromatography either on columns or during TLC
development. Therefore, it is advisable to perform chromatography in darkrooms
with enclosed columns. Whenever possible, solutions of carotenoids are kept on ice
during experimentation and only subjected to minimal heat when necessary. Gentle
heat (below 40°C) can be applied when reducing the volume carotenoid solutions
under nitrogen gas or rotary evaporation.

6. Other solvents such as diethyl ether or 10% diethyl ether in petroleum ether (boiling point, 40–60°C) can be used as the extraction solvent instead of chloroform.
However, in tissues where the lycopene content is high, chloroform is necessary for
effective extraction.

7. Routinely, petroleum ether (boiling point, 40–60°C) is used as the solvent for
recording spectra, but other solvents for which extinction coefficients are known
can be used (*see* **Note 8**).

8. Ethyl acetate is used as the solvent for sample application onto HPLC. In our experience, it maintains the carotenoids in solution at 8°C without evaporation over 12
h. Solvents such as acetone result in precipitation occurring in the sample, whereas chloroform evaporates.

References

1. Cerdá-Olmedo, E. (1989) Production of carotenoids with fungi, in *Biotechnology of
Vitamins, Pigments and Growth Factors* (Vandamme, E., ed.), Elsevier Applied
Science, Amsterdam, pp. 27–42.
2. Merz, U. (2000) *Business Report: The Global Market for Carotenoids,* Business
Communications Co., Norwalk, CT.
3. Sandmann, G., Albrecht, M., Schnurr, G., Knorzer, O., and Böger, P. (1999) The
biotechnological potential and design of novel carotenoids by gene combination in
Escherichia coli. Trends Biotechnol. **17,** 233–237.
4. Sandmann, G. (2001) Carotenoid biosynthesis and biotechnological application.
Arch. Biochem. Biophys. Acta **385,** 4–12.
5. Sutter, R. P. (1975) Mutations affecting sexual development in *Phycomyces
blakesleeanus. Proc. Natl. Acad. Sci. USA* **72,** 127–130.
6. Fraser, P. D., Truesdale, M., Bird, C. R., Schuch, W., and Bramley, P. M. (1993).
Carotenoid biosynthesis during tomato fruit development. Evidence for tissue specific gene expression. *Plant Physiol.* **105,** 405–413.
7. Fraser, P. D., Misawa N., Linden, H., Yamano, S., Kobayashi, K., and Sandmann, G.
(1992) Expression in *Escherichia coli,* purification and reactivation of the recombinant *Erwinia uredovora* phytoene desaturase. *J. Biol. Chem.* **267,** 19,891–19,895.
8. Davies, B. H. (1976) Carotenoids, in *Chemistry and Biochemistry of Plant Pigments,*
Vol. 2 (Goodwin, T. W., ed.), Academic Press, New York, pp. 38–165.

16

Insertional Mutagenesis in the Vitamin B_2 Producer Fungus *Ashbya gossypii*

María A. Santos, Laura Mateos, Karl-Peter Stahmann, and José-Luis Revuelta

Summary

The ascomycete *Ashbya gossypii,* a filamentous fungus, is a natural overproducer of vitamin B_2 and is currently exploited for the industrial production of this vitamin. Classical mutagenesis and selection of mutants showing improved production capacities have been routinely applied with success to the development of industrial strains of *A. gossypii* that overproduce vitamin B_2. However, this approach does not allow the subsequent isolation and identification of affected genes in organisms such as *A. gossypii* that lack a sexual life cycle, thus hampering further improvements of the strains on the basis of rational strategies derived from knowledge of the function of the genes involved in the process. Here, we describe an efficient in vitro *Himar1*-based transposition reaction of minitransposon elements on chromosomal *A. gossypii* DNA, which is subsequently reintroduced into *A. gossypii* by electrotransformation. Sequencing of the minitransposon insertion point in numerous transposition mutants and a limited screening of the insertional mutants demonstrate that this method can lead to the exhaustive mutagenesis of *A. gossypii,* allowing—for the first time—a genomic-scale mutational analysis of this biotechnologically important fungus.

Key Words: *Ashbya gossypii;* insertional mutagenesis; in vitro transposition; *Himar1;* riboflavin production.

1. Introduction

Many useful biological compounds are found in only small or trace quantities in their natural sources and are difficult or impossible to synthesize chemically. Driving the field of metabolic engineering is the hope that recombinant cells can serve as biosynthetic factories of this rich pool of natural substances *(1–4)*. Genes and biosynthetic enzymes can be purposefully modified to engineer more efficient pathways for the synthesis of natural substances and other novel metabolites *(3,4)*. However, reliance on under-

From: *Methods in Biotechnology, Vol. 18: Microbial Processes and Products*
Edited by: J. L. Barredo © Humana Press Inc., Totowa, NJ

standing the appropriate catalytic and regulatory mechanisms operating in the cell necessarily limits the production of compounds that can be synthesized in engineered organisms.

Riboflavin (vitamin B_2) is the precursor of the flavocoenzymes FMN and FAD, which are universally required for redox metabolic reactions in all organisms. Riboflavin is synthesized by many bacteria and by all fungi and plants, but it cannot be synthesized by vertebrates. Therefore, this compound is an essential nutrient for animals and humans. Microbial fermentation can be employed for the competitive production of vitamin B_2. The hemiascomycete *Ashbya gossypii,* a filamentous fungus, is a natural overproducer of this vitamin and is currently in use for industrial vitamin B_2 production *(5).*

Classical mutagenesis and selection of mutants showing improved production capacities have been routinely applied with success to the development of industrial strains of *A. gossypii* that overproduce vitamin B_2. However, this approach does not allow the subsequent isolation and identification of affected genes in organisms, such as *A. gossypii,* that lack a sexual life cycle. This hampers further improvements of the strains on the basis of rational strategies derived from knowledge of the function of the genes involved in the process. To circumvent this problem, transposons have been used to explore genetically determined processes in yeast and in other organisms. The insertion of a mobile genetic element into a gene or its controlling elements will destroy its function and could cause a detectable phenotype. Because the mutated gene is tagged by a known piece of DNA, the method has been named transposon tagging. Because transposon tagging has not been routinely established for fungi, the efficient transformation by homologous recombination of an organism provides an alternative method for generating insertion mutants.

A powerful asset for generating and identifying mutants of interest is the ability to generate large numbers of insertion alleles. Plasmid-based libraries of insertion alleles can be generated quickly and economically using transposons. For example, *A. gossypii* genomic DNA is mutagenized in vitro; transposon-mutagenized DNA is then shuttled back into *Ashbya,* either individually or *en masse,* where the insertion alleles substitute for their chromosomal copies by homologous recombination. Libraries of these alleles constructed with *mariner* family transposons, such as the transposable element *Himar1,* are advantageous because they always integrate into a TA target dinucleotide and hence exhibit less bias in target-site selection than do most other eukaryotic transposons, including the yeast transposon Ty1 *(6).* Moreover, purified *Himar1* transposase is sufficient to mediate in vitro transposition and does not need host-specific factors for transposase function *(7).* The use of mini-*Himar1* elements containing a selectable marker, the bifunctional kanamycin/G418 resistance marker *(G418)* for selection in both *Ashbya* and *Escherichia coli* as well as an *E. coli* plasmid origin of replication *(ColE1)* readily affords the identity of any mutagenized gene showing a phenotype of interest.

2. Materials

1. *Ashbya gossypii* ATCC 10895 wild-type strain (ATCC, Manassas, VA).
2. *Escherichia coli* strains BL21 (DE3), JM101, DH5α, and DH10 B.
3. Oligonucleotide primers: TpaseA: 5′-CCATATGGAAAAAAAGGAATTTCG-3′; TpaseB: 5′-CGGATCCGTTTTTATTATTCAACATAG-3′; reverse 5′-GAAACAGCTATGACCATG-3′; universal: 5′-TGTAAAACGACGGCCAGT-3′; ITR-L: 5′-CGTCGACCAATCATTTGAAGGTTGGTAC-3′; ITR-R 5′-CGTCGACTCGCTCTTGAAGGGAACTATG-3′; ColA: 5′-CACCATG-GTCTTCCGCTTCCTCGCTCAC-3′; ColB 5′-CACCATGGC AAAATCCCT-TAACGTGAG-3′; MR1: 5′-CGGATTACAGTCGACCAATC-3′; and MR2: 5′-CCGGATTACAGTCGACTCGC-3′ (*see* **Note 1**).
4. Restriction enzymes, RNase A, and T4 DNA ligase.
5. Agarose and DNA electrophoresis gel equipment.
6. Standard ligation mixture: 10 m*M* MgCl$_2$, 10 m*M* dithiothreitol (DTT), 50 m*M* Tris-HCl, pH 7.5, and 1 m*M* ATP.
7. Polymerase chain reaction (PCR) system, including thermal cycler, thermostable polymerase, and dNTPs.
8. Plasmids pBADMar1 and pmMar1 *(7)*.
9. Plasmid pET12 (Novagene, Madison, WI).
10. Plasmid pAG100 *(8)*.
11. Plasmid pBluescript SK (+) (Stratagene, La Jolla, CA).
12. Antibiotics: ampicillin, kanamycin, and G-418 disulfate.
13. LB medium: 10 g/L bacto-tryptone, 5 g/L bacto-yeast extract, 10 g/L NaCl, and 20 g/L agar. Adjust to pH 7.0 with 5 *N* NaOH. Autoclave at 121°C for 15 min.
14. LB–ampicillin: LB containing 100 µg/mL ampicillin.
15. LB–kanamycin: LB containing 50 µg/mL kanamycin.
16. LB–ampicillin–kanamycin: LB containing 100 µg/mL ampicillin and 50 µg/mL kanamycin.
17. IPTG (isopropyl-β-D-thio-galactopyranoside).
18. PMSF (phenylmethylsulfonyl fluoride).
19. Lysis buffer: 25 m*M* Tris-HCl, pH 7.8, 100 m*M* NaCl, and 1 m*M* PMSF.
20. Sonicator.
21. T buffer: 25 m*M* Tris-HCl, pH 7.8, 2 m*M* DTT, 100 m*M* NaCl, 5 m*M* MgCl$_2$, 10% glycerol (v/v), and 1 m*M* PMSF.
22. Sodium dodecyl sulfate–polyacrylamide gel electrophoresis (SDS-PAGE) equipment.
23. Yeast tRNA.
24. 0.5 *M* EDTA.
25. Electroporator and cuvets of 0.2 and 0.4 cm.
26. Whatman 3MM cellulose filters (Whatman, Kent, UK).
27. MA2 medium: 2% bacto-peptone, 0.2% yeast extract, 2% glucose, and 0.6% myo-inositol, pH 6.8.

28. Spheroplast buffer: 1.2 *M* sorbitol and 20 m*M* EDTA.
29. Zymolyase 20T (Seikagaku, Tokyo, Japan).
30. 3 *M* Sodium acetate, pH 7.0.
31. AT buffer: 50 m*M* potassium phosphate buffer, pH 7.0, and 25 mM DTT.
32. STM buffer: 270 m*M* sucrose, 10 m*M* Tris-HCl, pH 7.5, and 1 m*M* MgCl$_2$.
33. SPA medium: 0.3% soybean meal (w/v), 0.3% yeast extract (w/v), 0.3% malt extract (w/v), 2% corn step liquor (v/v), and 1% dextrose (w/v), pH 6.8.

3. Methods

The methods described outline (1) the expression and purification of active *Himar1* transposase, (2) the construction of functional minitransposons suitable for DNA tagging in *A. gossypii,* (3) the construction of plasmid-based libraries of insertion alleles by in vitro transposition, (4) the insertional mutagenesis mediated by transformation of *A. gossypii* and the isolation of mutants, and (5) the rescue of the insertion alleles from *A. gossypii* transformants and the identification of the insertion sites.

3.1. Expression and Purification of Active Himar1 Transposase

The construction of the expression plasmid for an active form of transposase Himar1p is described in **Subheadings 3.1.1.–3.1.6.** This contains (1) the description of the expression vector, (2) the amplification and cloning of the DNA fragment encoding *Himar1* transposase in the expression vector, and (3) the expression and purification of *Himar1* transposase, and (4) the analysis of the transposase activity.

3.1.1. pET-12a Expression Vector

pET-12a is a useful vector for the cloning and expression of recombinant proteins in *E. coli (9).* The vector, based on the T7 promoter/RNA polymerase system, allows sequences inserted into the unique *Nde*I and *Bam*HI cloning sites to be expressed under the control of the T7 promoter, which is not recognized by *E. coli* RNA polymerase. Therefore, virtually no expression occurs until a source of T7 RNA polymerase is provided. Recombinant pET-12a plasmids carrying the cloned sequence are transferred into expression hosts, such as *E. coli* BL21 (DE3) cells, containing a chromosomal copy of the T7 RNA polymerase gene under the IPTG-inducible *lacUV5* promoter, and expression is induced by the addition of IPTG. T7 RNA polymerase is very active and highly selective for the T7 promoter, allowing the expressed product to comprise more than 50% of the total cell protein a few hours after induction.

3.1.2. Amplification and Cloning of the DNA Fragment Encoding Himar1 Transposase

1. In a standard PCR reaction, amplify the 1.6-kb *Himar1* transposase coding region using 10 ng of pBADMar1 DNA as template (pBADMar1 is a plasmid kindly

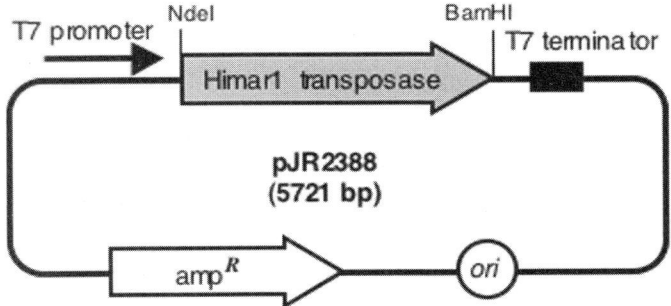

Fig. 1. Schematic of the *Himar1* tranposase expression plasmid pJR2388. This plasmid is derived from the expression vector pET-12a and contains a cDNA fragment encoding the *Himar1* tranposase under the transcriptional control of the T7 promoter.

donated by Dr. D. J. Lampe that contains a DNA fragment corresponding to the *Himar1* transposase gene) and the oligonucleotides TpaseA and TpaseB to add *Nde*I and *Bam*HI restriction sites upstream and downstream from the coding sequence, respectively.

2. Digest the cleaned PCR product with *Nde*I and *Bam*HI restriction enzymes and clone the DNA fragment into double-digested (*Nde*I + *Bam*HI) expression vector pET-12a using standard procedures (*10*) (*see* **Note 2**). This yields the expression plasmid pJR2388 (*see* **Fig. 1**).

3.1.3. Expression and Purification of Recombinant Transposase

The process involves the introduction of the expression plasmid for *Himar1* transposase in *E. coli* cells, the induction of protein expression, and the extraction and purification of the transposase.

3.1.3.1. TRANSFORMATION OF *E. COLI* AND INDUCTION OF PROTEIN EXPRESSION

1. Transform *E. coli* strain BL21 (DE3) competent cells with the expression plasmid pJR2388 using standard methods (*10*) and select the transformants onto LB–ampicillin plates
2. Inoculate 5 mL LB–ampicillin liquid medium with a single transformant colony and grow overnight at 37°C.
3. Inoculate 200 mL of LB–ampicillin medium with 50 μL of the overnight culture and incubate at 37°C in a shaker at 200 rpm until the culture reaches a density of 10^8 cells/mL. This takes about 4–5 h.
4. Induce the culture by adding 2 mL of 1 *M* IPTG. Shift the culture growth temperature to 28°C and incubate for an additional 2 h with vigorous shaking.

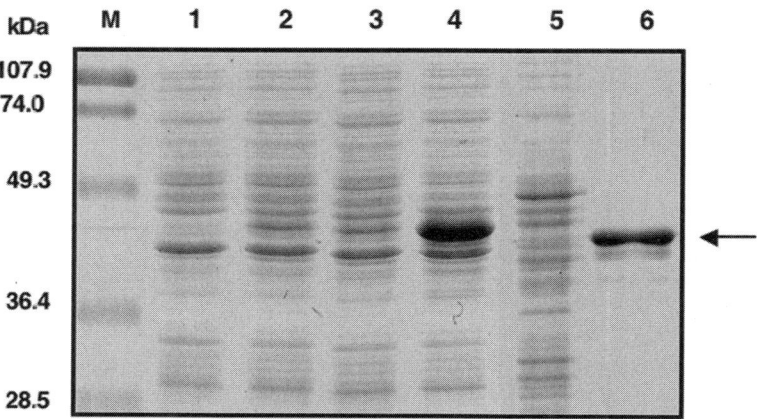

Fig. 2. Expression and purification of *Himar1* transposase. Crude cell extracts (5 µL) of *E. coli* strain BL21 (DE3) cells bearing control vector pET-12a (*lanes 1* and *2*) or *Himar1* transposase expression plasmid pJR2388 (*lanes 3* and *4*) from uninduced (*lanes 1* and *3*) or from IPTG-induced cultures (*lanes 2* and *4*) were resolved together with standard protein markers (*lane M*) in a 14% SDS-PAGE gel and stained with Coomassie blue. *Lanes 5* and *6* contain protein fractions corresponding to the purification process. *Lane 5* corresponds to the supernatant following disruption of bacterial cells expressing *Himar1* transposase, and *lane 6* corresponds to the purified *Himar1* transposase fraction after solubilization of protein precipitate in T buffer.

3.1.3.2. Protein Extraction and Purification

1. Harvest the cells by centrifugation at 5000*g* for 5 min and wash the pellet twice with 1 vol of cold, sterile water.
2. Resuspend the cell pellets in 10 mL of lysis buffer and disrupt the cells on ice by sonication using five bursts of 30 s.
3. Remove nonlysed cells by centrifugation at 5000*g* for 5 min at 4°C and transfer the cell lysate to a fresh tube.
4. Centrifuge at 10,000*g* for 15 min at 4°C. Transfer the supernatant containing the soluble protein fraction to a fresh tube and keep at 4°C for later analysis. Wash the precipitate, the insoluble protein fraction, twice with 20 mL of lysis buffer.
5. Solubilize the protein pellet by resuspension in 10 mL of fresh T buffer and vortex the solution gently for 30 s until a homogenous suspension is obtained. Keep the solution on ice for 30 min and vortex every 10 min. Add 10 mL of T buffer and repeat the process. The solution becomes clear when 65–70% of the protein is soluble.
6. Clear the protein solution by centrifugation at 10,000*g* for 10 min at 4°C and store the cleared supernatant containing active transposase at 4°C in 1-mL aliquots. Under these conditions, the *Himar1* transposase is active at least for 2 mo.
7. Analyze the induction process by SDS-PAGE (*see* **Fig. 2**).

3.1.4. Analysis of the Transposase Activity

The mechanism for the transposition of the *Himar1* transposon is similar to the mechanism proposed for the *mariner* elements *(7)*. Mariner transposons move through a DNA intermediate during transposition using a "cut-and-paste" mechanism, resulting in excision of the transposon from the original location and insertion at novel sites in the genome *(11,12)*. Two essential components are necessary in this cut-and-paste process: an active transposase functioning in trans and the inverted terminal repeats (ITRs) that flank the elements and are recognized and mobilized by the transposase. *Himar1* transposon always integrates into a TA target dinucleotide, which is duplicated upon insertion. Given these facts, the *Himar1* transposon can be exploited for random insertional mutagenesis in *A. gossypii* by inserting the *G418* selectable marker between the ITRs (resulting in the minitransposon A) and inducing its integration in the host chromosome DNA in vitro. As a preliminary step toward this goal, it is important to determine the efficiency of the purified *Himar1* transposase to mediate transposition in an in vitro assay and to determine that no host-specific factors are needed for the transposase function. To analyze the activity of the purified *Himar1* transposase (*see* **Subheading 3.1.3.2.**), we used the pBluescript SK (+) vector as host DNA in the transposition reaction. The insertion of minitransposon A into the ampicillin-resistant vector yields ampicillin/kanamycin-resistant plasmid molecules that can be efficiently recovered by transformation of *E. coli* cells with the transposition reaction mixture and selection of the transformants on media containing both antibiotics. For textual convenience, here we describe the analysis of the transposase activity, but the construction of the minitransposon A necessary for this analysis is described on the next subsection. The analysis of the transposase activity is performed as follows (*see* **Fig. 3**):

1. Purify the 2.2-kb *Apa*I–*Eco*RI DNA fragment containing the minitransposon A.
2. In a microfuge tube, mix 15 ng of minitransposon A donor DNA, 20 ng of pBluescript SK (+) target DNA, 2 µg of yeast tRNA, and 25 n*M* (approx 20 ng) of purified *Himar1* transposase in a final vol of 20 µL of T buffer (*see* **Note 3**). Run a control reaction by omitting the *Himar1* protein. Incubate for 1 h at 37°C and stop the reaction by the addition of 0.4 µL of 0.5 *M* EDTA.
3. Perform a phenol/chloroform extraction and recover the DNA by standard ethanol precipitation.
4. Dissolve the pellet in 10 µL of sterile water and use 1 µL to transform *E. coli* DH10B electrocompetent cells by standard methods *(10)*.
5. Plate two different dilutions (10^{-3} and 10^{-5}) from the transformation mixture onto LB–ampicillin plates and the rest of the transformation mixture onto LB plates containing ampicillin and kanamycin.
6. Count the number of colonies growing on each type of media. Transposase activity is measured by the transposition frequency, which we express as the ratio of doubly ampicillin kanamycin-resistant colonies to ampicillin-resistant colonies.

A

B

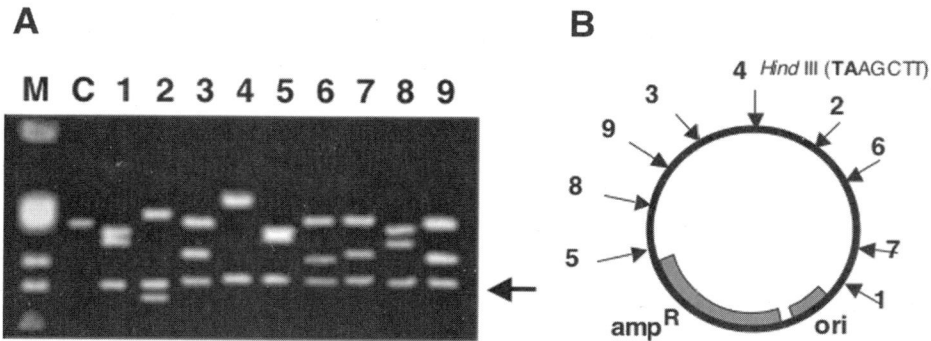

Fig. 3. The *Himar1* in vitro transposition reaction occurs randomly and efficiently into a plasmid target. (**A**) Restriction analysis of in vitro transposition products. DNA was isolated from doubly ampicillin kanamycin-resistant colonies generated by in vitro transposition assays using purified *Himar1* transposase and minitransposon A and pBluescript SK (+) as donor and target DNA, respectively. The DNA was digested with *Hind*III and fragments were separated by agarose gel electrophoresis and stained with ethidium bromide. Lane C: target plasmid pBluescript SK (+); lanes 1–9: transposition products isolated from nine independent ampicillin kanamycin-resistant colonies. A 921-bp fragment is diagnostic for insertion of the 2.3-kb minitransposon A sequence into the pBluescript SK (+) target plasmid. Two additional bands totaling 4.3 kb account for the digestion of the vector sequences at the *Hind*III site from the polylinker. (**B**) Localization of the insertion sites of minitransposon A in the transposition products analyzed. The exceptional restriction pattern shown by transposition product 4 can be explained in terms of the destruction of the *Hind*III site from the polylinker as a result of the integration of the minitransposon A into the TA dinucleotide present in (the *Hind*III) its recognition sequence. With the exception of the essential ampicillin resistance (amp^R) and replication origin (ori) regions, the locations of minitrasnposon insertion are evenly distributed across the target plasmid.

3.2. Construction of Minitransposons Specific for A. gossypii

Two specific minitransposons, A and R, have been constructed for use in *A. gossypii* (*see* **Fig. 4**). The first contains the inverted terminal repeats (5′ ITR and 3′ ITR) from the transposon *Himar1* flanking the dominant G418-resistance marker (*8*). In addition to the *G418* marker, the minitransposon R also contains the bacterial replication origin *ColE1*. To construct the minitransposon A:

1. Amplify the inverted terminal repeats by PCR using as template DNA of the plasmid pmMAR1 (*7*) and the oligonucleotides universal and ITR-L to amplify the 199-bp 5′ ITR fragment, and the reverse and ITR-R to amplify the 235-bp 3′ ITR fragment.

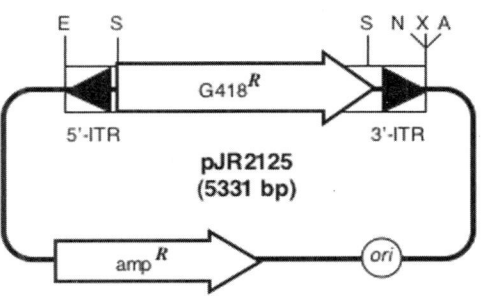
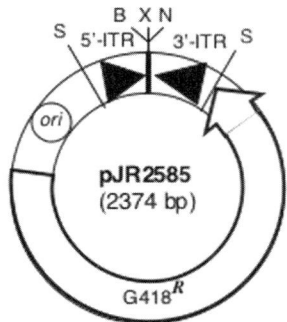

Fig. 4. Structure of the plasmids pJR2125 and pJR2585, which are donors of the minitransposons A and R, respectively. The most representative restriction sites (A, *Apa*I; B, *Bam*HI; E, *Eco*RI; N, *Not*I; S, *Sal*I; X, *Xba*I) are shown.

2. Digest the 5′ ITR fragment with *Apa*I and *Sal*I, and the 3′ITR fragment with *Eco*RI and *Sal*I. Clone both fragments into the pBluescript SK (+) vector previously digested with *Eco*RI and *Apa*I restriction enzymes.
3. Clone the G418-resistance marker, isolated as a 2.1-kb *Sal*I–*Sal*I fragment (*see* **Note 4**), into the unique *Sal*I site located between the 5′ ITR and the 3′ ITR sequences of the recombinant plasmid, pJR2125, obtained in **step 2.**

To provide replicative capacity to the minitransposon A, we constructed the minitransposon R as follows:

1. Amplify by standard PCR methods the *ColE1* origin of replication using as the template the plasmid pBluescript SK (+) and the oligonucleotides ColA and ColB. The oligonucleotides ColA and ColB are designed to add *Nco*I sites to both ends of the PCR product.
2. Release the 2.2-kb DNA fragment that encompasses minitransposon A from the plasmid pJR2125 by digestion with *Xba*I and circularize the purified DNA fragment by joining of its ends with T4 DNA ligase.
3. Heat-inactivate the T4 DNA ligase at 65°C for 5 min and dilute the sample twofold. In the same microfuge tube, linearize the circular DNA molecules by digestion with *Bsp*HI.
4. Clone the *ColE1* fragment previously digested with the *Nco*I restriction enzyme into the compatible *Bsp*HI site of minitransposon A and isolate the recombinant plasmid, minitransposon R, by bacterial transformation and selection on LB plates containing kanamycin and ampicillin.

3.3. Construction of a Genomic Library of Insertion Alleles by In Vitro Transposition

The process involves (1) the isolation of high-molecular-weight DNA from *A. gossypii,* (2) the insertional mutagenesis of the genomic DNA by an in vitro

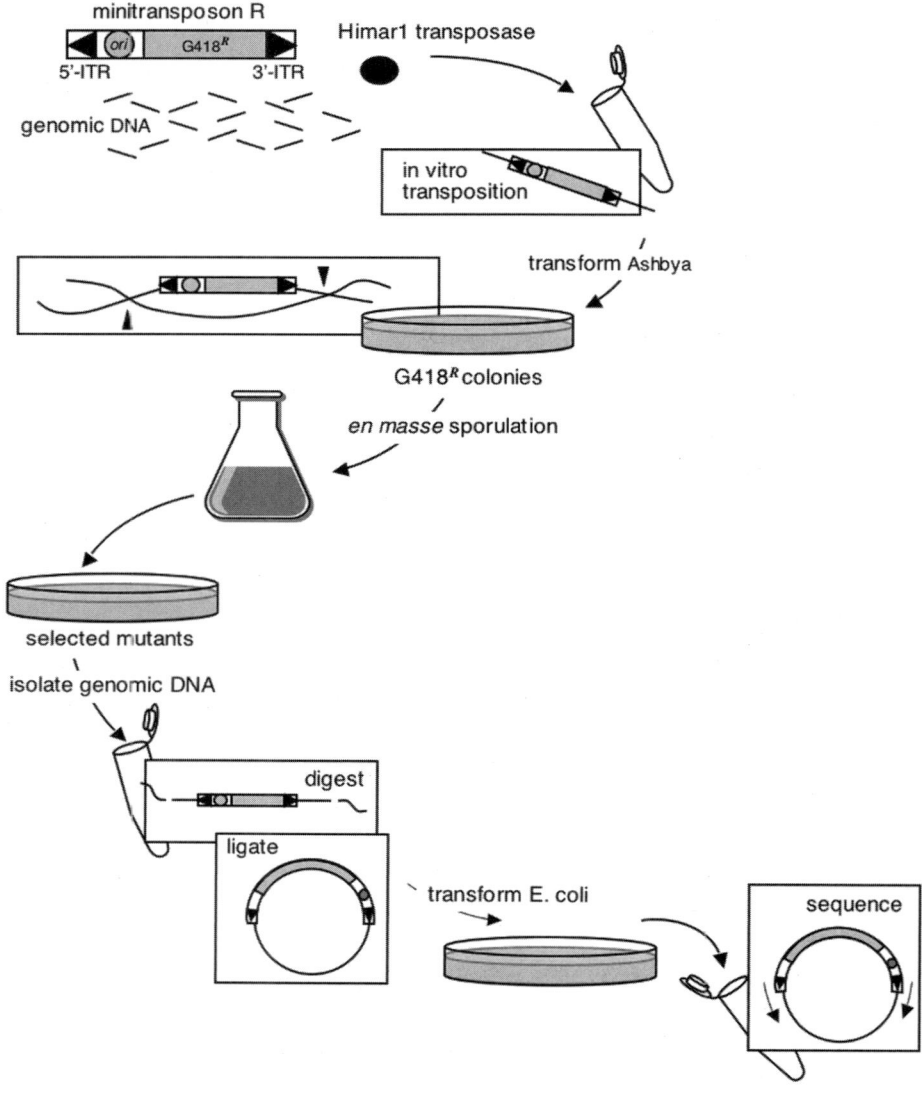

Fig. 5.

transposition reaction, and (3) the amplification of the mutagenized library; these procedures are described in **Subheadings 3.3.1.–3.3.3.** (*see* **Fig. 5**).

3.3.1. Isolation of High-Molecular-Weight DNA From A. gossypii

Because the ascomycete fungus *A. gossypii* shows mycelial growth with long branched hyphae, harvesting of the mycelium from liquid cultures is conveniently achieved by vacuum filtration through sterile Whatman 3MM cellulose filters. To obtain clean, high-molecular-weight genomic DNA, we use the following method:

1. Inoculate 200 mL of MA2 medium with a small piece of mycelium picked from a colony using sterile fine-tipped forceps. Incubate with vigorous shaking (200 rpm) at 28°C for 24 h.
2. Harvest the mycelia by vacuum filtration through sterile Whatman 3MM cellulose filters. Rinse the pad of mycelia twice with 100 mL of distilled, sterile water.
3. Transfer the mycelium to a 250-mL Erlenmeyer flask and resuspend in 20 mL of spheroplast buffer. Add 20 mg of Zymolyase 20T and incubate at 37°C with gentle shaking (50–60 rpm) until >95% are spheroplasted (30 min to 1 h).
4. Bring up the solution to 50 mM Tris-HCl, pH 7.5, 150 mM NaCl, 100 mM EDTA, and 0.5% SDS and lyse the spheroplasted cells by incubating at 65°C for 20 min.
5. Extract the spheroplast lysate with phenol/chloroform (1 : 1) until a clear interphase is observed. Transfer the aqueous phase to a clean, sterile, 200-mL Erlenmeyer flask and precipitate the nucleic acids by adding 2–3 vol 100% ethanol.
6. If the DNA is sufficiently concentrated, spool it onto a glass rod and place in a microfuge tube. If it cannot be spooled, decant the supernatant carefully and wash the precipitate twice with 20 mL of 80% ethanol.
7. Dissolve the precipitate nucleic acids in 1 mL of sterile water and degrade the RNA by adding 20 µl of a 10-mg/mL RNase A solution.
8. Incubate at 37°C for 30 min and precipitate the RNA-free DNA with 0.6 vol 2-propanol and 1/10 vol 3 M sodium acetate, pH 7.0.
9. Wash the precipitate twice with 1 mL 80% ethanol and resuspend the genomic DNA in 250 µL of water.

Fig. 5. Schematic representation of the in vitro mutagenesis technique and gene identification strategy. The minitransposon R is incubated with target DNA and *Himar1* transposase to perform an in vitro transposition reaction. The resulting mutagenized DNA is electrotransformed into *Ashbya*. The transformants are plated onto G418-selective medium. After growth, the transformant colonies are scraped from the plates and allowed to sporulate *en masse* in liquid G418-selective, sporulation medium. The spores are plated onto G418-selective medium and assayed for the phenotype of choice. Genomic DNA is isolated from the mutants of interest and digested with a restriction enzyme that does not cut within the minitransposon, thus leaving the *G418*-resistance marker and the *E. coli* origin of replication intact. These fragments are ligated intramolecularly to form a circular plasmid, which can be transformed into *E. coli* plated on LB–kanamycin. The plasmids can be prepared and the sequences flanking the minitransposon R insertion can be determined by sequencing from oligonucleotide primers MR1 and MR2, which are complementary to the right and left ends of the minitransposon, respectively. The identity of the gene can be determined by BLASTX or BLASTN searches of DNA databases.

3.3.2. Insertional Mutagenesis of Genomic DNA by In Vitro Transposition

Insertional mutants are generated in *A. gossypii* by means of shuttle mutagenesis. Transposon insertions are generated in restriction fragments of *Ashbya* genomic DNA; insertion alleles are subsequently introduced ("shuttled") into *Ashbya* by DNA-mediated transformation. Each fragment of genomic DNA, carrying a single minitransposon R insertion, will integrate at its corresponding genomic locus by homologous recombination (*see* **Fig. 5**). To generate target DNA fragments of discrete size for in vitro transposition, the genomic DNA is digested with a frequent-cutter restriction enzyme, such as *Pst*I. After the in vitro transposition reaction, the DNA molecules carrying an integrated minitransposon R are self-ligated to generate circular plasmids, which are able to replicate and amplify in *E. coli* (*see* **Fig. 5**).

3.3.2.1. IN VITRO TRANSPOSITION REACTION

1. Digest 100 µg of genomic DNA with *Pst*I. Linearize 10 µg of pJR2585 plasmid DNA by digestion with *Bam*HI and *Not*I. Check both digestions by electrophoresis on a 0.8% agarose gel. Extract the restricted DNAs with phenol/chloroform (1 : 1) and ethanol precipitate, and resuspend in distilled water using standard methods *(10)*.
2. Set up a transposition reaction by mixing 1 µg of the *Bam*HI–*Not*I DNA fragment containing the minitransposon R, 65 µg *Pst*I-cut *Ashbya* genomic DNA, 100 µg of yeast tRNA, and 12 n*M* (approx 50 µg) of *Himar1* transposase in a final vol of 1 mL of T buffer.
3. Incubate at 37°C for 1 h and stop the reaction by adding 20 µL of 0.5 *M* EDTA.
4. Extract the reaction once with phenol/chloroform (1 : 1) and ethanol precipitate the nucleic acids. Dissolve the DNA pellet in 250 µL of distilled water.

3.3.2.2. AMPLIFICATION OF THE MUTAGENIZED LIBRARY

The products of the transposition reaction consist of random, linear *Pst*I *A. gossypii* DNA fragments carrying an integrated minitransposon R that contains a kanamycin-selectable marker *(G418)* and a bacterial origin of replication *(ColE1)*. These linear molecules, which represent the entire *A. gossypii* genome, can be converted into a library of fully functional circular bacterial plasmids by self-ligation of their compatible *Pst*I ends. This allows the amplification of the minitransposon R-mutagenized library for its subsequent use in *A. gossypii* transformation.

1. Resuspend the DNA sample obtained in **Subheading 3.3.2.1., step 4** in 1.25 mL final vol of standard DNA ligase buffer and add 20 units of T4 DNA ligase. Incubate the self-ligation reaction at room temperature for 3 h.
2. Distribute the ligation reaction in two microfuge tubes and extract once with phenol/chloroform (1 : 1).

3. Recover the DNA by standard ethanol precipitation and dissolve the pellets in 100 μL of distilled water. Pool both samples and store at 4°C.

4. Thaw about 1100 μL of frozen *E. coli* DH10B electrocompetent cells and aliquot 50 μL into ice-cold sterile microfuge tubes. Pipet 5 μL of the mutagenized DNA library (**step 3**) into each microfuge tube, mix gently with the pipet tip, and incubate on ice for 1 min. Set up a negative transformation control without DNA.

5. Transfer the transformation mixtures to ice-cooled 0.2-cm electrode gap electroporation cuvets. Place the cuvets in the slot of the electroporation apparatus and deliver an electrical pulse of 25 μF, 2.5 kV, and 200 Ω. After the pulse, rapidly remove the electroporation cuvet and add 1 mL of LB medium at room temperature.

6. Transfer the electrotransformed cell mixtures to sterile microfuge tubes and incubate at 37°C with gentle shaking for 45 min. Centrifuge the tubes at 8000*g* for 30 s, discard the supernatant, and resuspend the cells in 100 μL LB medium.

7. Plate each sample onto LB–kanamycin plates and incubate overnight at 37°C. We typically obtain around 2000–5000 transformants colonies per plate.

8. Pool all the transformants in a single sample by scraping the plates with 5 mL of LB medium. Inoculate 500 mL of LB–kanamycin medium with the pool of transformants and incubate at 37°C for 6 h with vigorous shaking (200 rpm).

9. Harvest the cells by centrifugation and prepare plasmid DNA by standard alkaline lysis methods *(10)*. Resuspend the plasmid DNA precipitate in distilled water to obtain a concentration around 2–4 μg/μL.

3.4. Insertional Mutagenesis of A. gossypii and Isolation of Mutants

The transformation process in *A. gossypii* occurs mainly by homologous recombination. One-step gene disruption or gene replacement is efficiently achieved in this fungus by using a selectable marker that is inserted into a linear DNA fragment carrying the gene to be disrupted/replaced. The DNA flanking each side of the selectable marker recombines with the genome, the selectable marker being inserted into the target gene, thereby disrupting/replacing it. In order to convert the library of plasmid-borne fragments of *Ashbya* genomic DNA, carrying a single minitransposon R insertion, into a library of linear DNA molecules suitable for insertion-allele replacements, the minitransposon R-mutagenized plasmid library is relinearized by digestion with the same enzyme (*Pst*I) previously used for fragmentation of the genomic DNA (*see* **Subheading 3.3.2.1. step 1**).

The insertional mutagenesis process and the isolation of mutants include the transformation of *A. gossypii* and selection of the heterokaryotic, primary mutants (*see* **Subheading 3.4.1.**), selection of homokaryotic mutants (*see* **Subheading 3.4.2.**), and plasmid rescue and target gene identification (*see* **Subheading 3.4.3.**).

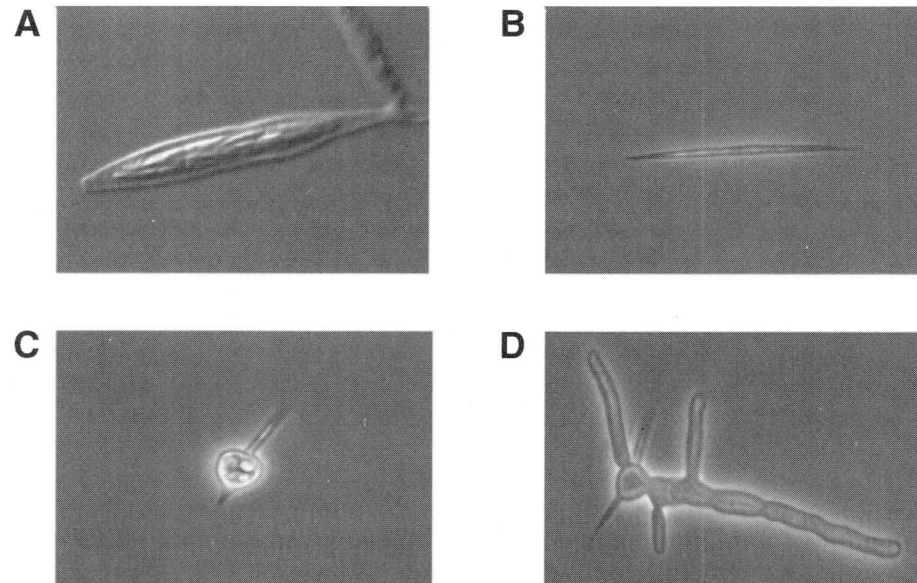

Fig. 6. Photomicrographs showing different stages of the germination of *A. gossypii* spores: **(A)** sporiferous sac containing eight asexual uninucleate spores; **(B)** nongerminated, needle-shaped spore; **(C)** swollen, germinative spore after 3–4 h growth; **(D)** germ cell showing short germinative hyphae after 12–14 h growth. This is the optimal stage for the electrotransformation of *Ashbya* germ cells.

3.4.1. Transformation of A. gossypii *and Selection of Heterokaryotic Mutants*

The fungus *A. gossypii* propagates by asexual, uninucleate spores, which form endogenously inside the hyphae and can be released either by autolysis or by enzymatic digestion of the cell wall with Zymolyase 20T. The uninucleate spores germinate to form a coenocytic multinucleate mycelium, thus repeating the life cycle asexually. Optimal transformation frequencies are obtained with recently germinated spores (i.e., when the germ cells are just beginning to emit germinative hyphae and contain around 20–30 nuclei) (*see* **Fig. 6**). The following method is designed to obtain a large number (appox 10,000) transformants:

1. Digest with *Pst*I to completion 200 µg of a 1-µg/µL DNA solution of the minitransposon R-mutagenized plasmid library (of **Subheading 3.3.2.2. step 9**).
2. Inoculate 250 mL of MA2 medium with 2.5 mL of a solution of fresh *A. gossypii* spores (approx 10^8 spores/mL) that has been stored at 4°C for less than 1 wk.

3. Incubate with shaking at 28°C for 12–14 h or until approx 75–90% of the germ cells show small germinative tubes.
4. Collect the germ cells by vacuum filtration through filter paper. Wash twice with 200 mL of sterile, distilled water. Resuspend the cells in 20 mL of AT buffer and incubate with shaking at 28°C for 30 min.
5. Centrifuge for 10 min at 8000*g* at 4°C. Remove the supernatant by gentle aspiration using an automatic pipet and resuspend the pellet in 20 mL of prechilled STM buffer.
6. Wash the cells twice with 20 mL of prechilled STM buffer and than resuspend in 1.5 mL of the same buffer. Aliquot 120 µL of the cell suspension into prechilled 0.4-cm electrode gap electroporation cuvettes.
7. Add to each cuvet 5 µL of *Pst*I-digested DNA solution of the minitransposon R-mutagenized library **(step 1)**. Pulse at 1.5 kV/cm, 100Ω, and 25 µF.
8. Immediately after the pulse, remove the electroporation cuvet and add 1 mL of MA2 medium at room temperature. Transfer the electroporated cell samples to microfuge tubes and centrifuge at 8000*g* for 5 min.
9. Remove about 800 µL of the supernatant by gentle aspiration using a micropipet and plate the rest onto MA2–agar plates. Incubate at 28°C for 4–6 h to allow the expression of the *G418*-resistance marker.
10. Pour 5 mL of overlay agar (0.4% agarose w/v) containing 1.5 µg/µL of G418 and maintained at 42–50°C onto each of the transformation plates. Incubate at 28°C for 2–4 d to allow growth of the transformant, G418-resistant colonies.

3.4.2. Clonal Selection of Homokaryotic Mutants

In the transformation of *A. gossypii,* only one nucleus of the 20–30 nuclei present in the germ cell results transformed. As a consequence, the transformant colonies consist of a heterokaryotic mycelium that contains a mixture of transformed and nontransformed nuclei. Homokaryotic cells bearing a single type of transformed nucleus are achieved by a process of clonal selection that involves the formation of uninucleate spores by sporulation of the primary, heterokaryotic transformant colonies and subsequent selection of the transformant spores in G418-selective medium. The selection of homokaryotic transformants able to display recessive phenotypes is performed as follows:

1. Pool all the transformants (approx 10,000) in a single sample by scraping the mycelia of the transformation plates with a sterile spatula and resuspend in 500 mL of SPA liquid sporulation medium containing G418 (100 µg/mL).
2. Incubate at 28°C with shaking (200 rpm) for 4–5 d. The formation of spores begins after 1–2 d of incubation and is completed by d 4 or 5.
3. Add 30 mL of a filter-sterilized aqueous solution containing 300 mg of Zymolyase 20T, and incubate at 37°C for 1 h with shaking (200 rpm).
4. Bring the cell suspension up to 0.03% Triton X-100 using a sterile aqueous stock solution to lyse the mycelial cells (*see* **Note 5**).

5. Distribute the cell lysate in four sterile 200-mL centrifuge tubes and collect the spores by centrifugation at 1000g for 10 min at room temperature.
6. Discard the supernatant and wash the precipitate three times with 100 mL of 0.03% Triton X-100.
7. Resuspend the spore precipitates in 15 mL (the smallest volume possible) of 0.03% Triton X-100 and keep at 4°C for short-term storage. For long-term storage, bring the spore solution up to 25% glycerol and store at –80°C.

3.4.3. Plasmid Rescue and Target Gene Identification

After the isolation of homokaryotic spores, the mutagenized *Ashbya* colonies can be assayed for any phenotype of choice. To determine the identity of any mutagenized gene, genomic DNA from the selected mutant colony is isolated and cut with a restriction enzyme that releases the complete minitransposon R and some of the flanking *Ashbya* genomic DNA regions. This DNA is then circularized by ligation and transformed into *E. coli,* where it can replicate—by virtue of the *ColE1* segment—and be selected for by the presence of the G418 marker. The sequence of the *A. gossypii* DNA into which the minitransposon R insertion occurred is then determined merely by sequencing the resulting plasmids using primers derived from the right or the left ends of the minitransposon (*see* **Fig. 5**). The sequence can be compared to the DNA sequence databases and the identity of the mutagenized gene determined by homology. In addition, the rescued DNA segments flanking the minitransposon can be used as probes to isolate the entire target gene from a genomic *A. gossypii* library.

1. Inoculate a single mutant colony in 50 mL of selective (MA2–G418) medium. Incubate with vigorous shaking (200 rpm) at 28°C for 24 h. Isolate genomic DNA from the mutant by scaling up the process described in **Subheading 3.3.1.** to 50 mL of culture.
2. Digest 10 µg of genomic DNA with a restriction enzyme that does not cut within the minitransposon R element (e.g., *Xho*I, *Bam*HI, *Not*I, *Eco*RI, *Pst*I, or *Xba*I). The total reaction volume should be large: at least 10 times greater than the volume of the genomic DNA added. Digest for at least 5 h.
3. Heat-inactivate the restriction enzyme at 65°C for 20 min and ethanol-precipitate DNA. If using an enzyme that cannot be heat-inactivated, extract the sample with phenol/chloroform before precipitation.
4. Resuspend DNA in 200 µL of standard ligation mixture. Add 1 µL of T4 DNA ligase and incubate overnight at 16°C.
5. Transform 2 µL of the ligation reaction into 100 µL electrocompetent *E. coli* DH10α cells and select on LB–kanamycin plates.
6. Prepare plasmids from several colonies. A *Hin*dIII digest should give an internal minitransposon R diagnostic band of 921 bp.
7. Perform sequencing reactions using the primers MR1 and MR2 for sequencing flanking regions from both ends of the minitransposon R.

8. Perform BLASTN and BLASTX searches on DNA and protein databases to determine the identity of the gene that has been disrupted. The *A. gossypii* genome has recently been sequenced and the sequence will become accessible to the public *(13)*.

4. Notes

1. The sequences underlined indicate restriction sites as follows: *Nde*I in the oligonucleotide TpaseA; *Bam*HI in TpaseB; *Sal*I in ITR-L and ITR-R, and *Nco*I in ColA and ColB.
2. The PCR products obtained by amplification with proofreading thermostable polymerases are occasionally difficult to clone directly into the blunt-ended pET-12a vector. These PCR fragments are efficiently cloned into pBluescript SK (+) digested with *Eco*RV and the recombinant plasmids readily selected by blue/white colony screening. The cloned PCR fragments are subsequently transferred to the expression vector pET-12a by standard cloning as restriction fragments obtained by digestion of the recombinant pBluescript plasmid with the appropriate restriction enzymes.
3. We calculated the protein concentrations based on the known extinction coefficient ($E = 53,510$) of the *Himar1* transposase *(14,15)*.
4. The *G418* marker consists of the coding sequence of the kanamycin-resistance gene, *kan*[r] *(16,17)*, under the transcriptional control of the strong promoter of the *TEF1* gene from *A. gossypii* *(8)*. The original *G418* marker present in the plasmid *pAG-100* contains single *Eco*RI and *Pst*I restriction sites that we destroyed by filling in of the restricted ends with the Klenow fragment of *E. coli* DNA polymerase I, following standard procedures *(10)*.
5. The *A. gossypii* spores are hydrophobic and form low-density aggregates in aqueous solutions that cannot be isolated by centrifugation. Addition of the detergent Triton X-100 (0.03%, v/v) to the spore solution prevents the formation of spore aggregates and allows their recovery by low-speed centrifugation.

Acknowledgments

The authors thank Dr. D. J. Lampe for providing plasmids pBADMAR1 and pmMAR1 and Dr. P. Phillipsen for plasmid pAG-100. Financial support by the Comisión Interministerial de Ciencia y Tecnología (CICYT, AGL2000-1537-C02-02) and BASF Aktiengesellschaft is gratefully acknowledged.

References

1. Nielsen, J. (1998) The role of metabolic engineering in the production of secondary metabolites. *Curr. Opin. Microbiol.* **1,** 330–336.
2. Stafford, D. E. and Stephanopoulos, G. (2001) Metabolic engineering as an integrating platform for strain development. *Curr. Opin. Microbiol.* **4,** 336–340.
3. Bailey, J. E. (1991) Toward a science of metabolic engineering. *Science* **252,** 1668–1681.

4. Bailey, J. E. (1999) Lessons from metabolic engineering for functional genomics and drug discovery. *Nat. Biotechnol.* **17,** 616–618.
5. Stahmann, K. P., Revuelta, J. L., and Seulberger, H. (2000) Three biotechnical processes using *Ashbya gossypii, Candida famata,* or *Bacillus subtilis* compete with chemical riboflavin production. *Appl. Microbiol. Biotechnol.* **53,** 509–516.
6. Craig, N. L. (1997) Target site selection in transposition. *Annu. Rev. Biochem.* **66,** 437–474.
7. Lampe, D. J., Churchill, M. E., and Robertson, H. M. (1996) A purified mariner transposase is sufficient to mediate transposition *in vitro. EMBO J.* **15,** 5470–5479. Erratum: *EMBO J.* 1997, **16,** 4153.
8. Steiner, S. and Philippsen, P. (1994) Sequence and promoter analysis of the highly expressed *TEF* gene of the filamentous fungus *Ashbya gossypii. Mol. Gen. Genet.* **242,** 263–271.
9. Studier, F. W. and Moffatt, B. A. (1896) Use of bacteriophage T7 RNA polymerase to direct selective high-level expression of cloned genes. *J. Mol. Biol.* **189,** 1113–1130.
10. Sambrook, J. and Russell, D. (2001) *Molecular Cloning: A Laboratory Manual,* 3rd ed. Cold Spring Harbor Laboratory Press, Cold Spring Harbor, NY.
11. van Luenen, H. G., Colloms S. D., and Plasterk, R. H. (1994) The mechanism of transposition of Tc3 in *C. elegans. Cell* **79,** 293–301.
12. Mizuuchi, K. (1992) Polynucleotidyl transfer reactions in transpositional DNA recombination. *J. Biol. Chem.* **267,** 21,273–21,276.
13. Brachat, S., Dietrich F. S., Voegeli, S., et al. (2003) Reinvestigation of the *Saccharomyces cerevisiae* genome annotation by comparison to the genome of a related fungus: *Ashbya gossypii. Genome Biol.* **4(R45),** 1–13.
14. Lampe, D. J., Grant, T. E., and Robertson, H. M. (1998) Factors affecting transposition of the Himar1 mariner transposon *in vitro. Genetics* **149,** 179–187.
15. Gill, S. G. and von Hippel, P. H. (1989) Calculation of protein extinction coefficients from amino acid sequence data. *Anal. Biochem.* **182,** 319–326.
16. Jiménez, A. and Davies, J. (1980) Expression of a transposable antibiotic resistance element in *Saccharomyces. Nature* **287,** 869–871.
17. Webster, T. D. and Dickson, R. C. (1983) Direct selection of *Saccharomyces cerevisiae* resistant to the antibiotic G418 following transformation with a DNA vector carrying the kanamycin-resistance gene of Tn903. *Gene* **26,** 243–252.

17

Improved Polysaccharide Production Using Strain Improvement

Thomas P. West

Summary

A strain improvement procedure is outlined for the isolation of mutant strains from the bacterium *Pseudomonas* sp. ATCC 31461 that are capable of enhanced production of the polysaccharide gellan compared to its parent strain. With gellan having a number of industrial applications, the isolation of such mutant strains could prove valuable to its large-scale production. The mutant strains are isolated using a procedure that involves chemical mutagenesis and screening for resistance to the antibiotic ampicillin, with possible mutants being selected based on their mucoid appearance on the antibiotic-containing solid medium. The mutant strains were identified by following protocols using gravimetric determinations to measure the concentrations of gellan and cellular biomass. The viscosimetric properties of the mutant strain were compared to its parent strain by measuring the viscosities of their whole-culture medium and crude polysaccharide using procedures outlined.

Key Words: Gellan; polysaccharide; mutant; isolation; ampicillin; biomass; viscosity.

1. Introduction

The polysaccharide gellan is synthesized extracellularly by the bacterium *Pseudomonas* sp. ATCC 31461 *(1–3)*. The applications of this bacterial polysaccharide include its use as a thickening agent or as an agar substitute *(4,5)*. Gellan is an anionic heteropolysaccharide with a tetrasaccharide structure that has been shown to contain 20% glucuronic acid, 60% glucose, and 20% rhamnose *(6,7)*. The rate of gelation of this polysaccharide is increased when monovalent and divalent cations are added *(8–11)*. Although glucose was initially utilized as the carbon source for gellan production by *Pseudomonas* sp. ATCC 31461 *(1,2)*, alternative carbon sources have been observed to support polysaccharide production by ATCC 31461 *(12–15)*.

In addition to the structure of gellan, prior studies have examined the physiological conditions that promote gellan production *(16–19)*. It has been shown

From: *Methods in Biotechnology, Vol. 18: Microbial Processes and Products*
Edited by: J. L. Barredo © Humana Press Inc., Totowa, NJ

that hydrolyzed soybean protein is an excellent nitrogen source for gellan synthesis *(16)*. The optimal pH of gellan production by ATCC 31461 is at neutrality *(19)*. Although mutagenesis of cells from the closely related species *Xanthomonas campestris* has been found to be effective in isolating mutants exhibiting increased production of the polysaccharide xanthan *(20)*, little research has focused on the isolation of mutants from *Pseudomonas* sp. ATCC 31461 that exhibit elevated polysaccharide synthesis. It is known that a spontaneous variant strain isolated from a rifampicin-resistant strain was able to synthesize slightly higher gellan levels than its parent strain *(21,22)*. Only by understanding the biosynthetic pathway by which gellan is produced will it be possible to construct mutants having an enhanced ability to elaborate the polysaccharide *(3)*. Recently, a procedure was devised to screen ampicillin-resistant mutant strains for possible enhanced polysaccharide production *(23)*. This procedure was based on the finding that gellan formation might share common nucleotide sugar pathway steps with peptidoglycan synthesis in this pseudomonad *(22)*. The procedure to isolate mutant strains of *Pseudomonas* sp. ATCC 31461 that exhibited enhanced polysaccharide production and to partially characterize the properties of the isolated mutant strains are outlined relative to the growth conditions utilized, the cell mutagenesis procedure, the protocol for the isolation of mutants exhibiting elevated gellan production, the gravimetric methods for gellan and biomass determinations, and the determination of whole-culture medium viscosity and crude polysaccharide viscosities. By analyzing the gellan and biomass determinations as well as the viscosity measurements of whole-culture medium viscosity and crude polysaccharide for each possible mutant isolated, it is possible to learn how similar polysaccharide production by the potential mutant is compared to its parent strain producing authentic gellan.

2. Materials

2.1. Growth Conditions

1. *Pseudomonas* sp. ATCC 31461 (ATCC, Manassas, VA).
2. Hydrolyzed soybean meal Soytone (Difco Laboratories, Detroit, MI).
3. Nutrient broth (Difco).
4. Constant-temperature incubator.
5. Autoclave.
6. Glucose solution: sterile 50% (w/v).
7. Minimal medium: 0.05% K_2HPO_4, 0.01% $MgSO_4 \cdot 7H_2O$, 0.09% hydrolyzed soybean meal, 1 mL/L salt solution, and 3% glucose.
8. Salt solution: 0.18% $MnCl_2 \cdot 4H_2O$, 0.248% $FeSO_4 \cdot 7H_2O$, 0.028% H_3BO_3, 0.003% $CuCl_2 \cdot 2H_2O$, 0.002% $ZnCl_2$, 0.007% $CoCl_2 \cdot 2H_2O$, 0.002% $NaMoO_4 \cdot 2H_2O$, and 0.21% sodium citrate $\cdot 2H_2O$.
9. Rotary shaker.

2.2. Mutagenesis

1. Ethylmethane sulfonate (Sigma, St. Louis, MO).
2. Spectrophotometer.
3. Test tubes (16 × 125 mm).
4. NaCl solution: sterile 0.85% (w/v).
5. Sodium thiosulfate solution: sterile 6% (w/v).

2.3. Isolation of Mutants Exhibiting Elevated Gellan Production

1. 0.2-μm Disposable syringe filters.
2. Ampicillin solution: sterile 10 mg/mL (w/v).
3. Sterile flat toothpicks.
4. Sterile 24% (w/v) sucrose solution.
5. Sterile 2-mL serum bottles.
6. Sterile 13-mm slotted stoppers.

2.4. Gellan Determination

1. 0.45-μm HVLP filters: diameter of 25 mm (Millipore, Bedford, MA).
2. Water bath.
3. Marbles.
4. 1 N KOH.
5. Trichloroacetic acid solution: 50% (w/v); store at 4°C.
6. Filtration unit: 25-mm fritted glass filter holder; stopper; 250-mL vacuum filtering flask; 15-mL glass funnel; clamp.
7. Oven.

2.5. Biomass Determination

1. 0.45-μm HVLP filters: diameter of 47 mm (Millipore).
2. Filtration unit: 47-mm fritted glass filter holder; stopper; 1-L vacuum filtering flask; 250-mL glass funnel; clamp.

2.6. Determination of Whole-Culture Medium and Crude Polysaccharide Viscosity

1. Rotary viscometer Cannon Model LV2000 with small-sample adaptor (Cannon, State College, PA).
2. Silicone viscosity standard for rotary viscometers, Cannon RT1000 (Cannon).

3. Methods

The methods described indicate (1) the growth conditions utilized, (2) the cell mutagenesis procedure, (3) the protocol for the isolation of mutants exhibiting elevated gellan production, (4) the gravimetric method for gellan determination, (5) the gravimetric method for biomass measurement, and (6) the determination of whole-culture medium and crude polysaccharide viscosities.

3.1. Growth Conditions

3.1.1. Strain and Medium

1. Use *Pseudomonas* sp. ATCC 31461 as the parent strain in the procedure because it has been shown to synthesize authentic gellan *(1)*.
2. Streak this pseudomonad strain for isolation on nutrient agar plates and grow for 72 h at 30°C. Store the plates at 4°C until use (*see* **Note 1**).
3. Grow the parent and mutant strains in minimal medium *(16)* containing a modified salt solution *(9)*.
4. Add the carbon source glucose (3%) separately to the medium after steam sterilization using a sterile 50% (w/v) stock glucose solution.

3.1.2. Culture Conditions

1. Inoculate 20 mL of nutrient broth in a sterile 125-mL Erlenmeyer flask from a nutrient agar plate of strain ATCC 31461 using an inoculating loop and shake the culture in a rotary shaker at 250 rpm for 48 h at 30°C.
2. Inoculate batch minimal medium cultures (25–50 mL) in 250-mL Erlenmeyer flasks with 10^6 cells/mL of the nutrient broth culture.
3. Shake the batch cultures at 250 rpm in a rotary shaker at 30°C over a period of 72 h.

3.2. Mutagenesis

1. Inoculate a nutrient broth culture (20 mL) of strain ATCC 31461 in a sterile 125-mL Erlenmeyer flask and shake culture at 250 rpm at 30°C.
2. Remove samples of the medium to measure the growth of each culture. Monitor the growth of each culture by following its optical density (OD_{600}) with a spectrophotometer and use the culture for mutagenesis once an OD_{600} of 0.30 is observed.
3. Remove 4.95 mL from the culture into a sterile, capped 16-mm × 125-mm test tube. Remove 0.05 mL of culture and place in a sterile, capped 16-mm × 125-mm test tube containing 4.95 mL of sterile 0.85% NaCl for measuring cell survival. Dilute this sample to 10^{-3}, 10^{-4}, 10^{-5}, and 10^{-6} in sterile 0.85% NaCl. Spread 0.1 mL of the 10^{-3}, 10^{-4}, 10^{-5}, and 10^{-6} untreated bacterial cells onto nutrient agar plates. Incubate plates for 72 h at 30°C (*see* **Note 2**).
4. Add 0.05 mL of ethylmethane sulfonate to the tube containing 4.95 mL culture and vigorously mix the contents (*see* **Note 2**).
5. Place the tube in a 30°C incubator without shaking for 60–120 min *(24)*.
6. Remove 0.6 mL from the tube after 60–120 min and add to a sterile 125-mL Erlenmeyer flask containing 17.4 mL of nutrient broth. Also, remove 0.05 mL from the tube and place in a sterile, capped 16-mm × 125-mm test tube containing 4.95 mL of sterile 0.85% NaCl for measuring cell survival. Dilute this sample to 10^{-3}, 10^{-4}, 10^{-5}, and 10^{-6} in sterile 0.85% NaCl. Spread 0.1 mL of the 10^{-3}, 10^{-4}, 10^{-5}, and 10^{-6} treated bacterial cells onto nutrient agar plates. Incubate plates for 72 h at 30°C (*see* **Note 2**).

7. Aerate the flask containing the mutagenized bacterial cells by shaking it at 250 rpm at 30°C for 48 h.

8. After 72 h, count the number of colonies on the nutrient agar plates that the dilutions of the untreated and treated cells were spread (*see* **Note 2**).

3.3. Isolation of Mutants Exhibiting Elevated Gellan Production

3.3.1. Isolation Protocol

1. Filter sterilize a 10-mg/mL solution of ampicillin using a 0.2-μm syringe filter.
2. Steam-sterilize nutrient agar and add the appropriate volume of sterile ampicillin to provide a final concentration of 100 mg/L in the medium. Pour the agar plates and allow them to dry overnight. Store the plates at 4°C in a refrigerator until use.
3. Spread approx 10^7 mutagenized cells onto nutrient agar plates containing 100 mg/L ampicillin. Incubate the plates for 72 h at 30°C.
4. Visually inspect the colonies on the plates for their degree of polysaccharide formation. Pick colonies that have a highly mucoid appearance onto nutrient agar plates containing ampicillin (100 mg/L). Streak plate for isolation and incubate plate for 72 h at 30°C.
5. Store plates with potential mutants in a refrigerator at 4°C until screened for polysaccharide production.

3.3.2. Screening of Possible Mutants

1. Inoculate ATCC 31461 and each potential mutant using an inoculating loop from their respective plates into 25 mL of nutrient broth in sterile 125-mL Erlenmeyer flasks. Shake the cultures at 250 rpm for 48 h at 30°C.
2. Inoculate 25 mL of minimal medium cultures containing 3% glucose in sterile 125-mL Erlenmeyer flasks with 10^6 cells of the 48-h nutrient broth cultures (approx 0.25 mL of the 48-h culture). Shake the cultures for 48 h at 30°C with aeration (250 rpm).
3. Determine the concentration of gellan after 48 h of growth as described in **Subheading 3.4.** for each culture.
4. Retain those potential mutants able to elaborate higher gellan concentrations than ATCC 31461 after 48 h of growth on nutrient agar plates containing ampicillin (100 mg/L) for further study.
5. Compare gellan and biomass production over a period of 72 h by the potential mutants and ATCC 31461 after completing the initial screening.
6. Inoculate 50 mL of minimal medium in sterile 250-mL Erlenmeyer flasks with nutrient broth cultures of the potential mutants and ATCC 31461. Grow three cultures of each potential mutant and ATCC 31461 for 72 h at 30°C with aeration (250 rpm).
7. Remove samples (3 mL) from each culture at 0, 24, 48, and 72 h of growth. Process the samples for gellan and biomass production as described in **Subheadings 3.4.** and **3.5.**
8. Identify a mutant strain, such as strain EGP-2, if the strain exhibits enhanced gellan production compared to ATCC 31461 (*see* **Table 1**) while its biomass production remains comparable to its parent strain after 72 h of growth (*see* **Table 2**).

Table 1
**Polysaccharide Production by Parental Strain *Pseudomonas* sp. ATCC 31461
and Mutant Strain EGP-2**

	Polysaccharide concentration produced by strain	
Fermentation time (h)	ATCC 31461	EGP-2
0	0.00 ± 0.00	0.00 ± 0.00
24	1.40 ± 0.12	1.49 ± 0.04
48	1.73 ± 0.07	2.36 ± 0.08
72	1.89 ± 0.33	3.33 ± 0.00

Notes: The strains were grown on glucose as a carbon source in medium over a period of 72 h during which polysaccharide levels were monitored. The polysaccharide levels are expressed as g/L, where each result represents the mean of three separate determinations ± SD.

Table 2
**Biomass of the Parent Strain *Pseudomonas* sp. ATCC 31461
and Mutant Strain EGP-2**

	Cell weight of each strain	
Fermentation time (h)	ATCC 31461	EGP-2
0	0.00 ± 0.00	0.00 ± 0.00
24	0.70 ± 0.07	0.56 ± 0.02
48	0.49 ± 0.05	0.85 ± 0.04
72	0.84 ± 0.24	0.91 ± 0.04

Notes: Cells weights were measured over a period of 72 h after being grown in a medium containing glucose as the sole carbon source. Cell weight levels are given as g/L, where each result indicates the mean of three independent trials ± SD.

9. For short-term storage, streak the identified mutant strains for isolation on nutrient agar plates containing ampicillin (100 mg/L) at 30°C for 72 h. Store the plates at 4°C in a refrigerator until use.
10. For storage longer than 2 wk, grow the identified mutant strains in nutrient broth cultures containing ampicillin (100 mg/L). Dilute the cultures in half with 24% sucrose and lyophilize in 2-mL serum bottles using slotted stoppers. Store the stoppered vials in a freezer at –20°C.

3.4. Gellan Determination

1. Remove 3-mL samples from each culture and add to 16 × 125-mm test tubes.
2. Cap tubes with marbles and incubate for 15 min at 100°C.

3. Add 0.04 mL of 1 *N* KOH to each tube to bring medium pH to 10.0 and mix vigorously (*see* **Note 3**).
4. Incubate tubes at 80°C for 10 min.
5. Add 0.0112 mL of 50% trichloroacetic acid directly into the medium in each tube, which adjusts its pH to 7.15. Incubate at 25°C for 10 min (*see* **Note 3**).
6. Centrifuge treated medium at 20,400*g* for 30 min at 25°C. Save the cell pellet for the biomass determination (*see* **Note 4**).
7. Collect the resultant supernatant. Add the supernatant (1.5 mL) to a 16 × 125-mm test tube and mix vigorously with 4.5 mL of ice-cold 95% ethanol.
8. Place the mixture in a freezer at –18°C for 20 min to allow the polysaccharide to precipitate.
9. Collect the precipitated polysaccharide by filtration on preweighed 0.45-μm filters (25 mm in diameter).
10. Dry the filters to constant weight at 80°C and reweigh to calculate the gellan concentration.

3.5. Biomass Determination

1. Wash the cell pellet obtained following the centrifugation of the medium with water (5 mL) and centrifuge at 20,400*g* for 30 min at 25°C.
2. Suspend the cells in water and collect by filtration on preweighed 0.45-μm filters (47 mm in diameter).
3. Dry the filters to constant weight at 80°C in an oven and reweigh to calculate the biomass concentration.

3.6. Determination of Whole-Culture Medium and Crude Polysaccharide Viscosity

3.6.1. Determination of Whole-Culture Medium Viscosity

1. Inoculate 50-mL cultures of glucose minimal medium in 250-mL Erlenmeyer flasks with 10^6 cells from 48-h nutrient broth cultures of each mutant strain and ATCC 31461. Aerate the cultures (done in triplicate) by shaking them at 250 rpm for 72 h at 30°C.
2. Remove 12 mL of culture medium at 0, 24, 48, and 72 h. Place the medium into the sample chamber of the small-sample adaptor of the viscometer (*see* **Note 5**).
3. Determine the viscosity of the culture medium at 23°C after 0 and 24 h of growth using spindle TL5 spindle at speed of 6 rpm (*see* **Note 5**).
4. Measure the viscosity of the culture medium at 23°C after 48 and 72 h of growth using spindle TL6 at speed of 6 rpm (*see* **Note 5**).
5. Compare the viscosity of the culture medium of each mutant strain, such as strain EGP-2, with that of ATCC 31461 over the period of 72 h (*see* **Table 3**).

3.6.2. Determination of Crude Polysaccharide Viscosity

1. Use the whole-culture medium (12 mL) removed from the cultures after 72 h. Place the medium into a 16 × 125-mm test tube and cap the tube.

Table 3
Culture Medium Viscosity

	Viscosity of culture broth	
Fermentation time (h)	ATCC 31461	EGP-2
0	2 ± 0	2 ± 0
24	532 ± 23	338 ± 23
48	2302 ± 177	3546 ± 123
72	2309 ± 19	3706 ± 70

Notes: The viscosity of the culture medium was determined for a period of 72 h in the glucose minimal medium cultures of the parent strain *Pseudomonas* sp. ATCC 31461 and mutant strain EGP-2. Viscosity is expressed as centipoises, where each result represents the mean of three independent trials ± SD.

Table 4
Viscosity of the Crude Polysaccharide Produced by the Strains

Strain	Viscosity	% Change
ATCC 31461	193 ± 27	100
EGP-2	185 ± 22	96

Notes: The viscosity of the crude polysaccharide produced by the parent strain *Pseudomonas* sp. ATCC 31461 was compared to the polysaccharide synthesized by mutant strain EGP-2 grown on glucose for 72 h. Each value is expressed as centipoises and has been normalized to 1 g/L polysaccharide. The values represent the mean of three separate trials ± SD.

2. Place the test tubes in a 100°C water bath for 15 min.
3. Add 1 *N* KOH (0.16 mL) to each tube to bring the pH of the medium to 10. Vortex each tube vigorously.
4. Incubate the tubes at 80°C for 10 min. Add 50% trichloroacetic acid (0.0448 mL) to the medium and vortex vigorously. Incubate tubes at 25°C for 10 min.
5. Pour contents of each tube into a centrifuge tube and spin contents at 20,400*g* for 30 min at 25°C. Save clarified supernatant and pour into the sample chamber of the small-sample adaptor (*see* **Note 5**).
6. Using the TL5 spindle at a speed of 1.5 rpm, determine the viscosity of the crude polysaccharide produced by the strains after 72 h (*see* **Note 5**).
7. Compare the viscosity of the crude polysaccharide produced by each mutant strain, such as strain EGP-2, after 72 h with the viscosity of the polysaccharide produced by ATCC 31461 after 72 h by dividing the polysaccharide viscosity by its gellan concentration (*see* **Table 4**). The values should be similar.

4. Notes

1. *Pseudomonas* sp. ATCC 31461 should be streaked for isolation biweekly on nutrient agar plates to ensure that the organism remains viable. It is important to steam sterilize cultures of *Pseudomonas* sp. ATCC 31461 and isolates from this strain at the end of experiments as a biosafety precaution.

2. By comparing the bacterial cell concentrations prior to and after mutagenesis, a survival rate can be calculated. A typical survival rate after treatment of the bacterial cells with ethylmethane sulfonate is 2.2%. Use ethylmethane sulfonate with caution because it is a mutagen. Wear disposable gloves when handling ethylmethane sulfonate and properly dispose of material in contact with it. Sodium thiosulfate (6%) can be used to inactivate ethylmethane sulfonate *(25)*.

3. During the alkali treatment of the culture medium samples and the subsequent neutralization of the alkali-treated samples with trichloroacetic acid, the medium should be thoroughly mixed because of the small volume of base or acid being added. Also, the neutralized medium should be centrifuged within a few hours if possible because prolonged exposure to the acid (such as overnight) appeared to greatly reduce the ability of ethanol to precipitate the polysaccharide.

4. After 72 h of growth, it might be necessary to heat the treated medium at 100°C in a water bath to ensure complete separation of the polysaccharide from the bacterial cells. It might also be necessary to use a spatula to help in separating the polysaccharide from the cell pellet after centrifugation.

5. The viscometer should have been calibrated at 23°C with a viscosity standard that is commercially available. In addition, it is important that the viscosity reading shown on the meter on the instrument has stabilized. It might be necessary to dilute the culture medium collected after 48 or 72 h from 1.2-fold to 2-fold with minimal medium to measure its viscosity using the spindles and speeds given.

Acknowledgments

This work is published as paper 3374, Journal Series, South Dakota Agricultural Experiment Station. This work was supported by funds from the South Dakota AES and from US Department of Agriculture Grant No. 99-35501-8419. The expert technical assistance of Beth Reed in helping to develop the protocols was greatly appreciated.

References

1. Kang, K. S., Veeder, G. T., Mirrasoul, P. J., Kaneko, T., and Cottrell, I. W. (1982) Agar-like polysaccharide produced by a *Pseudomonas* species: production and basic properties. *Appl. Environ. Microbiol.* **43,** 1086–1091.

2. Anson, A., Fisher, P. J., Kennedy, A. F. D., and Sutherland, I. W. (1987) A bacterium yielding a polysaccharide with unusual properties. *J. Appl. Bacteriol.* **62,** 147–150.

3. Sa-Correia, I., Fialho, A. M., Videira, P., Moreira, L. M., Marques, A. R., and Albano, H. (2002) Gellan gum biosynthesis in *Sphingomonas paucimobilis* ATCC

31461: Genes, enzymes and exopolysaccharide production engineering. *J. Ind. Microbiol. Biotechnol.* **29**, 170–176.

4. Lin, C. C. and Casida, L. E., Jr. (1984) GELRITE as a gelling agent in media for the growth of thermophilic microorganisms. *Appl. Environ. Microbiol.* **47**, 427–429.

5. Omoto. T., Uno, Y., and Asai, I. (1999) The latest technologies for the application of gellan gum. *Prog. Colloid Polym. Sci.* **114**, 123–126.

6. Jansson, P.-E., Lindberg, B., and Sandford, P. A. (1983) Structural studies of gellan gum, an extracellular polysaccharide elaborated by *Pseudomonas elodea*. *Carbohydr. Res.* **124**, 135–139.

7. O'Neill, M. A., Selvendran, R. R., and Morris, V. J. (1983) Structure of the acidic extracellular gelling polysaccharide produced by *Pseudomonas elodea*. *Carbohydr. Res.* **124**, 123–133.

8. Chandrasekaran, R., Radha, A., and Thailambal, V. G. (1992) Roles of potassium ions, acetyl and L-glyceryl groups in native gellan double helix: an X-ray study. *Carbohydr. Res.* **224**, 1–17.

9. Manna. B., Gambhir, A., and Ghosh, P. (1996) Production and rheological characteristics of the microbial gum gellan. *Lett. Appl. Microbiol.* **23**, 141–145.

10. Tang, J., Lelievre, J., Tung, M. A., and Zeng, Y. (1994) Polymer and ion concentration effects on gellan strength and strain. *J. Food Sci.* **59**, 216–220.

11. Tang, J., Tung, M. A., and Zeng, Y. (1995) Mechanical properties of gellan gels in relation to divalent cations. *J. Food Sci.* **60**, 748–752.

12. Pollock, T. J. (1993) Gellan-related polysaccharides and the genus *Sphingomonas*. *J. Gen. Microbiol.* **139**, 1939–1945.

13. Dlamini, A. M. and Peiris, P. S. (1997) Production of exopolysaccharide by *Pseudomonas* sp. ATCC 31461 *(Pseudomonas elodea)* using whey as fermentation substrate. *Appl. Microbiol. Biotechnol.* **47**, 52–57.

14. West, T. P. and Strohfus, B. (1998) Effect of carbon source on exopolysaccharide production by *Sphingomonas paucimobilis* ATCC 31461. *Microbiol. Res.* **153**, 327–329.

15. Fialho, A. M., Martins, L. O., Donval, M.-L., et al. (1999) Structures and properties of gellan polymers produced by *Sphingomonas paucimobilis* ATCC 31461 from lactose compared with those produced from glucose and from cheese whey. *Appl. Environ. Microbiol.* **65**, 2485–2491.

16. West, T. P. and Strohfus, B. (1998) Effect of complex nitrogen sources upon gellan production by *Sphingomonas paucimobilis* ATCC 31461. *Microbios* **94**, 145–152.

17. West, T. P. and Strohfus, B. (1999) Effect of yeast extract on gellan production by *Sphingomonas paucimobilis* ATCC 31461. *Microbios* **97**, 85–93.

18. West, T. P. and Fullenkamp, N. A. (2000) Ability of casamino acids to support gellan production by *Sphingomonas paucimobilis* ATCC 31461. *Microbios* **102**, 89–101.

19. West, T. P. and Fullenkamp, N. A. (2001) Effect of culture medium pH on bacterial gellan production. *Microbios* **105**, 133–140.

20. Rodríguez, H. and Aguilar, L. (1997) Detection of *Xanthomonas campestris* mutants with increased xanthan production. *J. Ind. Microbiol.* **18**, 232–234.

21. Fialho, A. M., Monteiro, G, A., and Sá-Correia, I. (1991) Conjugal transfer of recombinant plasmids into gellan gum-producing and non-producing variants of *Pseudomonas elodea* ATCC 31461. *Lett. Appl. Microbiol.* **12,** 85–87.

22. Martins, L. O., Fialho, A. M., Rodrigues, P. L., and Sá-Correia, I. (1996) Gellan gum production and activity of biosynthetic enzymes in *Sphingomonas paucimobilis* mucoid and non-mucoid variants. *Biotechnol. Appl. Biochem.* **24,** 47–54.

23. West, T. P. (2002) Isolation of a mutant strain of *Pseudomonas* sp. ATCC 31461 exhibiting elevated polysaccharide production. *J. Ind. Microbiol. Biotechnol.* **29,** 185–188.

24. Watson, J. M. and Holloway, B. W. (1976) Suppressor mutations in *Pseudomonas aeruginosa. J. Bacteriol.* **125,** 780–786.

25. Lindegren, G., Hwang, Y. L., Oshima, Y., and Lindegren, C. C. (1965) Genetical mutants induced by ethyl methanesulfonate in *Saccharomyces. Can. J. Genet. Cytol.* **7,** 491–499.

18

Use of the *Morganella morganii phoC* Gene as Reporter in Bacterial and Yeast Hosts

Stefania Cresti, Cesira L. Galeotti, Serena Schippa, Gian Maria Rossolini, and Maria C. Thaller

Summary

In this chapter are described the applications of the *Morganella morganii phoC* gene, encoding a molecular class A nonspecific phosphatase, as a reporter both in prokaryotic *(Escherichia coli)* and in eukaryotic *(Saccharomyces cerevisiae)* systems. The activity of PhoC can be detected by means of simple, sensitive, and relatively inexpensive tests in either qualitative (histochemical) or quantitative (liquid) assays. The methods are suitable to monitor gene expression and to determine transcriptional activity and the inducibility of a promoter or other regulatory elements. The intrinsic properties of phoC could be also exploited for studies on secretion, or in applications where the host cell should be recovered still viable after the reporter assay for further analysis. Indeed, in *S. cerevisiae,* the PhoC protein is secreted very efficiently into the culture medium and its activity can be easily assessed without interfering with the growth conditions required by the experimental approach of choice.

Key Words: *Saccharomyces cerevisiae; Escherichia coli;* yeast; bacteria; reporter gene; phosphatase; eukaryotic microorganism; prokaryote; phosphatase assay.

1. Introduction

Reporter gene systems have been developed and are widely used as a tool to indirectly measure the in vivo activity of transcriptional regulatory sequences. Reporter genes usually encode a product that is easy to be monitored *(1–7)*. The most popular reporter gene for bacteria is the *Escherichia coli lacZ* gene encoding β-galactosidase, which can be easily assayed using either soluble (e.g., *o*-nitrophenyl-galactopyranoside) *(8)* or insoluble (e.g., 5-bromo-4-chloro-3-indolyl-β-D-galactopyranoside) chromogenic substrates *(9)*. The same gene is a widely used reporter system also for the yeast *Saccharomyces cerevisiae,* together with the *cat* gene, encoding chloramphenicol acetyltransferase *(10)*. However, the assay in this case is relatively cumbersome *(11)*.

From: *Methods in Biotechnology, Vol. 18: Microbial Processes and Products*
Edited by: J. L. Barredo © Humana Press Inc., Totowa, NJ

The *Morganella morganii* PhoC protein is a molecular class A nonspecific phosphatase that can be efficiently produced and secreted when its gene *(phoC)* is expressed in *E. coli (12)*. Compared to *lacZ,* which can only be used with *E. coli* hosts carrying *lacZ* mutations, *phoC* can be used with any *E. coli* host *(12,13)*, and its activity can be easily detected qualitatively or quantitatively using insoluble or soluble chromogenic substrates that are relatively inexpensive. Moreover, the functional features of this enzyme, including tolerance to a broad range of temperatures and pH as well as to phosphate and cation concentration *(12)*, add to the interest of *phoC* as a reporter gene, widening its possible range of hosts and applications. Indeed, preliminary experiments demonstrated that PhoC can be produced and efficiently secreted in the yeast *S. cerevisiae* without undergoing degradation and that its activity can be easily detected and assayed with different substrates and in different growth conditions. *S. cerevisiae* has been used extensively as a model organism for the study of the most diverse biological processes and it represents an important model system for heterologous protein production. The wealth of methodological approaches available for this microorganism, however, is not matched with the availability of many different reporter genes useful for easy detection of promoter function and signal peptide efficiency in secretion.

The aim of this chapter is to describe the methods for using the *M. morganii phoC* as a reporter gene in *E. coli* and in *S. cerevisiae*. Two models will be discussed: (1) an analysis of the influence of the *Neisseria meningitidis GNA33* signal peptide sequence on gene expression will be used to illustrate the techniques related to the use of *phoC* as a reporter gene in *E. coli (see* **Subheading 1.1.**) and (2) the use of *phoC* as a reporter gene in *S. cerevisiae* will be illustrated through the analysis of the transcriptional activity of the yeast *UAS_{GAL}–CYC1* hybrid promoter *(14)*, during the construction of a promoter probe plasmid vector to be employed in yeast *(see* **Subheading 1.2.**)

1.1. Analysis of the Influence of a Signal Peptide Sequence on the Overall Heterologous Expression in E. coli

In a genome-wide screening for *N. meningitidis* serogroup B antigens, it was found that different forms of GNA33, a lipoprotein with homology to the *E. coli* murein transglycosylase MltA, were expressed at considerably different levels in a heterologous system (T7 expression system in *E. coli*). The analyzed forms were found to be different as a result of the presence or absence of a signal peptide sequence. The *GNA33 ORF* devoid of the N-terminal 20 amino acids that correspond to the signal peptide sequence gave the highest expression level, whereas the full-length form was poorly expressed. In order to evaluate the possible influence of the leader peptide on the overall heterologous expression, both wild and mutated *GNA33* signal peptides have been fused with the *phoC* reporter gene.

1.2. Use of phoC *as the Reporter Gene in the Construction of a Yeast Promoter-Probe Vector*

The yeast *S. cerevisiae* has been a popular model system for large-scale gene expression studies and it has several remarkable advantages, such as the availability of the entire genomic sequence, the relatively small size of genome, and a large body of genetic information *(15)*.

A better knowledge of the efficiency of its promoters and the possibility to engineer them by creating new hybrid promoters, aimed to solve specific biotechnological problems, is highly desirable. A system was therefore defined for the study of DNA sequences showing promoter activity in the yeast *S. cerevisiae* using the heterologous reporter enzyme PhoC of *M. morganii,* which is efficiently secreted by the yeast host. One promoter-probe vector was constructed, designed to screen and analyze yeast promoters. The system has been evaluated by analyzing the transcriptional activity of the well known UAS_{GAL}–*CYC1* hybrid promoter *(14)*.

2. Materials

2.1. Construction of Plasmids (E. coli)

1. pET21b(+) plasmid (Novagen Inc., Madison, WI) *(16)*.
2. *Morganella morganii* genomic DNA.
3. Primer 5′PhoC: 5′-gccgc**catatg**AAGAAGAATATTATCGCCGG-3′, which added *Nde*I restriction site (bold) to the sequence.
4. Primer 3′PhoC: 5′-gccgc**ctcgag**TTATTTCTGTGATTTTTGTGCAAATTCC-3′, which added *Xho*I restriction site (bold) to the sequence.
5. Primer 33Pho1: 5′-CTGTACGGCATCGCCGCCGCCATCCTCGCCGCCGC-GATCCCGGCCGGCAAGATGCC-3′.
6. Primer 33Pho2: 5′-gcc**catatg**AAAAAATACCTATTCCGCGCCGCCCTGTACGG CATCGCCGCCGCC-3′. An *Nde*I site (bold) was created at the 5′ end of the resulting polymerase chain reaction (PCR) fragment.
7. Primer *phoC*-FWD: 5′-gggga**G**CGATCCCGGCAGGCAACGAT-3′.
8. Primer *phoC*-REV: 5′-GGG**aagctt**ATTTCTGTGATTTTTGTGCAA-3′. A *Hind*III site (bold) was created downstream of the stop codon of *phoC*.
9. Primer *URA3*-FWD: 5′-AACCTAGAGGCCTTTTGATGTTA-3′.
10. Primer *URA3*-REV: 5′-GCGGCTC**GAGCTCggatcc**TAATAACTGATATAATT AAAT-3′. A *Bam*HI site (bold) was introduced at the 3′ end of *URA3*.
11. Restriction endonucleases, T4 DNA ligase.
12. Agarose and electrophoresis equipment.
13. Equipment for plasmid preparation.

2.2. Construction of Plasmids (S. cerevisiae)

1. YEpsec1 expression vector *(16,17)*.
2. Material described in **Subheading 2.1., steps 2–6.**

2.3. Transformation

1. Equipment for sterilization by filtration (0.22 μm).
2. Ampicillin stock solution: 100 mg/mL in sterile water. Ampicillin can be harmful by inhalation, ingestion, or skin absorption. Wear safety glasses and gloves.
3. Equipment for plasmid preparation.
4. Restriction endonucleases and DNA sequencing equipment *(18)*.
5. Luria–Bertani (LB) agar medium: 10 g/L tryptone, 5 g/L yeast extract, 15 g/L agar, and 10 g/L NaCl. Heat to dissolve and adjust the pH to 7.0 with 5 *N* NaOH (approx 0.2 mL). Sterilize by autoclave and cool to 50°C *(18)*.

2.3.1. Transformation of E. coli

E. coli strain BL21(DE3) (F⁻*ompT [lon] hsdS$_B$* (r$_B$⁻m$_B$⁻); an *E. coli* B strain with DE3, a λ-prophage carrying the T7 RNA polymerase gene *(16)*.

2.3.2. Transformation of S. cerevisiae (19)

1. *E. coli* strain HB101 (*F–, thi-1, hsdS20 (rB–, mB–), recA13, ara-14, leuB6, proA2, lacYl, galK2, rpsL20 (strr), xyl-5, mtl-1, supE44–*) *(20,21)*.
2. *S. cerevisiae* KD53B strain (*his*1, *ura*3-52).
3. YNB (10X) stock solution: 67 g/L yeast nitrogen base (w/o amino acids). Sterilize by filtration (0.22 μm) and store at 4°C (*see* **Note 1**).
4. Dextrose stock solution: 200 g/L of dextrose. Sterilize by filtration (0.22 μm) and store at room temperature; stable for several months.
5. L-Histidine stock solution: 10 mg/mL (w/v) L-histidine; Sterilize by filtration (0.22 μm) and store in the dark at 4°C; stable for several months. L-Histidine can act as a skin, eye, or respiratory irritant. Wear safety glasses and gloves and use in adequate ventilation.
6. SD/H medium (synthetic dextrose/histidine): sterilize 80 mL of agar 2% (w/v) by autoclaving for 20 min at 15 psi, cool to 54–56°C in a water bath, and aseptically add 10 mL of YNB 10X, 10 mL of dextrose stock solution, and 0.5 mL of histidine stock solution.
7. PEG 4000 stock solution: 400 g/L (w/v) of polyethylene glycol molecular weight 4000 in distillated water. Sterilize by filtration (0.22 μm) and store at 4°C; stable for several months. PEG 4000 can act as a skin, eye, or respiratory irritant. Wear safety glasses and gloves and use in adequate ventilation.
8. LiCl stock solution: 100 m*M* lithium chloride anhydrous in distilled water. Sterilize by filtration (0.22 μm) and store at 4°C; stable for several months. Lithium chloride can act as a skin, eye, or respiratory irritant. Wear safety glasses and gloves and use in adequate ventilation.

2.4. Qualitative Test With the Phenolphtalein Diphosphate–Methyl Green Technique

1. PDP stock solution: 20% (w/v) solution of phenolphtalein diphosphate and tetra-sodium salt (Sigma–Aldrich Chemie Gmbh, Steinheim, Germany) in distilled water. Sterilize by filtration (0.22 μm) and store at –20 °C; once frozen, the stock

solution is stable for several months (*see* **Note 2**). PDP can act as a skin, eye, or respiratory irritant; avoid exposure to dust.
2. MG stock solution: 5 mg/mL of methyl green (Sigma–Aldrich Chemie Gmbh). Sterilize by filtration (0.22 μm) and store in the dark at 4°C. MG stock solution is stable for several months (*see* **Note 3**). MG is irritating to skin, eye, and respiratory tract. Wear safety glasses and gloves and use in adequate ventilation.

2.4.1. Plate Screening in E. coli

1. IPTG stock solution: 100 m*M* isopropyl-β-D-thio-galactopyranoside (IPTG). IPTG can be harmful by inhalation, ingestion, or skin absorption. Wear safety glasses and gloves and use in adequate ventilation.
2. Tryptose phosphate agar medium (Difco, Detroit, MI).
3. TPMG medium (tryptose–phenolphtalein–methyl green): add 5 mL of PDP stock solution and 5 mL of MG stock solution to 0.5 L of tryptose phosphate agar medium. Sterilize by autoclaving and cool to 50°C.

2.4.2. Plate Screening in S. cerevisiae

1. YPG agar medium: 10 g/L yeast extract, 20 g/L peptone, 20 g/L agar, 20 g/L galactose, 5 g/L KH$_2$PO$_4$, and 0.5 g/L K$_2$HPO$_4$. Heat to dissolve and sterilize by autoclaving (*see* **Note 4**).
2. YPG/PDP medium: add to 100 mL of YPG agar sterile medium, melted and cooled to 50°C, 1 mL of PDP stock solution (2 mg/mL final concentration), and 1 mL of MG stock solution (0.05 mg/mL final concentration).

2.5. Quantitative Assays of PhoC Activity: Induction of the Cultures and Preparation of the Samples

2.5.1. Induction of the Cultures and Preparation of the Samples (E. coli)

1. LB broth: 10 g/L tryptone, 5 g/L yeast extract, and 10 g/L NaCl. Heat to dissolve and adjust the pH to 7.0 with 5 *N* NaOH (approx 0.2 mL). Sterilize by autoclaving (*18*).
2. Ampicillin stock solution (*see* **Subheading 2.3.**).
3. IPTG stock solution (*see* **Subheading 2.4.1.**).
4. EDTA solution: 0.5 *M* EDTA, pH 8.0. Add 186.1 g of disodium ethylenediaminetetraacetate 2H$_2$O to 800 mL of distilled water. Dissolve by stirring vigorously on a magnetic stirrer and adjust the pH to 8.0 with NaOH (about 20 g of NaOH pellets). Adjust the volume to 1 L with distilled water. Aliquot and sterilize by autoclaving. EDTA is harmful if swallowed; it can be harmful if inhaled or through skin contact and it is an eye, respiratory tract, and skin irritant. Wear safety glasses and use in adequate ventilation.

2.5.2. Induction of the Cultures and Preparation of the Samples (S. cerevisiae)

1. YP (yeast extract peptone) broth (2X): 20 g/L yeast extract, 40 g/L peptone, 10 g/L KH$_2$PO$_4$, and 1 g/L K$_2$HPO$_4$. Sterilize by filtration (0.22 μm).

2. YPD broth: add 10 mL of dextrose stock solution and 40 mL of sterile distilled water to 50 mL of sterile YP broth (2X).
3. Galactose stock solution: 20% (w/v) galactose. Sterilize by filtration (0.22 μm) and store at room temperature; stable for several months.
4. YPG broth: add 10 mL of galactose stock solution and 40 mL of sterile distilled water to 50 mL of sterile YP broth (2X).
5. Dextrose stock solution: 20% (w/v) dextrose. Sterilize by filtration (0.22 μm) and store at room temperature; stable for several months.

2.6. Quantitative Assays of PhoC Activity by Means of the pNPP Phosphatase Test

1. *p*NPP stock solution: 45.7 mg/mL *p*-nitrophenyl phosphate disodium salt hexahydrate (*p*NPP) in distilled water. Store at –20°C. Defrost only the necessary aliquots, avoiding repeated thawing-and-freezing cycles. Stable for several months. *p*NPP is toxic by inhalation, in contact with skin, or if swallowed, and it is irritating to eyes, respiratory system, and skin. Wear safety gloves.
2. 2 *M* Sodium acetate–acetic acid buffer, pH 5.5: dissolve 272.2 g of Na–acetate trihydrate in 80 mL of distilled water. Adjust the pH to 5.5 with glacial acetic acid. Adjust the volume to 100 mL. Na–acetate trihydrate can be harmful by ingestion, inhalation, or skin absorption; it can act as an irritant. Wear safety glasses. *Caution:* Glacial acetic acid is strongly corrosive and causes serious burns and indicates lachrymotion. Wear nitrile gloves and use fume hood.
3. 2 *M* NaOH: 40 g of sodium hydroxide pellets and 80 mL of distilled water. Dissolve by stirring on a magnetic stirrer. Adjust the volume to 100 mL with distilled water. NaOH is very corrosive, causes severe burns, and can cause serious permanent eye damage. It is very harmful by ingestion and is harmful by skin contact or by inhalation of dust. Handle in adequate ventilation; wear safety glasses and neoprene or polyvinyl chloride (PVC) gloves.

3. Methods

The methods described outline (1) the construction of plasmids (*see* **Subheadings 3.1.** and **3.2.**), (2) the transformation procedure (*see* **Subheading 3.3.**), (3) the description of the PDP-MG histochemical plate screening for PhoC activity (*see* **Subheading 3.4.**), (4) the induction of cultures and preparation of samples (*see* **Subheading 3.5.**), and (5) the analysis of phosphatase activity by means of the *p*-nitrophenyl phosphate (*p*NPP) test (*see* **Subheading 3.6.**).

3.1. phoC as a Reporter Gene in E. coli: The Construction of Plasmids

The construction of expression plasmids for the study of *GNA33* signal peptide on heterologous protein expression in *E. coli* is described in **Subheadings 3.1.1.–3.1.4.** This includes (1) the description of the expression vector, (2) the construction of plasmids pET33*phoC* and pET*phoC*$_{wt}$ designed to compare the influence of $GNA33_{wt}$ and $phoC_{wt}$ signal peptide sequences on the expression of the *phoC* gene, and (3) the construction of *GNA33* mutants, with the objective of achieving a better understanding of the role played by the *GNA33* signal peptide sequence on regulation.

3.1.1. Expression Vector

The pET21b(+) plasmid is an expression vector derived from pBR322 that contains the T7 promoter and terminator *(16)*. The selectable marker is ampicillin. The presence on the plasmid of the whole *lacI* gene, coding the transcriptional repressor, helps to keep the gene expression tightly regulated. A schematic of pET21b(+) is shown in **Fig. 1A.**

3.1.2. Construction of Plasmid pETphoC_wt

The wild-type *phoC* gene was amplified from *M. morganii* genomic DNA by using the primers 5′PhoC and 3′PhoC, that added *NdeI* and *XhoI* restriction sites to the sequence The PCR product was digested with the same endonucleases, gel-purified, and inserted into the *NdeI/XhoI* sites of pET21b(+) to obtain pET*phoC_wt*, to be employed as the positive control (*see* **Fig. 1B**).

3.1.3. Construction of Plasmid pET33phoC

In order to evaluate the influence of the *GNA33* signal peptide sequence on the heterologous expression in *E. coli,* the 690-bp sequence coding for the mature portion of the *M. morganii* PhoC protein were fused to the signal sequence-encoding 60 bp, creating the *GNA33–phoC* fusion.

Two PCR steps were employed. In the first step, the 33Pho1 and 3′PhoC primers were used. The sequence immediately downstream of the signal peptide cleavage site, 5′-ATCCCGGCAGG-3′, was changed to 5′-ATCC**CGGC-CGG**-3′ in order to create an *EagI* site (**bold**). The amplicon so obtained was then used as a template for the second PCR step, which was performed using the primers 33Pho2 and 3′PhoC. An *NdeI* site was created at the 5′ end of the resulting PCR fragment. The PCR product was digested with *NdeI* and *XhoI*, gel-purified, and inserted into the *NdeI/XhoI* sites of the pET21b(+) vector to obtain pET33*phoC* (*see* **Fig. 1C**).

3.1.4. Construction of GNA33 Mutants

To gain a better understanding of the role of this sequence in regulation, several mutants of the *GNA33* signal peptide sequence have been obtained and characterized. To generate mutations (at nucleotide and/or amino acid level) in the *GNA33* signal peptide sequence, several degenerated DNA linkers were synthesized and inserted into the *NdeI/EagI* sites of the pET33*phoC* vector (*see* **Fig. 1D**).

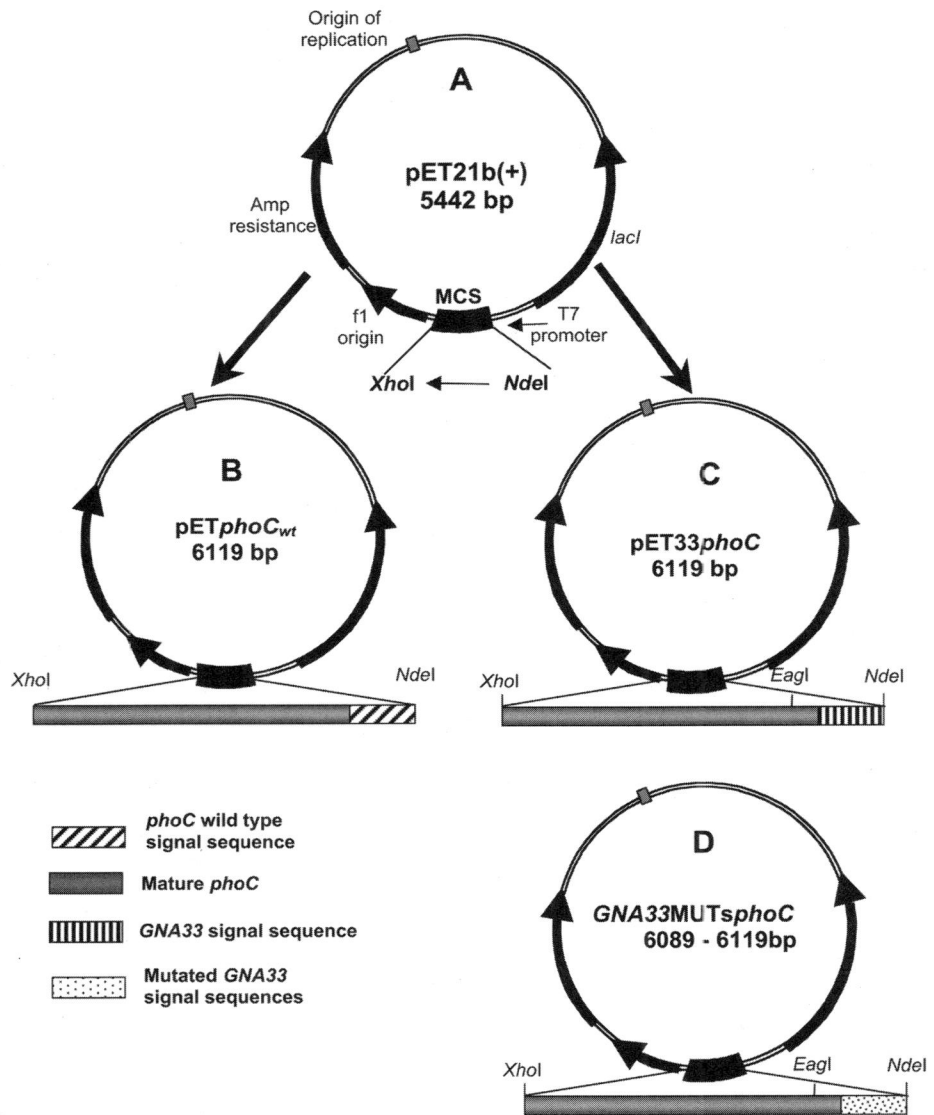

Fig. 1. **(A)** Schematic of the pET21b(+) vector adapted from Novagen. **(B)** *Nde*I/*Xho*I insert creating the pET*phoC*wt plasmid. **(C)** *Nde*I/*Xho*I insert creating pET33*phoC*. A unique *Eag*I site was created. **(D)** Several *GNA33* mutants were created by inserting synthetic linkers between *Nde*I and *Eag*I sites.

3.2. phoC *as a Reporter Gene in* S. cerevisiae: The Construction of Plasmids

The use of *phoC* as a reporter gene in the eukaryotic yeast host *S. cerevisiae* to analyze the activity of the UAS_{GAL}–*CYC1* hybrid yeast promoter is described in **Subheadings 3.2.1.–3.2.3.** This include (1) the description of the expression vector YEpsec1, (2) the construction of plasmid YEp*phoC,* obtained by fusing the *phoC* reporter gene with the signal peptide sequence of the *Kluyveromyces lactis* killer toxin in the YEpsec1 expression vector, and (3) the deletion of the promoter, to obtain a promoter-probe vector (YEp*phoC*ΔP).

3.2.1. Yeast Expression Vector YEpsec1

The plasmid YEpsec1 *(17)* is a multicopy "yeast episomal plasmid," derived from pEMBLyex2 *(22)* by adding the *K. lactis* k1 (killer toxin) signal peptide sequence. It is a shuttle vector that replicates both in *S. cerevisiae* and in *E. coli* because of the presence of the yeast replication origin of the 2μ plasmid, together with both the replication origins of plasmid pMB1 and of the f1-phage. The selection markers are *URA3* for *S. cerevisiae* and *amp* and *leuB* for *E. coli.* In this vector, the "upstream activating sequence" of the *GAL1*–*GAL10* genes *(UAS_{GAL})* is located 5′ to the TATA box of the *CYC1* promoter to form a hybrid promoter. The transcription is, therefore, induced by galactose and repressed by glucose. Immediately downstream of the hybrid promoter, there is the sequence encoding the signal peptide of the *K. lactis* killer toxin (k1), followed by a multiple cloning site harboring unique restriction sites (*see* **Fig. 2A**).

3.2.2. Construction of YEpphoC

In order to study its suitability as a reporter gene in *S. cerevisiae,* the *M. morganii phoC* gene was fused with the yeast signal peptide sequence of the *K. lactis* k1. The fragment encoding the mature protein was amplified from *M. morganii* genomic DNA, with the primers *phoC*-FWD and *phoC*-REV designed to place the fragment in frame with the signal sequence and to add a *Hind*III restriction site downstream of the stop codon of *phoC.* The amplicon (702 bp) was then restricted with the endonuclease *Hind*III, gel-purified, and inserted into the *Sma*I/*Hind*III sites of the YEpsec1 vector (*see* **Fig. 2B**).

3.2.3. Promoter-Probe Vector Construction

To obtain a promoter-probe vector, the region spanning the UAS_{GAL} and *CYC1* sequences was deleted from YEp*phoC* by a *Stu*I/*Sac*I digestion. As the *Stu*I site in YEpsec1 is located within the *URA3* gene, the fragment (449 bp) removed from the 3′ end of *URA3* was replaced with an amplicon obtained using the primers *URA3*-FWD and *URA3*-REV. In this step, a *Bam*HI cloning site (bold) was introduced at the 3′ end of *URA3.* The amplicon was restricted

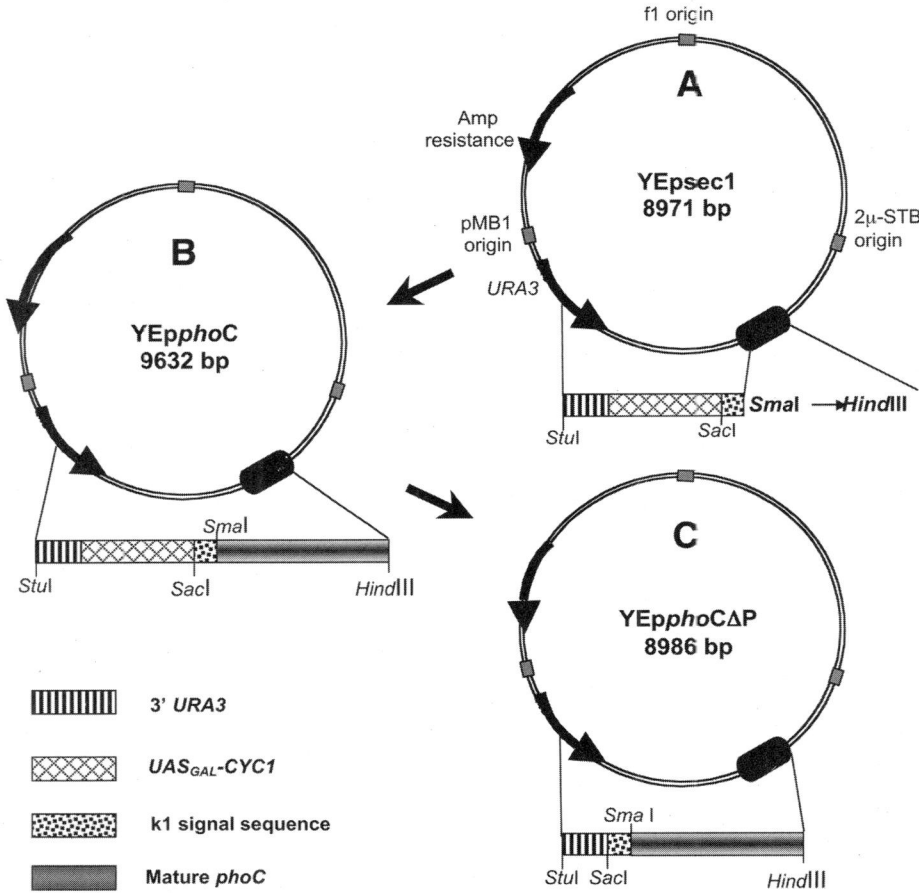

Fig. 2. **(A)** Schematic of the YEpsec1 vector. **(B)** *Sma*I/*Hind*III insertion of the *phoC* fragment encoding the mature PhoC protein, to obtain the YEp*phoC* plasmid. **(C)** Deletion of the *UAS*GAL–CYC1 region to create the promoter-probe vector (YEp*phoC*ΔP); a *Bam*HI unique site was introduced at the 3′ end of *URA3* for cloning purposes.

with *Stu*I and *Sac*I endonucleases, gel-purified, and cloned in the YEp*phoC* vector, digested with *Stu*I and *Sac*I, to obtain the YEp*phoC*ΔP vector (*see* **Fig. 2C**).

3.3. Transformations

All of the DNA manipulations were performed by standard recombinant DNA methods *(18)* to construct the expression plasmids and are not described here in detail because of space limitations.

3.3.1. Transformation Procedures in E. coli

The plasmids were used to transform *E. coli* BL21(DE3) *(20,21)* cells. Plasmids pET*phoC*$_{wt}$ and pET21b(+) were used as positive and negative controls, respectively. The mutant signal peptide sequences produced by the amplification with degenerate primers are described in **Fig. 3B.**

1. Transform *E. coli* competent cells by standard methods *(18)*.
2. Plate on LB agar medium with ampicillin (100 µg/mL) and incubate overnight at 37°C.
3. Select single colonies and grow them overnight in LB broth with ampicillin.
4. Perform plasmid preparations and check plasmids by restriction mapping.

3.3.2. Transformation Procedures in S. cerevisiae

All of the constructed plasmids were amplified in the *E. coli* HB101 strain and used to transform the yeast strain *S. cerevisiae* KD53b. The plasmid YEp*phoC* was used to follow the activity of the induced or repressed UAS_{GAL}–$CYC1$ promoter. The plasmid YEp*phoC*ΔP was used to verify the absence of PhoC activity after deletion of the promoter.

1. Transform *E. coli* HB101 by standard methods *(18)*.
2. Plate on LB agar medium containing ampicillin (100 µg/mL) and incubate overnight at 37°C.
3. Check plasmids by restriction enzyme digestion and DNA sequencing.
4. Prepare SD/H medium, mix it by swirling (avoid to produce bubbles), and pour it into the plates.
5. Transform *S. cerevisiae* KD53B strain *(19)* (*see* **Notes 5** and **6**).
6. Plate transformants on SD/H medium (*see* **Note 7**).

3.4. Qualitative Plate Screening of PhoC Activity With the PDP-MG Technique

The PDP-MG technique was originally devised as a histochemical screening to detect phosphatase-producing clones in bacterial genomic libraries *(22,23)* and has been since used successfully for several purposes related to *in situ* detection of phosphatase activity *(13,24)*. This simple and effective method can be used for qualitative assays with both bacteria and yeast, as illustrated in **Subheadings 3.4.1.** and **3.4.2.**

3.4.1. Plate Screening in E. coli

1. Prepare the TPMG medium, mix well, and pour into the plates (*see* **Note 8**).
2. Streak *E. coli* BL21(DE3) transformants on TPMG plates containing IPTG (final concentration = 0.5 m*M*).
3. Incubate the plates at 37°C until colonies appear.

A

```
                10          20          30          40          50          60
33-L1a   ATGAAGAAGT ACCTTTTCAG CGCCGCC... .......... .......... ..........
33-L1d   ATGAAAAAAT ACTTTTTCCG CGCCGCC... .......... .......... ..........
33-L1f   ATGAAAAAAT ATCTCTTTAG CGCCGCCCTG TACGGCATCG CCGCCGCCAT CCTCGCCGCC
GNA33wt  ATGAAAAAAT ACCTATTCCG CGCCGCCCTG TACGGCATCG CCGCCGCCAT CCTCGCCGCC

33-L1a   MKKYLFSAA..........
33-L1d   MKKYDFRAA..........
33-L1f   MKKYLFSAALYGIAAAILAA
GNA33wt  MKKYLFRAALYGIAAAILAA
```

```
                10          20          30          40          50          60
33-S1e   ATGAAAAAAT ACCTATTC.. .......... ......ATCG CCGCCGCCAT CCTCGCCGCC
33-S1c   ATGAAAAAAT ACCTATTCCG AGCTGCCCAA TACGGCATCG CCGCCGCCAT CCTCGCCGCC
33-S1b   ATGAAAAAAT ACCTATTCCG GGCCGCCCAA TACGGCATCG CCGCCGCCAT CCTCGCCGCC
33-S1i   ATGAAAAAAT ACCTATTCCG CGCCGCTTTG TACGGGATCG CCGCCGCCAT CCTCGCCGCC
GNA33wt  ATGAAAAAAT ACCTATTCCG CGCCGCCCTG TACGGCATCG CCGCCGCCAT CCTCGCCGCC

33-S1e   MKKYLF......IAAAILAA
33-S1c   MKKYLFRAAQYGIAAAILAA
33-S1b   MKKYLFRAAQYGIAAAILAA
33-S1i   MKKYLFRAALYGIAAAILAA
GNA33wt  MKKYLFRAALYGIAAAILAA
```

GNA33wt = GNA33 signal peptide sequence

B

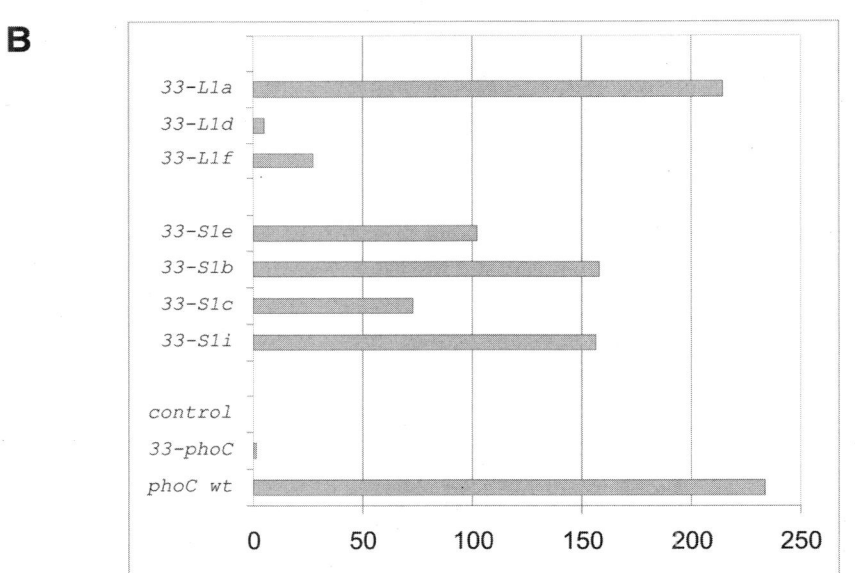

Fig. 3. (**A**) Alignment of DNA sequences and deduced amino acid sequences of mutated *GNA33* signal peptide sequences with the wild type *(GNA33wt)*. Mutations are highlighted in gray. (**B**) *p*NPP assay for the analysis of PhoC activity of both wild-type *(phoCwt)* and mutated *(33-phoC) GNA33* signal peptide. *phoCwt* was used as a positive control. PhoC activity was determined in cell samples from liquid cultures of *GNA33* signal peptide mutants.

4. Inspect the plates for the presence of green-stained colonies.
5. Compare the color intensity, which is directly related to PhoC activity, to both the positive and negative controls (*see* **Notes 9** and **10** and **Fig. 4**).

3.4.2 Plate Screening in S. cerevisiae

1. Prepare YPG/PDP medium, mix by swirling (avoiding to produce bubbles), and pour into the plates.
2. Select single isolated colonies of *S. cerevisiae* KD53b transformants (YEp*pho*C) from SD/H agar, and streak them on YPG/PDP agar medium.
3. Incubate the plates at 30°C until colonies appear.
4. Inspect the plates for the presence of the green stained colonies, surrounded by a green halo, that indicates activation of the promoter (*see* **Fig. 4**).

3.5. Quantitative Assays of PhoC Activity: Induction of the Cultures and Preparation of the Samples

In both the *E. coli* and yeast systems, the protein production needs to be induced. In *E. coli* BL21(DE3), the transcription is initiated from the pET21b(+) T7 promoter by means of an IPTG inducible T7 RNA polymerase. In *S. cerevisiae,* the analyzed promoter is induced by galactose. As a consequence, the phosphatase test must be performed on induced cultures.

3.5.1. Induction of E. coli Cultures and Preparation of the Samples

1. Pick an isolated colony of the clone to be tested and inoculate a 5-mL solution of LB broth (with ampicillin 100 µg/mL).
2. Grow overnight at 37°C.
3. Dilute the culture in 20 mL of freshly prepared LB broth plus ampicillin, to obtain an optical density (OD_{600}) equal to 0.1.
4. Incubate at 37°C until an OD_{600} equal to 0.5 is reached.
5. Induce expression by adding 200 µL of a 100 mM solution of IPTG (final concentration = 1 mM).
6. After 2–3 h of induction, collect the cultures.
7. Measure the OD_{600} and centrifuge a volume of culture equal to $1/OD_{600}$ mL (15,000g for 5 min) (*see* **Notes 11** and **12**).
8. Suspend the cells in 50 µL of sterile H_2O and add 10 µL of 0.5 M EDTA (*see* **Note 13**).
9. Incubate at room temperature for 30 min.

3.5.2. Induction of phoC Expression in Liquid Cultures of S. cerevisiae and Preparation of the Samples

1. Prepare YPD broth.
2. Pick a single isolated green colony of *S. cerevisiae* KD53b (YEp*pho*C) from YPG/PDP agar and inoculate a 5-mL volume of YPD broth.
3. Grow overnight at 30°C, with shaking.

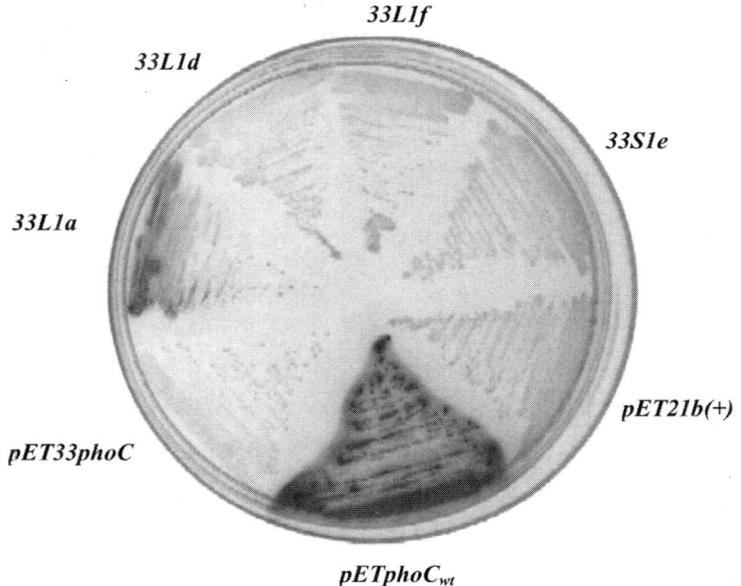

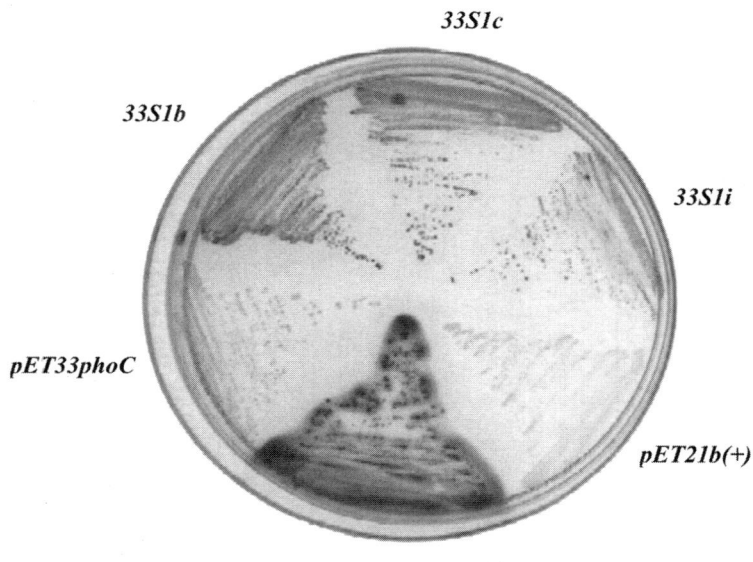

Fig. 4.

4. Prepare the YPG broth.
5. Dilute the culture 1 : 100 in YPG broth.
6. Collect 1.5 mL of culture and use 1 mL to measure the OD_{600} against YPG broth.
7. Centrifuge the remaining 0.5 mL of culture at 12,000g for 5 min and freeze the supernatant.
8. Place the culture at 30°C, with shaking.
9. Collect 1.5 mL of yeast culture every 6 h; measure the OD_{600} value determined as described in **step 4** and prepare a sample of supernatant as described in **step 5.**
10. After 24 h, split the culture into two and add dextrose to a final concentration of 2% (w/v) to one of the two samples.
11. Collect samples from the two cultures as described in **step 7.**

3.6. pNPP Phosphatase Assays

The *M. morganii* PhoC phosphatase activity was assayed using *p*NPP as a substrate and determining the amount of released *p*-nitrophenol (*p*NP) by measuring the $OD_{414.5}$ *(25)*. To avoid the need of performing a protein determination for each phosphatase assay, the data were expressed as nanomoles of *p*NP liberated per minute per milliliter of a culture with an OD_{600} value equal to 1 (*see* **Note 14**).

3.6.1. Quantitative Phosphatase Assay in E. coli Cultures

1. Prepare a test tube for each *E. coli* transformant to be tested by adding 50 µL of 2 *M* Na-acetate–acetic acid buffer, pH 5.5, and 840 µL of distilled water to the 50 µL of cell suspension (*see* **Note 15**).
2. Prepare a "blank" tube by adding 50 µL of 2 *M* Na-acetate–acetic acid buffer, pH 5.5, to 900 µL of distilled water.
3. Pre-equilibrate the samples at 37°C for 15 min.
4. Initiate the reaction by the addition of 50 µL of 100 m*M* *p*NPP to each tube (*see* **Note 16**).
5. Incubate at 37°C for 15 min.
6. Stop the reaction by adding 2 mL of 2 *M* NaOH.

Fig. 4. TPGM plate qualitative screening for the analysis of both wild-type and mutated *GNA33* signal peptide sequence. The positive control was pET*phoC*$_{wt}$ and the negative one was pET21b(+). The cells transformed with pET33*phoC* do not show any PhoC activity and are fully comparable to the negative control. PhoC activity is, on the other hand, evident in the positive control. These observations confirm that the negative influence of *GNA33* on gene expression is confined to the 60 bp fused to the reporter gene. Some of the analyzed mutants (33S1b, 33S1c, and 33S1i) show a weak positive reaction, whereas the modification introduced in the 33L1a mutant has significantly improved the heterologous expression, as compared to wild-type *GNA33*.

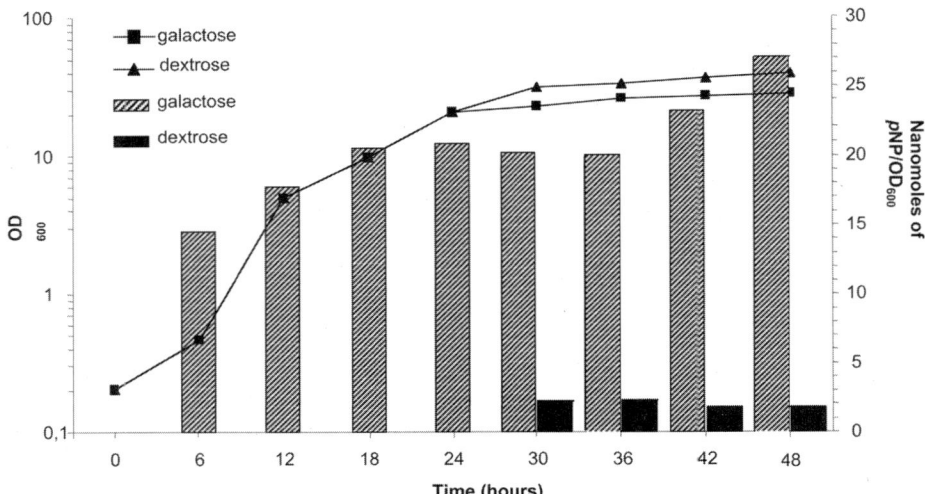

Fig. 5. The activity of the UAS_{GAL}–$CYC1$ promoter was analyzed by measuring the activity of the PhoC phosphatase in culture supernatants. Striped bars indicate the PhoC activity in a yeast culture grown in galactose medium, whereas black bars represent the PhoC activity found in the same culture after the addition of glucose-containing medium. The growth curves of the culture in galactose medium with and without feeding of glucose are also shown.

7. Calculate the nanomoles of pNP released, according to the formula ($OD_{414.5} \times 10^9/t \times 18,472) \times (Va/10^3)$, where $OD_{414.5}$ is the value of the observed absorbance at 414.5 nm, t is the testing time (15 min), 18,472 is the molar extinction coefficient of pNP (M^{-1} cm^{-1}), and Va is the total volume of the assay (3 mL).

The results of quantitative assays with wild and mutated *GNA33* signal peptide sequences are shown in **Fig. 3.**

3.6.2. Quantitative Phosphatase Assay in S. cerevisiae Cultures

1. Gently thaw the supernatants on ice.
2. Prepare a "blank" tube by adding 50 μL of 2 *M* Na-acetate–acetic acid buffer, pH 5.5, to 900 μL of distilled water.
3. Use 50 μL of each supernatant to measure phosphatase activity as described in **Subheading 3.6.1.** Calculate the released nanomoles of pNP according to the formula described in **Subheading 3.6.1, step 7.** Calculate the rate (released nanomoles/OD_{600}) in order to normalize the results to the number of cells present in the culture at the sampling time.

The results of quantitative assays with *S. cerevisiae* KD53B (YEp*phoC*) are shown in **Fig. 5.**

4. Notes

1. YNB medium can be purchased also without ammonium sulfate; in this case, prepare the YNB 10X using 1.7 g of YNB w/o ammonium sulfate and 5 g of ammonium sulfate.
2. A critical point for a good result of the histochemical testing is the PDP salt form. It must be the tetrasodic salt. Disodic or other salts do not work.
3. The quality of methyl green is a critical point. Some commercial preparations yield poor results, most probably as a result of a high content of impurities (especially methyl blue).
4. The phosphate buffer is added to the medium in order to prevent the expression of yeast phosphatases.
5. Several methods are available to transform yeast cells, but the LiCl protocol is simpler than the spheroplast procedure and it is, therefore, often preferred even if its efficiency is lower.
6. When the LiCl method is chosen, remember that it can cause the appearance of "petites" (respiratory-deficient yeast cells). Avoid picking these cells for further experiments, choosing instead the rather large colonies that are respiratory competent (phenotype "grande"). To distinguish "grande" from "petite," yeast cells can be plated onto agar YP with different carbon sources; in the presence of 2% (w/v) glucose, both "grande" and "petite" strains are able to grow, but when the carbon source is 3% (w/v), glycerol "petite" mutants cannot grow. Finally, when both glucose 0,05% (w/v) and glycerol 3% (w/v) are present, the petite cells start to grow but stop as soon as the glucose has been consumed, forming white and small (max 1–2 mm) colonies.
7. Prepare 1 : 10 dilution from each transformation tube, in sterile and distilled water. Plate 1 mL from each dilution on two SD plates. Sterilize a bent glass rod by dipping it into ethanol and then in the flame of a Bunsen burner. Let the rod cool and rub it on the inner side of the plate cover. When the rod has cooled to room temperature, very gently spread the transformed cells over the surface of the agar plate. Do not let the plate surfaces dry by a prolonged spreading: such a procedure could damage the transformed cells. Instead, leave the plates at room temperature until the liquid is absorbed. Placing the plates in a laminar flow cap with the lid slightly opened will be useful also to minimize the risk of bacterial contamination.
8. TPMG plates can be stored at 4°C for at least 2 mo without losing their potential for detection of the PhoC activity.
9. Sometimes, a very high expression of the *phoC* gene causes a "white colony" phenotype: white colonies surrounded by a deep green halo.
10. If a too heavy inoculum is applied on TPMG plates, the confluent bacterial growth causes medium consistent alkalinization and the methyl green turn colorless. In this case, the PhoC activity is recognizable by the diffuse pink-purple color resulting from the released phenolphthalein.
11. This volume of culture contains the same number of cells as 1 mL of a culture with an $OD_{600} = 1$.

12. When the cultures are collected after 2–3 h of induction, the possible secretion of PhoC in the medium or the possible lysis of cells are normally not relevant. However, if a prolonged incubation is necessary, the possible presence of PhoC activity in the culture supernatant should be checked.
13. This preliminary incubation should inhibit some *E. coli* phosphatases that could interfere with the test. It is especially useful when the test is performed on whole cells or on crude cellular extracts.
14. A culture of *E. coli* with an $OD_{600}=1$ corresponds to approx 1×10^9 cells/mL. A culture of *S. cerevisiae* KD53B with $OD_{600}=1$ corresponds to approx 1×10^7 cells/mL.
15. Crude bacterial extracts can also be prepared to perform phosphatase assays and evaluate PhoC activity in comparison to total protein amount. However, *p*NPP molecules are small enough to enter the bacterial cell. It is, therefore, possible to perform the test directly on whole-cell suspensions and compare the activities of suspensions with the same turbidity.
16. If many samples must be tested at same time, it is convenient to start the reaction at regular time intervals (e.g., 30 s) and stop them in the same order. The use of a timer can be helpful.

References

1. An, G., Hidaka, K., and Simonovitch, L. (1982) Expression of bacterial beta-galactosidase in animal cells. *Mol. Cell Biol.* **2,** 1628–1632.
2. Chalfie, M., Tu, Y., Euskirchen, G., Ward, W. W., and Prasher, D. C. (1994) Green fluorescent protein as a marker for gene expression. *Science* **263,** 802–805.
3. Gorman, C. M., Moffat, L. F., and Howard, B. H. (1982) Recombinant genomes which express chloramphenicol acetyltransferase in mammalian cells. *Mol. Cell Biol.* **2,** 1044–1051.
4. Jefferson, R. A., Burgess, S. M., and Hirsh, D. (1986) beta-glucuronidase from *Escherichia coli* as a gene-fusion marker. *Proc. Natl. Acad. Sci. USA* **83,** 8447–8451.
5. Olsson, O., Koncz, C., and Szalay, A. A. (1988) The use of the *luxA* gene of the bacterial luciferase operon as a reporter gene. *Mol. Gen. Genet.* **215,** 1–9.
6. Ow, D. W., Wood, K. W., De Luca, M., De Wet, J. R., Helinski, D. R., and Owell, S. H. (1986) Transient and stable expression of the firefly luciferase gene in plant cells and transgenic plants. *Science* **234,** 856–859.
7. Wildt, S. and Deuschle, U. (1999) *cobA,* a red fluorescent transcriptional reporter for *Escherichia coli,* yeast, and mammalian cells. *Nat. Biotechnol.* **17,** 1175–1178.
8. Lapage, S. P., Efstratiou, A., and Hill, L. R. (1973) The ortho-nitrophenol (ONPG) test and acid from lactose in Gram-negative genera. *J. Clin. Pathol.* **26,** 821–825.
9. Horwitz, J. P., Chua, J., Curby, R. J., et al. (1964) Substrates for cytochemical demonstration of enzyme activity. I. Some substituted 3-indolyl-beta-D-glycopyranosides. *J. Med. Chem* **7,** 574–575.
10. Mount, R. C., Jordan, B. E., and Hadfield, C. (1996) Reporter gene systems for assaying gene expression in yeast. *Methods Mol. Biol.* **53,** 239–248.

11. Alipour, H., Eriksson, P., Norbeck, J., and Blomberg, A. (1999) Quantitative aspects of the use of bacterial chloramphenicol acetyltransferase as a reporter system in the yeast *Saccharomyces cerevisiae*. *Anal. Biochem.* **270,** 153–158.

12. Thaller, M. C., Berlutti, F., Schippa, S., Lombardi, G., and Rossolini, G. M. (1994) Characterization and sequence of PhoC, the principal phosphate-irrepressible acid phosphatase of *Morganella morganii*. *Microbiology* **140,** 1341–1350.

13. Thaller, M. C., Berlutti, F., Schippa, S., Selan, L., and Rossolini, G. M. (1998) Bacterial acid phosphatase gene fusions useful as targets for cloning-dependent insertional inactivation. *Biotechnol. Prog.* **14,** 241–247.

14. Guarente, L., Yocum, R. R., and Gifford, P. (1982) A *GAL10-CYC1* hybrid yeast promoter identifies the *GAL4* regulatory region as an upstream site. *Proc. Natl. Acad. Sci. USA* **79,** 7410–7414.

15. Cherry, J. M., Adler, C., Ball, C., et al. (1998) SGD: *Saccharomyces* Genome Database. *Nucleic Acids Res.* **26,** 73–79.

16. Studier, F. W. and Moffatt, B. A. (1986) Use of Bacteriophage T7 RNA polymerase to direct selective high-level expression of cloned genes. *J. Mol. Biol.* **189,** 113–130.

17. Baldari, C., Murray, J. A., Ghiara, P., Cesareni, G., and Galeotti, C. L. (1987) A novel leader peptide which allows efficient secretion of a fragment of human interleukin 1 beta in *Saccharomyces cerevisiae*. *EMBO J.* **6,** 229–234.

18. Sambrook, J. and Russell, D. W. (eds.) (2001) *Molecular Cloning: A Laboratory Manual,* Cold Spring Harbor Laboratory Press, Cold Spring Harbor, NY.

19. Gietz, R. D. and Woods, R. A. (2002) Transformation of yeast by lithium acetate/single-stranded carrier DNA/polyethylene glycol method, in *Guide to Yeast Genetics & Molecular Cell Biology* (Christine Guthrie, G. R. F., ed), Academic–Hardback, London, pp. 87–96.

20. Bolivar, F. and Backman, K. (1979) Plasmids of *Escherichia coli* as cloning vectors. *Methods Enzymol.* **68,** 245–267.

21. Boyer, H. W. and Roulland-Dussoix, D. (1969) A complementation analysis of the restriction and modification of DNA in *Escherichia coli. J. Mol. Biol.* **41,** 459–472.

22. Cesareni, G. and Murray, J. A. (1987) Plasmid vectors carrying the replication origin of philamentous single-stranded phages. *Genet. Eng.* **9,** 135–154.

23. Riccio, M. L., Rossolini, G. M., Lombardi, G., Chiesurin, A., and Satta, G. (1997) Expression cloning of different bacterial phosphatase-encoding genes by histochemical screening of genomic libraries onto an indicator medium containing phenolphthalein diphosphate and methyl green. *J. Appl. Microbiol.* **82,** 177–185.

24. Thaller, M. C., Berlutti, F., Schippa, S., Iori, P., Passariello, C., and Rossolini, G. M. (1995) Heterogeneous pattern of acid phosphatases containing low-molecular-mass polypeptides in members of the family *Enterobacteriaceae. Int. J. Syst. Bact.* **45,** 255–261.

25. Neumann, H. and van Vreedendaal, M. (1967) An improved alkaline phosphatase determination with *p*-nitrophenyl phosphate. *Clin. Chim. Acta* **17,** 183–187.

19

Gene Expression Arrays in Food

Bart Weimer, Yi Xie, Lan-szu Chou, and Adele Cutler

Summary

Gene expression in microbes can be detected with an oligonucleotide-based DNA macroarray. The DNA macroarray is composed of short (22–24-mer) oligonucleotide probes immobilized onto a nylon membrane via polyinosine tails. An indirect high-density labeling method was used to effectively incorporate biotin into the nucleic acids. The hybridization signals were detected with a chemiluminescence-based method and digitized with a desktop scanner. The utility of this protocol was demonstrated by profiling the expression of 375 metabolically related genes in *Lactococcus lactis* ssp. *lactis* IL1403 during heat, acid, and osmotic stresses. The macroarray accurately detects known stress responses in lactococci. Additionally, expression profile changes of a commercial starter during cheddar cheese ripening can also be profiled.

Key Words: Gene expression; DNA macroarray; DNA microarray; expression arrays; metabolomics; functional genomics; amino acid metabolism; peptide hydrolysis.

1. Introduction

A simple, low-cost oligonucleotide-based filter array protocol in the format of a DNA macroarray is described. The primary features in this protocol are its macroarray format and use of oligonucleotides as probes. The main difference between DNA microarrays and macroarrays is the distance between spots (pitch size). The pitch sizes in DNA macroarrays are in the range of millimeters, instead of micrometers, as in DNA microarrays. The large pitch size eliminates the need for a high-resolution DNA microarray scanner. This DNA macroarray is an ideal choice for arrays with low to medium probe densities, such as metabolic-pathway-focused arrays or bacterial genome arrays. In contrast to cDNA probe arrays that are commonly used in DNA microarrays, oligonucleotide probes are more specific for the target molecules. This difference is critical for examining the biological relevance of the gene expression differences and minimizes false-positive results. Oligonucleotide arrays can differentiate the expression of genes with >90% identity (*1*).

From: *Methods in Biotechnology, Vol. 18: Microbial Processes and Products*
Edited by: J. L. Barredo © Humana Press Inc., Totowa, NJ

Fundamentally, the purpose of genome-wide expression profiling is similar to single gene expression analysis: to determine differential expressions between the control and treatments. However, the method used in statistical analysis of genome-scale expression data is much different from single gene expression analysis because of the massive number of data, the experimental noise, and the potential for a non-normal data distribution. Hence, care must be taken during data analysis steps. Without a firmly grounded analysis technique, biological conclusions derived from DNA array data are misleading at best and false at the worst.

2. Materials

2.1. Enzyme, Chemicals, and Kits

1. Terminal transferase (New England BioLabs, Beverly, MA).
2. dITP (Roche Applied Science, Indianapolis, IN).
3. Lysozyme solution: 50 mg/mL lysozyme (Sigma, St. Louis, MO) and 50 mM EDTA, pH 8.0.
4. RNAqueous kit (Ambion Inc., Austin, TX).
5. RNeasy Mini kit (Qiagen, Valencia, CA).
6. Superscript™ II (Invitrogen, Carlsbad, CA).
7. Random Primer (Invitrogen).
8. RNaseOUT™ (Invitrogen).
9. aa-dNTP stock mixture 20X: 10 mM dATP, dGTP, dCTP, 4 mM dTTP (Invitrogen), 6 mM aminoallyl-dUTP (Ambion Inc.), and 0.1 mM ddATP (Sigma).
10. RNaseH (Epicentre, Madison, WI).
11. Qiaquick-PCR purification kit (Qiagen).
12. PBS (phosphate-buffered saline): phosphate buffer, pH 7.2.
13. Sulfo-NHS-LC-biotin (Pierce, Rockford, IL).
14. SSC 20X: 175.3 g/L of NaCl, and 88.2 g/L of sodium citrate. Adjust to pH 7.0 with a few drops of 1 M HCl.
15. Pre/hybridization solution: 5X SSC, 0.1% N-laurosylsarcosine, 0.02% sodium dodecyl sulfate (SDS), and 1X blocking reagent (VectorLab, San Diego, CA).
16. Washing solution 2X: 2X SSC and 0.01% SDS.
17. Washing solution 0.5X: 0.5X SSC and 0.01% SDS.
18. North2South Hybridization and Detection Kit (Pierce).
19. mRNA transcripts of RCP1, XCP2, LTP4, and LTP6 genes from *Arabidopsis thaliana* (Stratagene, La Jolla, CA).

2.2. Instruments and Accessories

1. Manual arrayer (VP Scientific, San Diego, CA).
2. Positively charged Nylon membrane (Millipore, Bedford, MA).
3. GFA filter paper (Whatman International Ltd., Maidstone, UK).
4. Hybridization oven.
5. Desktop microcentrifuge.
6. Water bath.

7. Thermocycler.
8. Chemiluminescence film (Roche Applied Science).
9. Desktop scanner.
10. Stomacher.
11. Desktop scanner (capable of scanning 16-bit gray-scale image).

2.3. Software

1. ImageJ (http://rsb.info.nih.gov/ij/).
2. SpotFinder, an ImageJ plugin (www.stat.usu.edu/~adele/spotfinder/index.htm).
3. Significance Analysis of Microarrays (SAM), a Microsoft Excel addon (http://www.stat.stanford.edu/~tibs/SAM/index.html).

3. Methods

The methods outlined are described as array construction (*see* **Subheading 3.1.**), RNA collection (*see* **Subheadings 3.2.** and **3.3.**), cDNA production and labeling (*see* **Subheading 3.4.**), hybridization (*see* **Subheading 3.5.**), and data analysis (*see* **Subheading 3.6.**). Additional resources for array construction and data analysis are available commercially as well. Specific insights for the difficult steps, to increase your success, are listed in the Notes.

3.1. Design and Fabrication of an Oligonucleotide-Based DNA Macroarray

3.1.1. Oligonucleotide Probes Designing

A DNA macroarray containing a single probe per gene was designed based on the published *Lactococcus lactis* ssp. *lactis* IL1403 genome sequence *(2)* (*see* **Note 1**). A set of 384 oligonucleotide probes (including controls) was designed based on length (22–24-mer), melting temperature (63–65°C), GC percentage (40–60%), the absence of significant secondary structures, and at least four mismatches with any sequence in the genome. The probe targets were selected based on their roles in protein degradation pathways and carbohydrate, energy, fatty acid, nucleic acid, and amino acid metabolisms *(2)*. The nine control probes include four positive controls designed from the spiked *Arabidopsis thaliana* genes and five negative controls, including an empty spot, a random 15-mer probe, a probe designed from reverse complementary strand of dnaK, and probes from the plasmid-encoded genes *prtP* and *prtM*. Probes were synthesized by Sigma-Genosys (Austin, TX).

3.1.2. Oligonucleotide Probe Tailing

Add a polyinosine (polyI) tail to each oligonucleotide using a terminal transferase. Each tailing reaction mixture is composed of 100 pmol of oligonucleotide, 100 nmol of dITP, 20 U of terminal transferase, 5 mM of $CoCl_2$, 0.2 M

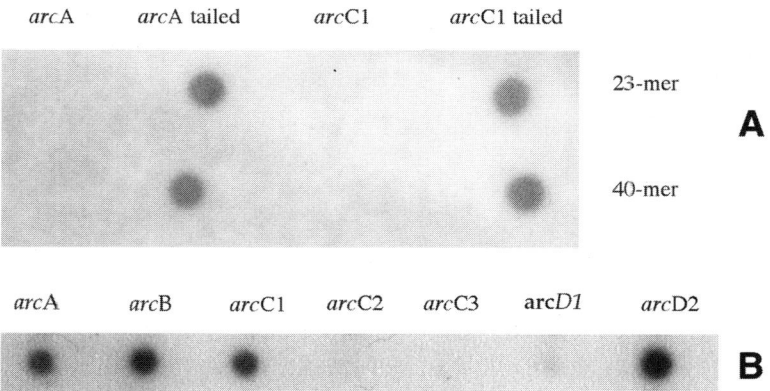

Fig. 1. (**A**) Use of tailing to reduce the length of the probe and (**B**) specificity of particular probes.

of potassium cacodylate, 0.25 mg/mL of bovine serum albumin (BSA), and 25 m*M* Tris-HCl, pH 6.6. Incubate the tailing reaction mixtures at 37°C for 2 h. This modification to the oligonucleotide probes is critical, because it allows us to use shorter probes while maintaining the probe specificity (*see* **Fig. 1**).

3.1.3. Array Printing

1. Spot tailed oligonucleotides in duplicate onto a positively charged Nylon membrane with a manual array device.
2. Fix spotted oligonucleotides by baking the membrane at 80°C for 30 min.
3. Store the printed arrays at room temperature until used.

3.2. Recovery of Bacteria From Laboratory Media or Food

For bacteria grown in laboratory media, a centrifugation at 16,000*g* for 2 min on a desktop microcentrifuge is usually sufficient to collect the sample bacteria from the liquid media. (One DNA macroarray experiment usually requires approx 10^8–10^9 cfu [colony-forming units] of bacteria.) To recover bacteria from complex media like food or cheese, an effective extraction step has to be conducted (*see* **Note 2**).

1. Grind 20 g of cheddar cheese in a motorized mill, add 80 mL of double-distilled water (ddH$_2$O), and mix with a stomacher for 2 min.
2. Filter the mixture with a GFA filter paper to remove fat and other particles and collect the filtrate.
3. After centrifugation at 5000*g* for 5 min at 4°C, proceed for RNA extraction as described in **Subheading 3.3.**

3.3. Total RNA Isolation

3.3.1. Spike in Control mRNAs (Optional)

Add the mRNA transcripts (1 ng each) of *RCP1* and *XCP2* genes from *A. thaliana* before the total RNA isolation and use as the positive controls for RNA extraction.

3.3.2. Using Ambion RNAqeous Kit to Extract Bacterial Total RNA

1. Resuspend the pellet in lysozyme solution and incubate for 10 min at 25°C.
2. After centrifugation at 5000*g* for 1 min, resuspend the pellet in 300 µL of the lysis/binding solution supplied in the RNAqueous kit.
3. Add an equal volume of 64% ethanol to the lysate. Apply the lysate/ethanol mixture to the RNAqueous Filter Cartridge and centrifuge for 1 min. If the filter clogs, apply an additional equal-volume mixture of lysis/binding solution and 64% ethanol to the filter. Discard the fluid that flows through the filter.
4. Wash the filter with 700 µL of wash solution 1 supplied in the RNAqueous kit. Centrifuge for 1 min and discard the fluid that flows through the filter.
5. Wash the filter twice with 500 µL of wash solution 2/3 supplied in the RNAqueous kit. Centrifuge for 1 min and discard the fluid that flows through the filter.
6. Apply 50 µL elution solution (preheated to approx 70–80°C) to the center of the filter. Recover the elutate by centrifugation for 30 s. Repeat with additional preheated 50 µL elution solution. The final total volume is 100 µL.

3.3.3. Cleanup Extracted RNA With Qiagen RNeasy Mini Kit

1. Add 350 µL Qiagen RNeasy Buffer RLT to the total RNA solution of **Subheading 3.3.2.**
2. Add 250 µL ethanol (96–100%) and mix thoroughly by repeated pipetting. Immediately apply the sample to the RNeasy mini column. Centrifuge for 15 s at 10,000*g* in a benchtop microfuge and discard the fluid that flows through the filter.
3. Transfer the RNeasy column into a new collection tube. Pipet 500 µL buffer RPE onto the RNeasy column. Centrifuge for 15 s at 10,000*g* and discard the fluid that flows through the filter. Repeat this step with an additional 500 µL of buffer RPE and centrifuge for 2 min at 10,000*g*. Discard the fluid that flows through the filter.
4. Place the RNeasy column into a new 2-mL collection tube and centrifuge in a microcentrifuge at full speed for 1 min.
5. Transfer the RNeasy column into a new 1.5-mL collection tube. Pipet 50 µL RNase-free water directly onto the RNeasy column silica-gel membrane. Centrifuge at 10,000*g* for 1 min. Repeat this step with an additional 50 µL of RNase-free water. The final total volume will be 100 µL.

3.3.4. Concentrate RNA With LiCl Precipitation

1. Mix the RNA solution with 50 µL LiCl PPT solution supplied in the RNAqueous kit, and incubate at –80°C for at least 30 min.

2. Centrifuge in a microfuge at full speed for 15 min. Carefully remove the supernatant. Wash pellet with equal volume of cold 70% ethanol, recentrifuge, and
   remove the supernatant.
3. Air-dry the pellet for 5–10 min. Do not let the pellet dry completely.
4. Completely dissolve the RNA pellet in 10 µL of RNase-free water.

3.4. Total RNA Labeling With Biotin

Biotin was incorporated into the cDNA with reverse transcription and an
indirect labeling procedure.

3.4.1. Spike in Control mRNAs (Optional)

Spike the mRNA transcripts (1 ng each) of genes *LTP4* and *LTP6* from *A.
thaliana* into the purified total RNA mixture before the RNA labeling reaction
and use as the positive controls for labeling reaction.

3.4.2. Reverse Transcription With Superscript II

1. Prepare priming reaction mixture by adding together 1 µL random primer (3
   µg/µL), 1 µL 20X aa-dNTP, and 100 µg of total RNA. Bring up the final volume
   to 12 µL with RNase-free water (*see* **Note 3**).
2. Heat the mixture to 65°C for 5 min and snap-cool on ice for 5 min.
3. Collect the contents of the tube by brief centrifugation and add 4 µL of 5X first-
   strand buffer, 2 µL of 0.1 *M* dithiothreitol (DTT), and 1 µL RNaseOUT (40 U/µL).
   Mix the contents gently.
4. Add 1 µL of Superscript II (200 U/µL) and mix by pipetting. Incubate the mixture
   at 25 C for 10 min.
5. Incubate the mixture at 42°C for 50 min.
6. Inactivate Superscript II by heating to 70°C for 15 min and cool to room temperature.
7. Add 0.2 µL RNase H (20 U/µL) and incubate 37°C for 20 min.
8. Add 100 µL buffer PB to the above reaction mixture and then apply the mixture to
   QIAquick Spin Column. Centrifuge at 10,000*g* for 1 min and discard flowthrough.
   Wash the column with 0.75 mL 80% ethanol twice. Centrifuge at 10,000*g* for an
   additional 1 min to get rid of the last bit of the ethanol and then elute the purified
   amino-labeled single-strained cDNA (aa-cDNA) with 50 µL PBS buffer, pH 7.2,
   twice. The total elution volume will be 100 µL.

3.4.3. Labeling the aa-cDNA With Biotin

1. Warm Sulfo-NHS-LC-biotin vial to room temperature.
2. Prepare the stock solution in PBS buffer, pH 7.2, at 10 µg/µL. Immediately add 20 µL
   into the purified amino-labeled single-strained cDNA solution from **Subheading 3.4.2.**
3. Incubate the mixture at 25°C for 1 h.
4. Quench the reaction with 6 µL of 4 *M* hydroxylamine solution for 15 min at room
   temperature.
5. Clean up the reaction again with the QIAquick-PCR purification kit. Briefly, add
   600 µL buffer PB and then apply the mixture to QIAquick Spin Column.

Centrifuge at 10,000g for 1 min and discard flowthrough. Wash with 0.75 mL buffer PE twice. Centrifuge at 10,000g for an additional 1 min to get rid of the last bit of ethanol and then elute the biotin-labeled ssDNA with 50 µL Buffer EB twice. The total elution volume is 100 µL.

3.5. Hybridization and Detection

3.5.1. Hybridization and Stringent Wash

1. Prehybridize the DNA macroarrays for at least 1 h at 52°C in a 15-mL Pre/Hybridization solution.
2. At the mean time, heat the biotin-labeled cDNA mixture from **Subheading 3.4.3.** at 100°C for 5 min and snap-cool on ice for 5 min.
3. After prehybridization, mix 10 mL of freshly prepared Pre/Hybridization buffer with labeled DNA targets and add to the hybridization tube to replace the prehybridization solution.
4. After hybridization at 52°C overnight, wash the membrane twice at 50°C for 5 min with a 2X washing solution.
5. Wash the membrane twice with a 0.5X washing solution at 50°C for 15 min.

3.5.2. Detection of Hybridization Signals With Noth2South Hybridization and Detection Kit

1. Equilibrate the North2South blocking buffer and wash buffer to room temperature.
2. Add blocking buffer to generously cover the membrane (0.25 mL/cm² membrane), and incubate at room temperature for 1 h.
3. Add streptavidin–HRP conjugate to the solution at 1 : 300 final dilution and further incubate at room temperature for 15 min.
4. Wash the membrane four times with 1X wash buffer at room temperature. Each wash lasts for 5 min.
5. Incubate the membrane in equilibration buffer (0.25 mL/cm² membrane) for 5 min at room temperature.
6. Prepare chemiluminescent substrate working solution by mixing equal volumes of the luminol/enhancer solution and stable peroxide solution. Add this working solution on the surface of the membrane and incubate for 5 min at room temperature.
7. Drain and blot membrane briefly on clean filter paper. Transfer the moist membrane to a plastic wrap, and carefully remove any trapped air bubbles.
8. Place the membrane into a film cassette and expose the blot to film for 60 min at room temperature; however, this could also be done with an imager if available (*see* **Note 4**). Representative DNA macroarray images acquired from stress treatments of *L. lactis* IL1403 and from starter cultures during cheddar cheese ripening are shown in **Figs. 2** and **3,** respectively.

3.6. Data Analysis of Gene Expression

1. Digitize the exposed films into 16-bit TIFF images with a desktop scanner.
2. Process the images with an ImageJ plugin, called SpotFinder (*3,4*).

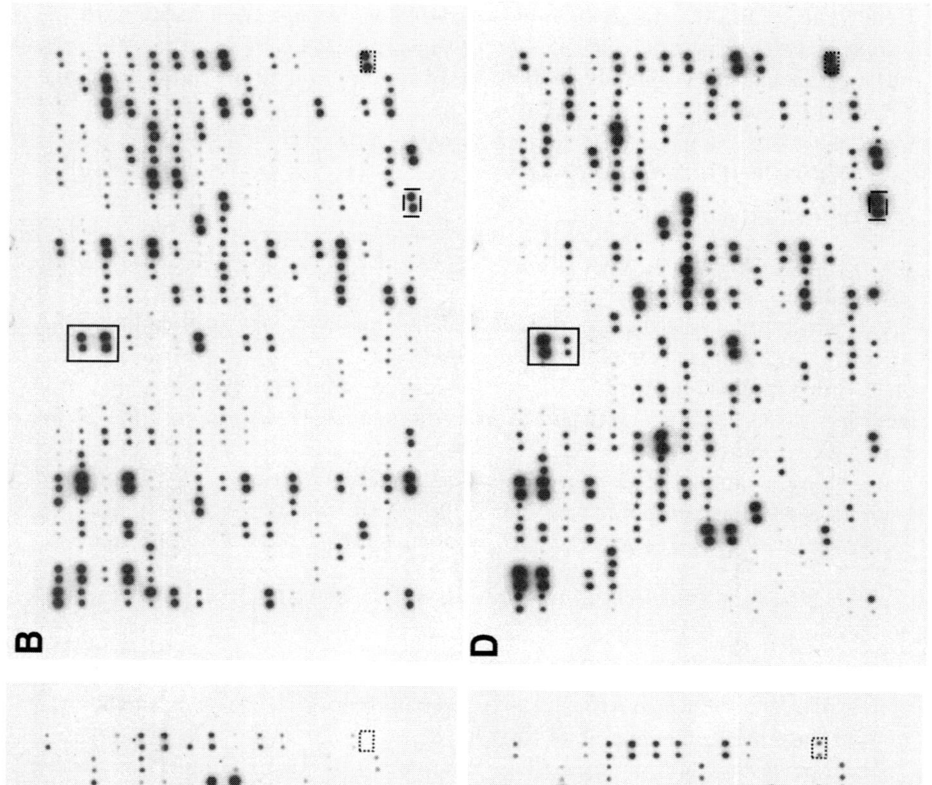

B

D

A

C

Fig. 2.

3. After background correction, calibrate the data from different experiments using an adapted lowess-based procedure *(5)* with a reference dataset generated by averaging the expression intensity of each gene over the control and treatment conditions.
4. Calculate the statistical significance of differential gene expression with a statistical package SAM *(6)* (*see* **Note 5**).

4. Notes

1. Although it is possible to design a large set of oligonucleotide probes manually, automated oligonucleotide design software is preferred and, in some cases, essential. Several free computer programs are specifically designed for this purpose, such as OligoArray (http://berry.engin.umich.edu/oligoarray2/), Probesel (http://www.zaik.uni-koeln.de/bioinformatik/arraydesign.html), and ArrayOligo-Selector (http://arrayoligosel.sourceforge.net/). The oligonucleotide probe set used in this report for *L. lactis* ssp. *lactis* IL1403 was designed with our custom software. It is available as a service for designing of other custom sets of oligonucleotide probes at the Center for Integrated BioSystems (http://www.usu.edu/biotech/).
2. Obtaining high-quality intact RNA is the first and often the most critical step in gene expression profiling. Care must be taken to avoid RNA degradation by RNases during the total RNA extraction. Bacterial mRNA molecules usually have very short half-lives, on the order of a few minutes. Certain genes can be induced or repressed during handling and processing of samples, leading to variable expression levels of these genes. To better preserve the bacterial mRNA abundance levels in the sample, it is often worthwhile to apply RNA stabilization reagents, such as the RNAprotect Bacteria Reagent (Qiagen).
3. Although the aminoallyl-dUTP level in the dNTP pool works reasonable well for many bacteria, some optimization might be needed to achieve even higher biotin incorporation during the total RNA labeling step. This is particularly important for the situation where the bacterial concentration is small or the extraction efficiency is low. In the some extreme cases, dTTP in the dNTP pool might be completely replaced by aminoallyl-dUTP.
4. Because of the limited dynamic range of the X-ray film, the intensities from the DNA array often have a nonlinear relationship with the mRNA abundance levels, which, in turn, makes any biological inference difficult. For example, the same fold changes in the high signal range might not have the same meaning as it does in low range

Fig. 2. Examples of expression profiles of stress responses in *L. lactis* spp. *lactis* IL1403 with a DNA macroarray. **(A)** expression profiles in control condition; **(B)** expression profiles after heat shock; **(C)** expression profiles after acid shock; **(D)** expression profiles after osmotic shock. The single-line box represents (1) *lys*Q (Lys-specific permease), (2) *lys*S (Lys-tRNA synthase); the long dash box represents *ych*H (acetyltransferase); and the short dash box represents *dna*K (heat-shock protein).

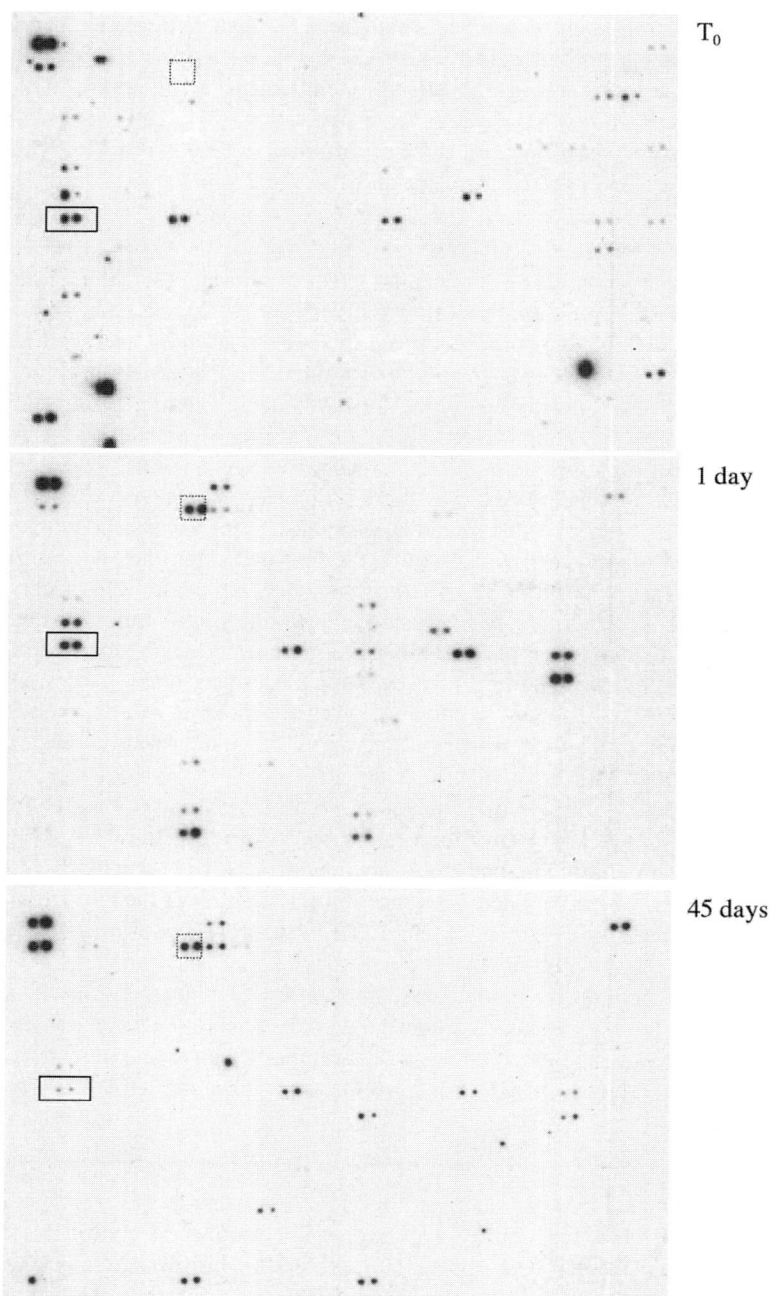

Fig. 3. Expression profile in cheddar cheese at T_0, 1 d and 45 d. The solid-line box is *arc*D1 (Arg/Orn antiport); the dashed-line box is *cit*F (citrate synthase; E.C. 4.1.3.6).

because of signal nonlinearity. Although other free or commercial software can be used for spot detection, a follow-up data linearization step might be required *(4)*. SpotFinder software is specially designed to overcome this problem. With a model-based spot-detection algorithm, this software achieves spot detection and signal linearization in a single step. It is advisable to use this image processing program on the array images derived from chemiluminescence-based or radioactive-based detection.

5. The statistical procedures based on individual gene analysis will result in massive parallel comparisons. This, in turn, will lead to large numbers of false-positive results because of the "multiple testing" problem *(7)*. For example, in an array with 1000 spots, suppose that there are 50 corresponding genes expressed that would be truly different between the control and treatment. With a Type I error rate at 0.05 per comparison, we would expect to detect $50+0.05(1000-50) = \sim 98$ differentially expressed genes; that is, almost 50% of the detected genes would be false positives. There are several statistical software that are designed to handle this issue. SAM is best known for its ability to minimize false positives while maintaining the sensitivity (reduces false negatives). In order to achieve a low false-positive rate, we adopted stringent criteria to assign significance, such as low false discovery rate 1% and a minimal twofold change in expression levels.

Acknowledgments

We thank the Dairy Management, Inc., the Western Dairy Center, the President's Fellowship, and Gandhi Fellowship of Utah State University for support.

References

1. Ramakrishnan, R., Dorris, D., Lublinsky, A., et al. (2002) An assessment of Motorola CodeLink microarray performance for gene expression profiling applications. *Nucleic Acids Res.* **30,** 30.
2. Bolotin, A., Wincker, P., Mauger, S., et al. (2001) The complete genome sequence of the lactic acid bacterium *Lactococcus lactis* ssp. *lactis* IL1403. *Genome Res.* **11,** 731–753.
3. Xie, Y. (2002) Computational design and data analysis in DNA macroarray, MS thesis, Utah State University, Logan, UT.
4. Xie, Y., Cutler, A., and Weimer, B. (2002) Linearization of DNA macroarray data, The 10th International Conference on Intelligent System for Molecular Biology, Edmonton, Canada.
5. Yang, Y. H., Dudoit, S., Luu, P., and Speed, T. P. (2001) Normalization for cDNA microarray data. *Proc. SPIE* **4266,** 19.
6. Tusher, V. G., Tibshirani, R., and Chu, G. (2001) Significance analysis of microarrays applied to the ionizing radiation response. *Proc. Natl. Acad. Sci. USA* **98,** 5116–5121.
7. Dudoit, S., Yang, Y. H., Callow, M. J., and Speed, T. P. (2002) Statistical methods for identifying differentially expressed genes in replicated cDNA microarray experiments. *Statist. Sin.* **12,** 111–139.

20

Optimization of Proteome Analysis for Wine Yeast Strains

Tammi L. Olineka, Apostolos Spiropoulos, Paula A. Mara,
and Linda F. Bisson

Summary

Proteome analysis is a powerful tool for the investigation of yeast in their natural habitats. Two-dimensional gel electrophoresis has been used extensively in proteomics, but reproducibility is notoriously difficult to achieve with this technology. An optimized procedure allowing high reproducibility for the analysis of wine strains of *Saccharomyces cerevisiae* and *Saccharomyces bayanus* has been developed and the critical procedures leading to irreproducibility identified. The steps most impacting gel reproducibility are sample preparation (including accurate protein concentration determination), pouring of the gels, staining, and spot detection. Each of these steps has been optimized for the analysis of the proteome of commercial and native isolates of *Saccharomyces*. Careful attention must also be paid to other steps in the procedure as well.

Key Words: Yeast; proteome; 2D gel electrophoresis; SDS-PAGE; *Saccharomyces;* wine strains.

1. Introduction

Proteomics is the investigation of the expressed protein complement of the genome of an organism. Given the diversity of the physical, chemical, and biochemical properties of proteins, proteome analysis is far more complex than the related transcriptome analyses, which are based on common properties of nucleic acid polymers. The proteome will vary depending on environmental and growth conditions and the specific genotype of the strain; thus, accurate quantitation of relative protein amounts and reproducibility of the protein profile are critical in these studies. Although both transcriptome and proteome technologies provide useful data on the physiology of an organism and its response to environmental change, they present different and complimentary kinds of information *(1)*. Transcriptome analyses define the steady-state levels of mRNA for the condition selected and indicate the transcriptional activity of the organism. Proteome analysis allows determination of the actual protein content and permits

From: *Methods in Biotechnology, Vol. 18: Microbial Processes and Products*
Edited by: J. L. Barredo © Humana Press Inc., Totowa, NJ

visualization of modified protein forms and of active protein degradation. However, proteins of low abundance are challenging to detect.

Our work focuses primarily on the comparative analysis of nonproliferative cells of the yeast *Saccharomyces*. This yeast is being investigated in its natural environment—the fermentation of grape juice. The majority of the yeast life-span in this environment is spent in one of several nonproliferative stages. Our observations confirm those of others that there is not a good correlation between mRNA and protein levels under nongrowing conditions *(2,3)* in contrast to the good correlation found in actively growing cells *(4)*. Under nongrowing conditions, posttranslational regulation is significant, necessitating analysis of the proteins themselves. Further, it is challenging to make mRNA preparations of the requisite purity and stability from yeast cells grown in grape juice, limiting the use of transcriptome technologies in the analysis of yeasts from their native environment. Using published protocols as a starting point, we have optimized several aspects of yeast proteome analysis for use with commercial and native isolates of *Saccharomyces*.

Proteome analysis is commonly performed using two-dimensional polyacrylamide gel electrophoresis (2D PAGE) *(5–7)*. In the first dimension, proteins are separated from complex protein mixtures based on their isoelectric points. During electrophoresis, proteins migrate across a pH gradient until they achieve a neutral charge, which occurs at their isoelectric point. Because the proteins have a neutral charge at this point, further migration does not occur. The first-dimension gels are then loaded onto a second dimension that separates proteins according to molecular weight and Stokes radius. The proteins are then visualized by staining with silver *(8)* or Coomassie blue *(9)*. Protein gels can then be scanned into a gel analytical software package and spots of interest detected. Once areas of interest have been selected, the proteins can be identified using a technique such as peptide mass fingerprinting *(10–12)*. Proteome analysis is best conducted with organisms for which the entire sequence of the genome is known, as is the case for *Saccharomyces (13)*. This permits unequivocal identification of protein spots and is essential for the analysis of organisms from native growing conditions where contaminating microbes might be present.

Proteome analysis via 2D sodium dodecyl sulfate (SDS)–PAGE has been called a "black art" because of the numerous steps involved and the difficulty with reproducibility and quantitation *(14)*. Indeed, careful attention to detail must be paid at all times. However, it is possible to develop procedures and quality control measures that elevate reproducibility to a statistically acceptable range. In general, it is important to use electrophoresis-grade reagents, including ultrapurified water (18 meg Ω). Reagents should be fresh and dated upon receipt so that the shelf life can be monitored. Volumetric flasks can be used to assure accurate volume measurements. Where possible, premixed commercial solutions should be purchased. Many are available from Genomic Solutions Inc. (Ann Arbor, MI), and we have had better reproducibility with these reagents than with making the solutions from individual components. There are steps that need to be optimized for each type of condition to be evaluated. For example, cell breakage is a critical variable in the analysis of yeast strains. Laboratory strains in rich-media conditions break more readily than commercial or native isolates of *Saccharomyces*. In general, exponentially growing cells break more readily than stationary or nonproliferating cells. Cells cultivated in minimal medium are more difficult to break than those in rich media. In our analyses, a synthetic grape juice medium is used. Cells grown in this medium behave as if they are grown in minimal media.

In our experience, better reproducibility is attained using poured rather than precast gels. It is not the chemistry of the precast gels that is problematic but, rather, the exposure of the gels to extremes of temperature during shipping, which might not be detectable until the gels are run. Because multiple researchers in our laboratory must run gels that are ultimately to be compared to each other, it was important to develop a standard, complete protocol and to identify the steps in the procedures most impacting reproducibility and sensitivity. Methods were also evaluated for compatibility with subsequent protein extraction and fractionation for identification by peptide mass fingerprinting. This chapter presents the following methods: (1) cell cultivation, (2) cell extract preparation and protein determination, (3) pouring of first- and second-dimension gels, (4) running of the gels, and (5) gel staining and drying. A sample of the separation achieved is shown in **Fig. 1.**

2. Materials

All chemicals were ACS grade unless indicated otherwise.

2.1. Strain and Growth Conditions

1. *Saccharomyces cerevisiae* or *Saccharomyces bayanus* (ATCC, Manassas, VA).
2. Bottle-top filters, 0.45 µm.
3. Ergo stock: Add 62.5 mg ergosterol, 6.25 mL Tween-80, and 95% ethanol to make a final volume of 25 mL. Store at 4–7°C (*see* **Note 1**).
4. Solution 1: To 250 mL of autoclaved deionized water add 110 g D-fructose, 110 g D-glucose, and 4 mL of ergo stock (swirl).
5. Solution 2: To 200 mL of autoclaved deionized water, add 6.0 g L(+)tartaric acid, 3.0 g L(–)malic acid, and 0.5 g citric acid (anhydrous).
6. Solution 3: To 200 mL of autoclaved deionized water, add 1.7 g yeast nitrogen base (YNB) without amino acids or ammonium sulfate (Difco Laboratories, Detroit, MI), 2.0 g vitamin assay Casamino acids, 6.0 mg myo-inositol, 0.2 g CaCl$_2$ (anhydrous), 0.8 g L-arginine HCl, 1.0 g L-proline, 0.1 g L-tryptophan, and 1.0 g ammonium phosphate.

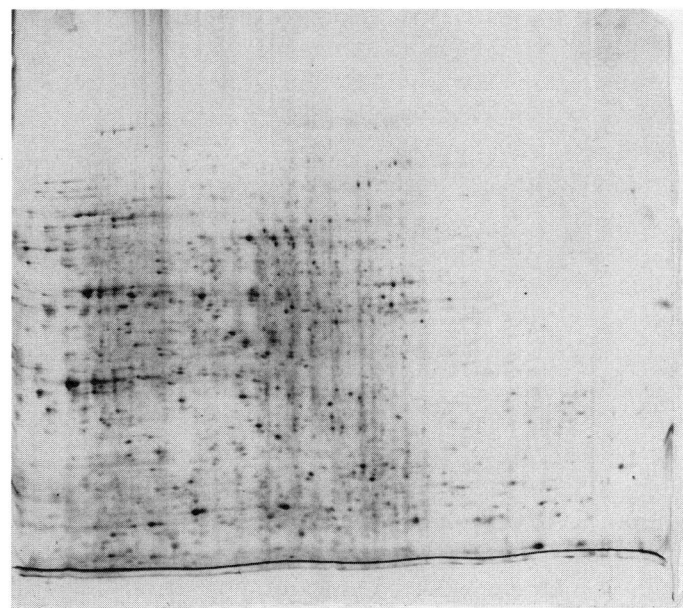

Fig. 1. Typical 2D gel separation. A commercial strain of *S. cerevisiae* was grown in the synthetic juice medium and harvested at the time of transition from active growth to the stationary phase.

7. Triple M synthetic (grape juice) medium: combine solutions 1, 2, and 3 and bring pH to 3.25 using 3 *N* KOH. Bring the volume up to 1 L and then filter-sterilize through a 0.45-µm bottle-top filter (*see* **Note 2**). The final composition is 110 g/L D-glucose, 110 g/L D-fructose, 10 mg/L ergosterol, 1 mL/L Tween-80, 6 g/L L(+)tartaric acid, 3.0 g/L L(−)malic acid, 0.5 g/L citric acid (anhydrous), 1.7 g/L YNB without amino acids or ammonium sulfate (Difco Laboratories), 2.0 g/L Casamino acids, 6 mg/L myo-inositol, 0.2 g/L $CaCl_2$, 0.8 g/L arginine-HCl, 1.0 g/L proline, 0.1 g/L tryptophan (not present in Casamino acids), and 1.0 g/L ammonium phosphate (*see* **Note 3**).
8. YPD medium: 20 g yeast extract, 10 g peptone, 20 g glucose, and 20 g agar. Bring volume to 950 mL with deionized water. Mix separately 20 g glucose in 50 mL deionized water. Sterilize each solution by autoclaving and then mix to final volume of 1 L. For liquid medium, omit agar.
9. Clinitest tablets (Bayer, Elkhart, IN).

2.2. Protein Extraction

1. Beadbeater.
2. Speed-vac.

3. Glass beads 0.5 mm, nitric acid washed.
4. Solution A: 0.5 mL Tris-HCl, pH 8.0 (1 M stock), 0.1 mL MgCl$_2$ (0.5 M stock), 0.2 mL CaCl$_2$ (0.1 M stock), 0.5 μL RNase A, and 1 μL DNase 1. Bring to 1.0 mL with 18 meg Ω water. Store at –20°C. Add 1 μL protease cocktail before use.
5. Thiourea/urea lysis buffer: 1.522 g thiourea, 4.204 g urea, 0.4 g of 3-[(3-cholamidopropyl)-dimethylammonio]-1-propane sulfonate (CHAPS), 0.1 g dithiothreitol (DTT), and 200 μL carrier ampholytes, pH 3.0–10.0. Bring volume to 10 mL with 18 meg Ω water. Store as 1-mL aliquots at –20°C. Add 1 μL protease cocktail per milliliter prior to use.
6. Sample buffer I: 0.3 g SDS, 3.088 g DTT, 0.444 g Tris-HCl, and 0.266 g Tris base. Bring to 100 mL with 18 meg Ohm water. Store as 1.5-mL aliquots at –70°C. The pH of the solution should be 8.0.
7. Protease cocktail: 1 mg/mL leupeptin, 1 mg/mL pepstatin, 10 mg/mL N-tosyl-L-phenylalanine chloromethyl ketone (TPCK), and 231 mg/mL 4-(2-aminoethyl) benzene-sulfonyl fluoride (AEBSF). Store as 100-μL aliquots at –70°C.

2.3. Determination of Protein Concentration Using a Modified Amido Black Method

1. 47-mm Nitrocellulose filter, 0.45-μm pore size.
2. Whatman filter paper 3MM (Whatman, Kent, UK).
3. Disposable glass test tubes (12 × 75 mm).
4. Bovine serum albumin (BSA): 1 mg/mL BSA in 50 mM Tris-HCL, pH 7.5.
5. Stain solution: 9 mL methanol (100%), 2 mL glacial acetic acid (concentrated), 9 mL of 18 meg Ω water, and 0.2 g amido black.
6. Destain solution: 90 mL methanol (100%), 2 mL glacial acetic acid (concentrated), and 8 mL 18 meg Ω water.
7. Eluant: 2.5 mL NaOH (1 N stock), 1 mL EDTA (5 mM stock), 50 mL ethanol (100%), and 46.5 mL 18 meg Ohm water.
8. Trichloroacetic acid (TCA) 50%: 55 mL of 90% TCA and 45 mL of 18 meg Ω water.
9. TCA 6%: 6.5 mL of 90% TCA and 93.5 mL of 18 meg Ohm water.

2.4. Proteomics

Products were purchased from Genomic Solutions unless stated otherwise.

1. Investigator™ 2D electrophoresis system (Genomic Solutions, Ann Arbor, MI).
2. 18-cm × 1-mm Threaded glass capillary tubes.
3. Glass plates.
4. 1-mm Glass spacers (Genomic Solutions).
5. Hamilton syringe (Genomics Solutions).
6. IEF extrusion syringe (Genomics Solutions).
7. Genomic Systems analytical casting solution, snap-cap vial *(15)*: 9.5 M urea, 2.0% Triton X-100, 4.1% acrylamide, and 5 mM CHAPS. Store at –20°C.
8. Ampholytes 3–10.
9. Ammonium persulfate (APS), 10%.

10. NaOH, 1 *N*.
11. Phosphoric acid, 85%.
12. Overlay buffer: 3.0 g urea, 0.2 mL Triton X-100, 0.25 mL ampholytes (40% stock), and 0.77 g DTT. Add 18 meg Ω water to a final volume of 100 mL.
13. Cathode (upper) buffer: 10 mL of 10 *N* NaOH and 990 mL of 18 meg Ω water. Degas solution for 10 min prior to use.
14. Anode (lower) buffer: add 1.4 mL 85% phosphoric acid to 2 L of 18 meg Ω water.
15. Equilibration buffer: 0.3 *M* Tris base, 0.75 *M* Tris-HCl, 3.0% SDS, 0.77 g DTT, and 0.01 g bromophenol blue. Bring to 100 mL with 18 meg Ω water.
16. Glycine gel solution: 300 mL duracryl, 30% with 0.65% BIS (*N,N'*-methylene bis-acrylamide) (premixed solution, Genomic Solutions) *(15)*, 250 mL glycine slab buffer, 418 mL of 18 meg Ω water, 10 mL of 10% SDS (filtered through 0.45 µm), 2.5 mL of 10% APS (fresh), and 500 µL *N,N,N',N'*-tetramethylethylenediamine (TEMED).
17. Glycine slab gel buffer: 130.8 g Tris base and 66.3 g Tris-HCl; mix into 500 mL 18 meg Ω water. Bring to 1 L with 18 meg Ω water. The pH should be 8.7–8.8. Adjust pH with concentrated NaOH or glacial acetic acid, if necessary.
18. SDS overlay buffer: 0.5% agarose, 5 g/L SDS, 72.1 g/L glycine, 15 g/L Tris base, and a few crystals of bromophenol blue.
19. Saturated isobutanol solution, 50%.
20. Running buffer (Tris/glycine/SDS): 25 m*M* Tris base, 192 m*M* glycine, and 0.1% SDS. Use 18 meg Ω water.
21. Standard sealing solution: running buffer containing 0.5% agarose and bromophenol blue (enough to color solution).
22. Sealing solution for plates and spacers: 1.5% agarose in 18 meg Ω water with bromophenol blue (enough to color solution).
23. Loading paper and protein standard (Amersham Biosciences, Uppsala, Sweden).

2.5. Silver Staining

1. Staining trays: white plastic photodeveloper trays, one per gel.
2. Large (2.5 gal) zip-closure bags to fit over trays.
3. Cellophane sheets (Genomic Solutions).
4. Blotter paper sheets.
5. Orbital shaker.
6. Gel dryer attached to a vacuum pump.
7. Latex gloves.
8. Fixer solution: 50% methanol and 12% acetic acid. Use 18 meg Ω water.
9. Wash 1: 50% ethanol.
10. Developer: 0.5 mL formaldehyde (37%), 60 g sodium carbonate, and 4 mg sodium thiosulfate; fill to 300 mL with 18 meg Ω water.
11. Sodium thiosulfate, 0.02%.
12. Silver nitrate, 0.1%.
13. Methanol, 30%.
14. Glycerol, 3%.

2.6. Detection and Gel Analysis

1. Flatbed scanner (with transparency adapter) or Typhoon (Amersham Biosciences).
2. Phoretix software (Nonlinear Dynamics, Newcastle, UK).

3. Methods

3.1. Strain and Growth Conditions

3.1.1. Yeast Strain Handling

We are investigating the yeast *Saccharomyces* under native growth conditions of grape juice. For this purpose, we use natural vineyard isolates or isolates from active dry commercial preparations. In all cases, individual colonies are isolated, streaked to purity, given an accession number, and stored in 15% glycerol at –70°C. This is necessary, as many commercial preparations contain a mixture of genetically related strains that might display differences in protein expression patterns. When evaluating reproducibility of methodology, it is important to use known genetically identical isolates.

3.1.1.1. PREPARATION OF STOCK CULTURES

1. Prepare a stock solution of 60% glycerol and sterilely filter through a 0.45-μm bottle-top filter.
2. Grow an overnight culture (12–18 h) at 25°C and 120 rpm using a well-isolated colony. Use 10–50 mL of YPD liquid medium.
3. Combine 1 part glycerol to 3 parts culture and mix well.
4. Place aliquots of yeast stock culture in 2-mL, sterile, screw-cap cryo-tubes and store at –70°C.

3.1.1.2. REGROWING CULTURES FROM STOCKS

1. Remove cryo-tube from freezer and place on ice. Do not let tube thaw.
2. Using a sterile loop or toothpick, remove some of frozen yeast stock and streak onto a YPD plate.
3. Immediately return cryo-tube to –70°C. Do not use a stock culture tube more than three times.
4. Incubate plate at 30°C for 48 h or until well-isolated colonies appear. Evaluate colonies for uniformity of morphology. If variants are seen, select another stock culture tube and repeat process.

3.1.2. Triple M Synthetic Medium

Grape juice varies dramatically in composition and is an undefined medium. Yeast strains behave differently, depending on the composition of the juice or must (juice plus skins and seeds). Therefore, in the majority of our studies, we use a synthetic juice medium, called "must minimal medium" or "Triple M" *(16,17)*. This medium is low in pH, as is natural grape juice, and high in tartrate, malate, and sugar (glucose and fructose). It also contains ample other macro nutrients and micronutrients and factors needed to promote ethanol tolerance. In its "complete" form, native and commercial isolates readily utilize all of the sugar present, converting it to ethanol and carbon dioxide. However, like native grape juice, the medium is somewhat chemically unstable and tartrate will precipitate. If this occurs, the medium should not be used, as other nutrients are removed in the precipitation process. Unlike natural grape juice, standard Triple M does not contain any phenolic compounds, but these can be added as needed for the study.

3.1.3. Strain Cultivation

An isolated colony is used to inoculate 50 mL of Triple M and grown overnight at 30°C and 200 rpm; then, 250mL of Triple M is inoculated to an optical density (OD_{580}) of 0.03. Samples and optical densities of the culture are taken until the fermentation is complete (sugar content less than 0.5%; checked with Clinitest tablets for residual glucose level).

3.2. Yeast Protein Extraction

The preparation of the crude protein extract is one of the most critical steps in proteome analysis. Environmental and strain background factors influence the amount of breakage of the cells. It is important to assess cell breakage under the microscope and to aim for equivalent levels of disruption. We routinely see 70–80% breakage of the cells. Because the yeast life cycle is asynchronous and budding cells break more readily than nonbudding cells, it is also useful to determine if nonbudded cells have ruptured. It is important to remember that these technologies evaluate the average expression profile of the population of cells. Selective breakage might give a false picture of the population as a whole; therefore, it might also be useful to evaluate the percentages of cells in different stages of the growth cycle before and after breakage.

In working with crude extracts of *Saccharomyces,* it is also important to consider steps that might activate native proteases inside of the cell *(18,19)*. The vacuole contains several proteases that, once released and activated, can degrade proteins in the extract. This degradation is prevented initially by the antiprotease peptides present in the cytoplasm, but the protease–peptide interaction can be disrupted by detergents and other treatments, leading to the activation of the proteases. For this reason, a protease-inhibitor cocktail is added to each solution used in the preparation of the yeast extract. Initially, it is important to monitor the effectiveness of protease inhibition by running a time-course analysis of the same extract and determining if degradation products appear.

1. Sample volume is determined by the OD_{580} of the culture. Place a volume equivalent to 2 OD units in a 2-mL microcentrifuge tube. For example, if the sample OD

is 0.5, 4 mL would be taken and two microfuge tubes would be used with the pellets pooled for washing in **step 2** (*see* **Note 4**).

2. Centrifuge all tubes for 30 s at max speed in the microfuge. Remove the supernatant using a pipettor. Add 2 mL of 18 meg Ω water, vortex briefly, and spin tubes for 30 s at max speed. Wash two times more under the same conditions. Pipet off supernatant. Store at –70°C.
3. Add 20 μL of solution A *(20)* to the yeast cell pellet. Add 50 mg of acid-washed glass beads (0.5 mm).
4. Bead beat each tube for 30 s; repeat five times with 3-min intervals on ice (*see* **Note 5**).
5. Incubate on ice for 10 min.
6. Add 50 μL thiourea/urea lysis buffer *(21)* to each tube.
7. Centrifuge at max speed for 10 min. Transfer the clear supernatant to a sterile microfuge tube.
8. Either store the clear supernatant at –70°C (overnight) or continue with **step 9.**
9. Add 200 μL of acetone to the 70-μL sample and incubate on ice for 2 h.
10. Centrifuge at max speed for 15 min in a microfuge.
11. Dry pellet in a Speed-vac for 5 min (*see* **Note 6**).
12. Add 20 μL of sample buffer I *(15)* (can be purchased from Genomic Solutions) and incubate on ice for 30–60 min.
13. Boil for 5 min and then briefly cool on ice.
14. Add 50 μL thiourea/urea lysis buffer (*see* **Note 7**).
15. Store at –70°C (for up to 6 mo).

3.3. Amido Black Protein Assay

For comparative analyses, it is critical that consistent amounts of protein be loaded onto the gels. We noticed significant problems with reproducibility that appeared to be consistent with a wide variation in protein loading. The linearity of several different protein determination methods using BSA as a standard was therefore evaluated, but with the standard processed via the acetone precipitation steps described earlier. Only one method showed a consistent linear response, the amido black protein assay *(22)*. Substances in our samples interfered with all other protein methods evaluated, including the popular Coomassie brilliant blue staining protocol recommended for other types of sample.

1. Prepare standards and samples in triplicate in 1.5-mL Eppendorf tubes. Standards consist of 0, 5, 10, 15, 20, and 25 μL of a solution of 1 mg/mL BSA in 50 m*M* Tris-HCl, pH 7.5. Samples should consist of 10–20 μL protein per tube. Mix standards and samples into 18 meg Ohm water to a final volume of 270 μL per tube (*see* **Note 8**).
2. Add 30 μL of 10% SDS in 1 *M* Tris-HCl, pH 7.5, to each tube and swirl to mix (do not vortex).
3. Add 100 μL of 50% TCA to each tube, vortex immediately, and let stand at room temperature for 5 min. Only add acid to samples for one filter (6–12 samples per filter) (*see* **Note 9**).

4. Place the nitrocellulose filter onto the filter apparatus and start the vacuum. Mark sample orientation on the filter paper with a pencil. Slowly pipet the contents of one tube onto one spot on the filter paper. Immediately add 300 µL of 6% TCA to the tube and pipet this wash onto the same location that the sample was loaded (*see* **Note 10**).
5. Change the pipet tip and repeat with the next tube.
6. Once 6–12 samples have been placed on the filter, soak the filter in stain solution for 10 min. The stain solution must be made on the day that it is used (make 10 mL at a time).
7. Rinse filter with distilled water.
8. Place filter in destaining solution for three successive destaining baths for a total of 10 min (*see* **Note 11**).
9. Dry filter on 3MM Whatman filter paper. Filters can be stored overnight or longer before elution.
10. Cut out spots and transfer each one to a labeled glass test tube (12 × 75 mm). Add 1.5 mL of eluent to each tube and incubate at room temperature for 20–30 min.
11. Vortex briefly and read the OD_{630}.
12. Generate a standard curve from BSA samples. Use this to convert absorbance readings from each sample to micrograms of protein per tube. Divide this by sample volume for micrograms of protein per microliter.

3.4. Pouring the Gels

3.4.1. First-Dimension Gels

3.4.1.1. PREPARING THE 1D GEL SOLUTION

1. Thaw out 1D analytical frozen casting solution for 1.5–2 h at room temperature or 30–45 min in a 37°C water bath.
2. Pour 1100 mL of 18 meg Ω water into the clean 1D casting cylinder. The water should be 20–25°C to avoid precipitation of urea. Make sure the cylinder base is level.
3. Add 350 µL of ampholyte stock solution to thawed 1D gel solution.
4. Add 40 µL of fresh 10% APS solution to thawed 1D gel solution.
5. Mix the solution completely by inverting the vial several times. Try to minimize bubbles when mixing (do not shake). Work quickly because gel polymerizes rapidly once APS is added.

3.4.1.2. CASTING THE 1D GELS

1. Place the threaded capillary tubes with the shrink wrap facing down into the vial of 1D gel solution. Tilt the vial gently to one side as the tubes are placed into the solution to avoid the formation of air bubbles (*see* **Fig. 2**).
2. Guide the small end of the 1D casting funnel over the tops of the tubes and press the snap-cap vial onto the assembly.
3. Slowly lower the snap-cap vial, tubes, and funnel combination into the casting cylinder so that the water entering through the lower hole in the funnel lays over the gel solution in the vial (*see* **Fig. 3**). Lower slowly to avoid mixing the solution.
4. Allow funnel and vial combination to settle at the bottom of the cylinder.

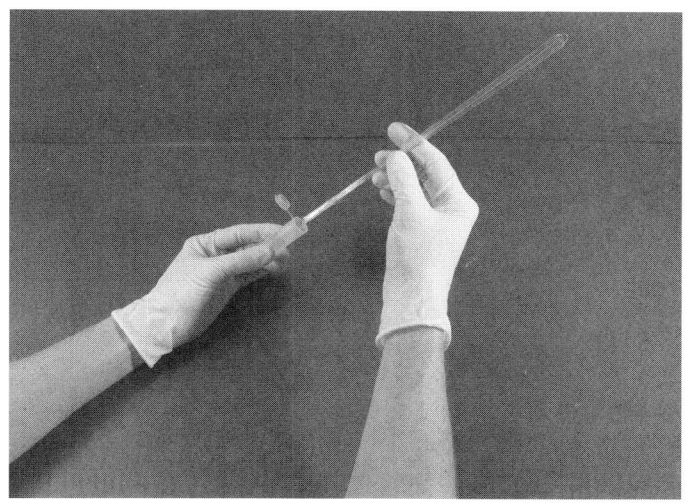

Fig. 2. Tilting the vial of 1D running gel solution to load the tubes.

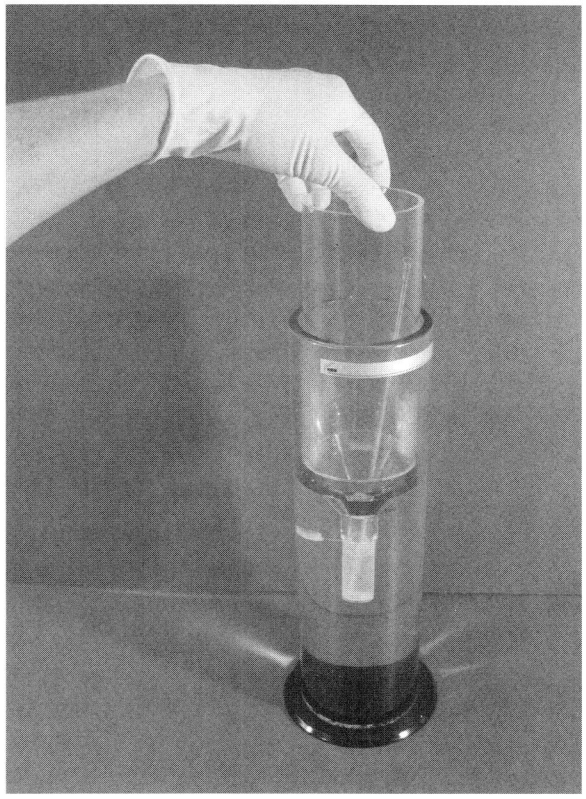

Fig. 3. Casting the 1D gels.

5. Let the gels polymerize for at least 1 h.
6. Slowly lift the funnel and vial combination from the casting cylinder. Allow the water to drain while lifting.
7. Separate the snap-cap vial and tubes from the funnel.
8. Carefully remove the tube bundle(s) and the attached polyacrylamide plug from the snap-cap vial.
9. Separate the tube bundles by pulling each bundle sideways from the other bundle(s). Do not lift the tube bundle upward. This can pull the gel out of the tubes.
10. Wrap the bundles in plastic wrap and store at 4°C for up to 2 mo.

3.4.2. Second-Dimension Gels

3.4.2.1. CLEANING THE PLATES AND SPACERS

1. Soak glass plates and glass spacers overnight in 18 meg Ω water with detergent designed for cleaning electrophoresis glassware. Rinse all plates and spacers thoroughly with 18 meg Ω water and allow to dry. Handle plates and spacers wearing clean gloves at all times.
2. Clean all plates and spacers with 95% ethanol (use Kimwipes).

3.4.2.2. ASSEMBLING THE SINGLE PLATES AND THE CASTING BOX

1. Clean the 2D casting box with 18 meg Ω water and Kimwipes (wear clean gloves).
2. Place the casting box on its back with the front opening facing up and the bottom of the box at the front of the bench.
3. Place a blue polyethylene film (with the longest side horizontally in the box, with 1 cm sticking out of the box). Fold this first sheet in half in order to see the height of the gel solution while pouring the gels.
4. Place a clean dry glass plate with the beveled side at the top and facing up onto your bench with paper towels underneath. Wipe the plate with 95% ethanol and dry with a Kimwipe.
5. Position the spacers flush with the top and sides of the glass plates.
6. Carefully place a second, clean, dry glass plate on the spacers with the top bevel facing down to form a V shape at the top of the plate assembly. This slot will receive the 1D gel. Press firmly on the top plate at the sides (where the spacers are located). Make sure the spacer is still flush at the bottom and the side of the glass plates.
7. Place the complete plate assembly into the casting box with the beveled edge at the top of the box. Place a blue polyethylene sheet on top of the plate assembly (the top should be about 1 cm above the plate assemblies) and repeat the above steps until you have made the number of plate assemblies you wish to pour (the casting box holds six plate assemblies). After all of the plate assemblies have been installed, place two additional spacers on the outside edge of the last assembly. These spacers ensure that pressure is applied to the edges instead of the center of the plates.
8. Fill the rest of the box with clean glass plates. Place blue polyethylene sheets between each plate. Leave enough room for one black foam spacer to bring the assemblies to the edge of the box. When pouring six gels (use 600 mL of gel solution), one extra glass plate, one set of white plastic spacers, and one foam spacer are needed.

9. Make sure that the O-ring is seated well. Clamp the front cover onto the casting box. Clamp opposing sides first and the bottom to ensure that the gasket is compressed uniformly. Make sure that the plates fit snugly but not tightly in the casting box.
10. Stand the casting box upright and level the box. Place the 2D casting cylinder in a ring stand so that the cylinder is at least 30 cm above the bench top.
11. Attach the silicone tubing to the casting cylinder and close it tightly with a clamp.

3.4.2.3. PREPARING THE ACRYLAMIDE CASTING SOLUTION

1. Mix the glycine gel solution with 18 meg Ω water, then degas for 10 min at 25 mg Hg. Six hundred milliliters are required to cast six 2D gels using the Genomic Solutions casting system.
2. Add the appropriate amount of SDS.
3. Immediately before pouring the gels, add the TEMED and APS to initiate gel polymerization (make sure that everything is set up before you do this step).
4. Swirl gently to mix.

3.4.2.4. CASTING SECOND-DIMENSION GELS

1. Pour the acrylamide casting solution into the casting cylinder. Allow bubbles to surface before opening the clamp.
2. Open the binder clamp and allow casting solution to flow through the silicone tubing into a beaker until the tubing is filled with liquid. Immediately connect the 2D casting cylinder to the casting box with the silicone tubing and control the rate of filling with the clamp. Clamp the tube once the acrylamide solution is 0.5 cm below the top of the beveled edge of the glass plate.
3. Using a pipet, carefully layer 0.5 mL of 50% isobutanol onto the surface of the gel in each plate assembly. Do not allow the butanol to mix with the acrylamide solution.
4. Drain remaining casting solution from the casting cylinder into a beaker. Plug the opening on casting box and disconnect the tubing for cleaning.
5. Dispose of the acrylamide after polymerization occurs.
6. After the gels have polymerized, place the casting box on its back. Release the casting box clamps and remove the front cover.
7. While wearing gloves, remove the foam spacer, plastic spacers, polyethylene sheets, and extra glass plates. Any excess polymerized acrylamide should be placed into the beaker.
8. Separate plate assemblies using a wide-blade spatula under the polyethylene sheets. Be very careful at this step. Do not twist the spatula or jab it into the plate assembly accidentally. Discard the polyethylene sheets and rinse the plate assemblies. Rinse the top of the gel plates with 18 meg Ω water and then place each one into a Zip lock bag and cover the gel with 2D upper running buffer.
9. Clean the box and equipment with detergent and 18 meg Ohm water.

3.5. Proteomics (Using the Genomic Solutions System)

The procedure for running the gels followed that recommended by the manufacturer of our 2D gel system, Genomic Solutions. The procedure can be found

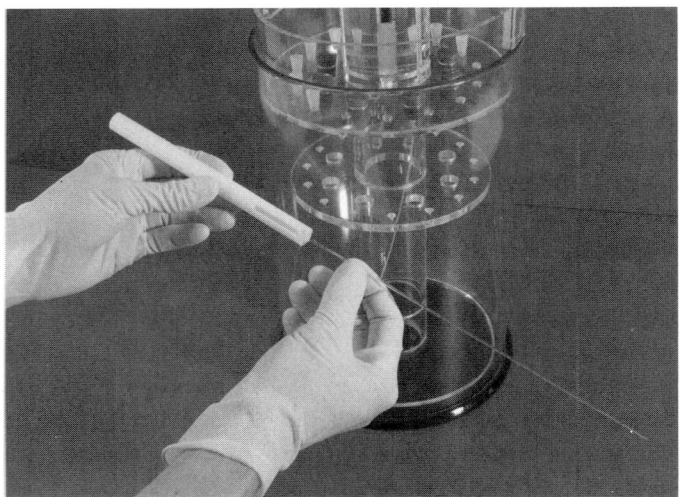

Fig. 4. Sliding grommets onto the top of a 1D gel for loading into the 1D running apparatus.

in their instruction manual and on-line (http://www.genomicsolutions.com) and is repeated here.

3.5.1. Running the First Dimension

3.5.1.1. Preparing the 1D Running System

1. Clean the 1D running tank with18 meg Ω water.
2. Pour 2 L of anode buffer into the lower reservoir of the running tank.
3. Lower the 1D running rack into the anode buffer.
4. Using a red (fine) Sharpie, mark the top of the gel in each tube.
5. Slide a grommet onto the top of each tube (*see* **Fig. 4**).
6. Place the tube with the grommet into the analytical grommet installer tool.
7. Carefully insert the tube and the grommet through the upper reservoir into one of the smaller holes marked 1–15 on the outer ring. Align the tube with the corresponding lower guide hole and slide the tube through the hole into the solution of the lower reservoir.
8. Carefully seat the grommet by using a gentle downward pressure on the grommet installer tool.
9. Repeat this procedure to install all of the tubes into the 1D running system.
10. Plug the eight larger holes in the center of the upper reservoir with preparative plugs. If fewer than 15 analytical tubes were installed in the previous steps, plug the remaining holes with the analytical plugs.
11. Fill the upper reservoir with 800 mL of degassed cathode solution. Make sure that there are no leaks.

3.5.1.2. DEBUBBLING THE TUBES

1. Fill the detachable syringe with the cathode solution.
2. Lower the needle to the gel surface of a tube and expel the cathode solution as you slowly bring the needle up to release all of the bubbles.
3. Repeat the process with all of the tubes.
4. Rinse the syringe with 18 meg Ω water.

3.5.1.3. LOADING THE OVERLAY BUFFER

1. Thaw sample overlay buffer.
2. Fill a 50-µL Hamilton syringe with a permanent 2 in. needle with sample overlay buffer (a disposable tip will introduce bubbles).
3. Lower the needle into the tube as far as it will go and expel 5 µL of overlay buffer. The overlay buffer is denser than the cathode solution and will settle onto the surface of the gel.
4. Repeat the process with all of the tubes.

3.5.1.4. LOADING SAMPLE ONTO FIRST DIMENSION

1. Once protein concentration has been determined, place an aliquot of the sample containing 35 µg of protein in a 500-µL tube. Bring volume up to 40 µL with thiourea/urea lysis buffer.
2. Rinse the 50-µL Hamilton syringe thoroughly with 18 meg Ω water. Draw up 40 µL of sample and slowly lower the syringe as far as it will go into the tube. Slowly expel the sample while raising the syringe (*see* **Note 12**).
3. Rinse the syringe with 18 meg Ω water 20–25 times and then load the next sample. Repeat this process.

3.5.1.5. PRERUNNING THE TUBE GELS

1. Put the lid onto the 1D box and plug the cable into the power supply (*see* **Note 13**).
2. Press FUNCTION SETUP. The following display will appear: Process (1P, 2F, 3S, 4B). Press 2 (for function 2F).
3. At # OF GELS, type in the number of gels being run.
4. MAXIMUM VOLTAGE (V): 250.
5. VHOLD (holding voltage): 125.
6. DURATION (h 0–99): 2 (h).
7. DURATION (m 0–59): 0 (min).
8. MAX CURRENT (µA): 110 (current per gel).
9. VOLT HOURS: 250.
10. Double check the parameters using FUNCTION SETUP.
11. Press START. When PROCESS (1P, 2F, 3S, 4B) appears, press 2.
12. An alarm will sound when the prerun is complete. This indicates that the designated volt-hours have elapsed. Press "." to silence the alarm. Once the alarm is disabled, the holding voltage (125 V) is maintained automatically.
13. Press STOP to turn off the holding voltage. When the PROCESS (1P, 2F, 3S, 4B) display appears, press 2 to turn off power to the tank.

3.5.1.6. PREFOCUSING THE TUBE GELS

1. Press FUNCTION SETUP; select 1 from the PROCESS (1P, 2F, 3S, 4B) display.
2. At # OF GELS, type in the number of gels being run.
3. MAX VOLTAGE (V): 2500.
4. DURATION (h 0–99): 1 (h).
5. DURATION (m 0–59): 0 (min).
6. MAX CURRENT (μA): 110.
7. Double check the parameters using FUNCTION SETUP.
8. Press START. When PROCESS (1P, 2F, 3S, 4B) appears, press 1.
9. An alarm will sound when the prefocus is complete. Press "." to silence the alarm. Prefocusing is complete when the voltage reaches maximum. This usually occurs within 60 min. If the maximum voltage is not reached within this time limit, prefocusing has not occurred.
10. Press STOP. When the PROCESS (1P, 2F, 3S, 4B) display appears, press 1 to turn off power to the tank.

3.5.1.7. FOCUSING THE TUBE GELS

1. Press FUNCTION SETUP; select 2 from the PROCESS (1P, 2F, 3S, 4B) display.
2. At # OF GELS, type in the number of gels being run.
3. MAX VOLTAGE (V): 2500.
4. VHOLD (holding voltage): 125.
5. DURATION (h 0–99): 22 (h).
6. DURATION (m 0–59): 30 (min).
7. MAX CURRENT (μA): 110 (current per gel).
8. VOLT HOURS: 55,000.
9. Double check the parameters using FUNCTION SETUP.
10. Press START. When PROCESS (1P, 2F, 3S, 4B) appears, press 2.
11. An alarm will sound when the prerun is complete. This indicates that the designated volt-hours have elapsed. Press "." to silence the alarm. Once the alarm is disabled, the holding voltage (125 V) is maintained automatically.
12. Press STOP to turn off the holding voltage. When the PROCESS (1P, 2F, 3S, 4B) display appears, press 2 to turn off power to the tank.

During isoelectric focusing, prepare equilibration buffer, upper and lower buffers for second dimension, and pour 2D gels. Preferably, pour the gels the night before or 6–7 h before the first dimension is completed. The 2D gels should sit in the gel box for a minimum of 2 h to make sure they are fully polymerized.

3.5.1.8. PREPARING 1D EQUILIBRATION BUFFER

1. Aliquot 100 mL of premixed equilibration buffer into a clean and sterile 100-mL bottle. Add 0.77 g of DTT and 0.1 g bromphenol blue to the premixed buffer. Mix the solution well.

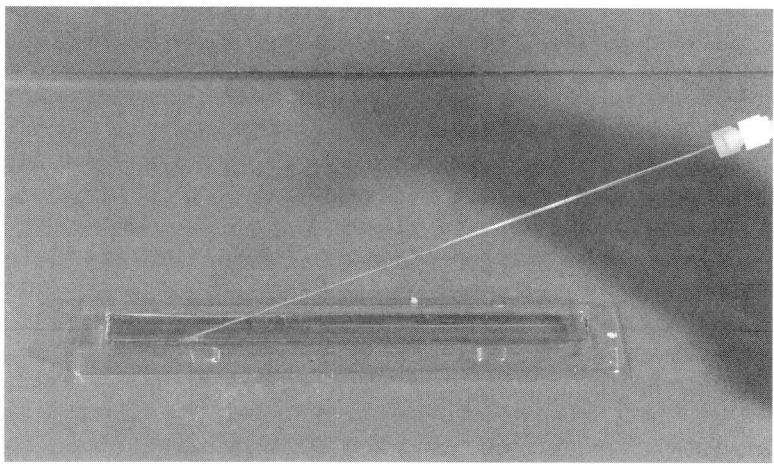

Fig. 5. Extruding the 1D gel.

2. Filter the solution through a 0.45-μm syringe filter. Put 10-mL aliquots into 15-mL polystyrene tubes. Freeze the aliquots at –20°C. To thaw frozen aliquots, let them stand at room temperature for about 2 h or put them into a 37°C water bath.

3.5.2. Setting Up a Second-Dimension Run

3.5.2.1. Preparing the Slab Gel Running Apparatus

1. Clean the 2D running tank and rinse thoroughly with 18 meg Ω water.
2. Place the running tank on the chiller plate with the stir bar and electrode posts on the left side. Connect the power cord to the AC power receptacle at the rear of the chiller.
3. Position a thermometer on the bottom of the running tank.
4. Close the drain valve on the tank by pressing the metal button on the valve until the hose connection pops out 3 mm and the O-ring shows. To open the valve, press the metal button and press the connection together.
5. Fill the lower chamber of the tank with 10 L (for five gels, 11 L for four gels) Tris/glycine/SDS running buffer (lower buffer can be used two to three times if kept chilled).
6. Turn on the chiller and allow buffer to chill overnight to 6–7°C.
7. Prepare 3 L of Tris/glycine/SDS running buffer (must be fresh every time) and chill (4°C) overnight.

3.5.2.2. Extruding the 1D Gels

1. Disconnect the 1D running system from the power supply.
2. Remove the cover of the 1D running assembly by pushing down on the tabs that pass through the cover. Push the tabs until the cover slips off of the inner rack.

3. Remove the tubes from the 1D running system using the grommet installer. Place the tubes on ice. Tubes can stay on ice up to 1 h. Load all gels onto the 2D gels within that hour (*see* **Note 14**).
4. Fill the 1-mL IEF extrusion syringe with 18 meg Ohm water.
5. Fill a holding tray with 1D gel equilibration buffer.
6. Insert the top end (with the thread flush with the top of the tube; it is the anode) of a tube into the gel extrusion adapter.
7. Turn the adapter by hand until it is tight (do not overtighten).
8. Exert constant pressure on the syringe plunger until the acidic end of the gel begins to move out of the tube. Decrease the pressure; slowly extrude the gel into the holding tray with the buffer (*see* **Fig. 5**).
9. Incubate the gel in the equilibration buffer for 2 min at room temperature and then proceed to load it onto a slab gel. Equilibrate each gel in a fresh tray of equilibration buffer. Do not reuse the equilibration buffer.
10. Repeat the above steps until all gels have been extruded, incubated in equilibration buffer, and then loaded onto slab gels.

3.5.2.3. LOADING THE TUBE GELS ONTO THE SECOND-DIMENSION SLAB GELS

1. Load the gel assemblies in the upper running chamber, sealing each into place with a Latex gasket. Fill unused slots with Latex gasket plugs. Use sealing solution for plates and spacers, which contains 1.5% agarose colored with bromphenol blue, to seal where the edge of each plate and spacer meet the gasket. Leaks can also be sealed with 1.5% agarose. Once all plates, gaskets, and plugs are in position, check for leaks with running buffer in the upper chamber (*see* **Note 15**).
2. Place one strip of loading paper with 10 µL of standard in the top left corner of the gel. Overlay the strip with 50 µL of standard sealing solution (running buffer containing 0.5% agarose and bromophenol blue). Pour off any running buffer that might be on the top of the gels before preparing the standards.
3. Use 2D gel running buffer in a squeeze bottle to fill the top of each slab gel.
4. Carefully lift the tube gel by the thread ends (with the anode on your right) (*see* **Fig. 6**).
5. Place the right end of the gel about 0.5 cm from the spacer. Trim the excess thread with scissors. Gently press the tube gel onto the slab gel with the gel installer (make sure there are no bubbles underneath the tube gel). The tube gel must be directly in contact with the slab gel.
6. Repeat **steps 1–5** until all of the gels are loaded.
7. Place the running rack into the 2D tank (the electrode on the left).
8. The running buffer in the lower chamber must reach the fill line (i.e., the electrode height) with the gel box in the tank.
9. Add running buffer to upper chamber to the fill line (approx 3 L).
10. Position the 2D cover with cable on the 2D running system with the cable connection on your left.
11. Seat the cover onto the running rack so the two connecting posts from the running rack and tank mate properly with the cable connection. If the cover is not completely down, no power will be supplied to the tank.

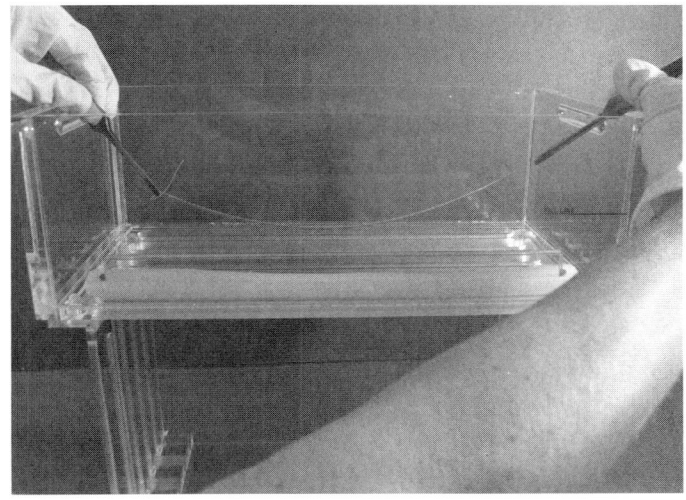

Fig. 6. Loading the 1D gel onto the second dimension.

3.5.2.4. FOCUSING SECOND-DIMENSION GELS

1. Press FUNCTION SETUP; select 3 from the PROCESS (1P, 2F, 3S, 4B) display.
2. SLAB OUTPUT SELECT (1L, 2R, 3B): select 1 for one running tank connected to the left slab output.
3. At # OF GELS, type in the number of gels being run.
4. MAX VOLTAGE (V): 500.
5. DURATION (h 0–99): 6 (h).
6. Double check the parameters using FUNCTION SETUP.
7. Press START. When PROCESS (1P, 2F, 3S, 4B) appears, press 3.
8. Record the temperature reading from the thermometer.
9. When the dye front is 1–2 cm from the bottom of the gel, the gel run is complete.
10. Press STOP. PROCESS (1P, 2F, 3S, 4B) will appear. Press 3 to stop the run.

3.5.2.5. REMOVING AND STORING FOCUSED SECOND-DIMENSION GELS

1. Remove the cover from the tank (*see* **Note 16**).
2. Drain the lower buffer into a container for reuse.
3. Prepare photographic developing trays with 200 mL of fixer solution for each gel.
4. Use the gasket puller to lift and remove the slab gaskets from the plate assemblies to drain the upper buffer.
5. Lift each plate assembly from the running rack.
6. Use a wide-blade spatula to separate a plate assembly by inserting and twisting gently between the two plates next to a spacer.

7. The gel will stick to one plate. Cut a notch at the top by the acidic end of the gel (*see* **Note 17**).

8. Carefully place each gel into a tray containing 200 mL of fixer. The gels can be stained at this time or stored overnight in the fixer solution. If the gels are stored overnight, pour in enough fixer to cover the gel and place the tray into a large labeled Zip lock bag and store in a refrigerator (*see* **Note 18**).

3.6. Silver Staining and Drying of Gels Using a Modified Blum Staining Method

One of the most challenging aspects of proteomics is staining of the gels following electrophoresis. In choosing a staining protocol, it is important to consider what type of further analysis will occur. If protein spots are to be excised, the proteins eluted, and subjected to further analysis, the staining protocol must be compatible with the downstream procedures to be used. We evaluated several different gel-staining protocols with the dual goals of obtaining a maximum number of spots but also being able to conduct peptide mass fingerprinting using matrix-assisted laser desorption ionization–time of flight (MALDI-TOF) mass spectrometry *(23,24)*. We found the method of Blum *(25–27)* to offer the best reproducibility in staining and optimum number of spots, but the method needed to be modified to be compatible with MALDI-TOF analysis. The modifications were to remove the formaldehyde from some of the solutions and to alter the time of some of the incubation periods. It is also important to monitor the staining process so that the gels can be removed if the background begins to stain too darkly. As with all previous procedures, it is important that fresh solutions be used (*see* **Note 19**).

3.6.1. Staining of the Gel

1. A minimum of 200 mL to a maximum of 400 mL of each solution should be used. For all incubation periods, the tray should be placed on an orbital shaker at 120 rpm (*see* **Note 20**).

2. Put gel in a tray with fixer for 1 h or longer on an orbital shaker, 120 rpm.

3. Remove fixer and put gel in wash solution 1 for three times for 20 min each. To change solutions, lift the developing tray from the shaker. Stabilize the gel by pressing it gently against the tray with gloved fingertips. Be sure to touch the gel only below the dye front or along the side with the BSA ladder because even "clean" gloves can leave fingerprints. Tilt the developing tray to pour off the solution. This step should only take a few seconds to pour off as much solution as possible.

4. Remove wash 1 and put gel in 0.02% sodium thiosulfate for 1 min. Rinse gel with three changes of 18 meg Ω water (20 s each).

5. Submerge gel in chilled 0.1% silver nitrate for 20 min at 4°C.

6. Rinse gel with two changes of 18 meg Ohm water (20 s each) (*see* **Note 21**).

7. Put gel in the developer solution (made fresh every time) and shake constantly. Prepare the sodium carbonate–sodium thiosulfate solution on the day it is used.

Add the formaldehyde shortly before using, within minutes if possible. The formaldehyde and developer should be used in a fume hood. As the solution becomes cloudy, replace with a new solution. Make sure all of the gels are developed for about the same time. Developing will take from 5 to 15 min, depending on the temperature of the gel and the developer solution (*see* **Note 22**).

8. When desired intensity is achieved, rinse gel for 20 s with 18 meg Ω water and drain.
9. Add fixer solution for 10 min.
10. Wash in 50% methanol for 20 min or longer (overnight is fine).
11. Place tray with gel and 50% methanol in a large zip-closure bag and store at 4°C or dry immediately (*see* **Note 23**).

3.6.2. Drying of Gels

1. Drain the fixer solution from the gels.
2. Soak the gel in 30% methanol for 30 min at 120 rpm.
3. Soak the gel in 3% glycerol for 30 min with shaking and drain.
4. While the gel is soaking, cut two pieces of blotter paper to fit gel dryer. Shortly before the end of the glycerol soak, completely wet two sheets of cellophane in 3% glycerol.
5. Place one sheet of wet cellophane in the center of the first sheet of blotter paper.
6. Carefully lift the gel from the tray, handling only on the edges that will not be analyzed. Slowly lower the gel onto the wet cellophane. Make sure that the cellophane extends beyond the gel in all directions.
7. Remove bubbles from between the gel and cellophane. Add 3% glycerol to facilitate bubble removal if necessary.
8. Carefully place the second sheet of wet cellophane on top of the gel. Use 3% glycerol to remove bubbles.
9. Place the second sheet of blotter paper on top of the cellophane.
10. Place this "gel sandwich" into the gel dryer and dry under vacuum for 2 h. Turn off the vacuum pump once the run is complete. Let the gel cool for 1 h or overnight before removing from the gel dryer.
11. Using clean gloves, open the blotter paper and trim the cellophane to within 0.5 cm from the edge of the gel.
12. Leave the cellophane attached to the gel and place in a clean zip closure bag.
13. Weight the gel down with books or glass plates to keep it flat until it is scanned.

3.7. Detection and Gel Analysis

Using a high-end flatbed scanner with a transparency adapter or a Typhoon, scan the gels 12-bit gray images at 300 dpi. The gels can be analyzed at this point. Phoretix version 5.1 by Nonlinear Dynamics is used by our lab (*see* **Note 24**).

4. Notes

1. Wear a mask and work in the fume hood for measuring and mixing ergosterol. Do not store ergo stock for more than 1 mo.
2. A bottle-top filter will only filter a max of 2 L of Triple M before clogging.

3. Make, sterilize, and use Triple M within 12 h. Autoclaving can enhance tartrate crystallization and is not recommended. Total assimilable nitrogen of this formulation is 433 mg/L. Nitrogen concentration is adjusted by changing ammonium phosphate and arginine HCl levels. For 208 mg/L nitrogen, decrease ammonium phosphate to 0.5 g/L and arginine HCl to 0.2 g/L. For 123 mg/L nitrogen, decrease ammonium phosphate to 0.1 g/L and arginine HCl to 0.2 g/L.

4. Make sure to wear (clean) gloves. Label tubes clearly and enter all codes into your lab notebook.

5. Be consistent with the bead-beating steps. Store glass beads and acetone at −20°C.

6. Do not dry pellet for longer than 5 min because protein concentration will be decreased.

7. The pellet might not resuspend completely, but it will resuspend after the freeze–thaw cycle in −70°C.

8. Use clean, sterilized, disposable tubes for best results, as reused tubes might contain protein that will be detected in the assay.

9. *Caution:* TCA is a strong acid.

10. Avoid excessive spreading and overlapping of spots. Samples with lower protein concentrations tend to spread more and those with high protein concentrations might plug the membrane.

11. Change the destaining solution when it becomes dark.

12. Careful sample loading is critical for reproducibility. If the sample is expelled too quickly, some of it will escape from the tube and the protein concentration loaded onto the gel will be unknown. Make sure that the syringe is rinsed well between samples.

13. Before starting a run, make sure there are no leaks around the grommets or plugs. Also make sure there are no bubbles in the 1D gel or in the tube above the gel.

14. Always extrude the gels in one direction so that you do not confuse the basic and acidic end. If you are right-handed, extruding toward your left is easiest. The 1D tubes are very fragile and must be handled with care. Using too much force to extrude the 1D gel will cause the tube gel to shoot off of the syringe and possibly break it. Record the order of the 1D gels in a lab notebook as they are placed on ice.

15. Do not use 2D gels with bubbles.

16. To avoid electric shock from the electrophoretic equipment, disconnect the power supply before draining the buffer or removing gels. Set the 2D upper running chamber onto a clean protective surface when loading and removing gels.

17. Keratins are everywhere (skin, hair, clothing, etc.) and will be detected by peptide mass fingerprinting. It is important that contamination of the gel surface be avoided.

18. Make sure that all of the equipment and tools used are cleaned thoroughly before storage.

19. Be careful with the gel; fingerprints can be left on the gel if you press too hard. Again, use clean, gloved hands (keratin contamination is a problem especially if proteins are to be subsequently identified by peptide mass fingerprinting).

20. Filter the developer before formaldehyde is added. Solutions containing formaldehyde must be made fresh the day of use. The Materials Safety Data Sheet for

formaldehyde should be consulted before using this reagent. Dispose of all waste solutions properly.

21. Use a glass container for storing silver nitrate solution.
22. **Step 8** should be stopped when gel is done developing (approx 6–8 min). The developer must be changed when it turns brown. All solution bottles and staining trays must be washed thoroughly after every run.
23. In high concentrations of methanol or ethanol, gels will shrink and turn opaque (white), but they will return to normal in the water wash.
24. The quality of the scan will affect the gel analysis. Consistency during scanning and subsequent analysis of the scanned gel will affect the final results.

References

1. Griffin, T. J., Gygi, S. P., Ideker, T., et al. (2002) Complementary profiling of gene expression at the transcriptome and proteome levels in *Saccharomyces cerevisiae. Mol. Cell. Proteomics* **1(4)**, 323–333.
2. Gygi, S. P., Rochon, Y., Franza, B. R., and Abersold, R. (1999) Correlation between protein and mRNA abundance in yeast. *Mol. Cell. Biol.* **19**, 1720–1730.
3. Mann, M. (1999) Quantitative proteomics? *Nat. Biotechnol.* **17**, 954–955.
4. Futcher, B., Latter, G. I., Monardo, P., McLaughlin, C. S., and Garrels, J. I. (1999) A sampling of the yeast proteome. *Mol. Cell. Biol.* **19**, 7357–7368.
5. Kaltschmidt, E. and Wittmann, H. G. (1970) Ribosomal proteins VII. Two-dimensional polyacrylamide gel electrophoresis for fingerprinting of ribosomal proteins. *Proc. Natl. Acad. Sci. USA* **67**, 1276–1282.
6. Laemmli, W. K. (1970) Cleavage of structural proteins during the assembly of the head of bacteriophage T4. *Nature* **227**, 680–685.
7. O'Farrell, P. H. (1975) High resolution two-dimensional electrophoresis of proteins. *J. Biol. Chem.* **250**, 4007–4021.
8. Switzer, R. C., Merril, C. R., and Shifrin (1979) A highly sensitive silver stain for detecting proteins and peptides in polyacrylamide gel. *Anal. Biochem.* **98**, 231–237.
9. Blackstock, W. P. and Weir, M. P. (1999) Proteomics: quantitative and physical mapping of cellular proteins. *Trends Biotechnol.* **17**, 121–127.
10. Shevchenko, A., Jensen, O. N., Podtelejnikov, A. V., et al. (1996) Linking genome and proteome by mass spectrometry: large scale identification of yeast proteins from two dimensional gels. *Proc. Natl. Acad. Sci. USA* **93**, 14,440–14,445.
11. Larsson, T., Norbeck, J., Karlsson, H., Karlsson, K.-A., and Blomberg, A. (1997) Identification of two-dimensional gel electrophoresis resolved yeast proteins by matrix-assisted laser desorption ionization mass spectrometry. *Electrophoresis* **18**, 418–423.
12. Sagliocco, F., Guillemot, J.-C., Monribot, C., et al. (1996) Identification of proteins of the yeast protein map using genetically manipulated strains and peptide mass fingerprinting. *Yeast* **12**, 1519–1533.
13. Mewes, H. W., Alberman, K., Bahr, M., et. al. (1997) Overview of the yeast genome. *Nature* **387(Suppl.)**, 7–8.

14. Abbott, A. (1999) A post-genome challenge: learning to read patterns of protein synthesis. *Nature* **402**, 715–720.
15. Genomic Solutions (1998) *Genomic Solutions™ Investigator™ 2D Electrophoresis System: Operating and Maintenance Manual,* Genomic Solutions, Ann Arbor, MI.
16. Schultz, M. and Kunkee, R. E. (1977) Formation of hydrogen sulfide from elemental sulfur during fermentation by wine yeast. *Am. J. Enol. Vitic.* **28**, 137–144.
17. Spiropoulos, A., Tanaka, J., Flerianos, I., and Bisson, L. F. (2000) Characterization of hydrogen sulfide formation in commercial and natural wine isolates of *Saccharomyces. Am. J. Enol. Vitic.* **51**, 233–248.
18. Rabilloud, T. (1996) Solubilization of proteins for electrophoretic analysis. *Electrophoresis* **17**, 813–829.
19. Schieltz, D. M. (1999) Preparing 2D protein extracts from yeast, in *2D Proteome Analysis Protocols, Vol. 112* (Link, A. J., ed.), Humana, Totowa, NJ, pp. 31–34.
20. Berckelman, T. and Stestedt, T. (1998) *2D Electrophoresis Using Immobilized pH Gradients: Principles and Methods,* Amersham Pharmacia, Piscataway, NJ, pp. 9–13.
21. Rabilloud, T. (1997) Improvement of solubilization of proteins in two-dimensional electrophoresis with immobilized pH gradients. *Electrophoresis* **18**, 307–316.
22. Weiss, K. C. and Bisson, L. F. (2001) Optimization of the amido black assay for determination of the proteins content of grape juices and wines. *J. Sci. Food Agric.* **81**, 583–589.
23. Nelson, R. W., McLean, M. A., and Hutchens, T. W. (1994) Quantitative determination of proteins by matrix-assisted laser desorption/ionization time-of-flight mass spectrometry. *Anal. Chem.* **66**, 1408–1415.
24. Weiss, K. C., Yip, T.-T., Hutchens, T. W., and Bisson, L. F. (1998) Rapid and sensitive fingerprinting of wine proteins by matrix-assisted laser desorption/ionization time-of-flight (MALDI TOF) mass spectrometry. *Am. J. Enol. Vitic.* **49**, 231–239.
25. Blum, H., Beier, H., and Gross, H. J. (1987) Improved silver staining of plant proteins, RNA and DNA in polyacrylamide gels. *Electrophoresis* **8**, 93–99.
26. Rabilloud, T., Grodard, V., Peltre, G., Righetti, P. G., and Ettori, C. (1992) Modified silver staining for immobilized pH gradients. *Electrophoresis* **13**, 264–266.
27. Rabilloud, T., Carpentier, G., and Tarroux, P. (1988) Improvement and simplification of low-background silver staining of proteins by using sodium dithionate. *Electrophoresis* **9**, 288–291.

21

Bacteriocin-Producing Strains in a Meat Environment

Frédéric Leroy and Luc De Vuyst

Summary

Bacteriocin-producing lactic acid bacteria are promising strains to be used as novel, functional starter cultures for the fermentation of foods, such as fermented sausages. In this chapter, a dual approach is described consisting of an evaluation of the kinetics of the cell growth, lactic acid production, and bacteriocin activity of a bacteriocin-producing *Lactobacillus* strain during in vitro laboratory sausage simulation as well as an *in situ* evaluation using a model sausage system consisting of Petri dishes filled with sausage batter. Both evaluations are complementary and permit one to assess the potential of a candidate strain as a novel starter culture, in particular with respect to its antibacterial bacteriocin activity toward *Listeria*.

Key Words: Bacteriocin; lactic acid bacteria; functional starter culture; food fermentation; fermented sausage; fermented meat; modeling; food safety; *Lactobacillus; Listeria*.

1. Introduction

Lactic acid bacteria represent a major constituent of the starter culture that is used to initiate the fermentation of meat products—in particular, fermented dry and semidry sausages. A recent trend is to be found in the use of functional lactic acid bacterium starter cultures to improve the food fermentation process *(1,2)*.

Generally, *Lactobacillus sakei* and *Lactobacillus curvatus* are used as main acidifiers for the production of European-style fermented sausage, whereas US-style fermented sausages usually rely on pediococci. The starter culture usually also contains *Micrococcaceae* for their role in the reduction of nitrate and for reasons of color stability and aroma production *(3)*. The use of bacteriocin-producing lactic acid bacteria offers perspectives to obtain an enhanced control over the fermentation process because of the improved competitiveness of the producer strain *(4)*. In addition, certain food-borne pathogens present in the meat (e.g., *Listeria monocytogenes*) are inhibited by the bacteriocin producer *(5,6)*. L.

From: *Methods in Biotechnology, Vol. 18: Microbial Processes and Products*
Edited by: J. L. Barredo © Humana Press Inc., Totowa, NJ

monocytogenes is of particular concern because of its high fatality rate and the zero-tolerance policy prevailing in some countries.

Bacteriocins from lactic acid bacteria can be defined as small peptides or proteins that display antibacterial activity toward closely related strains *(7)*. Bacteriocin-producing lactic acid bacterium strains that perform well under optimal laboratory conditions do not necessarily perform well under actual sausage-making conditions. This is ascribed to inactivation by meat proteases, to diffusion limitations *(8)*, or simply because of the incapability of the producer strain to perform well under sausage-making conditions *(9)*. Therefore, if the strain is to be used on an industrial level in meat fermentation, it is essential to investigate whether the bacteriocin producer is well adapted to the sausage environment. This can be performed in several ways: (1) in vitro testing of the functionality of the strain in an appropriate growth medium or a meat simulation medium under optimal conditions in a laboratory fermentor, (2) in vitro testing of the functionality of the strain in a meat simulation medium under meat simulation conditions in a laboratory fermentor, (3) *in situ* testing of the functionality of the strain in a model sausage system (Petri dishes, beaker sausage, etc.), and (4) *in situ* testing of the functionality of the strain in a pilot-plant, including challenge tests.

An optimal experimental design to test bacteriocinogenic starter or cocultures includes measurement of all relevant growth and bacteriocin production characteristics (cell counts, pH decrease, sugar consumption, metabolite production, bacteriocin activities) and incorporation of the necessary controls (nonbacteriocinogenic isogenic mutant strains, pathogenic indicator strains).

In this chapter, the application of sausage simulation techniques for the study and evaluation of bacteriocin-producing lactic acid bacterium strains will be discussed. Because of the difficulty of directly measuring bacteriocin activity in a meat batter, a dual approach is followed. An in vitro simulation (making use of a laboratory fermentor that is filled with a liquid meat simulation medium mimicking the water phase of the sausage) is useful for estimating bacteriocin production under sausage fermentation conditions quantitatively, whereas *in situ* experiments using a model sausage system (e.g., in Petri dishes) permit one to evaluate the actual antilisterial action. The latter method yields an undirect measurement of bacteriocin activity because it compares the antibacterial effect (on a bacteriocin-sensitive indicator strain) of a bacteriocin producer with a comparable bacteriocin-negative control strain and does not involve direct bacteriocin activity measurements. Hence, the two methods are complementary. In addition, model sausage systems such as beaker sausage and pilot-plant trials can be performed as well. The latter are laborious and costly but might be considered as a last step in the evaluation procedure. In any case, challenge tests have to be performed when one wants to use a selected strain in practice.

Lactobacillus sakei CTC 494, an isolate from Spanish naturally fermented sausage, is used here as a case study for the evaluation of its potential to inhibit the food-borne pathogen *Listeria (6,10–13)*. *Lb. sakei* is used instead of *L. sakei* to differentiate *Listeria* species *(L.)*. In principle, any interesting bacteriocin-producing lactic acid bacterium can be evaluated as a potential starter culture for sausage fermentation using this approach. Mainly, the first two methods will be described in detail. For model sausage systems and pilot-plant trials, the reader is referred to the literature *(4–6,14–18)*.

2. Materials

1. 10- to 15-L scale laboratory fermentor permitting *in situ* sterilization (e.g., Biostat C; B. Braun Biotech International, Melsungen, Germany) with computer control of pH, temperature, airflow, substrate feeding, and stirring (e.g., Micro-MFCS for Windows NT; B. Braun Biotech International).
2. High-performance liquid chromatograph (HPLC) with a column suitable for sugar and organic acid separations (e.g., a prepacked column RT 300-7,8 Polyspher OA KC; VWR International, Darmstadt, Germany), a differential refraction detector for detection, and appropriate software for integration and quantification (e.g., Millennium Chromatography Manager; Waters Corporation, Milford, MA).
3. pH electrode suitable for meat stabbing.
4. Stomacher.
5. Filtration bridge with membrane filters (0.45-µm-pore-size filters) to perform biomass measurements as cell dry mass (CDM).
6. Bacteriocin-producing lactic acid bacterium strain (e.g., *Lb. sakei* CTC 494).
7. Bacteriocin-sensitive indicator organism (e.g., *Listeria innocua* LMG 13568).
8. MRS medium: 10 g/L bacteriological peptone, 8 g/L Lab Lemco powder, 4 g/L yeast extract, 20 g/L glucose, 2 g/L dipotassium hydrogen phosphate, 5 g/L sodium acetate trihydrate, 2 g/L triammonium citrate, 0.2 g/L $MgSO_4·7H_2O$, 0.038 g/L $MnSO_4·H_2O$, and 1 mL/L Tween-80 *(19)*. Sterilize at 121°C for 20 min.
9. MSM (meat simulation medium), per 9 L: 290 g meat peptone (e.g., bacteriological peptone; Oxoid, Basingstoke, UK), 232 g meat extract (e.g., Lab Lemco powder; Oxoid), 55.5 g lactic acid, 550 g sodium chloride, 0.10 g sodium nitrite, 2.00 g $MgSO_4·7H_2O$, 0.38 g $MnSO_4·H_2O$, and 10 mL Tween-80. One liter of glucose solution (150 g/L) is added separately to obtain a final volume of 10 L (*see* **Subheading 3.1.2.**).
10. Sausage batter (e.g., Spanish-style fermented dry sausage): 800 g/kg lean pork, 200 g/kg back fat, 25 g/kg sodium chloride, 0.1 g/kg sodium nitrite, 0.3 g/kg potassium nitrate, 0.5 g/kg sodium ascorbate, 1.5 g/kg sodium pyrophosphate, 7.0 g/kg dextrose, 10.0 g/kg lactose, 10.0 g/kg skimmed milk powder, 10.0 g/kg sodium caseinate, 0.02 g/kg Ponceau 4R, 3.0 g/kg black pepper, and 50 g/kg water *(6,18)*. Use fresh, unspoiled meat.
11. Brain–heart infusion (BHI): 12.5 g/L calf brain solids, 5 g/L beef heart infusion solids, 10 g/L proteose peptone, 2 g/L glucose, 5 g/L sodium chloride, and 2.5 g/L disodium phosphate. Sterilize at 121°C for 20 min.

12. Palcam agar: 39 g/L Columbia blood agar base, 3 g/L yeast extract, 0.5 g/L glucose, 0.8 g/L aesculin, 0.5 g/L ferric ammonium citrate, 10 g/L mannitol, 0.08 g/L phenol red, and 15 g/L lithium chloride. Sterilize at 121°C for 20 min. Use a selective supplement to inhibit growth of bacteria other than *Listeria* (10.0 mg polymyxin B, 5.0 mg acriflavine hydrochloride, and 20.0 mg ceftazidime).
13. Agar plates for the determination of *Lb. sakei* CTC 494 and *L. innocua* LMG 13568 cell counts (respectively MRS medium and BHI) and 1.5% of agar. Sterilize at 121°C for 20 min.
14. Bacteriocin indicator plates: BHI and 1.5% of agar. Sterilize at 121°C for 20 min.
15. Soft agar overlays for bacteriocin indicator plates: BHI and 0.7% of agar. Sterilize at 121°C for 20 min and inoculate with the indicator organisms (e.g., *L. innocua* LMG 13568).
16. Sodium phosphate buffer, 50 m*M*, pH 6.5. Sterilize at 121°C for 20 min.
17. Physiological saline solution: 0.85% sodium chloride. Sterilize at 121°C for 20 min.

3. Methods

The methods described outline (1) the evaluation of the strain during an *in vitro* laboratory sausage simulation and (2) an *in situ* evaluation using a model sausage system in Petri dishes.

3.1. In Vitro Sausage Simulation

3.1.1. Maintenance and Propagation of the Strains

Maintain the bacteriocin-producing lactic acid bacterium and the bacteriocin-sensitive *Listeria* indicator strains at −80°C in MRS medium and BHI, respectively, both of which contain 25% (v/v) glycerol as a cryoprotectant (*see* **Note 1**). To obtain fresh cultures, propagate the strains twice in 10 mL of their respective media at 30°C for 12 h before experimental use.

3.1.2. Laboratory Fermentor Experiments

Perform the following steps to start the fermentation process:

1. Prepare the inoculum by inoculating a fresh culture of the lactic acid bacterium to 100 mL of sterile MRS medium and incubate overnight at 30°C (*see* **Note 2**).
2. Prepare 9 L of a MSM that mimics the water phase of a fermented sausage. Add the simulation medium to the fermentor and sterilize the vessel *in situ* at 121°C for 20 min. Use a computer-controlled laboratory fermentor of sufficient size (10 L working volume) to allow extensive sampling and to minimize the volume effect of sampling.
3. Sterilize a glucose solution (150 g in 1 L) separately and aseptically add it to the fermentor, bringing the total fermentation volume to 10 L (*see* **Note 3**).
4. Inoculate the fermentor (1% v/v).
5. Impose a pH profile to the fermentor and control the fermentation pH (to within 0.05 of the set point) by automatic addition of 10 *M* NaOH (*see* **Note 4**). The

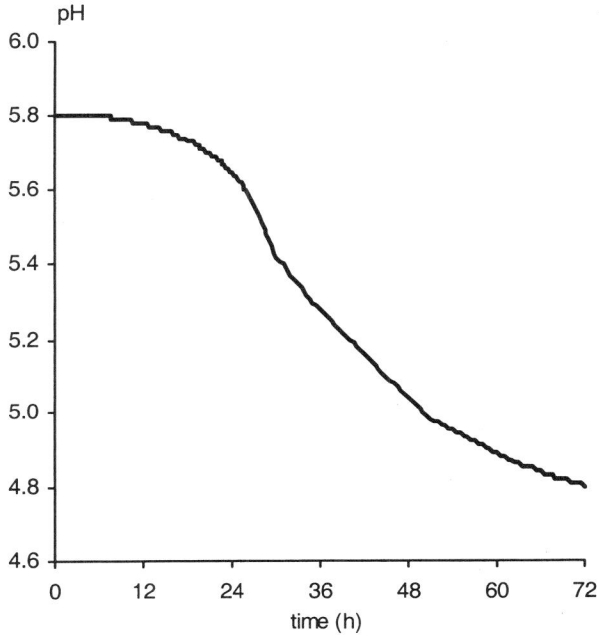

Fig. 1. Example of a pH profile that can be applied to the fermentor to simulate the meat buffering during European sausage fermentation (e.g., Belgian-style fermented sausage).

imposed pH profile will be dependent on the type of sausage studied but should bring about a pH decrease from 5.8–6.0 to 4.8–5.1 in 3 d (*see* **Fig. 1**).

6. Impose a water-activity profile to the fermentor to simulate the drying-in obtained during the fermentation by gradually pumping a sterile salt solution (140 g of sodium chloride in 500 mL of water) into the medium over a period of 3 d (0.12 mL/min). The latter corresponds with a decrease in water activity of 0.1 units, as commonly observed in sausage practice (from 0.958 to 0.948).
7. Control the temperature at a relevant sausage fermentation temperature (e.g., 20°C), preferably within 0.1°C of the setpoint.
8. Perform moderate agitation (50 rpm) of the medium to ensure its homogeneity.
9. Perform the experiments at least in duplicate for statistical reliability.

3.1.3. Sampling and Functionality Measurements

At regular time intervals, aseptically withdraw a sample (e.g., 50 mL) from the fermentor to determine cell growth, bacteriocin activity, organic acid concentration (in particular, lactic acid), and residual glucose concentrations. Determine cell growth as CDM by membrane filtration of a known volume of fermentation liquor, followed by washing of the filter with demineralized water and drying it overnight at 105°C (*see* **Note 5**). Perform microcentrifugation (13,000*g* for 10 min) to obtain cell-free samples, needed for the measurement of lactic acid and residual glucose concentrations by HPLC and for the estimation of bacteriocin activity levels by bioassay.

Perform HPLC as follows:

1. Pretreat cell-free samples with an equal volume of 20% (w/v) thrichloroacetic acid to precipitate proteins, perform microcentrifugation (13,000*g* for 10 min), and filter (0.2 µm) the solution into HPLC vials.
2. Inject 30 µL of the vials into the HPLC column and use a 0.005 *N* H$_2$SO$_4$ solution at a fixed flow rate of 0.4 mL/min as the mobile phase.

Determine the bacteriocin activity in the cell-free culture supernatant by bioassay, using a modified critical dilution method followed by spot-on-lawn *(20)*:

1. Prepare the overlay soft agar for the indicator plates by adding a fresh culture of *L. innocua* LMG 13568 with an optical density (OD$_{600}$) of 0.45- to 3.5-mL bottles containing sterile liquid BHI overlay agar (*see* **Note 6**).
2. Prepare a serial twofold dilution of the cell-free culture supernatant with sterile sodium phosphate buffer.
3. Spot the series with a micropipet (10 µL) onto the indicator plate, which is subdivided into distinct zones corresponding with the dilutions (*see* **Note 7**).
4. Incubate the overlaid agar plates overnight at 30°C (*see* **Note 8**).
5. Express bacteriocin activity in arbitrary units (AU), corresponding with 10 µL of the highest dilution causing a definite zone of inhibition on the lawn of the indicator strain. Activity of the *n*th dilution equals to $2^n \times 100$ AU/mL (*see* **Note 9** and **Fig. 2**).
6. Perform all analyses at least in duplicate.

3.1.4. Modeling of the Fermentation Data

If desired, the data obtained from the fermentations can be modeled. In this way, useful information (e.g., biokinetic parameters) can be obtained, such as the maximum specific growth rate or the specific bacteriocin production of the strain during batch fermentation. The latter parameters are useful for evaluation of the kinetic behavior of the strain or for strain comparison *(9,21)*.

Model the cell concentration [*X*] (in g CDM/L) as a function of time (*t*, in h) with the following equation:

$$dX/dt = \mu_{max}\, \gamma_i X$$

where μ_{max} is the maximum specific growth rate (h^{-1}) and γ_i is a dimensionless inhibition (subscript *i*) function.

For lactic acid bacteria, the logistic equation is frequently used, taking into account a maximum cell concentration for cell growth X_{max} (in g CDM/L) so that *(10,11)*:

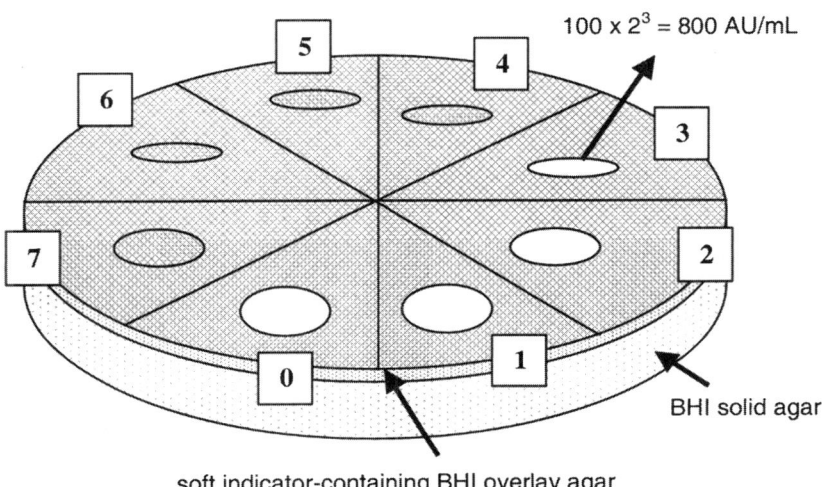

3rd is last clear inhibition zone:

100 x 2^3 = 800 AU/mL

BHI solid agar

soft indicator-containing BHI overlay agar

Fig. 2. Indicator plate for the estimation of bacteriocin activity, obtained by spotting 10 µL of a twofold dilution series and incubating overnight.

$$\gamma_i = 1 - X/X_{max}$$

This equation can be adapted for improved fitting of the growth data of lactic acid bacteria by introducing an inhibition exponent n to the equation *(22)*:

$$\gamma_i = (1 - X/X_{max})^n$$

Alternatively, the Monod equation can be used, but the latter is generally unable to model growth data of lactic acid bacteria because of the complexity of their nutrient requirements *(12)*:

$$\gamma_i = [N]/([N] + K_m)$$

where [N] is the limiting nutrient (in g/L) and K_m is the Monod constant.

A better and more realistic fit can be obtained by using a three-step inhibition function, the so-called nutrient-depletion model *(12)*:

$$\gamma_i = 1 \qquad\qquad\qquad\qquad\qquad\qquad \text{if } [X] < [X_1]$$
$$\gamma_i = 1 - I_1([X] - [X_1]) \qquad\qquad\qquad \text{if } [X_1] \leq [X] \leq [X_2]$$
$$\gamma_i = (1 - I_1([X_2] - [X_1]) - I_2([X] - [X_2]) \quad \text{if } [X] > [X_2]$$

Model the depletion of fermentable sugar [S] (in g/L) and its conversion into lactic acid [L] (in g/L) with the following equations:

$$d[S]/dt = -1/Y_{X/S}d[X]/dt$$
$$d[L]/dt = d[S]/dt$$

where $Y_{X/S}$ is the cell yield coefficient (in g CDM/g glucose), m_S is the cell maintenance coefficient (in g glucose/[gCDM × h]), and $Y_{L/S}$ is the yield coefficient for the production of lactic acid from substrate (in g lactic acid/g glucose).

Model the bacteriocin activity in the cell-free culture supernatant [B], expressed in mega-arbitrary units (MAU or 10^6 AU) per liter, as

$$d[B]/dt = k_B \, d[X]/dt - k_{inact}[X][B]$$

with $k_B = 0$ if $[X] < [X_B]_{min}$ or $[X] > [X_B]_{max}$; k_B is the specific bacteriocin production (in MAU/g CDM), k_{inact} is the apparent bacteriocin inactivation rate (in L/[gCDM × h]), and $[X_B]_{min}$ and $[X_B]_{max}$ are the minimum *(12)* and maximum *(22)* biomass concentration for bacteriocin production, respectively. The parameter k_{inact} indicates "apparent" inactivation of the bacteriocin because of adsorption of the bacteriocin molecules to the producer cells rather than true inactivation. $[X_B]_{min}$ reflects the biomass concentration at which enough induction factor is present to induce the bacteriocin production *(12)*. Bacteriocin biosynthesis is limited by a maximum biomass concentration for bacteriocin production $[X_B]_{max}$, at which a switch-off occurs. The latter has been observed for *Enterococcus faecium* RZS C5 *(22)*.

Use the Euler integration technique to solve the equations using spreadsheet software (e.g., Microsoft Excel) (*see* **Note 10**). Divide the fermentation time into small steps (e.g., 0.25 h) and calculate [X], [S], [L], and [B] for each moment. As an example, express cell growth at moment t_2 as a result of the previous moment t_1: $[X]_2 = [X]_1 + \mu_{max}\gamma_i [X]_1(t_2 - t_1)$.

Minimize the sum of squares between modeled values and experimental values by adjusting the biokinetic parameters for cell growth (μ_{max}, $[X_1]$, $[X_2]$, I_1, I_2), substrate consumption (m_S, $Y_{X/S}$), lactic acid production ($Y_{L/S}$), and bacteriocin activity (k_B, k_{inact}, $[X_B]_{min}$, $[X_B]_{max}$). Retain the parameters that result in the lowest sum of squares and, consequently, in the best fit.

3.2. In Situ *Evaluation Using Petri Dishes*

3.2.1. Model Sausages

1. Prepare the meat batter by mixing the sausage ingredients (e.g., according to a Spanish-style sausage recipe) *(6,18)*.
2. Temper the meat at $-1°C$ to $0°C$.
3. Use Petri dishes as casings and fill each dish with 100 g of meat batter.
4. Add the starter culture (10^6 CFU [colony-forming units]/g of meat batter) and the bacteriocin-sensitive *Listeria* indicator (10^4 CFU/g of meat batter).
5. Incubate the dishes at the sausage fermentation temperature (e.g., 20°C) for 3 d.

Perform the experiment at least in triplicate and calculate the average and standard deviation for the sampling data.

3.2.2. Sampling

1. Sample the model sausages at regular time intervals and use one dish per sample.
2. Measure the pH by inserting a pH electrode into the meat.
3. Aseptically take 10 g of meat from the center of the model sausage and homogenize it with 90 mL of sterile saline solution enriched with bacteriological peptone (1 g /L) for 1 min in a stomacher apparatus.
4. Determine total lactic acid bacteria counts on MRS agar by making a serial 10-fold dilution.
5. Incubate the plates anaerobically at 30°C for 48 h.

Measure the ability of the lactic acid bacteria to produce bacteriocins and, hence, to inhibit the sensitive indicator strain by direct antagonism *(23)*, as follows:

1. Prepare an indicator BHI overlay agar as described in **Subheading 3.1.3.**
2. Overlay the counting plates and incubate overnight at 30°C.
3. Count colonies that are causing a clear inhibition zone in the overlay.

Using the latter method, the competitiveness of the bacteriocin producer can be followed by comparing the counts of bacteriocin-producing colonies with total lactic acid bacterium counts (*see* **Note 11**). Measure bacteriocin production in the meat by monitoring the reduction of cell counts of the sensitive *Listeria* indicator on selective Palcam agar. Incubate the latter plates at 30°C for 48 h. Express *Listeria* and *Lactobacillus* counts as a function of time (*see* **Note 12**).

4. Notes

1. A suitable indicator strain can be obtained by screening the supernatant of the bacteriocin-producing culture toward a series of *Listeria* strains with the spot-on-lawn method (*see* **Subheading 3.1.3.**). For most bacteriocin-producing lactobacilli or pediococci with relevance to meat fermentation, it is relatively easy to find a sensitive *Listeria* strain. A *L. innocua* strain can be used as an indicator instead of *L. monocytogenes* for safety reasons. It has been shown that *L. monocytogenes* is more sensitive to the bacteriocin of *Lb. sakei* CTC 494 than *L. innocua (6)*. Also, the experiment is, in principle, perfectly suitable to test for any desired sensitive strain other than *Listeria*.
2. Choose the inoculation level and the incubation time so that the strain will be actively growing at the moment of inoculation of the fermentor. This will avoid long lag phases. An inoculation of 1% (v/v) and incubation during 12 h at 30°C will generally be acceptable.
3. Glucose is added separately after sterilization. This procedure is followed to avoid Maillard-type chemical reactions of glucose with nitrogen compounds during sterilization at 121°C, which would result in a decrease of its bioavailability.

4. The increase in volume and the amount of Na$^+$ added because of the addition of 10 M NaOH for pH control is negligible. As an alternative, NH$_4$OH can be used.

5. Other possibilities to monitor cell growth include optical density (OD$_{600}$) measurements with a spectrophotometer or classical plate counting. Both methods can be carried out in parallel. The former method gives an immediate idea of the stage of the fermentation process; the latter method reveals the living status of the culture and allows the detection of potential microbial contamination resulting from failed practice.

6. Sterilize the 3.5-mL BHI bottles in the autoclave at 121°C for 20 min. Let them cool down in a water bath to 50°C before inoculation with the indicator strain. Too hot BHI would inactivate the *Listeria* indicator, whereas too cold BHI would lead to solidification of the gel.

7. It is suggested to work in a laminar flow hood to avoid contamination of the indicator plates. It is also critical to avoid mechanical disturbance of the spotted liquid while drying into the agar to ensure sharp, circled inhibition zones. The number of dilutions to be spotted depends on the indicator strain and the activity of the bacteriocin, but normally eight dilutions should be sufficient.

8. Respect the incubation time of the indicator plates for all measurements and assure that the plates are not overincubated to avoid that growth of resistant indicator strains results in turbid inhibition zones.

9. To avoid variability in bacteriocin activity values resulting from subjective interpretation of the inhibition zones, all measurements should be made by the same person. This results in good reproducibility of the activity titres. For zones that are not entirely clear, take the intermediate activity between the highest dilution yielding a clear zone and the next dilution displaying lower clearness, or make appropriate dilutions.

10. Alternatively, if one is familiar with programming, software such as Matlab can be used. When working with Microsoft Excel, the solver function can be used as well to achieve a minimum of the sum of squares.

11. If desired, strain domination of the starter culture can be verified by testing randomly chosen colonies and by identifying the authenticity of the bacterial colonies by plasmid profile analysis according to the method of Anderson and McKay *(24)*. The latter research article offers a detailed overview of the protocol. Alternatively, repetitive polymerase chain reaction techniques can be performed *(25,26)*. In both cases, sample the colonies according to a progressive colony sampling plan based on the accumulated binomial distribution.

12. For comparison with a non-bacteriocin-producing strain (e.g., a classical sausage starter), repeat the experiment using a bacteriocin-negative strain of the same species instead of the bacteriocin producer. The use of an isogenic, nonbacteriocin-producing strain is recommended.

Acknowledgments

The authors acknowledge the Research Council of the Vrije Universiteit Brussel, the Fund for Scientific Research-Flanders (FWO), the Institute for the Promotion of Innovation by Science and Technology in Flanders (IWT), and in particular, the STWW project "Functionality of novel starter cultures in traditional fermentation processes." FL was supported by a grant of the FWO (post-doctoral fellowship).

References

1. De Vuyst, L. (2000) Technology aspects related to the application of functional starter cultures. *Food Technol. Biotechnol.* **38,** 105–112.
2. Leroy, F. and De Vuyst, L. (2004) Lactic acid bacteria as functional starter cultures for the food fermentation industry. *Trends Food Sci. Technol,* **15,** 67–78.
3. Campbell-Platt, G. and Cook, P. E. (eds.) (1994) *Fermented Meats,* Blackie Academic & Professional, London.
4. Vogel, R. F., Pohle, B. S., Tichaczek, P. S., and Hammes, W. P. (1993) The competitive advantage of *Lactobacillus curvatus* LTH1174 in sausage fermentations is caused by formation of curvacin A. *Syst. Appl. Microbiol.* **16,** 457–462.
5. Foegeding, P. M., Thomas, A. B., Pilkington, D. H., and Klaenhammer, T. R. (1992) Enhanced control of *Listeria monocytogenes* by *in situ*-produced pediocin during dry fermented sausage production. *Appl. Environ. Microbiol.* **58,** 884–890.
6. Hugas, M., Garriga, M., Aymerich, M. T., and Monfort, J. M. (1995) Inhibition of *Listeria* in dry fermented sausages by the bacteriocinogenic *Lactobacillus sake* CTC 494. *J. Appl. Bacteriol.* **79,** 322–330.
7. De Vuyst, L. and Vandamme, E. J. (eds.) (1994) *Bacteriocins of Lactic Acid Bacteria: Microbiology, Genetics and Applications.* Blackie Academic & Professional, London.
8. Stiles, M. E. and Hastings, J. W. (1991) Bacteriocin production by lactic acid bacteria: potential for use in meat preservation. *Trends Food Sci. Technol.* **2,** 247–251.
9. Leroy, F., Verluyten, J., Messens, W., and De Vuyst, L. (2002) Modelling contributes to the understanding of the different behaviour of bacteriocin-producing strains in a meat environment. *Int. Dairy J.* **12,** 247–253.
10. Leroy, F. and De Vuyst, L. (1999) Temperature and pH conditions that prevail during the fermentation of sausages are optimal for the production of the antilisterial bacteriocin sakacin K. *Appl. Environ. Microbiol.* **65,** 974–981.
11. Leroy, F. and De Vuyst, L. (1999) The presence of salt and a curing agent reduces bacteriocin production by *Lactobacillus sakei* CTC 494, a potential starter culture for sausage fermentation. *Appl. Environ. Microbiol.* **65,** 5350–5356.
12. Leroy, F. and De Vuyst, L. (2001) Growth of the bacteriocin-producing *Lactobacillus sakei* strain CTC 494 in MRS medium is strongly reduced due to nutrient exhaustion: a nutrient depletion model for the growth of lactic acid bacteria. *Appl. Environ. Microbiol.* **67,** 4407–4413.
13. Leroy, F. and De Vuyst, L. (2003) A combined model to predict the functionality of the bacteriocin-producing *Lactobacillus sakei* strain CTC 494. *Appl. Environ. Microbiol.* **69,** 1093–1099.
14. Berry, E. D., Liewen, M. B., Mandigo, R. W., and Hutkins, R. W. (1990) Inhibition of *Listeria monocytogenes* by bacteriocin-producing *Pediococcus* during the manufacture of fermented semidry sausage. *J. Food Prot.* **53,** 194–197.

15. Sabel, D., Yousef, A. E., and Marth, E. H. (1991) Behaviour of *Listeria monocytogenes* during fermentation of beaker sausage made with or without a starter culture and antioxidant food additives. *Lebensm. Wiss. Technol.* **24**, 252–255.

16. Schillinger, U., Kaya, M., and Lücke, F. K. (1991) Behaviour of *Listeria monocytogenes* in meat and its control by a bacteriocin-producing strain of *Lactobacillus sake*. *J. Appl. Bacteriol.* **70**, 473–478.

17. Campanini, M., Pedrazzoni, I., Barbuti, S., and Baldini, P. (1993) Behaviour of *Listeria monocytogenes* during the maturation of naturally and artificially contaminated salami: effect of lactic-acid bacteria starter cultures. *Int. J. Food Microbiol.* **20**, 169–175.

18. Callewaert, R., Hugas, M., and De Vuyst, L. (2000) Competitiveness and bacteriocin production of *Enterococci* in the production of Spanish-style dry fermented sausages. *Int. J. Food Microbiol.* **57**, 33–42.

19. de Man, J. C., Rogosa, M., and Sharpe, M. E. (1960) A medium for the cultivation of lactobacilli. *J. Appl. Bacteriol.* **23**, 130–135.

20. De Vuyst, L., Callewaert, R., and Pot, B. (1996) Characterisation and antagonistic activity of *Lactobacillus amylovorus* DCE 471 and large scale isolation of its bacteriocin amylovorin L471. *Syst. Appl. Microbiol.* **19**, 9–20.

21. Leroy, F., Degeest, B., and De Vuyst, L. (2002) A novel area of predictive modelling: describing the functionality of beneficial microorganisms in foods. *Int. J. Food Microbiol.* **73**, 251–259.

22. Leroy, F. and De Vuyst, L. (2002) Bacteriocin production by *Enterococcus faecium* RZS C5 is cell density limited and occurs in the very early growth phase. *Int. J. Food Microbiol.* **72**, 155–164.

23. Barefoot, S. F. and Klaenhammer, T. R. (1983) Detection and activity of lactacin B, a bacteriocin produced by *Lactobacillus acidophilus*. *Appl. Environ. Microbiol.* **45**, 1808–1815.

24. Anderson, D. G. and McKay, L. L. (1983) Simple and rapid method for isolating large plasmid DNA from lactic streptococci. *Appl. Environ. Microbiol.* **46**, 549–552.

25. Gevers, D., Huys, G., and Swings, J. (2001) Applicability of rep-PCR fingerprinting for identification of *Lactobacillus* species. *FEMS Microbiol. Lett.* **205**, 31–36.

26. De Vuyst, L., Avonts, L., Hoste, B., et al. (2004) The lactobin A and amylovorin L471 encoding genes are identical, and their distribution seems to be restricted to the species *Lactobacillus amylovorus* that is of interest for cereal fermentation. *Int. J. Food Microbiol,* **90,** 93–106.

22

In Vitro and In Vivo Interactions of Nonpathogenic Bacteria With Immunocompetent Cells

Eduardo J. Schiffrin, Nabila Ibnou-Zekri, Jean M. Ovigne, Thierry von der Weid, and Stephanie Blum

Summary

The intestinal microbiota is a postnatally acquired organ that is composed of a large diversity of bacterial genera and species and has an influence on the physiology of the host, both locally at the intestine and systemically. They perform important functions for the host and can, in turn, be modulated by environmental factors such as nutrition. Specific components of the intestinal microflora, including *Lactobacilli* and *Bifidobacteria,* have been associated with beneficial effects on the host, such as promotion of gut maturation and integrity, antagonisms against pathogens, and immune modulation. Beyond this, the microflora seems to play a significant role in the maintenance of intestinal immune homeostasis and prevention of inflammation. To date, the contribution of the intestinal epithelial cell in the first-line defense against pathogenic bacteria and microbial antigens has been recognized. However, the interactions of intestinal epithelial cells with commensal bacteria are less understood. This chapter intends to summarize some methods that can be used to understand the cellular and molecular mechanisms underlying bacterial modulation of the innate immune response of the host and their contribution to the homeostasis of the immune function.

Key Words: Intestinal microbiota; intestinal epithelial cells; intestinal mucosa; commensals; probiotics; innate immune response.

1. Introduction

1.1. The Intestinal Microbiota Origin and Complexity

Soon after birth, the mammalian intestine becomes heavily colonized by a complex microbiota. Neonatal intestinal colonization has been described, according to traditional bacteriological analysis of the stools, as a dynamic process characterized by a succession of dominant bacteria *(1)*. The type of diet has an influence on the process *(2)*; for example, in breast-fed babies, bifidobacteria are

From: *Methods in Biotechnology, Vol. 18: Microbial Processes and Products*
Edited by: J. L. Barredo © Humana Press Inc., Totowa, NJ

dominant and further diversity only appears after weaning (3). In contrast, formula-fed babies show a higher diversity of bacterial genera and species.

In the last years, the development of new methods is helping to overcome the ignorance of the composition and temporal dynamics of the intestinal microbiota. In fact, the development of nucleic-acid-based methods has significantly contributed to the study of complex bacterial ecosystems (4,5). The analysis of 16S rDNA from human fecal samples revealed that a large proportion of intestinal bacteria have escaped from description so far. The cultivation-independent PCR-TGGE (polymerase chain reaction–temperature gradient gel electrophoresis) or PCR-DGGE (denaturing gradient gel electrophoresis) analysis combined with measurements of ecological diversity can be applied for monitoring physiological changes of the microbiota because of age (6), diet (7), functional conditions, or alterations induced by antibiotics, immunosuppression, or inflammation (5,8).

Recent studies combining these technologies confirm that Bifidobacterium spp. are a dominant component of breast-fed babies from the first days of life. In addition, the new molecular tools also shed light on other major components of the neonatal microbiota such as streptococci, enterococci, Ruminococcus, and Clostridium (6,9). After weaning, major shifts in bacterial components occur, leading to the establishment of a mature type of microbiota, which is stable during the adult life. However, changes can be detected during infections, antibiotic treatments, immunosuppression, or inflammatory bowel disease (10).

Bacterial density and components of the microbiota vary in different parts of the gastrointestinal tract (GIT), showing increasing numbers and diversity from stomach to the mid-gut and distal intestine (11).

Multiple factors play a role in creating different ecological environments along the intestine. Some of them are the type of the mucins produced at different levels of the GIT mucosa, luminal pH, redox potential, and nutrient availability. These and other—yet to be identified—factors generate different habitats and ecological niches in different organs of the GIT.

The normal intestinal microbiota is composed of planktonic bacteria and microorganisms entrapped into the mucus layer overlying the intestinal epithelial cells (IECs) (12). Colonic mucins are "arranged" in a laminated structure alternating sialomucins and sulfomucins. Bacteria are consistently observed within the slowly moving outer layer (13), indicating a role of mucins in the colonization by commensals (14,15). Exceptionally, commensals such as the filamentous segmented bacteria, anchor into the epithelial tissue (13). Furthermore, the "tissular" arrangement of the microbiota in the mucus layer strongly suggests an integrated metabolic activity.

Recognition of specific receptors by commensal microorganisms on mucins or epithelial cells have been proposed, but evidence is rare. Some authors have described *Lactobacillus* adhesins that bind to small-intestine and gastric mucins *(16)*. In contrast, colonization of enteropathogens highly depends on binding to specific receptors on the mucosal surface, either on the IECs or in the mucous layer. Thus, it seems that colonization by commensals occurs in a more "promiscuous" manner.

1.2. Intestinal Commensals, Antipathogenic Activity, and Immune Homeostasis

1.2.1. Competition for Mucosal Receptors and Nonspecific Protection of the Epitheial Surface

There is growing evidence that the indigenous microbiota is important to protect the host against enteropathogens *(17)*. This has been demonstrated by the higher susceptibility of germ-free animals to infections *(18,19)*. Protection of the intestinal mucosa by the commensal bacteria cannot solely be attributed to competitive exclusion of pathogens *(20,21)*; it also implies other underlying mechanisms *(22)* such as metabolic activity and consumption of nutrients in a defined habitat, local production of metabolites *(23)* or antibacterial products *(24,25)*, and protection of epithelial cytoskeleton/tight junctions. These and more have been discussed in the context of colonization resistance to pathogens.

Finally, the intestinal mucus layer overlying the IEC is an anatomical barrier harboring most of the commensal bacteria. The production of intestinal mucus and the thickness of the mucus layer are stimulated by intestinal bacteria, specifically by certain probiotics *(26)*.

1.2.2. Commensal Protection by Immune Modulation of the Host

Another mechanism of host protection against enteropathogens is the modulation of innate (nonspecific) mucosal defenses by commensal bacteria. Although cell-to-cell interaction between commensals and the mucosal cells might be a rare event, a crosstalk between bacteria and immunocompetent cells in the mucosal microenvironment is likely to occur.

Bacterial products, cell wall components, or nucleotides (unmethylated CpG motifs) have been shown to be recognized by mammalian cells (often receptor mediated) and leading to distinct cellular responses. The initial responses to commensals or probiotic bacteria imply a wide array of cellular modifications in IECs and adjacent immune cells. The innate host response can define the type of the subsequent adaptive (antigen specific) immune response. Furthermore, the innate response to components of the microbiota probably plays a role in the maintenance of mucosal tissue homeostasis.

Although molecular basis of host–bacteria interactions are not fully understood, there is clear evidence that commensal bacteria stimulate the development of the mucosal immune system *(27)*. Gut commensal microbes play a role in increasing circulating antimicrobial antibodies and keeping Peyer's patches under permanent activation *(28)*.

Most of the intestinal IgA is reactive with indigenous bacteria *(29)*. Intestinal IgA antibodies are polyreactive or show high crossreactivity against bacterial motifs and they represent a first line of defense against enteropathogens, preventing bacterial attachment to specific intestinal receptors. Protection against pathogens based on IgA secretory activation by commensals has been experimentally demonstrated using segmented filamentous bacteria *(30)* and a combination of bacteroides and clostridium *(31)*.

The central role of the mucosal innate immune response induced by commensal bacteria has gathered increasing attention in the context of general host defenses and immunemodulation. The IEC is now considered an integral part of the mucosal innate response *(32)*. Initial in vitro studies emphasized IEC reaction to different pathogenic challenges *(33)*.

Recent advances in the characterization of microbial–epithelial interactions suggest that the innate epithelial response can also recognize and discriminate between different commensal bacteria. IEC gene activation resulting from commensal bacterial challenge can affect the physiology of nutrient absorption, mucosal defenses, and xenobiotic metabolism *(34,35)*. These observations confirm the front-line role of the IEC in the recognition of the intestinal microbiota. Thus, the IECs are not simply a passive barrier to infection but, rather, constitute a dynamic interphase, which can sense the highly changing environment of the gut and, together with adjacent immune cells, give origin to protective physiological responses.

During the last years, we have aimed to understand cellular and molecular basis of microbial–mucosal crosstalk in order to recognize distinct profiles of stimulatory capacity by commensal and probiotics.

Different in vitro cellular models have been developed, using different levels of complexity (single-cell or coculture-based models) to study nonpathogenic-derived signals and their characteristic responses elicited in human target cells (IECs, peripheral blood mononuclear cells, and dendritic cells). Human in vitro cocultures, produced by the cocultivation of human intestinal epithelial cell lines, such as CaCO-2 or HT-29 cells, and human peripheral blood mononuclear cells (PBMCs) using the transwell culture technique, have been proven useful models to investigate the molecular basis of microbial/epithelial interactions *(36)*. Finally, in vivo gnotobiotic mice (C3H) monocolonized by commensals have been used to study the effect of selected probiotic strains on host immune physiology.

2. Materials

1. Phosphate-buffered saline (PBS) 1X, pH 7.2 (Sigma AG, Basel, Switzerland).
2. Penicillin/streptomycin 10,000 IU/mL, 10,000 UG/mL (Sigma).

3. Dulbecco's modified Eagle's medium (DMEM) glutamine, high glucose (Amimed, Allschwil, Switzerland).
4. Fetal calf serum (FCS), (Amimed).
5. RPMI 1640 (Gibco–Invitrogen, Basel, Switzerland).
6. Nonessential amino acids (Sigma).
7. Human AB serum (Sigma).
8. Gentamycin (Gibco–Invitrogen).
9. Human CaCO-2 epithelial cells (ATCC, Manassas, VA).
10. 25-mm Cell culture insert, 0.4-µm nucleopore size (Becton Dickinson, Basel, Switzerland).
11. Six-well tissue culture plates (Costar, Cambridge, MA).
12. 24-Well tissue culture plate (Costar).
13. Ficoll–Hypaque 1077 (Sigma).
14. Human peripheral blood mononuclear cells purified from buffy coats (Blood Transfusion Center, Lausanne, Switzerland).
15. MultiCell-ERS electrode, voltmeter/ohmmeter (Milian, Geneva, Switzerland).
16. *Escherichia coli* LTH 634 (Institute of Food Technology, Hohenheim University, Stuttgart, Germany).
17. *Lactobacillus johnsonii* La 1 NCC533 (Nestlé Culture Collection; Lausanne, Switzerland).
18. *Lactobacillus sakei* LTH 681 (Institute of Food Technology).
19. *Lactobacillus paracasei* NCC2461 (Nestlé Culture Collection).
20. Brain–heart infusion (BHI) broth (Chemie Brunschwig AG, Basel, Switzerland).
21. LPS, *Escherichia coli* Serotype O111:B4 (Sigma).
22. Micro RNA Isolation Kit (Stratagene, La Jolla, CA).
23. Nucleospin kit (Macherey-Nagel; Oensingen, Switzerland)
24. MuLV reverse transcriptase (Perkin Elmer, Boston, MA)
25. *Taq* polymerase (Perkin Elmer)
26. dNTP nucleotides (Applied Biosystems, Rotkreuz, Switzerland)
27. Specific 3′ primers coding for human cytokines, such as tumor necrosis factor (TNF)-α, interleukin (IL)-1β, IL-8, MCP-1, interferon (IFN)-γ, IL-10, transforming growth factor (TGF)-β and β-actin.
28. Thermocycler (Perkin-Elmer)
29. Agarose gels (BioWhittaker Molecular Applications, Rockland, ME)
30. TBE: Tris-borate EDTA (Life Technologies AG, Basel, Switzerland)
31. TAE: Tris-acetate EDTA (Life Technologies AG)
32. Enzyme-linked immunosorbent assay (ELISA) for TNF-α, IL-1β, IL-10, IL-4, IFN-γ (Diaclone, Besançon, France).
33. C3H/n germ-free mice (Centre National de la Recherche Scientifique, Orleans, France).
34. MRS–agar plates (Chemie Brunschwig AG).
35. Glycerol (Merck, Darmstadt, Germany).
36. GasPak system (AnaeroGen, Hampshire, UK).
37. Polytron (Kinematica, Lausanne, Switzerland).

38. EDTA (Sigma).
39. Bovine serum albumin (BSA) (Sigma).
40. 0.1 *M* Phenylmethylsulfonyl fluoride (Sigma).
41. Soybean trypsin inhibitor (Sigma).
42. Maxisorp ELISA plates (Life Technologies AG).
43. Borate-buffered saline: 100 m*M* NaCl, 50 m*M* boric acid, and 1.2 m*M* $Na_2B_4O_7$, pH 8.2.
44. Biotinylated goat anti-mouse IgA or biotinylated goat anti-mouse IgG1 or IgG2 (Southern Biotechnology Associates Inc., Birmingham, AL).
45. Horseradish peroxidase-labeled streptavidin (KPL, Gaithersburg, MD).
46. TMB microwell substrate (KPL).
47. 1 *M* Phosphoric acid.
48. 3% Isofluorane (Merck).

3. Methods

3.1. Human In Vitro Coculture Model to Study Nonpathogenic Bacteria-Epithelial/Immune Cell Interactions

3.1.1. CaCO-2 Cell Culture

1. Seed human enterocytelike CaCO-2 cells at a density of 2.5×10^5 cells/mL on a 25-mm cell culture insert (0.4 µm nucleopore size).
2. Place the inserts into six-well tissue culture plates and culture for 18–22 d at 37°C and 10% CO_2 in DMEM (glutamine, high glucose) supplemented with 20% decomplemented FCS (56°C, 30 min), 1% MEM nonessential amino acids, 0.1% gentamycin, and 0.1% penicillin/streptomycin.
3. Change cell culture medium every second day until the cells are fully differentiated (d 21).
4. Determine continuously the transepithelial electrical resistance (TEER) in confluent CaCO-2 monolayers using a MultiCell-ERS electrode (voltmeter/ohmmeter).

3.1.2. Isolation of Human Peripheral Mononuclear Cells

1. Purify human peripheral blood mononuclear cells (PBMCs) from healthy volunteers from buffy coats using Ficoll–Hypaque gradient centrifugation (500*g*, 30 min).
2. Harvest PBMCs from the interface and wash five times with RPMI 1640.
3. Dilute at 2×10^6 cells/mL in RPMI 1640 containing 20% decomplemented human AB serum (56°C, 30 min) and gentamycin (150 µg/mL).

3.1.3. Bacteria and Culture Conditions

1. Grow aerophilic nonpathogenic *E. coli* LTH 634 in brain–heart infusion (BHI) broth at 37°C.
2. Grow *L. johnsonii* La 1 NCC 533, a human intestinal isolate, in MRS broth without acetate at 37°C.
3. Grow *L. sakei* LTH 681 isolated from fermented food in MRS broth without acetate at 30°C.

4. Harvest all bacteria by centrifugation (3000*g*, 15 min) after 24 h of cultivation, at stationary growth phase.
5. Wash the bacteria three times with PBS (1X, pH 7.2) and subsequently dilute to obtain final cell densities of 1×10^6 and 1×10^7 CFU/mL in RPMI 1640 medium.

3.1.4. CaCO-2/Leukocyte Coculture Model

1. Wash tissue culture inserts covered with CaCO-2 cell monolayers two times with prewarmed RPMI 1640 medium.
2. Transfer to six-well tissue culture plates.
3. Add 2 mL culture medium to the apical and basolateral compartment of the transwell cell culture system. In the CaCO-2/leukocyte cocultures, add 2 mL of freshly purified PBMCs (2×10^6 cells/mL) to the basolateral compartment of the culture plates.
4. To stimulate CaCO-2/leukocyte cocultures with nonpathogenic bacteria, challenge the apical surface of the CaCO-2 cell monolayers by the addition of 1×10^6 or 1×10^7 CFU/mL of either nonpathogenic *E. coli, L. johnsonii, L. sakei,* or LPS (1 µg/mL, Serotype O111:B4), respectively. Use pro-inflammatory cytokines, such as TNF-α (100 ng/mL) or IL-1β (1 µg/mL) as positive control. Culture medium without any stimulatory agent is recommended as a negative control, as some human PBMCs can be prestimulated. In the case of aerophilic *E. coli,* add gentamycin (150 µg/mL) to the apical medium to prevent bacterial outgrowth during the experiment.
5. To control the migration of immunostimulatory bacterial products, add LPS (0.1 µg/mL, 1 µg/mL, and 100 µg/mL; Serotype O111:B4) or gentamycin-treated bacteria (1×10^6 or 1×10^7 CFU/mL) to the cell culture inserts (0.4 µm pore size) in the absence of CaCO-2 cells. Measurement of the transepithelial electrical resistance (TEER, Ω cm^2) is a simple way to control the intact barrier function of CaCO-2 cell monolayers.
6. After stimulation for 6–36 h (37°C, 10% CO_2), depending on the cytokine/chemokine to be investigated, collect CaCO-2 cells from the inserts, wash with cold PBS (1X, pH 7.2) and lyse in denaturation buffer for total RNA extraction from Micro RNA Isolation Kit.
7. Store cellular lysates at –20°C.
8. For production of cytokines/chemokines, collect cell culture supernatants by aspiration from both apical and basolateral compartment (do not pool samples) and freeze at –20°C until further analysis. Note that the basolateral supernatant has to be centrifuged (350*g*) to avoid contamination with immune cells.

With this technical approach, our group recently demonstrated that IECs can recognize nonpathogenic bacteria in the presence of PBMCs. Furthermore, a characteristic response to a given nonpathogenic commensal strain can be observed distinguishing between two major cytokine/chemokine responses of intestinal epithelial cells (IECs). Nonpathogenic Gram-negative *E. coli* and certain Gram-positive *lactobacilli (L. sakei)* trigger a NF-κB-mediated inflammatory response

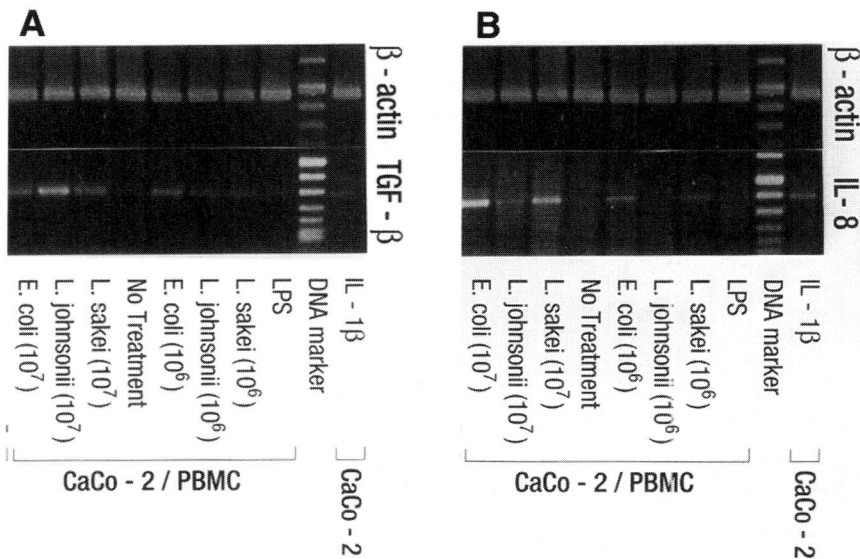

Fig. 1. (**A**) Upregulation of TGF-β mRNA in leukocyte-sensitized CaCO-2cells by *L. johnsonii* La1. Reverse transcription (RT)–PCR analysis of TGF-β-specific gene transcripts in CaCO-2 cells after stimulation of CaCO-2/leukocyte cocultures with nonpathogenic *E. coli*, *L. johnsonii* La1, and *L. sakei* (16 h, 10^6 and 10^7 CFU/mL). (**B**) Differential induction of IL-8 by nonpathogenic bacteria in leukocyte-sensitized CaCO-2cells. Determination of specific gene transcripts for IL-8 in CaCO-2 cells after stimulation of CaCO-2/leukocyte cocultures with nonpathogenic *E. coli*, *L. johnsonii* La1, and *L. sakei* (16 h, 10^6 and 10^7 CFU/mL). Controls for **A** and **B**: LPS (1 μg/mL), IL-1β (10 ng/mL), no treatment (medium). Results represent one of three independent experiments.

resulting in the production of TNF-α, IL-1β, IL-8, and MCP-1 (*see* **Fig. 1**). This initial epithelial pro-inflammatory reaction is only transient and self-limiting, as the PBMCs in the basolateral compartment are able to switch off the initial inflammatory response of the epithelium.

Other *Lactobacillus* strains, including *L. johnsonni* La1 and *L. gasseri* (both human intestinal isolates), induce immunoregulatory cytokines, such as TGF-β, in IECs without prior pro-inflammatory activation (*see* **Fig. 2**). TGF-β is produced by both immune and nonimmune cells and exhibits a broad range of functions, the most important being the modulation of immune responses. In the intestinal immune system, TGF-β displays an important role in the maintenance of intestinal barrier integrity and induction of oral tolerance.

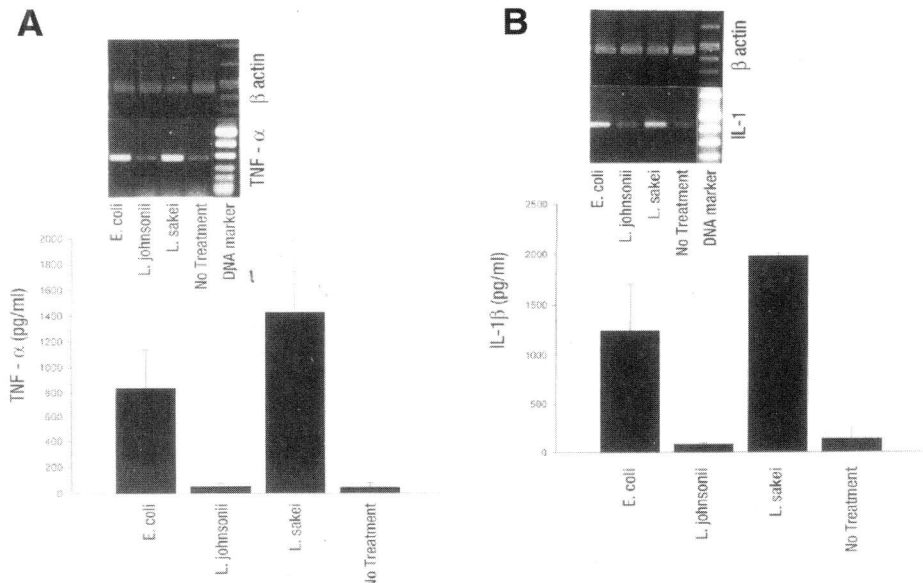

Fig. 2. Differential induction of pro-inflammatory cytokines TNF-α and IL-1β in CaCO-2/leukocyte cocultures by different nonpathogenic bacteria. Stimulation of CaCO-2/leukocyte cocultures with nonpathogenic *E. coli, L. johnsonii* La1, and *L. sakei* (16 h, 10⁶ CFU/mL). Secretion of TNF-α **(A)** and IL-1β **(B)** into the basolateral compartment was determined by ELISA technique (bar chart, pg/mL). RT-PCR analysis was used to determine the expression of TNF-α- **(A)** and IL-1β- **(B)** specific gene transcripts in CaCO-2 cells. Values are given as mean ± SD of triplicates.

3.1.5. RNA Extraction and Amplification by Reverse Transcription–Polymerase Chain Reaction

1. Isolate total RNA from CaCO-2 cells or PBMCs by the acid guanidinium thiocyanate/phenol/chlorophorm method using the Micro RNA Isolation Kit.
2. For semiquantitative reverse transcription–polymerase chain reaction (RT-PCR) analysis, incubate total RNA (0.5 µg), 1 mM of each dNTP, 2.5 U/µL MuLV reverse transcriptase, and specific 3′ primers coding for human cytokines, such as TNF-α, IL-1β, IL-8, MCP-1, IFN-γ, IL-10, TGF-β, and β-actin at 42°C for 30 min.
3. Perform PCR amplification in a total volume of 50 µL using *Taq* polymerase and specific cytokine primers using the following protocol: initial denaturation at 94°C for 1 min, 30 cycles of denaturation (94°C, 1 min), primer annealing (60°C, 1 min), and extension (72°C, 1 min).
4. Analyze PCR products by gel electrophoresis on 2% agarose gels containing 1X TBE, staining with ethidium bromide.

3.1.6. ELISA

Quantify TNF-α, IL-1β, IL-10, IL-4, and IFN-γ in cell culture supernatants by ELISA using the appropriate kit. Human in vitro IEC/PBMCs cocultures can be used to characterize early cellular events underlying the epithelial response to bacteria-derived signals. Both types of epithelial response against nonpathogenic bacteria indicate the importance of either a self-limiting or noninflammatory cellular immune response in the context of the antigen-rich intestinal environment. Thus, certain components of the enteric flora could contribute to maintain a low level of "physiological" intestinal inflammation, whereas others directly promote the production of immunoregulatory cytokines. These examples of an integrated epithelial/lamina propria response strongly suggest a role for the commensal microflora in homeostatic responses. Furthermore, these results provide direct evidence on the beneficial effects of specific probiotic strains on intestinal immune homeostasis (*see* **Note 1**).

3.2. Monocolonization of Rodents With Specific Commensals

3.2.1. Animals

1. Maintain 6-wk-old C3H/n germ-free mice in flexible film isolators.
2. Screen mice weekly to check germ-free status, sampling fecal samples and culturing them on MRS–agar plates under aerobic and anaerobic conditions.
3. At time 0, germ-free mice receive a single gavage of 10^9 CFU of bacteria (monoassociation), and colonization is readily detectable after 24 h in the feces. Control animals include ex-germ-free conventionlized animals by rearing germ-free mice with conventional mice, and conventional mice.
4. Sacrifice mice by cervical dislocation under 3% isoflurane anaesthesia.

3.2.2. Bacteria

L. johnsonii NCC533 and *L. paracasei* NCC2461 were originally isolated from human feces. *L. johnsonii* NCC533 belongs to the *L. acidophilus* group, is catalase negative, and produces DL-lactic acid. *L. paracasei* NCC2461 belongs to the *L. casei* group, is catalase negative, and produces L-lactic acid. Both strains grow well in vitro in MRS broth without acetate, at 37°C, under anaerobic conditions and in the presence of bile salts. Both strains resist simulated gastric juice conditions. For the preparation of gavages:

1. Grow the lactobacilli in MRS broth without acetate at 37°C, under anaerobic conditions in a GasPak system for 8 h (exponential growth phase).
2. Harvest bacteria by centrifugation at 4000g for 10 min.
3. Wash pelleted bacteria twice in sterile 0.9% NaCl and resuspend at 10^9 CFU in 0.2 mL, as estimated by plating dilutions.
4. Inoculate the gavage bacteria intragastrically with sterile 22-gage stainless-steel feeding needle.

3.2.3. Colonization and Translocation of Bacteria

1. Measure at different times after gavage with *L. johnsonii* or *L. paracasei* the translocation of the bacteria to Peyer's patches (PPs) and mesenteric lymph nodes (MLNs) and the density of colonization in the small intestine and colon. Use four mice per group and repeat the experiments three times. At each time-point, PPs are also sampled from control conventional mice to ensure that PP sampling is done sterily (PPs are found to be sterile in conventional mice).
2. Remove PPs and MLNs aseptically, wash five times in PBS and homogenize in 2 mL of sterile PBS–15% glycerol.
3. Plate 100 µL of each of the homogenates on MRS–agar plates. The final dilution on the plate is therefore 1 : 20, and the detection limit is 20 CFU/organ.
4. Flush intestinal colon contents with sterile PBS and mix.
5. After 5 min of sedimentation to remove large remains, collect bacteria from supernatants, wash, and resuspend in 10 mL of PBS–15% glycerol.
6. Plate 100 µL of the bacterial suspension on MRS–agar plates. Therefore, the initial dilution on the plate is 10^{-2} and the detection limit is 10^2 CFU/g.
7. Weight the luminal content in order to express bacterial levels in CFU per gram of luminal suspension.
8. Collect feces sterily from the rectum and resuspend in 5 mL of saline (10^{-2} dilution).
9. Plate serial dilutions of all samples on MRS–agar and incubate anaerobically at 37°C for 48 h.
10. Record colony counts.

3.2.4. Analysis of Intestinal Secretory IgA

1. Cut small intestines (including jejunum and ileum) into small pieces and homogenize mildly with a Polytron in 3 mL of PBS solution containing 0.01% soybean trypsin inhibitor, 36 mM EDTA, and 1% bovine serum albumin.
2. Collect supernatants after centrifugation at 2.000g for 10 min. at 4°C.
3. After adding 0.1 M phenylmethylsulfonyl fluoride, collect intestinal contents from the supernatants following ultracentrifugation at 10,000g for 20 min. at 4°C.
4. Freeze at –20°C.

3.2.5. Peyer's Patches Organ Culture

The method for PPs organ culture is the described by Klaasen et al. *(37)*.

1. Dissect PPs from the small bowel of mice and place in cold RPMI–10% fetal calf serum (FCS).
2. Wash the PPs three times with fresh medium.
3. Cut each PP in half with a sterile razor balade.
4. Place in a sterile flat-bottomed well of a 24-well plate in 2.0 mL of RPMI–10% FCS.
5. Culture for 10 d under 5% CO_2 at 37°C.

3.2.6. Antibacterial Antibodies Analysis

1. Detect *Lactobacillus*-specific immunoglobulins by coating bacteria crude lysate at 100 µg/mL in borate-buffered saline 3 h at 37°C on Maxisorp ELISA plates.
2. After four washes in PBS containing 0.05% Tween-20, block wells overnight at 4°C with PBS containng 20% FCS and 0.05% Tween-20.
3. Incubate samples for 4 h at 4°C.
4. After washing, add secondary biotinylated monoclonal antibodies (biotinylated goat anti-mouse IgA or biotinylated goat anti-mouse IgG1 cr IgG2) at a concentration of 0.5 µg/mL and incubate for 1 h at room temperature.
5. Wash wells, inoculate 1 µg/mL of horseradish peroxidase-labeled streptavidin, and incubate for 30 min at room temperature.
6. Wash the plates one last time and reveal with TMB microwell substrate.
7. Stop reactions with 1 *M* phosphoric acid and measure the optical density (OD_{450}).

With this methodology *(38)*, we have been able to show that for two different *Lactobacillus* species, in vitro cell systems that have shown similar behavior can induce a different immune stimulation in vivo (*see* **Note 2**). *L. johnsonii* was a stronger stimulus for IgA antibody production by PPs in culture and *L. johnsonii*-colonized mice produced higher levels of specific antibodies of the IgG2a and IgG1 isotypes with a predominance of the IgG1 isotype. *L. casei* induced lower levels of local and systemic antibodies with predominance in serum of the IgG2a isotype. The differential isotype induction by *L. casei* and *L. acidophilus* suggests a preferential Th1 and Th2 immune stimulation, respectively.

4. Notes

1. The in vitro epithelal/PBMCs coculture system has been useful to demonstrate the central role that immune cells play in the epithelial response to nonpathogenic bacteria and bacterial factors. Moreover, it has been shown that IECs recognize, in a distinct manner, different components of the commensal flora. As a consequence, it is possible to foresee the use of specific commensal/probiotcs for modification of the mucosal physiology.
2. The in vivo germ-free/gnotobiotc mouse model has permitted one to show that *Lactobacillus* strains with apparently similar properties in vitro could have distinct patterns of colonization and could induce diverse immune responses in vivo. Previous work has shown that colonization of germ-free mice with different combinations of commensal bacteria results in a rapid appearance of IgA+ plasma cells in the lamina propria *(39,40)*. It was apparent from those studies that certain components of the intestinal microbiota, such as segmented filamentous bacteria, had a major stimulatory capacity. IgA production in the intestine has become a major outcome to assess the effect of the microbiota on the immune system. IgA production at the mucosal sites could result from B1- or B2-lymphocyte secretory activity. B1 cells reside in the peritoneal cavity, differ from conventional B-cells (B2) in their phenotype, and are responsible for the production of multireactive antibodies that are not somatically

mutated and are directed against ubiquitous bacterial epitopes of the normal gut microbiota such as phosphoryl choline *(41)*. In contrast, B2-cells are replenished by bone marrow precursors, the antibody molecules involve somatic mutations, and the antigenic specificity is narrower than that of the B1 cells *(42)*. There is not still a clear indication whether one of the B-cell types is the main source of IgA antibodies stimulated by commensal bacteria *(43)*. The present methodology permits one to go beyond previous questions because the possibility to assess the type of humoral systemic response by the different strains of *Lactobacillus* gives a clear indication of the type of T-cell response generated by each bacterium (Th1 or Th2).

References

1. Machie, R. I., Sghir, A., and Gaskins, H. R. (1999) Developmental microbial ecology of the neonatal gastrointestinal tract. *Am. J. Clin. Nutr.* **69,** 1035S–1045S.
2. Edwards, C. A., Rumney, C., Davies, M., et al. (2003) A human flora-associated rat model of the breast-fed infant gut. *J. Pediatr. Gastroenterol. Nutr.* **37,** 168–177.
3. Martin, F., Savage, S. A. H., Parret, A. M., Gramet, G., Doré, J., and Edwards, C. (2000) Dynamics of colonisation of the colon in breast-fed infants. *Reprod. Nutr. Dev.* **40,** 180–185.
4. Tannock, G. W. (1999) Analysis of the intestinal microflora: a renaissance. *Antonie Leeuwenhoek* **76,** 265–278.
5. Vaughan, E. E., Schut, F., Heilig, H. G. H. J., Zoetendal, E. G., de Vos, W. M., and Akkermans, A. D. L. (2000) A molecular view of the intestinal ecosystem. *Curr. Issues Intest. Microbiol.* **1,** 1–12.
6. Favier, C. F., Vaughan, E. E., de Vos, W. M., and Akkermans, A. D. L. (2002) Molecular monitoring of succession of bacterial communities in human neonates. *Appl. Environ. Microbiol.* **68,** 219–226.
7. Tannock, G. W., Munro, K., Harmsen, H. J. M., Welling, G. W., Smart, J., and Gopal, P. K. (2000) Analysis of the fecal microflora of human subjects consuming probiotic products containing *Lactobacillus rhamnosus* DR 20. *Appl. Environ. Microbiol.* **66,** 2578–2588.
8. Zoetendal, E. G., Akkermans, A. D., and de Vos, W. M. (1998) Temperature gradient electrophoresis analysis of 16SrRNA from human fecal samples reveals stable and host-specific communities of active bacteria. *Appl. Environ. Microbiol.* **64,** 3854–3859.
9. Schwiertz, A., Gruhl, B., Lobnitz, M., Michel, P., Radke, M., and Blaut, M. (2003) Development of the intestinal bacterial composition in hospitalized preterm infnats in comparison with breast-fed, full-term infants. *Pediatr. Res.* **54,** 393–399.
10. Schiffrin, E. J. and Blum, S. (2002). Interactions between the microbiota and the intestinal mucosa. *Eur. J. Clin. Nutr.* **56(Suppl. 3),** S60–S64.
11. Marteau, Ph., Pochart, Ph., Doré, J., Béra-Maillet, C., Bernalier, A., and Corthier, G. (2001) Comparative study of bacterial groups within the human cecal and fecal microbiota. *Appl. Environ. Micorbiol.* **67,** 4939–4942.
12. Stoodley, P., Sauer, K., Davies, D. G., and Costerton, J. W. (2002) Biofilms as complex differentiated communities. *Annu. Rev. Microbiol.* **56,** 187–209.

13. Rozee, K. R., Cooper, D., Lam, K., and Costerton, J. W. (1982) Microbial flora of the mouse ileum mucous layer and epithelial surface. *Appl. Environ. Microbiol.* **43,** 1451–1463.

14. Matsuo, K., Ota, H., Akamatsu, T., Sugiyama, A., and Katsuyama, T. (1997) Histochemistry of the surface mucous gel layer of the human colon. *Gut* **40,** 782–789.

15. Deplancke, B. and Gaskins, H. R. (2001) Microbial modulation of innate defense: goblet cells and the intestinal mucus layer. *Am. J. Clin. Nutr.* **73,** 1131S–1141S.

16. Rojas, M., Ascencio, F., and Conway, P. L. (2002) Purification and characterization of a surface protein from *Lactobacillus fermentum* 104R that binds to porcine small intestinal mucus and gastric mucin. *Appl. Environ. Microbiol.* **68,** 2330–2336.

17. Kraehenbuhl, J. P. and Neutra, M. R. (1992) Molecular and cellular basis of immune protection of mucosal surfaces. *Physiol. Rev.* **72,** 853–879.

18. Berg, R. D. and Savage, D. C. (1975) Immune responses of specific pathogen-free and gnotobiotic mice to antigens of indigenous microorganisms. *Infect. Immun.* **11,** 320–329.

19. Collins, F. M. and Carter, P. B. (1978) Growth of salmonellae in orally infected germ-free mice. *Infect. Immun.* **21,** 41–47.

20. Ouwehand, A. C. and Conway, P. L. (1996) Purification and characterization of a component produced by *Lactobacillus* fermentum that inhibits the adhesion of K88 expressing *Escherichia coli* to porcine mucus. *J. Appl. Bacteriol.* **80,** 311–318.

21. Blomberg, L., Henriksson, A., and Conway, P. L. (1993) Inhibition of adhesion of *Escherichia coli* K88 to piglet ileal mucus by *Lactobacillus* ssp. *Appl. Environ. Microbiol.* **59,** 34–39.

22. Resta-Lennert, S. and Barrett, K. E. (2003) Live probiotics protect intestinal epithelial cells from the effects of infection with enteroinvasive *Esherichia coli* (EIEC). *Gut* **52,** 988–997.

23. Asahara, T., Nomoto, K., Shimizu, K., Watanuki, M., and Tanaka, R. (2001) Increased resistance of mice to *Salmonella* enterica serovar *typhimurium* infection by synbiotic administration of Bifidobacteria and transgalactosylated oligosaccharides. *J. Appl. Microbiol.* **91,** 985–996.

24. Lee, K. H., Jun, K. D., Kim, W. S., and Paik, H. D. (2001) Partial chatacterization of polyfermenticin SCD, a newly identified bacteriocin of *Bacillus polyfermenticus*. *Lett. Appl. Microbiol.* **32,** 146–151.

25. Boris, S., Jimenez-Díaz, R., Caso, J. L., and Barbes, C. (2001) Partial characterization of a bacteriosin produced by *Lactobacillus delbrueckii* subsp. Lactis UO004, an intestinal isolate with probiotic potential. *J. Appl. Microbiol.* **91,** 328–333.

26. Mack, D. R., Michail, S., Wei, S., McDougall, L., and Hollingsworth, M. A. (1999) Probiotics inhibit enteropathogenic *E. coli* adherence in vitro by inducing intestinal mucin gene expression. *Am. J. Physiol.* **276,** G941–G950.

27. Cebra, J. J. (1999) Influences of microbiota on intestinal immune system development. *Am. J. Clin. Nutr.* **69,** 1046S–1051S.

28. Berg, R. D. and Savage, D. C. (1975) Immune responses of specific pathogen-free and gnotobiotic mice to antigens of indigenous and nonindigenous microorganisms. *Infect. Immun.* **11,** 320–329.

29. Kramer, D. R. and Cebra, J. J. (1995) Early appearance of natural mucosal IgA responses and germinal centers in suckling mice developing in the absence of maternal antibodies. *J. Immunol.* **154**, 2051–2062.

30. Garland, C. D., Lee, A., and Dickson, M. R. (1982) Segmented filamentous bacteria in the rodent small intestine: their colonization of growing animals and possible role in host resistance to Salmonella. *Microb. Ecol.* **8**, 181–190.

31. Zachar, Z. and Savage, D. C. (1979) Microbial interference and colonisation of the murine gastrointestinal tract by *Listeria monocytogenes. Infect. Immun.* **23**, 168–174.

32. Mostov, K. and Zegers, M. (2003) Cell biology: just mix and patch. *Nature* **422**, 267–268.

33. Elewaut, D., DiDonato, J. A., Kim, J. M., Truong, F., Eckmann, L., and Kagnoff, M. F. (1999) NF-kappa B is a central regulator of the intestinal epithelial cell innate immune response induced by infection with enteroinvasive bacteria. *J. Immunol.* **163**, 1457–1466.

34. Hooper, L. V. and Gordon, J. I. (2001) Commensal host-bacterial relationships in the gut. *Science* **292**, 1115–1118.

35. Xu, J., Bjursell, M. K., Himrod, J., et al. (2003) A genomic view of the human–Bacteroides thetaiotaomicron symbiosis. *Science* **299**, 2074–2076.

36. Haller, D., Bode, C., Hammes, W. P., Pfeifer, A. M. A., Schiffrin, E. J., and Blum, S. (2000) Non-pathogenic bacteria elicit a differential cytokine response by intestinal epithelial cell/leukocyte co-cultures. *Gut* **47**, 79–87.

37. Klaasen, H. L. B. M., Van den Heijden, P. J., Stok, W., et al. (1993) Apathogenic, intestinal, segmented, filamentous bacteria stimulate the mucosal immune system of the mice. *Infect. Immun.* **61**, 303–306.

38. Ibnou-Zekri, N., Blum, S., Schiffrin, E. J., and von der Weid, T. (2003) Divergent patterns of colonization and immune response elicited from two intestinal *Lactobacillus* strains that display similar properties in vitro. *Infect. Immun.* **71**, 428–436.

39. Crabbe, P. A., Bazin, H., Eyssen, H., and Heremans, J. F. (1968) The normal microbial flora as a major stimulus for proliferation of plasma cells synthesizing IgA in the gut. The germ-free intestinal tract. *Int. Arch. Allergy Appl. Immunol.* **34**, 362–375.

40. Moreau, M. C., Ducluzeau, R., Guy-Grand, D., and Muller, M. C. (1978) Increase in the population of duodenal immunoglobulin A plasmocytes in axenic mice associated with different living or dead bacterial strains of intestinal origin. *Infect. Immun.* **21**, 532–539.

41. Bos, N. A., Bun, J. C. A., Popma, S. H., et al. (1996) Monoclonal immunoglobulin A derived from peritoneal B cells is encoded by both germ line and somatically mutated VH genes and is reactive with commensal bacteria. *Infect. Immun.* **64**, 616–623.

42. Bao, S., Beagley, K. W., Murray, A. M., et al. (1998) Intestinal IgA plasma cells of the B1 lineage are IL-5 dependent. *Immunology.* **94**, 181–188.

43. Thurnheer, M. C., Zuercher, A. W., Cebra, J. J., and Bos, N. A. (2003) B1 cells contribute to serum IgM, but not to intestinal IgA, production in gnotobiotic Ig allotype chimeric mice. *J. Immunol.* **170**, 4564–4571.

23

Volatile Sulfur Detection in Fermented Foods

Bart Weimer and Ben Dias

Summary

Detection of sulfur compounds is difficult in complex samples such as soil and food. The volatility of these compounds can be an advantage if a trapping method is used. This method must be inert for the highly reactive compounds to avoid artifacts resulting from sample preparation and analysis procedures. This chapter describes the use headspace analysis with cryofocusing to detect low levels (ppb) of volatile sulfur compounds produced via fermentation of milk. This method is relatively simple compared to the use of sophisticated instrumentation for the unique identification tools available. It is also useful for other sample types. Milk is a good sample of a difficult substrate that reacts with volatile compounds and is reflective of the level of difficulty for this analysis.

Key Words: Methanethiol; cheddar cheese; sulfur.

1. Introduction

Sulfur compounds are found in virtually every list of compounds associated with the flavor and aroma of many foods and a large majority of fermented foods, such as cheese and sausage. Included in the volatile sulfur compounds (VSCs) are methanethiol, methional, dimethyl sulfide (DMS), dimethyl disulfide (DMDS), dimethyl trisulfide (DMTS), carbonyl sulfide, and hydrogen sulfide *(1)*. Methanethiol, an influential aroma compound, has a very low odor-detection threshold of about 0.02 µg/mL in air *(2)*. It is a major contributor to the aroma of decaying meats *(3,4)* and has been characterized as putrid, fecal, and sulfurous. Other sulfur-containing compounds are associated with various flavors and aromas that are considered desirable at low concentrations (i.e., ppb). Dias and Weimer *(5,6)* found these compounds to be produced by bacteria and purified enzymes in pure culture and in cheese slurries by lactococci and brevibacteria. Other organisms also produce volatile sulfur compounds via various mechanisms that are controlled with the addition of methionine *(7)*.

From: *Methods in Biotechnology, Vol. 18: Microbial Processes and Products*
Edited by: J. L. Barredo © Humana Press Inc., Totowa, NJ

It is the result of the presence of these compounds in minute quantities in food that makes their analysis challenging. Usual concentration methods used for the analysis of other food volatiles are generally not applicable because of the reactive nature of these compounds. Many of these compounds are sensitive to heat and oxidation, which result in the production of artifacts during analysis. Hence, precautions need to be taken during sample extraction and handling to minimize these conversions.

Numerous techniques have been proposed for the extraction and concentration of volatiles in food. Headspace analysis is a quick and simple method for the analysis of volatiles requiring little sample preparation (8,9). Concentration of the headspace leads to improved detection of the volatile sulfur compounds. Cold trapping via cryofocusing is a method that has been used successfully for the analysis of sulfur compounds in various foods, and it has the advantage of having higher recoveries and fewer artifacts (8).

The use of sulfur selective detectors like the flame photometric detector (FPD) and chemiluminescent detectors are useful tools for the monitoring of sulfur compounds in complex matrices and are frequently used for the monitoring of sulfur compounds formed during fermentations (10). Little is known about the mechanism of production of VSCs in fermented foods. Two methods are postulated that include fermentation and chemical reactions via the Strecker reaction (11). Environmental conditions are important in the oxidation–reduction reactions that are also hypothesized to be involved in the production of the polysulfur compounds such as DMDS and DMTS. For example, methanethiol readily oxidizes to DMDS and DMTS under the appropriate conditions (11). Another example is the cheese environment. It has pH and oxidation–reduction conditions that change over time and are thought to be involved in the conversion between the molecular species.

2. Materials

1. Methanethiol (Sigma, St. Louis, MO).
2. Dimethyl sulfide (Sigma).
3. Dimethyl disulfide (Sigma).
4. Milk solution: 10% nonfat dry milk and 30% glycerol.
5. Ultrahigh-temperature processed milk (Gossners, Logan, UT).
6. Spectrophotometer.
7. Cuvets.
8. Elliker's growth medium (Difco, Detroit, MI).
9. Trypic soy broth (Difco).
10. L-Glucono-δ-lactone (GDL).
11. Liquid nitrogen.
12. Liquid-nitrogen container attached to the cyrofocusing unit.
13. Gas chromatographic (GC) equipment 17A Ver. 3 (Shimadzu Corp., Columbus, MD).

14. Shimadzu Class VP software (Shimadzu Corp.).
15. Centrifuge.
16. Centrifuge tubes (15 and 50 mL).
17. Shaking incubators at 25°C.
18. Static incubator at 32°C
19. Headspace autosampler Tekmar 7000 (Tekmar Inc., Cincinnati, OH).
20. Headspace vials (Tekmar Inc.).
21. Autosampler for GC.
22. Cryofocusing unit for GC.
23. Flame photometric detector at 394 nm.
24. Helium gas.
25. Argon gas.
26. Vortex.
27. SPB1-Sulfur column (Supelco, Bellefonte, PA).
28. Fused silica capillary column SPB1-sulfur 30 m × 0.32 mm inner diameter, 100% polydimethylpolysiloxane coated, 4 μm film thickness (Supelco).
29. Ice bath.
30. Water bath.
31. *Lactococcus lactis* ssp. *cremoris* S2 *(5)*.
32. *Brevibacterium linens* BL2 *(5)*.
33. 0.05 *M* Potassium phosphate buffer, pH 7.2.

3. Methods

The demonstration of the method for sulfur detection is described in an artificial cheese system known as cheese slurry. This demonstrates the production of VSCs via fermentation and highlights the use of headspace analysis in a model system with complicated microbial ecology and chemistry. The description is divided into bacterial growth (*see* **Subheading 3.1.**), slurry preparation (*see* **Subheading 3.2.**), and headspace sampling and analysis (*see* **Subheading 3.3.**).

3.1. Growth of Bacteria

3.1.1. Growth of L. lactis *ssp.* cremoris S2

1. Grow *L. lactis* ssp. *cremoris* S2 in Elliker's broth at 30°C and freeze in milk solution.
2. *B. linens* BL2 was grown in tryptic soy broth with aeration at 25°C and frozen in sterile tryptic soy broth containing 30% glycerol.
3. Store at –70°C until further use.
4. Before each use, thaw the frozen stock cultures and grow in their respective media at their respective temperatures for two transfers.
5. Inoculate *L. lactis* ssp. *cremoris* S2 into 2% ultrahigh-temperature processed milk and incubated at 30°C for 14 h. This was used to inoculate the cheese milk.
6. Grow *B. linens* BL2 in tryptic soy broth containing 0.25% L-methionine at 25°C for 36 h with aeration by shaking at 250 rpm.

7. Harvest cells by centrifugation (5000*g* for 10 min at 4°C), wash twice with 0.05 *M* potassium phosphate buffer, pH 7.2, and adjust to an optical density (OD_{600}) of 0.35 in 0.05 *M* potassium phosphate, pH 7.2.
8. Inoculate (5%) the washed cells were into the milk.

3.1.2. Growth of B. linens BL2

1. Grow *B. linens* BL2 in tryptic soy broth with aeration at 25°C and frozen in sterile tryptic soy broth containing 30% glycerol.
2. Store at –70°C until further use.
3. Before each use, thaw the frozen stock cultures and grow in their respective media at their respective temperatures for two transfers.
4. Grow *B. linens* BL2 in tryptic soy broth containing 0.25% L-methionine at 25°C for 36 h with aeration by shaking at 250 rpm.
5. Harvest cells by centrifugation (5000*g* for 10 min at 4°C), wash twice with 0.05 *M* potassium phosphate buffer, pH 7.2, and adjust to an OD_{600} of 0.35 in 0.05 *M* potassium phosphate, pH 7.2.
6. Inoculate (5%) the washed cells were into the milk.

3.2. Model Cheese Preparation

A model cheese was made as an aseptic cheese slurry system *(12)*.

1. Preacidify pasteurized milk containing 2% milk fat with 10% acetic acid to a pH of 6.3 and pour into sterile cheese slurry vats (1 L per vat).
2. Heat-treat the milk by heating to 75°C, holding at 75°C for 5 min, and rapidly cooling to 30°C for 15 min using an ice bath.
3. Acidify the cheese milk by inoculation with a 1% inoculum of *L. lactis* ssp. *cremoris* S2 and 0.70% GDL with 5% *B. linens* BL2.

3.3. Static Headspace Sampling and Analysis

3.3.1. Headspace Sampling

1. Transfer the cheese slurry (5 g) to sterile 22-mL headspace vials sealed with a Teflon lined silicon septum (*see* **Note 1**). Prepare duplicate vials for each treatment, at all time-points.
2. Flush vials with argon for 5 min while agitating using a vortex mixer set on high and subsequently incubate at 25°C.
3. Each day, load vials on to a headspace autosampler.
4. Heat samples to 70°C and allow equilibrating for 30 min before being mixed for 5 min at a power setting of 5.
5. Cryofocus a 2-mL injection at –70°C onto the top of an SPB1-sulfur column.
6. Make an injection into the GC by increasing the temperature of the cryofocusing unit to 200°C for 1 min.

3.3.2. GC Analysis

Volatile sulfur compound in the cheese slurry headspace are separated and detected using a GC system fitted with a FPD operating in the sulfur mode at

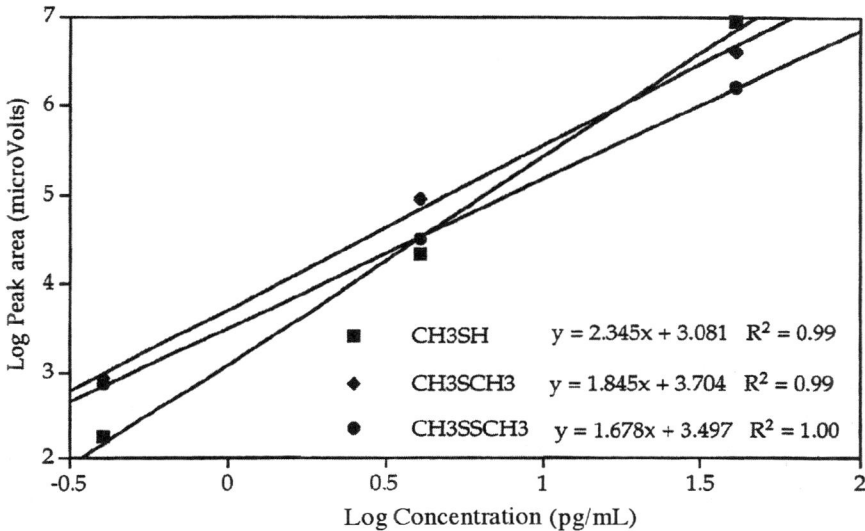

Fig. 1. Standard curves and linear regression equations for selected VSCs.

394 nm. A 30-m × 0.32-mm-inner diameter, 100% polydimethylpolysiloxane-coated (4 μm film thickness), fused silica capillary column is inserted into the GC system. Helium is used as the carrier gas at a flow rate of 1 mL/min. The initial column temperature of 30°C is held for 1 min, increased to 200°C at the rate of 10°C/min, and maintained for 10 min.

3.3.3. Data Analysis

The peak area of the eluting compounds is determined using electronic integration software and the data are analyzed on a personal computer using Shimadzu Class VP software. Compounds are identified by comparing retention times of authentic compounds to the retention times of compounds in the samples. The standard curves are obtained by adding increasing amounts of methanethiol followed by detection with the GC. The response is plotted against the concentration, followed by linear regression to determine the equation for the relationship (*see* **Fig. 1**) *(13)*. VSCs are quantified by preparing standard curves with authentic compounds (*see* **Fig. 1** and **Note 2**). The detection limit of the FPD response is 0.4 pg/mL, with a linear range of 0.4–40 pg/mL, for methanethiol, DMS, and DMDS. Chromatograms of slurry headspace are provided as an example of the production of these compounds by *L. lactis* S2 and *B. linens* BL2 (*see* **Fig. 2**).

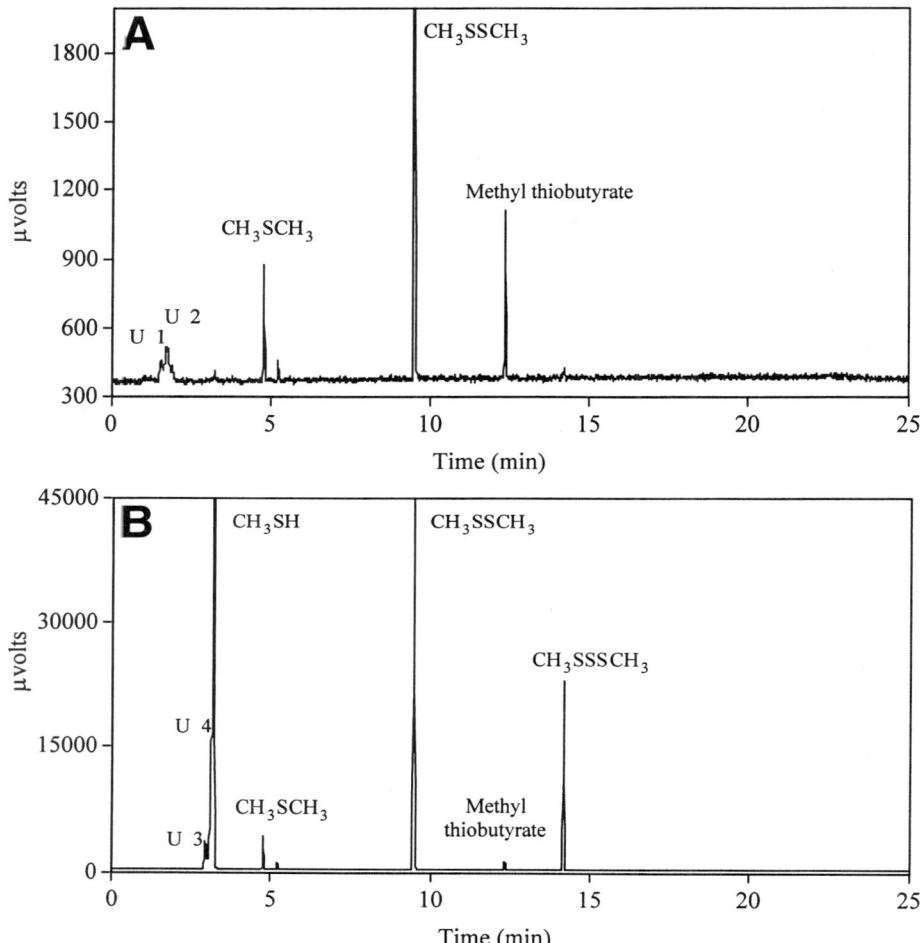

Fig. 2. Chromatograms from a cheddar cheese slurry headspace after 4 d of incubation at 25°C. The slurry is made by direct acidification with GDL containing *B. linens* BL2 (**A**) cr *L. lactis* ssp. *cremoris* S2 containing *B. linens* BL2 (**B**). U indicates unknown sulfur-containing compounds that were observed during the fermentation.

4. Notes

1. Methanethiol solutions can be obtained from several sources or is prepared by bubbling methanethiol gas through chilled oil. It is important to check the concentration of methanethiol in these solutions via titration with iodine. This is critical for the accurate quantification of these compounds because of their extreme volatility.

2. Care should be taken to make sure that all surfaces that the sample comes in contact with are inert and will not react with the sulfur compounds. Vial and septa selection is important; it is preferable to use Teflon-lined silicon septums. The system should be checked periodically for carryover by loading an empty vial into the headspace sampler and checking for unexpected compounds in the control vials.

References

1. Cliff, A. J. and Law, B. A. (1990) Peptide composition of enzyme-treated Cheddar cheese slurries, determined by reverse phase high performance liquid chromatography. *Food Chem.* **36,** 73–80.
2. Fazzalari, F. A. (ed.) (1978) *Composition of Odor and Taste Threshold Value Data,* ASTM, Baltimore, MD.
3. Lee, M., Smith, D. L., and Freeman, L. R. (1979) High-resolution gas chromatographic profiles of volatile organic compounds produced by microorganisms at refrigerated temperatures. *Appl. Environ. Microbiol.* **37,** 85–90.
4. Thomas, C. J. and McMeekin, T. A. (1981) Production of off odours by isolates from poultry skin with particular reference to volatile sulphides. *J. Appl. Bacteriol.* **51,** 529–534.
5. Dias, B. and Weimer, B. (1998) Conversion of methionine to thiols by lactococci, lactobacilli, and brevibacteria. *Appl. Environ. Microbiol.* **64,** 3320–3326.
6. Dias, B. and Weimer, B. C. (1999) Production of volatile sulfur compounds in Cheddar cheese slurries. *Int. Dairy J.* **9,** 605–611.
7. Weimer, B., Seefeldt, K., and Dias, B. (1999) Sulfur metabolism in bacteria associated with cheese. *Antonie van Leeuwenhoek* **76,** 247–261.
8. Reglero, G., Herraiz, T., and Herraiz, M. (1990) Direct headspace sampling with on column thermal focusing in capillary gas chromatography. *J. Chromatogr. Sci.* **28,** 221–224.
9. Reineccius, G. A. and Anandaraman, S. (1984) Analysis of volatile flavors in food constituents and residues, in *Food Constituents and Food Residues: Their Chromatographic Determination* (Laurence, J. F., ed.), Marcel Dekker, New York, pp. 98–198.
10. Mistry, B. S., Reineccius, G. A., and Jasper, B. L. (1994) Comparison of gas chromatographic detectors for the analysis of volatile sulfur compounds in foods, in *Sulfur Compounds in Foods* (Mussinan, C. J. and Keelan, M. E., eds.), American Chemical Society, Washington, DC, pp. 8–21.
11. Mussinan, C. J. and Keelan, M. E. (1994) Sulfur compounds in foods: an overview, in *Sulfur Compounds in Foods* (Mussinan, C. J. and Keelan, M. E., eds.), American Chemical Society, Washington, DC, pp. 1–7.
12. Roberts, M., Wijesundar, C., Bruinenberg, P. G., and Limsowtin, G. K. Y. (1995) Development of an aseptic cheese curd slurry system for cheese ripening studies. *Aust. J. Dairy Technol.* **50,** 66–69.
13. Kim, S. C. and Olsen, N. F. (1989) Production of methanethiol in milk fat-coated microcapsules containing *Brevibacterium linens* and methionine. *J. Microencapsul.* **6,** 799–811.

24

Bioleaching

Analysis of Microbial Communities Dissolving Metal Sulfides

Axel Schippers and Klaus Bosecker

Summary

Bioleaching is the oxidation of insoluble metal sulfides to metal sulfates by Fe(II) and/or sulfur-oxidizing *Bacteria* and *Archaea*. Metals can be dissolved by submerged leaching using fine-grained material that is suspended in the leach liquid. To analyze bioleaching communities, cells should be enumerated and their metal sulfide oxidation rate should be measured. For enumeration of total cell counts, the acridine orange direct counting (AODC) method using epifluorescence microscopy is described. To enumerate viable Fe(II) and sulfur-oxidizing bacteria, the most-probable-number (MPN) cultivation technique using selective media is described. Bioleaching activity can be measured by microcalorimetry, and from the heat output values, the biological metal sulfide oxidation rate can be calculated. Control microcalorimetric measurements also allow the determination of the chemical metal sulfide oxidation rate.

Key Words: Bioleaching; submerged leaching; metal sulfide oxidation; AODC; MPN; microcalorimetry; *Acidithiobacillus; Leptospirillum; Acidimicrobium; Sulfobacillus; Acidianus; Sulfolobus; Metallosphaera; Ferroplasma.*

1. Introduction

Acidophilic metal-sulfide-oxidizing microorganisms are increasingly used in commercial bioleaching operations to recover metals such as Au, Cu, Ni, Zn, U, and Co *(1–5)*. In sulfidic mine waste, bioleaching microorganisms oxidize pyrite and other metal sulfides and produce toxic metals containing acid rock drainage (ARD) *(6–9)*. Different acidophilic Fe(II) and/or sulfur-compound-oxidizing *Bacteria* and *Archaea* are involved in metal sulfide oxidation. The most important species are *Acidithiobacillus* (formerly *Thiobacillus*) *ferrooxidans, Acidithiobacillus thiooxidans, Acidithiobacillus caldus, Leptospirillum ferrooxidans, Leptospirillum ferriphilum, Leptospirillum thermoferrooxidans, Acidimicrobium ferrooxidans, Sulfobacillus acidophilus,*

From: *Methods in Biotechnology, Vol. 18: Microbial Processes and Products*
Edited by: J. L. Barredo © Humana Press Inc., Totowa, NJ

Sulfobacillus thermosulfidooxidans, Acidianus brierleyi, Sulfolobus metallicus,
Metallosphaera sedula, Ferroplasma acidiphilum, and *Ferroplasma acidar-*
manus. Cultivation techniques such as most-probable-number (MPN) *(5,6)* and
overlaid solid media *(10,11)*, as well as molecular ecological techniques such
as fluorescence *in situ* hybridization (FISH) *(7–9,12,13)*, immunological and
polymerase chain reaction (PCR)-based methods *(14–17)* have been applied to
analyze microbial communities dissolving metal sulfides. Here, we describe (1)
how to take samples and to detach microorganisms from solid particles, (2) the
determination of total cell counts by acridine orange direct counting (AODC)
using epifluorescence microscopy, and (3) the enumeration of viable mesoacidophilic Fe(II) and sulfur-oxidizing bacteria by the MPN technique using selective liquid media. Additionally, we describe (4) a method to quantify the
bioleaching activity using microcalorimetry and to calculate the corresponding
microbial metal sulfide oxidation rate *(5,6,18–20)*. Finally, (5) submerged
leaching for metal dissolution is shown *(21)*.

2. Materials

2.1. Sampling and Detachment of Microorganisms

1. Sterile boxes or dark flasks with a volume of at least 100 mL.
2. Rotary shaking device.
3. Glass vessels, 250 mL.
4. Thermometer (–10°C to 100°C).

2.2. Total Cell Counts by AODC

1. A fluorescence microscope fitted with a minimum 50-W mercury vapor lamp, a
   wide-band interference filter set for blue excitation, an Achroplan ×100/1.25 oil
   Ph3 ∞/0.17 objective lens, a Plan Neofluar ×40/0.75 dry Ph3 ∞/0.17 objective lens,
   and ×10 oculars, one equipped with a grid.
2. Acridine orange solution: 0.1% acridine orange and 100 mM acetate buffer, pH 4.7.
   Sterilize by filtration.
3. Sterile pipets and tubes.
4. Black polycarbonate membrane filters, 0.2 μm, 25 mm in diameter.
5. Sterile filter unit.
6. Sterile glass slides and cover slides.

2.3. Viable Cell Counts by MPN

1. Leathen medium *(22)* for enrichment of acidophilic Fe(II) oxidizers: 0.15 g/L
   $(NH_4)_2SO_4$, 0.05 g/L KCl, 0.5 g/L $MgSO_4 \cdot 7H_2O$, and 0.05 g/L K_2HPO_4. Adjust
   to pH 3.5 with H_2SO_4 or NaOH. After sterilization in an autoclave, add 1 mL/L of
   1% (w/w) $Ca(NO_3)_2 \cdot 4H_2O$ solution, sterilized by filtration, and 10 mL/L of 10%
   (w/w) $FeSO_4 \cdot 7H_2O$ solution acidified to pH 2.0 with sulfuric acid and sterilized
   by filtration (*see* **Note 1**). Fill 4.5 mL of medium in each previously sterilized tube
   (21 per sample) using a sterilized dispenser.

2. Starkey medium *(23)* for enrichment of acidophilic sulfur oxidizers: 0.3 g/L $(NH_4)_2SO_4$, 3.5 g/L K_2HPO_4, 2.5 mL/L of 10% (w/w) $CaCl_2 \cdot 2H_2O$, 5.0 mL/L of 10% (w/w) $MgSO_4 \cdot 7H_2O$, and 1.0 mL/L of 1% (w/w) $Fe_2(SO_4)_3$. Adjust to pH 3.5 with H_2SO_4 or NaOH. After sterilization in an autoclave, fill approx 0.5 g of elemental sulfur (three times steam-sterilized) in each previously sterilized tube (21 per sample) and 4.5 mL of medium in each tube using a sterilized dispenser.
3. Rotary shaker at 30°C equipped for tubes.
4. Microscope with phase contrast.

2.4. Determination of Bioleaching Activity and Calculation of Pyrite Oxidation Rates

1. Thermal Activity Monitor Thermostat type 2277 (Thermometric, Järfälla, Sweden) equipped with 4-mL Ampoule Micro Calorimetric Units (type 2277-201) or 20 mL Ampoule Micro Calorimetric Units (type 2230-000).
2. Glass ampoules for ampoule cylinders.

2.5. Metal Dissolution

1. Rotary shaker (120 rpm) or magnetic stirrer.
2. Sterile Erlenmeyer flasks (100 mL or 500 mL) sealed with metallic caps.
3. Sterile pipets.
4. Crushed sulfide ore (particle size < 0.2 mm), metal sulfide containing industrial waste or soil.
5. Leathen medium without Fe(II).
6. Diluted sulfuric acid (pH 2.0), sterilized by filtration.
7. Nonbleeding pH-indicator strips or pH-meter.
8. Test strips for the detection and semiquantitative determination of metal ions.
9. Acidophilic Fe(II)-oxidizing bacteria or archaea (e.g., *Acidithiobacillus ferrooxidans*).
10. Mixture of methanol : formaldehyde (9 : 1).
11. Microscope with phase contrast.

3. Methods
3.1. Sampling and Detachment of Microorganisms

Collect at least 10 g of representative solid sample material from bioleaching reactors, mine waste, soils, sediments, and so forth, in sterile boxes or dark flasks with a volume of at least 100 mL to provide enough air for the microorganisms to survive the transport. To avoid light and drying out, close the boxes or flasks and keep them at ambient temperature in the dark. Temperatures below 10°C can inactivate the bioleaching microorganisms. Perform the microbiological analysis within a few days to avoid a considerable change of the bioleaching community. Measure the temperature at the sampling site because the measurement of microbial activity should be done at the *in situ* temperature.

To detach microorganisms from the surface of the solid material, put 10 g of solid sample material in 100 mL of Leathen medium without Fe(II) in a 250-mL glass vessel and shake at 130 rpm for 2 h at room temperature. After settling of coarse solid particles, immediately use the liquid phase for total and viable cell counting.

3.2. Total Cell Counts by AODC

The total number of cells is determined by AODC using epifluorescence microscopy:

1. Add 0.5 mL of acridine orange solution to 4.5 mL of the liquid phase (*see* **Subheading 3.1.**).
2. After shaking, incubate for 3 min, shake again, and remove the stained cells and fine solid particles on a 0.2-μm black polycarbonate membrane filter, in a filter device.
3. Flush off excess dye from the membrane by rinsing with sterile deionized water.
4. Put the moist membrane filter on a glass slide and cover with a cover slide. To avoid drying out, start counting the cells immediately by using a fluorescence microscope (*see* **Note 2**).
5. Count only bacterially shaped fluorescing objects and calculate the total cell numbers per gram of solid material. The detection limit for bacterial cells is usually 1 × 10^5 cells per gram.

3.3. Viable Cell Counts by MPN

The number of viable cells of acidophilic Fe(II) and sulfur oxidizers is quantified by MPN:

1. Dilute the liquid phase (*see* **Subheading 3.1.**) in the ratio 1 : 10 in Leathen medium without Fe(II) and continue in 10-fold steps to 10^{-6}.
2. Inoculate with sterile pipets 0.5 mL of each 10-fold dilution tube into three parallel tubes containing 4.5 mL of a selective medium to enrich Fe(II) (Leathen medium) or sulfur-oxidizing microorganisms (Starkey medium) (*see* **Note 3**).
3. Incubate the inoculated tubes on a rotary shaker at 120 rpm for 4–6 wk in the dark at 30°C to enrich mesophilic microorganisms (*see* **Note 3**).
4. Acidophilic Fe(II)-oxidizing bacteria (*Acidithiobacillus ferrooxidans*-like or *Leptospirillum*-like) are considered to be present if the Leathen medium turns reddish brown. Growth of acidophilic sulfur-oxidizing bacteria (*Acidithiobacillus thiooxidans*-like) is indicated by high acid production in Starkey medium (pH below 2.0).
5. Identify bacteria by microscopy and differentiate between the two mesoacidophilic Fe(II) oxidizers microscopically. Analyze the highest positive dilutions for rods (indicating *Acidithiobacillus*) or curved and vibrioid-shaped cells (indicating *Leptospirillum*).
6. Calculate the viable cell numbers per gram of solid material by using MPN tables (http://www.cfsan.fda.gov/~ebam/bam-a2.html#tab1) (*24–26*).

3.4. Determination of Bioleaching Activity and Calculation of Pyrite Oxidation Rates

For the determination of microbial metal sulfide oxidation activity, the heat output of solid samples is microcalorimetrically measured using a thermal-activity monitor equipped with ampoule cylinders.

1. Put 10 g of a solid sample in a 20-mL glass ampoule (or 2 g in a 3-mL or 4-mL glass ampoule). Take care to have a representative mixture of fine and coarse particles. The size of the solid particles is restricted to the diameter of the ampoule mouth (*see* **Note 4**).
2. Close the glass ampoule containing solid material and air (*see* **Note 5**) tightly with a sealed cap using the tool provided by the company selling the microcalorimeter (*see* **Note 6**).
3. Insert the closed ampoule in the ampoule cylinder. After thermal equilibration for 30 min at the *in situ* temperature of the sampling site (0–90°C) (*see* **Note 7**), the heat output resulting from microbial and chemical oxidation of metal sulfides is recorded in microwatts for about 2–4 h until a stable value is obtained.
4. Measure the heat output caused only by chemical metal sulfide oxidation in a control experiment. Remove the cap and add 1 mL of chloroform to the solid sample. After 24 h, remove all chloroform by vacuum evaporation, close the ampoule with a new sealed cap tightly, and again measure the heat output (chemical oxidation). The difference between the two values is the microbial metal sulfide oxidation activity (in µW).
5. If pyrite is the dominant metal sulfide in the sample, calculate the chemical, the microbial, and/or the combined pyrite oxidation rate, because this rate correlates with the heat output. According to Rohwerder et al. *(18)*, a complete oxidation of FeS_2 to Fe(III) and sulfate produces a reaction energy of – 1546 kJ/mol (*see* **Note 8**):

$$FeS_2 + 3.75O_2 + 0.5H_2O \rightarrow Fe^{3+} + 2SO_4^{2-} + H^+, \qquad \Delta_f H^0 = -1546 \text{ kJ/mol} \quad (1)$$

Calculate the pyrite oxidation rate r (µg/kg/s) using the transformed reaction energy value of – 1.546 kJ/mmol, the molecular mass of pyrite of 0.12 kg/mol, the measured heat output a (µW), and the sample weight w (g) by the following equation:

$$r \text{ (µg/kg/s)} = 1/{-1.546} \text{ (mmol/kJ)} \times 0.12 \text{ (kg/mol)} \times a \text{ (µW)} \times 1/w \text{ (1/g)} \quad (2)$$

3.5. Metal Dissolution

Submerged leaching using fine-grained sulfidic material that is suspended in the leach liquid and kept in motion by shaking or stirring is the easiest and most efficient way to dissolve metals by microorganisms. High aeration and accurate monitoring and control favor the growth and the activity of the microorganisms *(21)*.

1. Depending on the total amount of solid sampling material, use 100- or 500-mL conical flasks sealed with metallic caps (*see* **Note 9**), add 0.5–5.0 g of fine-grained

sulfidic material and 50 mL or 100 mL Leathen medium without iron, inoculated with 5 mL or 10 mL of a well-grown bacterial or archaeal culture, respectively (*see* **Note 10**).

2. Adjust the pH in the leach suspension to 2–3 by adding sterile sulfuric acid.
3. After settling of the solid particles, take a small volume (1–2 mL) for counting the amount of bacterial or archaeal cells and for determining the initial concentration of dissolved metals.
4. Replace the sample volume by sterile culture medium without iron and determine the weight of the flask.
5. For control experiments (chemical leaching), replace the microbial inoculum by the same amount of a methanol/ formaldehyde mixture.
6. Incubate the Erlenmeyer flasks at 30°C on a shaking table (approx 120 rpm) or use a stirring system (*see* **Note 11**).
7. At recorded intervals (e.g., every second day), remove 1–2 mL of the leaching liquid for pH measurement, enumeration of bacterial or archaeal cells, and the determination of solubilized metals. Before sampling, determine the weight of the flask and readjust any loss caused by evaporation by adding sterile diluted sulfuric acid (pH 2.0). Take the liquid samples from the supernatant after settling of the solid material for about 5 min.
8. Replace the sample volume by sterile culture medium without Fe(II). Again, determine the weight of the flask before it is replaced on the shaker.
9. Continue the submerged leaching process for approx 3 wk, until the metal concentration and pH value in the supernatant remain constant.

4. Notes

1. Do avoid oxidation of Fe(II) and precipitation of Fe(III), add a few drops of sulfuric acid to $FeSO_4 \cdot 7H_2O$ before you add water and adjust the pH.
2. If you prefer to count later, let the membrane filter dry for a few minutes after rinsing in the filter unit. Afterward, the membrane filter is mounted in a minimum of paraffin oil between a glass slide and a cover slide. Before microscopic analysis, the membrane filters can be stored at −20°C for several months.
3. Enrichment of other species of Fe(II) and/or sulfur oxidizers (e.g., moderate acidophiles or thermophiles) requires different media and/or different incubation temperatures. Relevant references are available in the literature.
4. The glass ampoule shall not be completely filled with solid material to provide enough air for the microorganisms to be active.
5. The microbial and/or the chemical metal sulfide oxidation activity is determined under full aeration. In the natural environment (e.g., within mine waste dumps), the oxygen partial pressure can be lower than in the atmosphere, which can reduce the oxidation rate.
6. An ampoule not tightly sealed might give a wrong value because of heat loss by the evaporation of water.
7. The temperature dependence of the bioleaching activity above 10°C can be described by the Arrhenius equation (*19,20*).

8. The value is calculated for standard conditions (25°C). Reaction energy values for other metal sulfides than pyrite can be calculated from tables in the literature *(27)*.

9. If cotton plugs are used instead of metallic caps, bioleaching could be less efficient because of limited supply of O_2 and CO_2.

10. Heat sterilization of the solid sampling material should be renounced until it has been shown that no chemical or mineralogical changes take place in the substrate material during sterilization. At the beginning of the leaching test, the pulp density should be in the range of 1–5%. In the course of the leaching experiment, the pulp density can be increased step by step as long as the pH value in the leach suspension remains in the range 2.0–3.0.

11. If a magnetic stirring system is used, magnetic ore particles will stick to the stirrer or the plastic cover of the stirrer might be destroyed if a high pulp density is used. In this case, the results of metal dissolution have to be considered cautiously.

References

1. Brandl, H. (2001) Microbial leaching of metals, in *Biotechnology* (Rehm, H.-J. and Reed, G., in cooperation with Pühler, A., and Stadler, P., eds.), Wiley–VCH, Weinheim, Vol. 10, pp. 191–224.

2. Rawlings, D. E. (2002) Heavy metal mining using microbes. *Annu. Rev. Microbiol.* **56,** 65–91.

3. Rohwerder, T., Jozsa, P.-G., Gehrke, T., and Sand, W. (2002) Bioleaching, in *Encyclopedia of Environmental Microbiology* (Bitton, G., ed.), Wiley, New York, Vol. 2, pp. 632–641.

4. Rawlings, D. E., Dew, D., and du Plessis, C. (2003) Biomineralization of metal-containing ores and concentrates. *Trends Biotechnol.* **21,** 38–44.

5. Sand, W., Hallmann, R., Rohde, K., Sobotke, B., and Wentzien, S. (1993) Controlled microbiological in-situ stope leaching of a sulphidic ore. *Appl. Microbiol. Biotechnol.* **40,** 421–426.

6. Schippers, A., Hallmann, R., Wentzien, S., and Sand, W. (1995) Microbial diversity in uranium mine waste heaps. *Appl. Environ. Microbiol.* **61,** 2930–2935.

7. Bond, P. L., Druschel, G. K., and Banfield, J. F. (2000) Comparison of acid mine drainage microbial communities in physically and geochemically distinct ecosystems. *Appl. Environ. Microbiol.* **66,** 4962–4971.

8. Edwards, K. J., Bond, P. L., Gihring, T. M., and Banfield, J. F. (2000) An archaeal iron-oxidizing extreme acidophile important in acid mine drainage. *Science* **287,** 1796–1799.

9. Schrenk, M. O., Edwards, K. J., Goodman, R. M., Hamers, R. J., and Banfield, J. F. (1998) Distribution of *Thiobacillus ferrooxidans* and *Leptospirillum ferrooxidans* for generation of acid mine drainage. *Science* **279,** 1519–1522.

10. Johnson, D. B. (1995) Selective soild media for isolating and enumerating acidophilic bacteria. *J. Microbiol. Methods* **23,** 205–218.

11. Okibe, N., Gericke, M., Hallberg, K. B., and Johnson, D. B. (2003) Enumeration and characterization of acidophilic microorganisms isolated from a pilot plant stirred-tank bioleaching operation. *Appl. Environ. Microbiol.* **69,** 1936–1943.

12. Bond, P. L. and Banfield, J. F. (2001) Design and performance of rRNA targeted oligonucleotide probes for in situ detection and phylogenetic identification of microorganisms inhabiting acid mine drainage environments. *Microb. Ecol.* **41**, 149–161.

13. Pernthaler, J., Glöckner, F.-O., Schönhuber, W., and Amann, R. (2001) Fluorescence *in situ* hybridization (FISH) with rRNA-targeted oligonucleotide probes. *Methods Microbiol.* **30**, 207–226.

14. Wulf-Durand, P. de., Bryant, L. J., and Sly, L. I. (1997) PCR-mediated detection of acidophilic, bioleaching-associated bacteria. *Appl. Environ. Microbiol.* **63**, 2944–2948.

15. Jerez, C. A. (1997) Molecular methods for the identification and enumeration of bioleaching microorganisms, in *Biomining: Theory, Microbes and Industrial Processes* (Rawlings, D. E., ed.), Landes Bioscience, Austin. TX.

16. Goebel, B. M. and Stackebrandt, E. (1994) Cultural and phylogenetic analysis of mixed microbial populations found in natural and commercial bioleaching environments. *Appl. Environ. Microbiol.* **60**, 1614–1621.

17. Pizarro, J., Jedlicki, E., Orellana, O., Romero, J., and Espejo, R. T. (1996) Bacterial populations in samples of bioleached copper ore as revealed by analysis of DNA obtained before and after cultivation. *Appl. Environ. Microbiol.* **62**, 1323–1328.

18. Rohwerder, T., Schippers, A., and Sand, W. (1998) Determination of reaction energy values for biological pyrite oxidation by calorimetry. *Thermochim. Acta* **309**, 79–85.

19. Elberling, B., Schippers, A., and Sand, W. (2000) Bacterial and chemical oxidation of pyritic mine tailings at low temperatures: *J. Cont. Hydrol.* **41**, 225–238.

20. Rohwerder, T., Kahl, A., Wentzien, S., and Sand, W. (1999) Microcalorimetric determination of bioleaching activity and temperature dependence, in *Biohydrometallurgy and the Environment Toward the Mining of the 21st Century* (Amils, R. and Ballester, A., eds.), Elsevier, Amsterdam, Part A, pp. 749–758.

21. Bosecker, K. (1987) Microbial Leaching, in *Fundamentals of Biotechnology* (Präve, P., Faust, U., Sittig, W., and Sukatsch, A., eds.), Verlag Chemie, Weinheim, pp. 661–683.

22. Leathen, W. W., McIntyre, L. D., and Braley, S. A. (1951) A medium for the study of the bacterial oxidation of ferrous iron. *Science* **114**, 280–281.

23. Starkey, R. L. (1925) Concerning the physiology of *Thiobacillus thiooxidans,* an autotrophic bacterium oxidizing sulphur under acid conditions. *J. Bacteriol.* **10**, 135–162.

24. Alexander, M. (1965) *Methods of Soil Analysis,* Agronomy Series No. 9, ASA, Madison, WI, Part 2, pp. 1467–1472.

25. de Man, J. C. (1983) MPN tables, corrected. *Eur. J. Appl. Biotechnol.* **17**, 301–305.

26. McCrady, M. H. (1915) The numerical interpretation of fermentation-tube results. *J. Infect. Dis.* **17**, 183–212.

27. Wagman, D. D., Evans, W. H., Parker, V. B., et al. (1982) NBS tables of thermodynamic properties. *J. Phys. Chem. Ref. Data* **11(Suppl. 2)**.

25

Use of PhoN Phosphatase to Remediate Heavy Metals

Marion Paterson-Beedle and Lynne E. Macaskie

Summary

Heavy metals are removed by *Serratia* sp. via the activity of a cell-bound, atypical, PhoN-type acid phosphatase enzyme, which liberates HPO_4^{2-} from a suitable organic phosphate donor, with the stoichiometric precipitation of heavy-metal cations (M^{2+}) as insoluble $MHPO_4$ at the cell surface. The present chapter describes the production of *Serratia* sp. cells (and biofilm) using different growth conditions (i.e., in media containing glycerol and lactose as the carbon source and as batch and continuous cultures). Also, it describes methods of immobilization of cells by entrapment, using polyurethane foam as the matrix, and by covalent coupling, using silanized ceramic Raschig rings and glutaraldehyde as a crosslinking agent. The removal of U from solution using a packed-bed reactor system is described as well as the removal of Ni^{2+} or Co^{2+} using a hybrid bioaccumulative and chemisorptive mechanism, known as "microbially enhanced chemisorption of heavy metals."

Key Words: *Serratia; Citrobacter;* acid phosphatase; lactose; glycerol; biofilm; immobilization; reactor; uranium; nickel; cobalt.

1. Introduction

The production of a cell-bound, atypical, PhoN-type acid phosphatase enzyme by a *Citrobacter* sp. (NCIMB 40259) now identified as a *Serratia* sp. *(1)* has been well documented *(2–4)*. Heavy metals are removed by *Serratia* sp. via the activity of the phosphatase enzyme, which liberates HPO_4^{2-} from a suitable organic phosphate donor, with the stoichiometric precipitation of heavy-metal cations (M^{2+}) as insoluble $MHPO_4$ at the cell surface. Previous studies have shown that *Escherichia coli* bearing the corresponding cloned *phoN* gene from *Salmonella typhimurium* also promoted the formation of metal hydrogen phosphate on the cell surface *(5)*.

A model for phosphatase-mediated uranyl ion accumulation, with exocellular deposition of polycrystalline hydrogen uranyl phosphate ($HUO_2PO_4 \cdot 4H_2O$ [HUP]) that is identical (by X-ray diffraction analysis) to the structure of HUP

From: *Methods in Biotechnology, Vol. 18: Microbial Processes and Products*
Edited by: J. L. Barredo © Humana Press Inc., Totowa, NJ

prepared by inorganic crystal growth has been well published *(6,7)*. This cell surface-localized enzyme system is robust, functioning between temperatures of 2°C and 45°C, over a pH range from 4.5 to 8.5 and tolerating elevated metal concentrations. The metal phosphate deposits remain associated with the biomass, forming large accumulations (e.g., 9 g of uranium per gram of dry weight biomass, with steady-state metal removal of above 90% *(8)*. Also, the effective biomineralization of HUP from uranium mine waters has been demonstrated *(9)*.

Among heavy metals, nickel is particularly recalcitrant to bioremediation using the standard approaches of biosorption and bioaccumulation *(10)*. However, a hybrid bioaccumulative and chemisorptive mechanism, described as "microbially enhanced chemisorption of heavy metals" (MECHM) *(10,11)* has been used for the effective removal of nickel from dilute aqueous solution. Two approaches have been tested for the bioremediation of Ni^{2+} using *Serratia* sp. cells *(10–13)*. The first approach was based on a two-step process: (1) enzymatically mediated generation of a polycrystalline "host lattice" (priming layer) (i.e., HUP bound to the surface of the *Serratia* sp. cells) and (2) cation-exhange intercalation of Ni^{2+} into the interlamellar spaces of HUP *(10–13)*. "Chemical" HUP is an established ion-exchange material, with the intralamellar mobile protons freely exchangeable for other ions such as Na^+ and Ni^{2+} *(14)*. The second approach was to cochallenge immobilized *Serratia* sp. cells in a packed-bed reactor with a Ni^{2+} and UO_2^{2+} solution in the presence of glycerol 2-phosphate (phosphate donor for phosphate release and metal bioprecipitation), giving a sustained removal of both metals *(10)*.

Nuclear wastes contain not only uranium but also fission and activation products like ^{90}Sr, ^{137}Cs, and ^{60}Co. It has been shown that *Serratia* sp. biofilm in a continuous cochallenge system (i.e., Sr^{2+} or Cs^+ in the presence of uranyl ion and glycerol 2-phosphate) was able to promote the removal of approx 58% of Cs^+ and 50–56% of Sr^{2+} from dilute aqueous flows containing 1 mM of the metals *(15)*. However, and in contrast to Ni^{2+} (above), this technique was not able to remove Co^{2+} from solution, but using cells precoated with HUP as bioinorganic ion-exchangers, it was possible to remove the Co^{2+} until the reactor reached saturation *(15)*.

It is known that the phosphatase activity of *Serratia* sp. cells can vary greatly according to the pregrowth conditions. For efficient and economic industrial operation, it is essential to have a reproducible product of high phosphatase activity. Large-scale biomass production with high phosphatase activity has been achieved using a lactose-based minimal medium and an air-lift fermentor operated in the chemostat mode with steady output of continuously grown culture *(16)* or using a fed-batch culture system *(17)*. The phosphatase enzyme remains functional in resting cells, facilitating the use of immobilised pregrown cells in reactors.

The presence of supports during cell growth in continuous culture results in the "self-immobilization" of *Serratia* sp. cells as a biofilm. Inexpensive materials of large surface area are generally chosen (e.g., reticulated foams, ceramic Raschig rings, and porous glass). Alternatively, cells can be grown in liquid culture, the biomass concentrated, and chemically attached to suitable supports or physically entrapped by a solid matrix. The immobilized cells can be used in packed-bed reactors for continuous metal removal.

The aim of this chapter is to describe the production of *Serratia* sp. cells and biofilm using glycerol or lactose as the carbon source, the immobilization of cells by the methods of entrapment and covalent coupling, the preparation of packed-bed reactors for removal of uranium from acid solution, and the removal of nickel and cobalt by the MECHM method.

2. Materials

2.1. Preparation of Planktonic Cells and Biofilm

1. *Citrobacter* sp. strain (NCIMB 40259) now identified as a *Serratia* sp. *(1)* (licensed by Isis Innovation, Oxford, UK). *Serratia* sp. is classified as category 1 under the Advisory Committee on Dangerous Pathogens (i.e., a biological agent unlikely to cause human disease). Good microbiological practice is recommended for all work with such microorganisms.
2. Nutrient agar (Oxoid, Basingstoke, UK).
3. Glycerol minimal medium: 12.0 g/L Tris-HCl, 0.62 g/L KCl, 0.96 g/L $(NH_4)_2SO_4$, 0.67 g/L glycerol 2-phosphate, 0.063 g/L $MgSO_4 \cdot 7H_2O$, 0.32 mg/L $FeSO_4 \cdot 7H_2O$, and 2.0 g/L glycerol *(17)*. Dissolve the Tris-HCl in deionized water and reduce the pH to 7.2–7.5 with 2 M HCl. Add and dissolve all of the other components sequentially and adjust the pH of the medium to 7.2 with 2 M HCl or 2 M NaOH. Add deionized water to the required volume (e.g., for 1 L of medium, add deionized water to 980 mL), sterilize, and cool. Dissolve the glycerol in deionized water (e.g., for 1 L of medium, dissolve 2 g of glycerol in 20 mL of deionized water), sterilize, and cool. Add the glycerol solution aseptically to the medium prior to inoculation (*see* **Notes 1–3** and **Fig. 1**). Caution: KCl is irritant; $FeSO_4 \cdot 7H_2O$ is harmful; HCl is corrosive; NaOH is corrosive.
4. 250-mL and 500-mL Erlenmeyer flasks and sponge or cotton wool bungs.
5. 250-mL and 500-mL Erlenmeyer flasks containing an outlet at the bottom of the flask with approx 15 cm of silicone tubing attached, a glass connector (male), and a clip hose clamp (*see* **Fig. 1**).
6. Lactose minimal medium: 12.0 g/L Tris-HCl, 0.62 g/L KCl, 0.96 g/L $(NH_4)_2HPO_4$, 0.063 g/L $MgSO_4 \cdot 7H_2O$, 0.32 mg/L $FeSO_4 \cdot 7H_2O$, and 2.13 *(17)* or 0.6 g/L *(16)* lactose for batch or carbon-limiting continuous culture, respectively. Also, add 20 μL/L polypropylene glycol to continuous culture medium to control foaming. Follow the same preparation procedure described for glycerol minimal medium (*see* **Notes 1–3** and **Fig. 1**). *Caution:* KCl is an irritant; $FeSO_4 \cdot 7H_2O$ is harmful; HCl is corrosive; NaOH is corrosive.

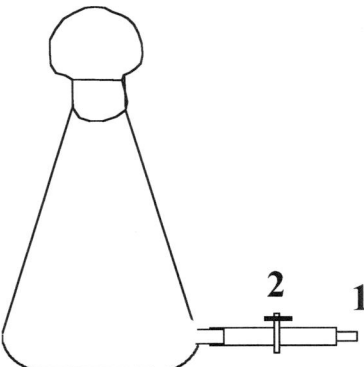

Fig. 1. A schematic of an Erlenmeyer flask used for the addition of cell culture or sterile lactose solution to large volumes of sterile medium (over 3 L): (1) glass connector (male); (2) clip hose clamp.

7. Supports for biofilm growth: polyurethane reticulated foam Filtren TM30 (Recticel, Wetteren, Belgium) and ceramic Raschig rings based on aluminum oxide (product no. SA5539, length = 5 mm, external diameter = 5 mm, internal diameter = 2 mm, median pore diameter = 1.8–50 µm or product no. SA6575, length = 7 mm, external diameter = 6.6 mm, internal diameter = 2 mm, pore size = approx 100 Å) (Norton Chemical Process Products Corp., Akron, OH).

8. 70% (v/v) Ethanol in water. *Caution:* Ethanol is highly flammable.

9. Rotary shaker (*see* **Note 4**).

10. 10-L Autoclavable bottle for large batch culture (*see* **Fig. 2** and **Note 4**).

11. Air-lift fermentor for continuous culture, heating mantle, thermometer, flow meter, peristaltic pump 101U (Watson Marlow, Falmouth, UK) (*see* **Fig. 3**). *Caution:* The apparatus must be adequately supported with clamps.

12. Silicone tubing (internal diameters: 4 mm, 6 mm, and 9 mm), tubing connectors, and cable ties (length = 112 mm and width = 2.41 mm).

13. Stainless-steel wire cage for ceramic Raschig rings support (approx 25 cm × 2 cm) (*see* **Note 5**).

14. Cotton and needle to prepare strands of polyurethane reticulated foam.

15. Supply of compressed air (*see* **Note 6**).

16. 20-L Autoclavable bottles with wide neck, screw top containing three outlets (inflow to fermentor, addition of lactose to medium, and air filter) (*see* **Fig. 2**).

17. 20-L Autoclavable bottles with wide neck, screw-top containing two outlets (outflow from fermentor and air filter).

18. Timer, 50-mL and 2-L measuring cylinders.

19. Isotonic saline: 8.5 g/L NaCl.

20. Centrifuge (capacity for six × 500-mL bottles with caps) (*see* **Note 7**).

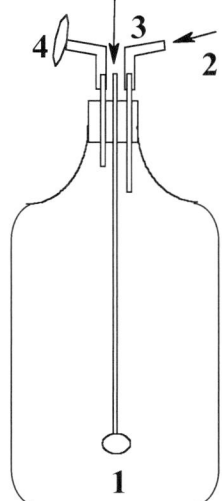

Fig. 2. A schematic of a vessel for a batch culture [(1) sintered glass; (2) culture inlet; (3) air inflow; (4) air filter] or for medium for the air-lift fermentor [(1) remove the sintered glass and leave the tubing immersed in the medium; (2) lactose solution inlet; (3) medium outflow; (4) air filter].

2.2. Assay of Phosphatase Activity (18,19)

1. 12 mg/mL *p*-Nitrophenyl phosphate in water. Can be stored for several days at 4°C in the dark.
2. 0.2 *M* 3-(*N*-Morpholino)propanesulfonic acid (MOPS)–NaOH buffer, pH 7.0. *Caution:* MOPS is irritant; NaOH is corrosive.
3. 0.2 *M* NaOH.
4. 20-mL Glass tubes.
5. Spectrophotometer, 2-mL cuvets ($10 \times 4 \times 45$ mm), water bath (30°C), timer, and vortex mixer.

2.3. Preparation of Immobilized Cells

2.3.1. Entrapment

1. Support for immobilization of cells: Hypol HP 3000, a polyurethane prepolymer for production of hydrophilic foams, containing less than 9% of toluene diisocyanate (TDI) (Hampshire Chemical Corp., Lexington, MA, now supplied by Dow Chemicals Co., Cheshire, UK) (*see* **Note 8**). *Caution:* TDI is irritant and a suspected carcinogen.
2. Isotonic saline: 8.5 g/L NaCl.

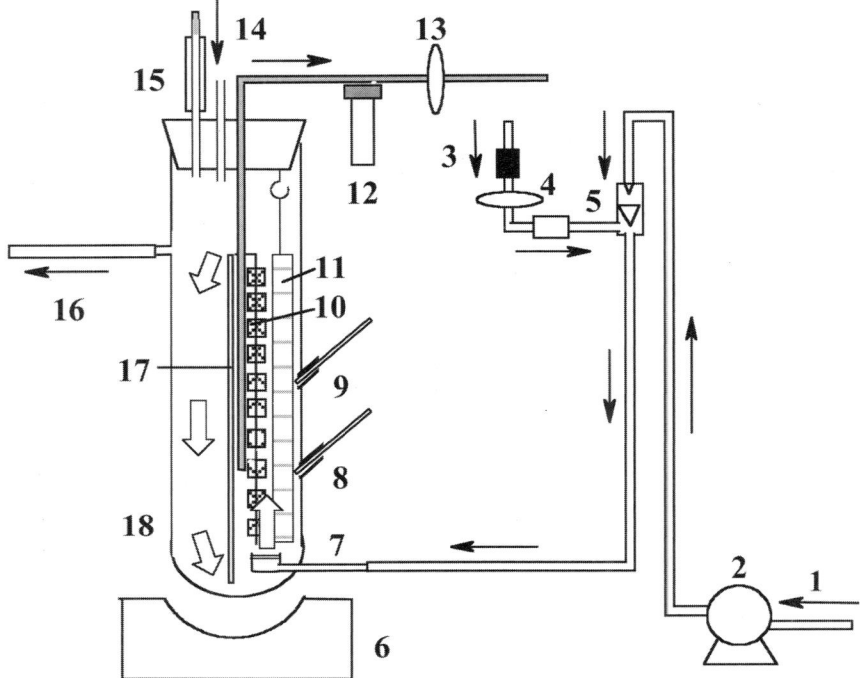

Fig. 3. Air-lift fermentor for simultaneous production of planktonic cells and biofilm. The dimensions of the fermentor are length = 45 cm and internal diameter = 10 cm, giving a volume of approx 2.5 L. Medium circulation is achieved via the central partition (length = 32 cm) with a U-shaped slot at the bottom, allowing rapid flow of medium between the up side (air-lift) and down side. (1) Sterile medium inflow; (2) peristaltic pump; (3) air inflow (*see* **Note 4**); (4) charcoal filter, air filter, and flow meter (1.8 L/min); (5) medium/air mixing chamber; (6) heating mantle to maintain constant temperature; (7) sintered glass inlet port; (8) thermometer port; (9) extra port (e.g., for a dissolve oxygen electrode, if required); (10) strands of polyurethane reticulated foam; (11) cage with ceramic Raschig rings; (12) sample collection vessel; (13) air filter; (14) culture inlet containing a glass connector (female); (15) exit air filter with condenser; (16) outflow port; (17) glass central partition; (18) fermentor body. Open arrows are the direction of airflow in the fermentor. Filled arrows are direction of (1) medium, (14) culture, and (3) air flow and circulation.

2.3.2. Covalent Coupling to Supports

1. Support for immobilization of cells: Ceramic Raschig rings based on aluminum oxide (product no. SA5539, length = 5 mm, external diameter = 5 mm, internal diameter = 2 mm, median pore diameter = 1.8–50 μm; Norton Chemical Process Products Corp.).

2. Nitric acid. Use only in a chemical fume hood. *Caution:* Nitric acid is oxidizing and corrosive.
3. 10% (w/v) 3-Aminopropyltriethoxy silane (APTS) in toluene. *Caution:* APTS is corrosive; toluene is highly flammable and harmful.
4. 0.1*M* Sodium phosphate buffer, pH 7.0.
5. 5% (v/v) Glutaraldehyde in 0.1 *M* sodium phosphate buffer, pH 7.0. *Caution:* Glutaraldehyde is harmful and dangerous for the environment.
6. 0.1 *M* Sodium acetate buffer containing 1 *M* NaCl, pH 4.5.
7. Acetone. *Caution:* Acetone is highly flammable and irritant.
8. Refluxing apparatus (a round-bottomed flask fitted with a standard water condenser) and heating mantle. *Caution:* The apparatus must be adequately supported with clamps.

2.4. Preparation of Reactor Packed-Bed Systems

2.4.1. Bioaccumulation of Hydrogen Uranyl Phosphate

1. *Serratia* sp. cells immobilized onto a support.
2. Cylindrical glass columns (e.g., internal diameter = 1.5 cm, column height = 9 cm) or Perspex columns.
3. Silicone tubing (internal diameter = 2 mm), rubber bungs (size 15); glass tubing (external diameter = 3 mm, length = 5 cm), clip hose clamps, tubing connectors, and cable ties (length = 112 mm, width = 2.41 mm).
4. Peristaltic pump 505U or 313S with a microcassette pumphead and polyvinyl chloride (PVC) pump tubing (color code: orange/yellow, internal diameter = 0.50 mm) (Watson-Marlow, Falmouth, UK).
5. Reactor: *see* **Fig 4.**
6. Uranium solution: 0.5 m*M* $UO_2(NO_3)_2 \cdot 6H_2O$, 20 m*M* MOPS–NaOH, 2 m*M* sodium citrate buffer (*see* **Note 9**), and 5 m*M* glycerol 2-phosphate, pH 6.0–7.0. *Caution:* Uranyl nitrate is very toxic, dangerous for the environment, oxidizing, and should be treated as a potential carcinogen. Commercially obtained uranyl nitrate is depleted uranium (i.e., [238]U). The specific activity of depleted uranium is 0.0000005 Ci/g (18,500 Bq/g) of contained uranium (*see* **Note 10**).

2.4.2. Intercalation of Nickel or Cobalt Into Hydrogen Uranyl Phosphate

1. *Serratia* sp. cells immobilized onto a support.
2. Reactor (as described in **steps 2–4** of **Subheading 2.4.1.**) primed with uranyl phosphate (*see* **Note 10**).
3. Nickel solution: 0.5 m*M* or 1 m*M* $Ni(NO_3)_2 \cdot 6H_2O$, pH 5.8–6.1. *Caution:* Nickel nitrate is harmful and oxidizing.
4. Cobalt solution: 1 m*M* $Co(NO_3)_2 \cdot 6H_2O$, pH 6.0. *Caution:* Cobalt nitrate is toxic and oxidizing.

2.4.3. Cocrystallisation of Nickel and Hydrogen Uranyl Phosphate

1. *Serratia* sp. cells immobilized onto a support.
2. Reactor as described in **steps 2–4** of **Subheading 2.4.1.**

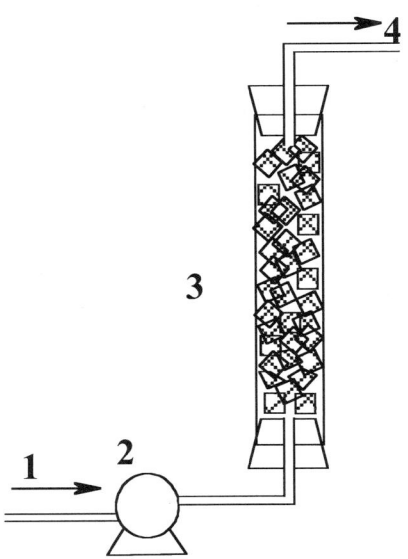

Fig. 4. A schematic of a reactor packed-bed system: (1) metal inflow solution, (2) peristaltic pump, (3) glass column containing cells immobilized onto a support (e.g., polyurethane reticulated foam coated with biofilm), and (4) outflow solution.

3. Mixed metal solution: 0.5 mM UO$_2$(NO$_3$)$_2$·6H$_2$O, 0.5 mM Ni(NO$_3$)$_2$·6H$_2$O, 20 mM MOPS–NaOH, 2 mM sodium citrate buffer (*see* **Note 9**), and 5 mM glycerol 2-phosphate, pH 7.0 (*see* **Note 10**).

2.5. Spectrophotometric Assay for Uranium (20)

1. 0.15% (w/v) Arsenazo III (2,2′[1,8-dihydroxy-3,6-disulfo-2,7-naphthalene-bis(azo)] dibenzenearsonic acid) in water. *Caution:* Arsenazo III is toxic and dangerous for the environment. Allow the solution to stand for 1 h and then filter to remove the residual solid. The solution is stable for a few months at ambient temperature.
2. 0.1 M UO$_2$(NO$_3$)$_2$ · 6H$_2$O stock solution in water. *Caution:* Uranyl nitrate is very toxic, dangerous for the environment, oxidizing, and should be treated as a potential carcinogen. Commercially obtained uranyl nitrate is depleted uranium (i.e., [238]U). The specific activity of depleted uranium is 0.0000005 Ci/g (18,500 Bq/g) of contained uranium (*see* **Note 10**).
3. Standard solution: 1 mM UO$_2$(NO$_3$)$_2$·6H$_2$O. Prepare working solution on the day of use by dilution of the stock solution.
4. 0.1M HCl. *Caution:* HCl is corrosive.
5. Spectrophotometer and 2-mL cuvets ($10 \times 4 \times 45$ mm).

2.6. Spectrophotometric Assay for Nickel (21)

1. 0.02% (w/v) of 2-(5-Bromo-2-pyridylazo)-5-diethylaminophenol (2-Br-PADAP reagent) in ethanol.
2. Acetate buffer, pH 5.5.
3. Ethanol. *Caution:* Ethanol is flammable.
4. 1% (w/v) of $Na_2S_2O_3$ in water.
5. $0.1 M$ $Ni(NO_3)_2 \cdot 6H_2O$ stock solution in water. *Caution:* Nickel nitrate is harmful and oxidizing.
6. Standard solution: 1 mM $Ni(NO_3)_2 \cdot 6H_2O$. Prepare working solution on the day of use by dilution of the stock solution.
7. Spectrophotometer and 2-mL cuvets ($10 \times 4 \times 45$ mm).

2.7. Spectrophotometric Assay for Phosphate (22)

1. Glassware must be phosphate-free (*see* **Note 11**).
2. 2.5% (w/v) Sodium molybdate solution in 1.67 M H_2SO_4. *Caution:* $Na_2MoO_4 \cdot 2H_2O$ is irritant; H_2SO_4 is corrosive.
3. 1 M HCl. *Caution:* HCl is corrosive.
4. Stannous chloride stock solution: 1.5 g of stannous chloride in 2.5 mL of concentrated HCl. Prepare 2 d prior to analysis to allow time for the stannous chloride to dissolve (allow to stand in a safe place). Prepare stock solution monthly. *Caution:* $SnCl_2$ and HCl are corrosive.
5. Stannous chloride working solution: 0.25% (v/v) of stannous chloride stock solution in 1 M HCl. Prepare working solution immediately prior to analysis.
6. Standard solution: 5 mM disodium hydrogen orthophosphate.
7. Disposable 2-mL cuvets ($10 \times 4 \times 45$ mm) (e.g., Sarstedt, Nümbrecht, Germany).
8. Disposable 1-mL plastic Pasteur pipets.
9. Spectrophotometer.

2.8. Spectrophotometric Assay for Cobalt (23)

1. 0.2 M Citric acid solution.
2. Phosphate–boric acid buffer: 6.2 g of boric acid, 35.6 g of disodium phosphate dehydrate, and 500 mL of 1 M NaOH in a total volume of 1 L.
3. 0.2% (w/v) Nitroso-R salt (1-nitroso-2-naphthol-3,6-disulfonate, disodium salt) in water. The solution undergoes no change for months if kept in the dark, but slowly decomposes in daylight. *Caution:* Nitroso-R is harmful.
4. 0.1 M $Co(NO_3)_2 \cdot 6H_2O$ stock solution. *Caution:* Cobalt nitrate is toxic and oxidizing.
5. Standard solution: 1 mM $Co(NO_3)_2 \cdot 6H_2O$. Prepare working solution on the day of use by dilution of the stock solution.
6. Concentrated nitric acid. Use only in a chemical fume hood. *Caution:* Nitric acid is oxidizing and corrosive.
7. Disposable 2-mL cuvets ($10 \times 4 \times 45$ mm).
8. Water bath (100°C), vortex mixer, spectrophotometer.

3. Methods

3.1. Preparation of Planktonic Cells and Biofilm

3.1.1. Batch Culture for Production of Planktonic Cells

1. Grow *Serratia* sp. cells for 48 h at 30°C on nutrient agar plates. Using a sterile loop, inoculate one loopful of bacterial cells into a 250-mL sterile Erlenmeyer flask containing 50 mL of glycerol or lactose minimal medium (*see* **Note 4**). Incubate the culture at 30°C on a rotary shaker for 16–18 h at 250 rpm.
2. Inoculate 5 mL of the overnight culture to a 500-mL sterile Erlenmeyer flask containing 100 mL of glycerol or lactose minimal medium. Incubate the culture at 30°C on a rotary shaker for 16–18 h at 250 rpm.
3. Inoculate the culture obtained in **step 2** to a 10-L sterile flask containing 3 L of medium. Incubate at 30°C using a water-bath with forced aeration (*see* **Fig. 2**) (*see* **Note 4**).
4. Harvest cells by centrifugation at 12,164g (*see* **Note 7**) at 4°C for 20 min and wash twice in isotonic saline. Suspend cells in isotonic saline and store in aliquots at 4°C up to 2 d.

3.1.2. Continuous Culture for Simultaneous Production of Planktonic Cells and Biofilm

1. Wash the support (approx 400 ceramic Raschig rings or approx 400 cubes of polyurethane foam) to be used for biofilm growth in deionized water. Transfer the support to a 2-L bottle, immerse in deionized water, autoclave, sterilize, and cool. Drain the support (*see* **Note 12**).
2. If polyurethane foam is used for biofilm growth, prepare 13 strands containing 25 cubes of polyurethane foam on each strand. Tie all of the strands around the central glass partition (*see* **Fig. 3**). If ceramic Raschig rings are used for biofilm growth, fill the stainless-steel wire cage with approx 300 rings (*see* **Fig. 3**).
3. Introduce the central glass partition into the fermentor. If polyurethane foam is used as a support, the strands should be located on the upside of the fermentor. If ceramic raschig rings are used, the cage should be hung on the upside (*see* **Fig. 3**). Put glass wool in the outlet of the condensers and inlet of air (i.e., glass connectors in between the flow meter and the mixing chamber), cover all outlets with foil and autoclavable tape, put on clip hose clamps and cable ties on all connections, and sterilize the fermentor, making sure that expanding air and steam can escape via the filters.
4. Prepare 40 L of minimal medium (lactose 0.6 g/L) using two 20-L bottles (*see* **Fig. 2**). Each bottle should contain one outlet with approx 50 cm of silicone tubing (internal diameter = 4 mm) and a glass connector (male) for the medium, one outlet with an air filter attached, one inlet with approx 10 cm of silicone tubing (internal diameter = 6 mm), and a glass connector (female) for the addition of lactose solution (*see* **Fig. 1**). Cover all outlets with foil and autoclavable tape and sterilize, making sure that expanding air and steam can escape via the filters (*see* **Note 3**).
5. Follow **step 1** of **Subheading 3.1.1.** Inoculate 5 mL of the overnight culture to a sterile 500-mL Erlenmeyer flask (with an outlet at the bottom of the flask with

approx 15 cm of silicone tubing attached, a male glass connector, and a clip hose clamp; *see* **Fig. 1**) containing 100 mL of 0.6 g/L lactose minimal medium. Incubate the culture at 30°C on a rotary shaker for 16–18 h at 250 rpm.

6. Connect the air inlet of the fermentor to the supply of air (approx 1.0 L/min.) (*see* **Fig. 3**). Make sure that all of the clip hose clamps are "open". Pump the medium into the fermentor until it reaches the level of the outflow. Control the temperature of the fermentor at 30°C. Aseptically inoculate (*see* **Note 2**) 100 mL of the overnight culture to the air-lift fermentor, increase aeration (approx 1.8 L/min; *see* **Note 13**), and leave as a batch culture under aeration for 24 h (*see* **Note 14**).

7. To remove a sample from the fermentor, connect a 50-mL plastic syringe to the outlet containing the sample collection vessel (*see* **Fig. 3**), pull the plunger and the sample vessel will fill with culture. Measure the phosphatase activity (*see* **Subheading 3.2.**), OD_{600}, and pH of the culture.

8. Pump the medium continuously into the fermentor for 6 d at a dilution rate (*D*) of 0.113 h^{-1} (*see* **Notes 15** and **16**). For a scaled-down system, the flow rate should be set to the same value of *D* (*see* **Note 16**).

9. Harvest the cells from the outlet by centrifugation at 12,164*g* at 4°C for 20 min and wash twice in isotonic saline. Suspend cells in isotonic saline and store in aliquots at −70°C until needed (*see* **Note 17**).

10. Remove the strands containing the polyurethane foam and/or the cage containing the ceramic rings coated with biofilm and suspend them from a glass rod in a 2-L glass measuring cylinder containing approx 200 mL of isotonic saline to maintain a humid atmosphere and cover with foil. Store at 4°C until needed.

11. Sterilize the fermentor prior to washing it (*see* **Note 18**).

3.2. Phosphatase Activity Assay (18,19)

Phosphatase specific activity is determined by measuring the liberation of *p*-nitrophenol (*p*NP) from *p*-nitrophenyl phosphate (*p*NPP) as described by Bolton and Dean *(18)*. One unit of enzyme activity is defined as 1 nmol of *p*NP liberated per minute per milligram bacterial protein. To avoid the need for performing a protein determination for each phosphatase assay, a calibration was generated between the turbidity of the bacterial suspension (OD_{600}) and the amount of protein present (measured by the method of Lowry) *(19)*. There was no significant difference in the calibration under different batch growth conditions and the conversion factor of 276 µg protein per unit of OD_{600} of 1 mL suspension was determined *(19)*.

1. Transfer 0.2 mL of cell suspension to three test tubes (A, B, and C).
2. Add 1.8 mL of MOPS–NaOH buffer, pH 7.0, to each test tube.
3. Pre-equilibrate samples at 30°C for 15 min.
4. Add 5 mL of 0.2 *M* NaOH to one test tube (C) to stop phosphatase activity.
5. Initiate the reaction by the addition of 0.4 mL of 12 mg/mL *p*NPP solution to test tube B, set the timer, and then mix.
6. Add 0.4 mL of 12 mg/mL *p*NPP solution to test tube C and mix.

7. Wait a few minutes, until tube B shows a yellow color, and then stop the reaction by the addition of 5 mL of 0.2 M NaOH. Vortex mix contents of test tubes B and C. Measure the absorbance of the suspension B at 410 nm against the reagent blank C.
8. Vortex mix the suspension in tube A and measure the absorbance at 600 nm against deionized water.
9. Calculate the phosphatase specific activity (nmol of pNP liberated per minute per milligram of bacterial protein) using the following equation:

$$\text{Phophatase specific activity} = \frac{A_{410} \times 10^9}{t \times 18{,}472} \times \frac{7.4}{10^3} \times \frac{1}{0.552 \times A_{600}} \tag{1}$$

where A_{410} is the absorbance of cell suspension B (against reagent blank C) at 410 nm, t is the time of reaction (in min), A_{600} is the absorbance of cell suspension A (against deionized water) at 600 nm, 18,472 is the molar extinction coefficient (M^{-1} cm^{-1}) for the above-described conditions, and 0.552 is the protein conversion factor (*see* **Note 19**).

3.3. Preparation of Immobilized Cells

There are two broad types of immobilization technique: (1) entrapment, where the microorganisms are caught in the interstices of porous or fibrous materials or physically restrained within or by a solid matrix and (2) attachment, where the microorganisms adhere to solid surfaces by self-adhesion or chemical bonding.

3.3.1. Entrapment

Polyurethanes constitute a group of polymers with highly versatile properties and, therefore, have been used as supports for the entrapment of *Serratia* sp. cells (*see* **Note 20**).

Several stages can be distinguished in the polyurethane foam production process *(24)*. Characteristic reactions take place at every stage (*see* **Fig. 5**). In the first and second stages, the reaction of isocyanate with hydroxyl groups begins to form urethane bonds to increase the viscosity of the reactant mix. At the same time, water reacts with isocyanate groups with evolution of CO_2, the formation of urea derivatives, and an increase in volume of the reaction mixture. As a result of these exothermic reactions, the temperature of the mixture increases. In the third stage, CO_2 evolution ceases, the reactions with hydroxyl groups come to completion, and the urethane and urea groups react with excess isocyanate groups to form branched or crosslinked groups.

1. Mix 34 g of Hypol HP 3000 polyurethane prepolymer with 25 g fresh (wet) weight of cells (i.e., the pellet [not suspended in isotonic saline]) obtained from the centrifugation of **Subheading 3.1.2., step 9** (*see* **Note 21**).
2. Allow the mixture to foam and cure for at least 30 min at ambient temperature.
3. Cut the foam containing entrapped cells into cubes of $7 \times 7 \times 7$ mm^3 and wash four times with 800 mL of isotonic saline. Store in isotonic saline at 4°C.

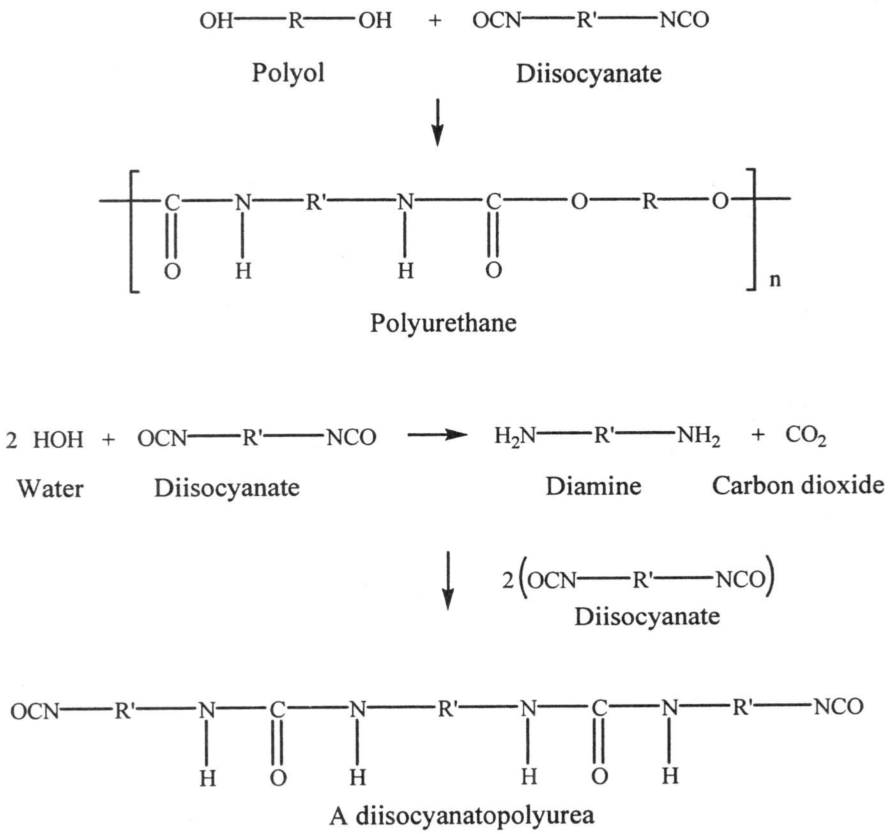

OH——R——OH   +   OCN——R'——NCO

Polyol   Diisocyanate

$$\downarrow$$

$$\left[\begin{array}{c}
\text{C} \quad \text{N} \quad \text{R'} \quad \text{N} \quad \text{C} \quad \text{O} \quad \text{R} \quad \text{O} \\
\| \quad | \qquad\qquad | \quad \| \\
\text{O} \quad \text{H} \qquad\qquad \text{H} \quad \text{O}
\end{array}\right]_n$$

Polyurethane

2 HOH   +   OCN——R'——NCO $\longrightarrow$ H$_2$N——R'——NH$_2$ +   CO$_2$

Water   Diisocyanate   Diamine   Carbon dioxide

$$\downarrow \quad 2\left(\text{OCN——R'——NCO}\right)$$
Diisocyanate

$$\text{OCN——R'——N—C—N——R'——N—C—N——R'——NCO}$$
$$\qquad\qquad\quad |\ \ \|\ \ |\qquad\quad\ |\ \ \|\ \ |$$
$$\qquad\qquad\quad \text{H}\ \ \text{O}\ \ \text{H}\qquad\quad \text{H}\ \ \text{O}\ \ \text{H}$$

A diisocyanatopolyurea

Fig. 5. Chemistry of polyurethane production.

3.3.2. Covalent Coupling to Supports

The immobilization of cells on solid supports by covalent coupling usually leads to very stable preparations with extended active life when compared with immobilized cell preparations obtained with other coupling methods (i.e., physical adsorption and ionic binding).

Inorganic supports, because of their physical properties, are suitable for industrial use and offer several advantages over their organic counterparts: high mechanical strength, thermal stability, resistance to microbial attack, easy handling, excellent shelf life, and easy regenerability.

The surface of most inorganic supports (e.g., glass, metal oxides, sand) is mainly composed of oxide and hydroxyl groups that provide a mildly reactive surface for activation and protein binding. The relative inertness of the inorganic

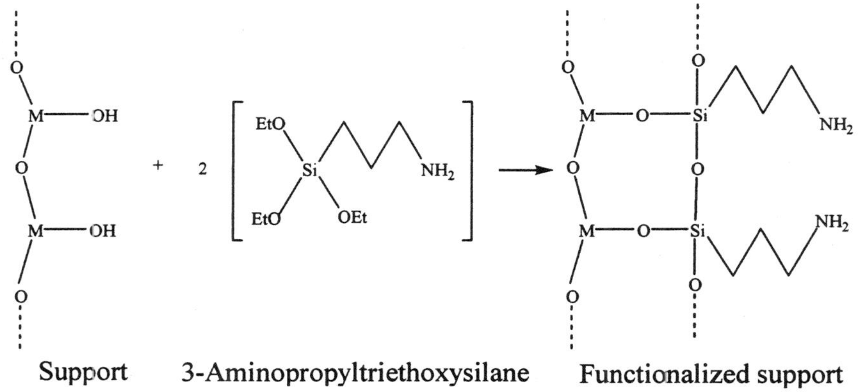

Fig. 6. Silanization of inorganic support (e.g., ceramic Raschig rings) using 3-amino-propyltriethoxysilane.

supports makes their derivatization difficult. However, this can be accomplished by an inorganic and/or crosslinked organic coating. One of the most popular derivatization methods of inorganic materials is the silanization method, which involves use of trialkoxy silane derivatives containing an organic functional group *(25)*. Coupling of these reagents (e.g., 3-aminopropyltriethoxy silane) to the support takes place presumably by displacement of the alkoxy residues on the silane, hydroxyl groups, or the oxidized surface of the inorganic support to form metal–*O*–Si linkages (*see* **Fig. 6**).

Supports containing amino groups (e.g., silanized support in **Fig. 6**) can be derivatized using glutaraldehyde, as shown in **Fig. 7.** One aldehyde group reacts with the amino group of the support to form a Schiff's base, whereas the second aldehyde group couples with an amino moiety of a ligand, thus achieving linkage of the ligand (protein) to the support (*see* **Fig. 7**). However, the nature of the reaction of glutaraldehyde with proteins and amino supports is not fully understood. It is known that commercial solutions of glutaraldehyde could contain α,β-unsaturated oligomers of glutaraldehyde; such oligomers would be formed by aldol condensation process via elimination of water (*see* **Fig. 8**). Therefore, coupling of protein amino groups to "glutaraldehyde" would take place via a Michael-type condensation reaction (i.e., conjugate-addition of an amine to the ethylenic double bond of an α,β-unsaturated carbonyl compound to give an alkylamino derivative (*see* **Fig. 8**) *(26)*. This mechanism explains the stability of the bond, which cannot be the result of a simple Schiff's base formation.

1. Treat 300 g of ceramic Raschig rings in a 2-L beaker with concentrated HNO₃ (i.e., enough to cover the rings) (*see* **Subheading 2.3.2., step 2**) for 4 h, mixing every 30

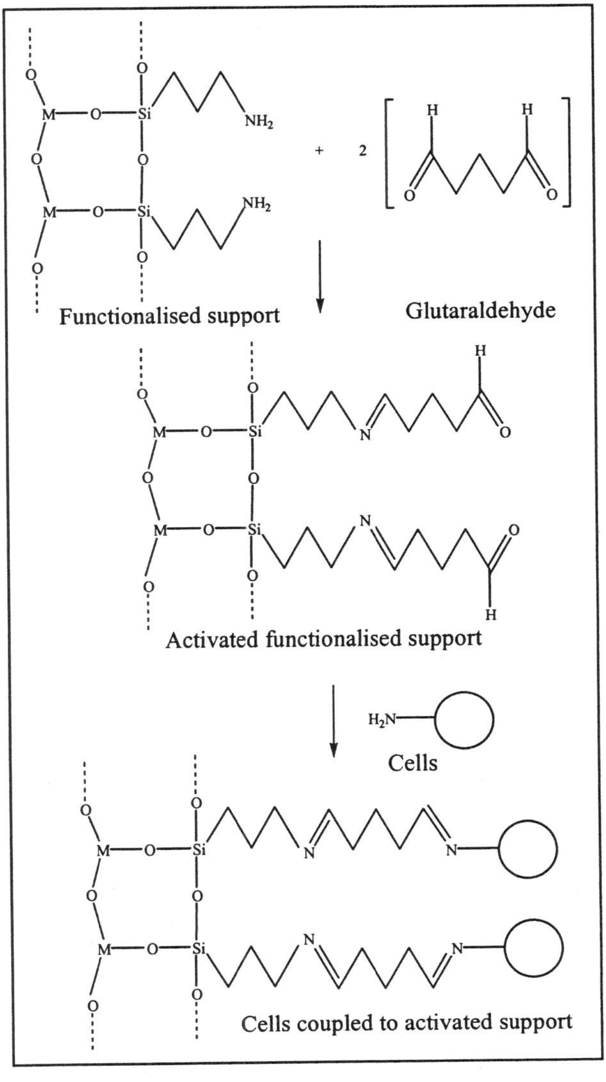

Fig. 7. Activation of amino bearing supports by glutaraldehyde and coupling of cells.

min. Drain the supports, wash them five times with deionized water (or until the pH of the washings is in the range of 5.0–6.0), and dry at 170°C in a furnace for 16 h.
2. Silanize the nitric acid treated rings by refluxing in a boiling water bath with 450 mL of 3-aminopropyltriethoxy silane solution for 50 min. Wash rings five times with

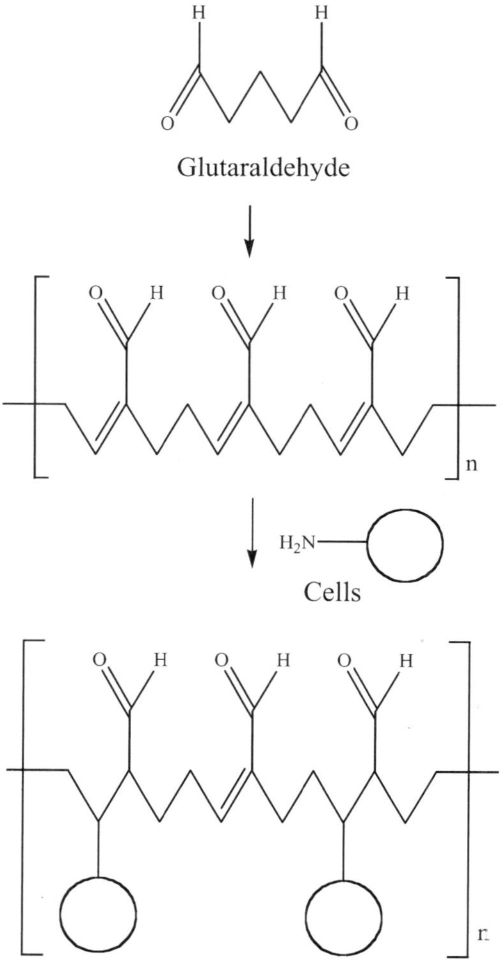

Fig. 8. Conjugate addition of amino groups to ethylene double bonds of α,β-unsaturated oligomers contained in commercial aqueous glutaraldehyde solutions.

250 mL of toluene, wash five times with 200 mL of acetone, and dry in a furnace oven at 115°C for 16 h. *Caution:* allow all acetone to evaporate before putting in oven.

3. Treat 130 g of silanized support with 724 mL of 5% glutaraldehyde solution for 1 h. Filter off the excess glutaraldehyde, wash the rings three times with 300 mL deionized water, and then wash three times with 300 mL of 0.1 *M* phosphate buffer, pH 7.0.

4. Suspend 49 g fresh (wet) weight of the cell pellet obtained in **step 9** of **Subheading 3.1.2.** (approx 85% moisture content; *see* **Note 21**) in 200 mL of 0.1 *M* phosphate

buffer, pH 7.0, and add to the aminopropyl-functionalized and glutaraldehyde-activated support. Allow the cells to react with the support for 15 h at 4°C. Wash the rings containing cells twice with 300 mL of 0.1 M acetate buffer, pH 4.5, containing 1 M sodium chloride, twice with 300 mL of deionized water, and twice with 300 mL 0.1 M phosphate buffer. Store the cells immobilized on to the rings in 0.1 M phosphate buffer, pH 7.0, at 4°C.

5. Wash the immobilized cells with isotonic saline prior to packing the reactor (*see* **Subheading 3.4.**) to eliminate the presence of phosphate.

3.4. Preparation of Packed-Bed Reactor Systems

3.4.1. Bioaccumulation of Hydrogen Uranyl Phosphate

1. Pack a cylindrical glass column (e.g., length = 9 cm, internal diameter = 1.5 cm, working volume = approx 13 mL) with immobilized *Serratia* sp. cells (e.g., polyurethane reticulated foam coated with biofilm [88 cubes, 5×5×5 mm³] [*see* **Subheading 3.1.2.**], or cubes of cells entrapped in Hypol foam [*see* **Subheading 3.3.1.**], or cells covalently coupled to Raschig rings [*see* **Subheading 3.3.2.**]).
2. Challenge the reactor (*see* **Fig. 4**) with uranium solution (**step 6 of Subheading 2.4.1.**), with an upward flow (approx 10 mL/h) via an external peristaltic pump and at ambient temperature (*see* **Note 22**).
3. Take samples (approx 1 mL) from the outflow and measure the uranium (*see* **Subheading 3.5.**) and phosphate (*see* **Subheading 3.7.**) contents and pH. If required, store samples at −20°C prior to analysis.

3.4.2. Intercalation of Nickel or Cobalt Into Hydrogen Uranyl Phosphate

1. Prepare a reactor as described in **Subheading 3.4.1., step 1** (*see* **Fig. 4**).
2. Prime the reactor with uranium solution (**Subheading 2.4.1., step 6**) for 24 h, with an upward flow via an external peristaltic pump, at ambient temperature. Set the flow rate (mL/h) to give approx 50% removal of the input uranium (*see* **Note 23**).
3. Take samples (approx 1 mL) from the outflow and measure the uranium contents (*see* **Subheading 3.5.**).
4. Estimate the amount of HUP in the reactor by calculation: difference between total U_{in} and total U_{out} for the duration.
5. Challenge the reactor primed with HUP with nickel or cobalt solution (**Subheading 2.4.2., steps 3** and **4**) at a flow rate of 10 mL/h until the column reaches saturation (appearance of Ni^{2+} or Co^{2+} in the column outflow).
6. Take samples (approx 1 mL) from the outflow and measure the nickel (*see* **Subheading 3.6.**) or cobalt (*see* **Subheading 3.8.**) and phosphate (*see* **Subheading 3.7.**) contents and pH. If required, store samples at −20°C prior to analysis.

3.4.3. Cocrystallization of Nickel and Hydrogen Uranyl Phosphate

1. Prepare a reactor as described in **step 1** of **Subheading 3.4.1.** (*see* **Fig. 4**).
2. Challenge the reactor with mixed metal solution (**Subheading 2.4.3., step 3**) with an upward flow via an external peristaltic pump, at ambient temperature. Set the flow rate (mL/h) to give approx 50% removal of the uranium from solution (*see* **Note 24**).

3. Take samples (approx 1 mL) from the outflow and measure the nickel (*see* **Subheading 3.6.**) and phosphate (*see* **Subheading 3.7.**) contents and pH. If required, store samples at –20°C prior to analysis.

3.5. Spectrophotometric Assay for Uranium (20)

1. Transfer 30 μL of sample or standard to a cuvet.
2. Add 1.97 mL of 0.1 *M* HCl.
3. Add 0.1 mL of 0.15% (w/v aq.) of arsenazo III solution and mix well using a 1-mL plastic Pasteur pipet.
4. Measure the absorbance at 652 nm against a reagent blank.
5. Plot a standard curve, absorbance vs uranyl nitrate concentration, and derive the linear regression equation used to determine the uranium content (expressed as m*M* or mg/L) of the samples.

3.6. Spectrophotometric Assay for Nickel (21)

1. Transfer 50 μL of a sample or standard to a cuvet.
2. Add 0.2 mL of acetate buffer, pH 5.5, 2 mL of ethanol, 1.4 mL of deionized water, and 1 mL of 1% (w/v) of $Na_2S_2O_3$.
3. Add 0.4 mL of 0.02% (w/v) 2-Br-PADAP in ethanol reagent.
4. Measure the absorbance after 1 h at 560 nm against a reagent blank (the complex is stable for 24 h).
5. Plot a standard curve, absorbance vs nickel nitrate concentration, and derive the linear regression equation used to determine the nickel content (expressed as m*M* or mg/L) of the samples.

3.7. Spectrophotometric Assay for Phosphate (22)

1. Transfer 0.97 mL of deionized water to a disposable 2-mL cuvet.
2. Add 0.6 mL 2.5% (w/v) sodium molybdate in 1.67 *M* H_2SO_4.
3. Add 0.03 mL of sample or standard (up to 5 m*M* PO_4^{3-}).
4. Add 0.4 mL of freshly prepared stannous chloride working solution.
5. Mix solution well using a 1-mL disposable plastic Pasteur pipet.
6. Allow to equilibrate for 20 min.
7. Measure the absorbance at 720 nm against a reagent blank.
8. Plot a standard curve, absorbance vs disodium hydrogen orthophosphate concentration, and derive the linear regression equation used to determine the phosphate content (expressed as m*M* or mg/L) of the samples.

3.8. Spectrophotometric Assay for Cobalt (23)

1. Transfer 0.03 mL of sample or standard (containing 1–10 μg of Co^{2+}) to a test tube.
2. Add 0.25 mL of 0.2 *M* citric acid.
3. Add 0.3 mL of phosphate–boric acid buffer. The pH of the solution should be close to 8.0.
4. Add 0.125 mL of 0.2% (w/v) Nitroso-R salt solution while stirring.

Table 1
Sterilization Times for Liquids in Different Vessels in the Autoclave at 121–123°C

Vessel	Volume (mL)	Exposure time (min)
Erlenmeyer flask	50	12–14
Erlenmeyer flask	200	12–15
Erlenmeyer flask	1000	20–25
Erlenmeyer flask	2000	30–35
Bottle	9000	50–55

Source: **ref. 27**.

5. Cover the test tubes containing the samples and transfer to a water bath at 100°C and leave for 1 min.
6. Add 0.25 mL of concentrated HNO_3 and leave for a further 1 min at 100°C.
7. Cool the sample in the dark.
8. Add 0.325 mL of deionized water.
9. Measure the transmittance of the solution at 420 nm.
10. Plot a standard curve, transmittance vs cobalt nitrate concentration, and derive the equation used to determine the cobalt content (expressed as mM or mg/L) of the samples. Note that the resultant graph is exponential.

4. Notes

1. Use a sterile pipet to add small amounts of glycerol or lactose solution to the medium, if dispensed into smaller flasks, prior to inoculation. Where the medium volume is large (e.g., over 3 L), use a 250-mL or 500-mL Erlenmeyer flask as described in **Fig. 1**. Sterilize the flask containing the carbon source solution and add it aseptically to the medium (*see* **Note 2**).
2. To connect sterile glass connectors (male and female) aseptically, take the foil off one connector and immerse it in 70% ethanol solution, repeat for the other connector, withdraw both from the ethanol solution, shaking off excess ethanol, and connect them immediately.
3. The necessary duration of the sterilization process depends on the size (i.e., the heat capacity) of the apparatus and vessels to be sterilized; this can be derived from the data in **Table 1** (*27*).
4. Small culture volumes (up to 500 mL) can be aerated using a rotary shaker (250 rpm); culture volume should comprise no more than 15–20% of the nominal maximum volume of the flask. Larger volumes should be grown in containers with forced aeration (*see* **Fig. 2**): compressed air fed via an external supply, via a charcoal filter to remove any volatile oils, and an antibacterial filter (0.2 μm) for air sterilization.

5. Some stainless steels can be antibacterial because of leaching of small quantities of metals. Ideally, microbiological quality stainless steel should be used or, if unavailable, the material should be leached extensively in deionized water before use and then tested for possible toxicity effects in trial batch cultures.

6. Ideally, a heavy-duty compressor should be used (e.g., CompAir Auto Power, E10: 2 HP, maximum pressure 150 psig, 200 L recipient size [Oscott Air Ltd., Birmingham, UK]). If this is not available, a scaled-down version of the fermentor (length = 22.5 cm and internal diameter = 5 cm)—all other components scaled down to match (except the biomass support materials)—can be run using air supplied from an aquarium pump (e.g., Rena Air pump, type 301R [Aquarium Pharmaceuticals, Inc., Chalfont, PA]) available from most pet stores.

7. Efficient separation of cells is usually achieved using a fixed-angle rotor at a relative centrifugal force (RCF) of 12,164g for 20 min, at 4°C. The MSE "Mistral" 6X 1 L with a swing-out rotor at the manufacturer's recommended maximum RCF (7,004g) is marginally adequate with a spin time of 30 min.

8. Hypol 3000 reacts with water to make foam and releases carbon dioxide in the process. Avoid contamination of the stored product with water, as pressure from generated carbon dioxide can violently rupture a sealed container.

9. 20 mM MOPS–NaOH was originally used to control the pH of the solution. However, it can be excluded from the solution [i.e., 0.5 mM UO$_2$(NO$_3$)$_2$·6H$_2$O, 2 mM citrate buffer, and 5 mM glycerol 2-phosphate, pH 6.0–7.0] without changes to the final result.

10. Radioactive material. Make sure that all necessary safety precautions are adopted. Spill trays must be used. Wear protective clothing and gloves at all times. Take special care when handling stock solutions. Deal with spillages immediately. Define the work areas with radioactive warning tape. Dispose of in accordance with local guidelines on radioactive waste disposal.

11. The phosphate assay is very sensitive (limit = approx 1 ppm) and it is necessary to preclean glassware by immersing in 2–5% Decon 90 solution (Decon Laboratories Ltd., Hove, UK), soaking for 24 h and extensive rinsing in deionized water. It is usually convenient to reserve the glassware once cleaned. Alternatively, assays can be conveniently carried out in 2-mL disposable cuvets, which contribute negligible free PO$_4^{3-}$ (e.g., Sarstedt).

12. This eliminates from the support any toxic components that may affect the phosphatase activity.

13. The aeration rate is much lower than usually used for aerobic cultures. The suggested rate maintains the culture at above 90% of saturation of dissolved oxygen (DO$_2$) on the upside of the fermentor *(16)*. This value reduces with time, reaching approx 50% at the end of the fermentation.

14. The glass wool from the condenser might get wet, either during autoclaving or fermentation. If that is the case, quickly change the wet glass wool for dry sterile glass wool, using tweezers that have been dipped in 70% ethanol solution.

15. The pumps should be calibrated prior to the experiment. Test the same tubing that will be used for the experiment and determine the flow rates (mL/h) at the various pump

settings. Plot a curve, pump setting vs flow rate, and derive the linear regression equation used to determine what setting should be used for the required flow rate.

16. Flow rate (mL/h) = $D \times V$, where V (mL) is the working volume of the fermentor. If $V = 2200$ mL and $D = 0.113$ h^{-1} the flow rate will be 248 mL/h. To determine the flow rate (mL/h), measure the volume of the outflow from the fermentor in a fixed period of time. Adjust the pump setting accordingly.

17. The frozen cells maintain their activity when thawed.

18. To remove the cells from the sintered glass from the air-lift fermentor or batch culture, immerse it subsequently in HCl : water (1 : 1) for 24 h and in 5% (v/v) Decon 90 in water for 24 h or longer if necessary. Rinse in deionized water.

19. When the molar extinction coefficient is known (ε, M^{-1} cm^{-1}) and a fixed path length established (l, 1 cm), the concentration of an unknown amount of a substance (c, M) can be determined by measuring the absorbance of the substance and applying the Lambert–Beer law:

$$c = \frac{A}{\varepsilon \times l} \qquad (2)$$

Using Eq. 2, the nanomoles of pNP in the assay will be

$$\frac{A_{410} \times 10^9}{18{,}472} \times \frac{7.4}{10^3} \qquad (3)$$

Using the conversion factor 276 µg of protein per unit of A_{600} of 1-mL suspension, the amount of protein (mg) in a 2-mL cell suspension will be

$$A_{600} \times 0.276 \times 2 = A_{600} \times 0.552 \qquad (4)$$

Therefore, with Eqs. 3 and 4, the phosphatase specific activity (i.e., nmmol pNP liberated per mg protein per min) can be determined.

20. Previous studies have used *Serratia* sp. (*Citrobacter* sp.) cells immobilized in gel matrices *(8,11,22)*, but gels lack mechanical strength for large-scale operation and their production often utilizes toxic components (e.g., polyacrylamide gel). Other methods of entrapment using nontoxic matrices (e.g., alginate and chitosan) have been tested, but the matrices might dissolve after a month at 4°C, probably via enzymatic activity of the organism.

21. The water : prepolymer ratio should be approx 0.63 : 1. If 25 g of the cell paste contains approx 21.25 g of water (moisture content of approx 85%), the ratio 21.25 g water : 34 g Hypol will be approx 0.63 : 1. The moisture content of the cells was determined by drying the sample for 48 h at 39.8°C, under vacuum, in an Abderhalden drying pistol using dichloromethane as the refluxing solvent.

22. Using the conditions described, the uptake of uranium from solution is over 95% during 5 d (at least) and column activity is retained over many weeks of continuous operation.

23. The use of the $F^{1/2}$ value compensates for variations in column activity attributable to differences in phosphatase activity and ensures comparable HUP loadings and distributions per column.

24. Shredded polyacrylamide gel containing entrapped *Serratia* sp. cells packed in a column (internal diameter = 2.2 cm, length = 8 cm), challenged with uranyl ion and nickel (0.5 mM each) in the presence of glycerol 2-phosphate, with a flow rate set that removed 56.3% of the UO_2^{2+} and 27.3% of the Ni^{2+}. This removal was sustained at steady state over 42 h, corresponding to 25 column volumes *(10)*. Other possibilities to increase the Ni^{2+} removal would be (1) to aim for approx 98% of UO_2^{2+} removal for approx 50% of Ni^{2+} removal, (2) to challenge with 1 mM of UO_2^{2+} and 0.5 mM of Ni^{2+}, or (3) to put the residual Ni^{2+} through again with new UO_2^{2+}. It has been shown that challenging a reactor packed with polyurethane foam coated with *Serratia* sp. biofilm with 1 mM UO_2^{2+} and 1 mM Sr^{2+}, with a steady-state removal of U of over 92%, the corresponding removal of Sr^{2+} was in the range of 50–56% *(15)*.

Acknowledgments

This work was supported by the BBSRC (grant no. 6/E1464). The authors thank Recticel (Belgium) and Norton Chemical Process Products Corporation (USA) for the supports for immobilization of cells.

References

1. Pattanapipitpaisal, P., Mabbett, A. N., Finlay, J. A., et al. (2002) Reduction of Cr(VI) and bioaccumulation of chromium by Gram positive and Gram negative microorganisms not previously exposed to Cr-stress. *Environ. Technol.* **23,** 731–745.
2. Jeong, B. C. and Macaskie, L. E. (1995) PhoN-type acid phosphatase of a heavy metal-accumulating *Citrobacter* sp.: resistance to heavy metals and affinity towards phosphomonoester substrates. *FEMS Microbiol. Lett.* **130,** 211–214.
3. Jeong, B. C., Poole, P. S., Willis, A. C., and Macaskie, L. E. (1998) Purification and characterization of acid-type phosphatase from a heavy-metal-accumulating *Citrobacter* sp. *Arch. Microbiol.* **169,** 166–173.
4. Jeong, B. C. and Macaskie, L. E. (1999) Production of two phosphatases by *Citrobacter* sp. grown in batch and continuous culture. *Enzyme Microb. Technol.* **24,** 218–224.
5. Basnakova, G., Stephens, E. R., Thaller, M. C., Rossolini, G. M., and Macaskie, L. E. (1998) The use of *Escherichia coli* bearing a *phoN* gene for the removal of uranium and nickel from aqueous flows. *Appl. Microbiol. Biotechnol.* **50,** 266–272.
6. Macaskie, L. E., Empson, R. M., Cheetham, A. K., Grey, C. P., and Skarnulis, A. J. (1992) Uranium bioaccumulation by a *Citrobacter* sp. as a result of enzymatically mediated growth of polycrystalline HUO_2PO_4. *Science* **257,** 782–784.
7. Yong, P. and Macaskie, L. E. (1995) Enhancement of uranium bioaccumulation by a *Citrobacter* sp. *via* enzymatically-mediated growth of polycrystalline $NH_4UO_2PO_4$. *J. Chem. Technol. Biotechnol.* **63,** 101–108.
8. Macaskie, L. E. (1990) An immobilized cell bioprocess for the removal of heavy metals from aqueous flows. *J. Chem. Technol. Biotechnol.* **49,** 357–379.
9. Macaskie, L. E., Yong, P., Doyle, T. C., Roig, M. G., Diaz, M., and Manzano, T. (1997) Bioremediation of uranium-bearing wastewater: biochemical and chemical factors influencing bioprocess applications. *Biotechnol. Bioeng.* **53,** 100–109.

10. Bonthrone, K. M., Basnakova, G., Lin, F., and Macaskie, L. E. (1996) Bioaccumulation of nickel by intercalation into polycrystalline hydrogen uranyl phosphate deposited *via* an enzymatic mechanism. *Nat. Biotechnol.* **14,** 635–638.

11. Basnakova, G. and Macaskie, L. E. (1997) Microbially enhanced chemiosorption of nickel into biologically synthesized hydrogen uranyl phosphate: a novel system for the removal and recovery of metals from aqueous solutions. *Biotechnol. Bioeng.* **54,** 319–328.

12. Basnakova, G., Spencer, A. J., Palsgard, E., Grime, G. W., and Macaskie, L. E. (1998) Identification of the nickel uranyl phosphate deposits on *Citrobacter* sp. cells by electron microscopy with electron probe X-ray microanalysis and by proton-induced X-ray emission analysis. *Environ. Sci. Technol.* **32,** 760–765.

13. Basnakova, G., Finlay, J. A., and Macaskie, L. E. (1998) Nickel accumulation by immobilised biofilm of *Citrobacter* sp. containing cell-bound polycrystalline hydrogen uranyl phosphate. *Biotechnol. Lett.* **20,** 949–952.

14. Clearfield, A. (1988) Role of ion exchange in solid-state chemistry. *Chem. Rev.* **88,** 125–148.

15. Paterson-Beedle, M. and Macaskie, L. E. (2004) Removal of cobalt, strontium and caesium from aqueous solutions using native biofilm of *Serratia* sp. and biofilm pre-coated with hydrogen uranyl phosphate, in *Proceedings 15th International Biohydrometallurgy Symposium,* Athens, (Tsezos, M., et al. eds.), Elsevier, in press.

16. Finlay, J. A., Allan, V. J. M., Conner, A., Callow, M. E., Basnakova, G., and Macaskie, L. E. (1999) Phosphatase release and heavy metal accumulation by biofilm-immobilised and chemically-coupled cells of a *Citrobacter* sp. pre-grown in continuous culture. *Biotechnol. Bioeng.* **63,** 87–97.

17. Macaskie, L. E., Hewitt, C. J., Shearer, J. A., and Kent, C. A. (1995) Biomass production for the removal of heavy metals from aqueous solutions at low pH using growth-decoupled cells of a *Citrobacter* sp. *Int. Biodeterior. Biodegrad.* **35,** 73–92.

18. Bolton, P. G. and Dean, A. C. R. (1972) Phosphatase synthesis in *Klebsiella (Aerobacter) aerogenes* growing in continuous culture. *Biochem. J.* **127,** 87–96.

19. Jeong, B. C. (1992) Studies on the atypical phosphatase of a heavy metal accumulating *Citrobacter* sp. D Phil thesis, University of Oxford, UK.

20. Yong, P., Eccles, H., and Macaskie, L. E. (1996) Determination of uranium, thorium and lanthanum in mixed solutions using simultaneous spectrophotometry. *Anal. Chim. Acta* **329,** 173–179.

21. Wei, F.-S., Qu P.-H., Shen, N.-K., and Ying, F. (1981) Sensitive spectrophotometric determination of nickel(II) with 2-(5-bromo-2-pyridylazo)-5-diethylaminophenol. *Talanta* **28,** 189–191.

22. Yong. P. and Macaskie, L. E. (1997) Effect of substrate concentration and nitrate inhibition on product release and heavy metal removal by *Citrobacter* sp. *Biotechnol. Bioeng.* **55,** 821–830.

23. Onishi, H. (ed.) (1986) *Photometric Determination of Traces of Metals, Part IIA: Individual Metals, Aluminium to Lithium,* 4th ed., Wiley, New York, pp. 454–459.

24. Woods, G. (ed.) (1990) *The ICI Polyurethanes Book,* 2nd ed., Wiley, Chichester, pp. 27–29.

25. Cabral, J. M. S. and Kennedy, J. F. (1991) Covalent and coordination immobilisa-
    tion of proteins, in *Protein Immobilisation—Fundamentals and applications*
    (Taylor, R. F., ed.), Marcel Dekker, New York, pp. 73–138.
26. Goldstein, L. and Manecke, G. (1976) The chemistry of enzyme immobilisation, in
    Immobilised Enzyme Principles (Wingard, L. B., Jr., Katchalski-Katzir, E., and
    Goldstein, L., eds.), Academic, London, pp. 23–126.
27. Schlegel, H. G. (1993) *General Microbiology,* 7th ed., Cambridge University Press,
    Cambridge.

26

Safe Dispatch and Transport of Biological Material

Vera Weihs and Christine Rohde

Summary

The correct transport of biological material is one of the major aspects of biolegislation. The awareness of this fact has only quite slowly come into being. The demands on the export, packaging, and shipping of biological material are manifold and the number of possible mistakes is endless. On the other hand, to be in conformity with all regulations is easy under the premises that a person with a sound knowledge of all the relevant regulations feels responsible for the shipment. This chapter describes which laws, regulations, and restrictions have to be respected before a biological substance can be released (*see* **Note 1**). Before a living microorganism is offered for dispatch, whether it is a transborder supply or supply within a country, it has to be made sure that the organism does not fall into wrong hands so that dispatch might be illegitimate. According to the International Air Transport Association Dangerous Goods Regulations (IATA DGR) *(1)*, the sender of infectious substances has to be a trained person, as infectious substances are classified as dangerous goods in Class 6, Division 6.2. The regulations for shipping biological substances by postal mail are laid down by the Universal Postal Union (UPU). The transport of hazardous biological material—performed in compliance with these regulations—can be regarded as safe by using triple-packaging UN combination systems.

Key Words: Dispatch; transport; IATA Dangerous Goods Regulations; infectious substances; export; import; Universal Postal Union (UPU); WHO Risk Groups; EEC Dual Use Directive; triple packaging; Shipper's Declaration; EN 829; genetically modified organisms; diagnostic specimens.

1. Export and Import of Biological Material and Dispatch Within a State

As a general rule, biological material—at least pathogens or otherwise hazardous biological material—should never be sent to private persons. Only adequately trained staff in sufficiently equipped laboratories should receive microorganisms in order to guarantee safe handling and proper use. **Table 1** demonstrates which issues have to be clarified before any dispatch of biological material.

From: *Methods in Biotechnology, Vol. 18: Microbial Processes and Products*
Edited by: J. L. Barredo © Humana Press Inc., Totowa, NJ

Table 1
Issues to Be Clarified Before the Dispatch of Biological Material

Risk group	Destination	Notice	Action/proof	Kind of dispatch
1	Intranational	National laws like: Genetic Engineering Act, Plant Protection Act, Infectious Diseases of Animals Act	Permission/registration/authorization	Mail
	Foreign country	Within the EU: Plant Protection Act; Air mail might not be admitted; Act on the Control of War Weapons; Total embargo?	Only for civil use; **No dispatch**	Mail/air mail; Air freight
2	Intranational	National laws like: Infectious Diseases Act, Infectious Diseases of Animals Act, Plant Protection Act, Genetic Engineering Act, Act on the Control of War Weapons	Permission/registration/authorization/registered laboratory	Private carriers by road
	Foreign country	EEC Regulation for the Control of Exports of Dual-Use Goods; Act on the Control of War Weapons; Import or quarantine restrictions; Import permit permission; EU: Plant Protection Act	Only for civil use; Authorization; Only for civil use	Private carriers by road or air freight
	Non-OECD country	Embargo? →→→→→; Act dealing with foreign trade →→→→→	**No dispatch**; **No dispatch**	
3 and 4	See Risk Group 2	See Risk Group 2	See Risk Group 2	Private carrier

Table 2
Useful Websites

Biological safety	Export restrictions	Transport
http://www.who.int/en/	http://www.un.org/News/ ossg/sanction.htm	http://www.unece.org/ trans/danger/danger.htm
http://www.biodiv.org	http://binas.unido.org/binas	http://www.iata.org/index.htm
http://biosafety.ihe.be/ Menu/BiosEur1.html	http://www.biodiv.org	http://www.icao.int
	http://biosafety.ihe.be/ Menu/BiosEur1.html	http://www.upu.int
	http://www.opbw.org	
	http://www.dfat.gov.au/ isecurity/pd/pd_4_96/ pd9.html	

1.1. Export

The export of microorganisms is usually governed by national law. However, there is international or regional (e.g., in Europe) harmonization on the lists of microorganisms affected (*see* **Table 2**). The Biological and Toxin Weapons Convention (BTWC) states that the prohibition of the development, production, and stockpiling of chemical and biological weapons as well as their elimination will facilitate the achievement of general and complete disarmament under strict and effective international control (*see* **Table 2**). With few exemptions, organisms classified in Risk Group 1 (no or very low individual and community risk according to WHO definition) *(2)* are admitted for export without restrictions. Respective lists allocating organisms to the Risk Groups can, for example, be found in EEC Directive 2000/54/EC on the protection of workers from risks related to exposure to biological agents at work *(3)*. However, a total embargo would exclude even such organisms from export. Sanctions by the United Nations (*see* **Table 2**) might be in place and have to be observed.

More general restrictions are imposed by those laws and directives controlling the export of goods with a possible dual use, which means they could be used for civil or military purposes. A regulative example is EEC Council Directive 3381/94/EEC setting up a Community regime for the control of exports of goods with dual use *(4)*. Organisms categorized as potential biological weapons include the causative agents of anthrax (*Bacillus anthracis*, Risk Group 3), pestilence (*Yersinia pestis*, Risk Group 3), or toxin-producing strains of organisms like *Aspergillus fumigatus*, *Clostridium botulinum*, and

Staphylococcus aureus (all Risk Group 2). Other examples are certain plant (crop) pathogenic species and all those genetically modified microorganisms that meet the definition of the respective paragraph in the EU list of dual-use goods *(4)*.

1.2. Import

The import of biological material might be subject to the quarantine regulations of a given recipient country. These restrictions often depend on the special ecological or climatic situation of a certain country: It is free of a certain pathogen and it might be expected that an introduced microorganism could spread excessively. The United States only permit entry of animal cell cultures after a certain period of quarantine in connection with an import permit. Any individual case of import of microorganisms to Australia, Canada, and New Zealand requires an import permit. The European Community Commission Directive 97/3/EC *(5)* restricts the import of organisms harmful to plants and plant products and their spread within the Community. In Germany, comparable restrictions concerning, for example, *Synchytrium endobioticum,* the causative agent of potato cancer, or *Erwinia amylovora,* causing fire blight in fruit trees, are imposed according to the German Plant Protection Act *(6)*. Similarly, the German Infectious Diseases of Animals Imports Enactment *(7)* restricts the import of *Chlamydia psittaci,* causing psittacosis, or *Zymonema farcinimosus,* responsible for lymphatic vessel inflammation of one-hoofed animals. In other countries, similar laws exist.

1.3. Dispatch Within a State

Handling of certain kinds of biological material might be restricted to certain persons or registered laboratories having the required permissions or registrations. Pathogens or otherwise hazardous biological material should only be delivered after having verified that the recipient has the respective permissions to handle the material: (1) to handle human pathogens, (2) to handle animal pathogens, (3) to handle plant pathogens, and (4) to handle genetically modified organisms. It should be noted that also the intranational dispatch of certain agents listed as potential dual-use material might be regulated: A signed end-user certificate is possibly required. The best sources of information are the respective national export offices.

2. Transport

Living, potentially or definitely pathogenic biological material is sent all around the world for scientific, industrial, or teaching purposes. During transport, such material might present a menace or danger for those persons involved in the transportation chain who are all not trained with respect to working with biological substances (e.g., postal employees, secretaries, and others who could be exposed if an accident occurs). To prevent or to reduce the possibility of an inadvertent release of microorganisms, national and international laws and regulations for the transport and the packaging of the material exist (*see* **Table 2**).

2.1. Packaging of Biological Material

The principle of the safe packaging of biological material relies on the system of triple packaging. The system specifically designed for transport of infectious substances (Class 6, Division 6.2) serves as the only protection for people, animals, and the environment. Also, nonhazardous biological substances (allocated to WHO Risk Group 1) that are consequently not regulated by dangerous goods transport regulations should, of course, reach their destination in a vital and undamaged status, as fast as possible.

The system of triple packaging is as follows. The packaging must consist of (1) a watertight primary receptacle, (2) a watertight secondary packaging, and (3) an outer packaging. Absorbent material has to be placed between the primary and the secondary packaging. Several primary receptacles should be wrapped individually. There should be sufficient absorbent material present (e.g., cotton wool) to absorb the entire liquid if a leakage occurs. The outer packaging bears all the labels and documentation: (1) label the package containing infectious materials as "Infectious Substance" by using the correct label (*see* **Fig. 1**) and (2) label shipments containing noninfectious material as "Perishable Biological Substances" (*see* **Fig. 2**). Labels are commercially available or are already printed on ready-to-use packaging.

Figures 3 and **4** depict the schematic construction of the triple-packaging principle. UN certified combination packagings *(8)* as well as EN 829 systems *(9)* are commercially available and have to meet the respective specific performance tests. In comparison to EN 829 packaging, the UN combination packaging has a much higher strength and stability.

2.2. Transport by Mail

National postal services are merged in the Universal Postal Union (UPU). In the Letter Post Compendium *(10)*, guidelines are laid down for the postal transport and packaging of perishable biological substances and infectious material. It is differentiated between noninfectious organisms (Risk Group 1) and infectious organisms (Risk Groups 2, 3, and 4).

Generally, the exchange of biological substances by mail is restricted to countries whose postal administrations agree to accept biological material. The national postal services of several countries do not allow shipment (receipt or dispatch) of perishable biological substances by mail, either by surface or air *(11)*. Almost all postal administrations prohibit infectious substances like all

Fig. 1. Label for infectious substances.

Fig. 2. Label for noninfectious biological substances.

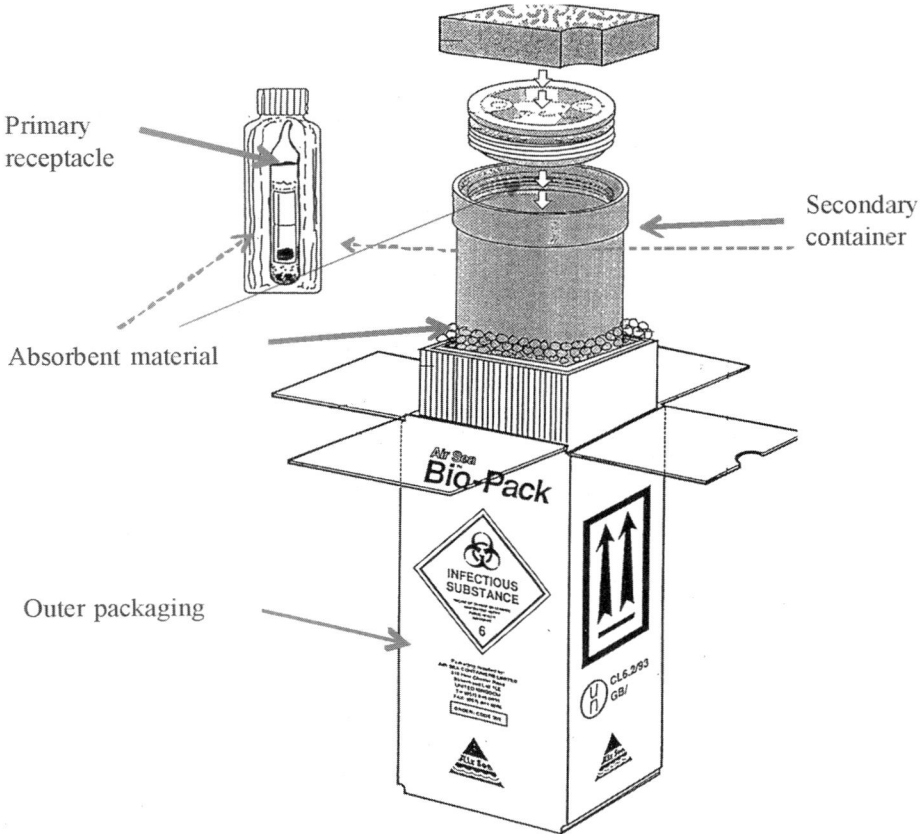

Primary receptacle

Secondary container

Absorbent material

Outer packaging

Fig. 3. Triple packaging for infectious substances according to UN Packing Instruction 620(IATA P1602).

other kinds of dangerous goods in the mail. Consequently, all postal administrations involved (senders, consignees, and all transit countries) would have to permit such a transport. As this is almost never possible, the only alternative of sending the material by freight and private courier has to be chosen. The national postal authorities or postal offices concerned give advice whether transport by mail is admitted (*see* **Table 2**). The following requirements have to be fulfilled in addition to those described in **Subheading 2.1:** (1) Indicate which organisms are shipped (e.g., on a delivery slip) between the outer and secondary container. (2) Label as "Lettre" to speed up transport and delivery (please note that any biological material must not be sent by postal parcel service).

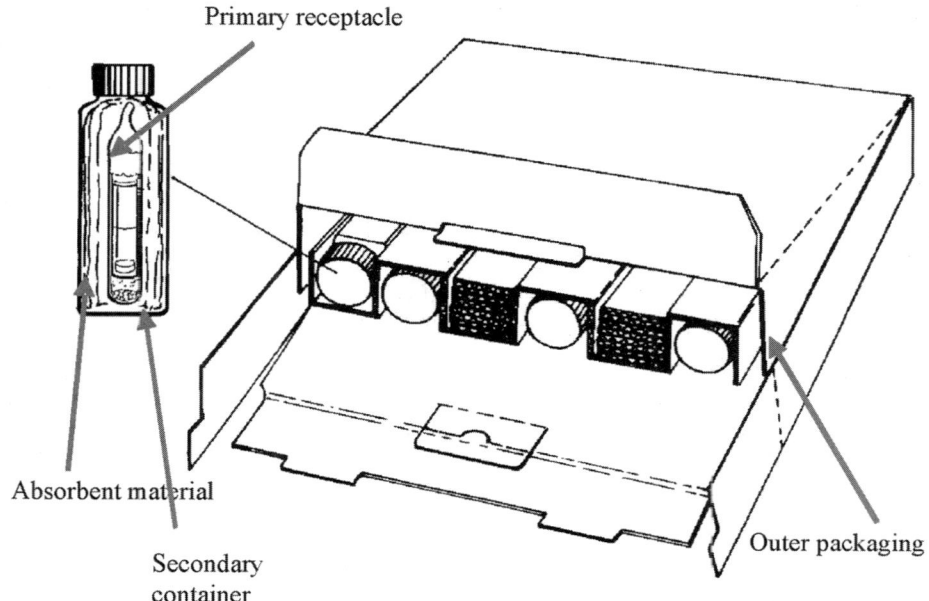

Fig. 4. Packaging according to standard EN 829 for noninfectious biological substances.

(3) Shipments of biological material into foreign countries might be subject to customs inspection. They must carry the green customs declaration C1 (*see* **Fig. 5**). Indicate number and type of material (e.g., 3 bacterial cultures, 2 plasmid preparations). The note "No commercial value" must be attached in case of free exchange between laboratories or any other supply free of charge. (4) If infectious substances are shipped by air mail (if permitted at all!), the IATA Dangerous Goods Regulations (DGR) *(1)* have to be followed and the Shipper's Declaration for Dangerous Goods must be filled in (*see* **Subheading 2.3.**).

2.3. Transport as Freight

If shipments containing biological material are not accepted in (air) mail, shipping by using private carrier service and (air) freight has to be chosen. The carrier will decide on the fastest mode of transport (road, rail, air, waterways) and will assist in filling in the necessary documents. The United Nations Committee of Experts for the Transport of Dangerous Goods established a system for the classification of all kinds of dangerous substances, called "Orange Book" *(8)*, with the aim of ensuring safe shipping for all modes of transport. The nine different classes include explosives, gases, flammable liquids, flam-

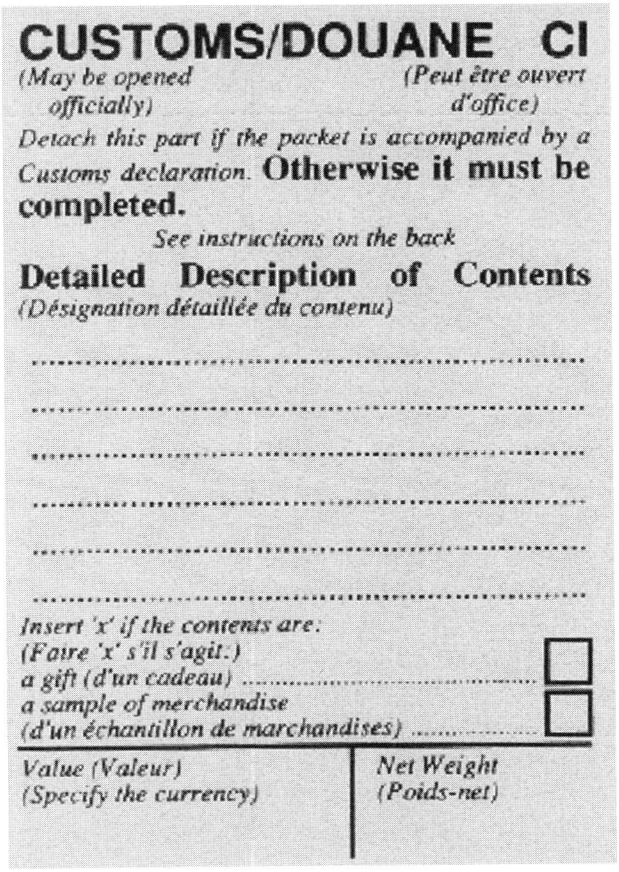

CUSTOMS/DOUANE CI

(May be opened   (Peut être ouvert
   officially)   d'office)

Detach this part if the packet is accompanied by a
Customs declaration. **Otherwise it must be completed.**

See instructions on the back

Detailed Description of Contents
(Désignation détaillée du contenu)

....

....

....

....

....

....

Insert 'x' if the contents are:
(Faire 'x' s'il s'agit:)
a gift (d'un cadeau) .... ☐
a sample of merchandise
(d'un échantillon de marchandises) .... ☐

Value (Valeur) (Specify the currency)	Net Weight (Poids-net)

Fig. 5. Label needed for customs clearance.

mable solids, oxidizing substances and organic peroxides, toxic and infectious substances, radioactive material, corrosives, and miscellaneous dangerous goods. The UN number, proper shipping name, Class, hazard label, packing group (not relevant for biological substances), and Packing Instruction (*see* **Table 3**) are important for shipping procedures. Infectious substances are classified in Class 6, Division 6.2. The UN numbers for infectious substances affecting humans and those affecting animals are UN 2814 and UN 2900, respectively. All consignments containing infectious substances (dangerous goods) are to be accompanied by the transport emergency card (available from courier services) and must be packed in UN-certified packagings.

For transport of infectious substances on the road or by rail the ADR (Accord Européen relatif au transport international des marchandises dangereuses par routes) *(12)* and COTIF/RID (regulations concerning the international carriage of dangerous goods by rail) *(13)*, respectively, apply in Europe. Rules for the international traffic of dangerous goods by inland waterways (ADN) *(14)* have been developed and the carriage of dangerous goods by sea is internationally regulated by the International Maritime Dangerous Goods Code (IMDG) *(15)*.

The International Civil Aviation Organization (ICAO) established guidelines for the transport of dangerous goods by air, implemented by the International Air Transport Association (IATA) (*see* **Table 3**). The regulations governing the transport by air are the most stringent but also the most clear. Air transport is playing a major role today, at least in case of international exchange of biological material. In this context, the IATA Infectious Substances and Diagnostic Specimens Shipping Guidelines *(16)* are especially concerned with the shipping of infectious biological material, whereas the transport of noninfectious biological material is not included. When sending infectious substances by air, a Shipper's Declaration for Dangerous Goods has to be filled in (in English). The Shipper's Declaration is a legal document and has to be signed by a trained person; general or job-specific training of a person involved in packing and shipping of dangerous goods is a precondition according to IATA Dangerous Goods Regulations (DGR), and certified training courses have to be attended by shippers of dangerous goods. The shipper is responsible for the correct documentation.

If the packaging contains infectious substances as the only dangerous good, the following labeling is necessary: (1) UN number plus technical (scientific) name of the infectious substance (UN 2814 or UN 2900, respectively); (2) name, address, and telephone number of sender; (3) name, address, and telephone number of recipient; (4) hazard label (*see* **Fig. 1**); (5) packaging orientation label (if it contains liquid material). Advance arrangements prior to dispatch are indispensable: The sender is urgently advised to contact the courier service when preparing a shipment. Also, the consignee has to be informed of the intended dispatch of infectious substances.

If noninfectious, perishable material is not accepted in air mail, it can be shipped by air freight. The Shipper's Declaration for Dangerous Goods is not applicable.

2.4. Transport of Genetically Modified Organisms

Genetically modified microorganisms have been altered through genetic engineering in a way that does not occur naturally. Five groups are differentiated when offered for air transport *(1)*:

Table 3
Relevant Extract From the IATA Dangerous Goods List

UN No.	Name and description	Class	Hazard label	Packing instruction	Max. net quantity per Package	
					Passenger aircraft	Cargo aircraft
1845	Carbon dioxide, solid (dry ice)[a]	9	Miscellaneous	904	200 kg	200 kg
2814	Infectious substance, affecting humans[a] (solid or liquid)	6.2	Infectious substance	602	50 g or mL	4 kg or L
2900	Infectious substance, affecting animals[a] (solid or liquid)	6.2	Infectious substance	602	50 g or mL	4 kg or L
3245	Genetically modified microorganisms *(acc. 3.6.2.1.2 (c))*	9	Miscellaneous	913	100 g or mL	100 g or mL
3373	Diagnostic specimens		None	650	4 kg or L	4 kg or L

[a] *See* **Note 2.**

1. Organisms being infectious to humans and/or animals. They are to be shipped as infectious substances according to IATA DGR, Class 6, Division 6.2, UN 2814 or UN 2900, following Packing Instruction 602.
2. Animals containing genetically modified microorganisms with infectious potential. They are not to be transported by air. Other modes of transport have to be chosen. Exemptions might be granted by the States concerned.
3. Organisms known or suspected to be dangerous to humans, animals, or the environment. They are not to be transported by air. Other modes of transport have to be chosen. Exemptions might be granted by the States concerned.
4. Organisms capable of altering animals, plants, or microbiological substances in a way that does not occur in nature but which do not meet the definition of an infectious substance. They are allocated to Class 9 (Miscellaneous Dangerous Goods), UN 3245. The applicable Packing Instruction (PI 913) is comparable to PI 602 for infectious material.
5. Genetically modified organisms which are not included under paragraphs 1 to 4. They are not regulated by IATA DGR and will be transported as Articles not Regulated as Dangerous Goods.

2.5 Transport of Diagnostic Specimens

The IATA DGR *(1)* define diagnostic specimens as any human or animal material including excreta, secreta, blood and its components, tissue and tissue fluids transported for diagnostic purposes (except living infected animals). Diagnostic specimens are assigned to UN 3373. (*See* **Note 1**; changes expected for 2005.) Packing Instruction 650 has to be followed. The Shipper's Declaration for Dangerous Goods is not to be used. Samples of diagnostic specimens without infectious substances like blood collected for blood transfusion are not subject to IATA DGR and, therefore, are not to be dispatched as dangerous goods.

3. Conclusion

As it has been shown, numerous steps have to be done before biological material can be shipped in compliance with the law (*see* **Note 1**). Please note that this chapter can only provide an overall picture on the complex topic of dispatch and transport. Special recurrent training courses have to be attended by personnel involved in shipment. By giving deep insight, the training courses also are invaluable fora for discussion and help.

4. Notes

1. Legislation concerning the dispatch and transport of infectious substances underlies constant changes. Organizations involved in shipping regulations can submit variations so that actuality in this chapter cannot be guaranteed. Therefore, it is necessary to keep oneself informed about recent changes before biological material is offered for transport.
2. If dry ice (carbon dioxide, solid) is used as a refrigerant to keep the biological material cold during transport, a special containment has to be used that allows constant release of carbon dioxide gas. There are special combination containments for infectious substances on the market that can keep the low temperature up to 3 d. Dry ice is a dangerous good (Class 9, miscellaneous dangerous goods) and can only be transported by freight. It has to be included in the documentation that accompanies the consignment and the latter has to be labeled accordingly.

References

1. International Air Transport Association (2004) *Dangerous Goods Regulations,* 45th ed., International Air Transport Association.
2. World Health Organization (1993) *Laboratory Biosafety Manual,* 2nd ed., World Health Organization, Geneva.
3. European Parliament (2000) Directive 2000/54/EC on the protection of workers from risks related to exposure to biological agents at work. OJ No. L262, pp. 21–45 of 18.09.2000.
4. European Parliament (1994) Council Directive 3381/94 setting up a Community regime for the control of exports of goods with dual use. OJ No. L 367, pp. 1–12 of 31.12.1994.
5. European Parliament (2000) Council Directive 2000/29/EC on protective measures against the introduction into the community of organisms harmful to plants or plant

products and against their spread within the community. OJ No. L. 169, pp. 1 of 10.07.2000.

6. German Plant Protection Act (1990). BGB1. I, p. 1221 of 28.06.1990.
7. German Infectious Diseases of Animals' Imports Enactment (1992). BGB1. I, p. 2467 of 23.12.1992.
8. United Nations (2001) *United Nations Recommendations on the Transport of Dangerous goods. Model Regulations,* 12th ed., United Nations, New York (known as the Orange Book).
9. CEN (Comité Européen de Normalisation) (1996) EN 829 In vitro diagnostic systems—transport packages for medical and biological specimens—requirements, tests.
10. Universal Postal Union (2000) Official compendium of information of general interest concerning the implementation of the Letter Post Services (1999 Beijing Convention and Letter Post Regulations).
11. Rohde, C. (1999) Shipping of infectious, non-infectious and genetically modified biological materials, international regulations. DSMZ—Deutsche Sammlung von Mikroorganismen und Zellkulturen, Braunschweig.
12. United Nations (2002) *European Agreement Concerning the International Carriage of Dangerous Goods by Road (ADR—Accord Européen relatif au transport international des marchandises dangereuses par routes),* United Nations, New York.
13. United Nations (2003) *Regulations Concerning the International Carriage of Dangerous Goods by Rail (RID),* TSO, London.
14. United Nations (2003) *European Agreement Concerning the International Carriage of Dangerous Goods by Inland Waterways (ADN—Accord Européen relatif au transport international des marchandises dangereuses par voies de navigation intérieures),* United Nations, New York.
15. International Maritime Organization (2002) *International Maritime Dangerous Goods Code (IMDG),* International Maritime Organization, London.
16. International Air Transport Association (2003) *Infectious Substances and Diagnostic Specimens Shipping Guidelines,* 4th ed., International Air Transport Association.

27

How to Deposit Biological Material for Patent Purposes

Vera Weihs

Summary

Patent protection can be granted for an invention that is new, of industrial value, and reworkable. To assure reworkability of an invention where biological material is involved, a sample should be deposited with an independent, recognized patent depositary. This chapter outlines in detail how a scientist renders the respective biological material available to third parties. Also demonstrated is how third parties can obtain the biological material that has been deposited for patent purposes.

Key Words: Legal protection; biotechnological inventions; invention reworkability; patent deposit; Budapest Treaty; International Depositary Authority; IDA; WIPO; strain release.

1. Deposit of Biological Material

If a scientist finds out something new that is not obvious for the average expert and which might be of industrial value, he has made an invention. In order to prevent others making commercial use of this invention without his consent, patent protection should be obtained. Patent protection guarantees the inventor some degree of exclusivity on the invention in order to be assured a reasonable profit and to justify the risks of development. In return, the inventor discloses the invention by giving an exact written description of the process (manifestation). The effect is that the knowledge is available to the public and others can build upon this knowledge. The manifestation gives the public the possibility to rework the invention after expiration of the period of patent protection.

In the area of biotechnology, where biological material is involved, words alone might be incapable of describing an invention sufficiently. In order to assure the reworkability of such an invention, a sample should be deposited at an independent place where such organisms will become so-called "patent strains." Because patent offices are not equipped for working microbiologically, independent and scientifically recognized culture collections are entrusted with the

From: *Methods in Biotechnology, Vol. 18: Microbial Processes and Products*
Edited by: J. L. Barredo © Humana Press Inc., Totowa, NJ

handling of the biological material to be deposited. The Budapest Treaty on the International Recognition of the Deposit of Microorganisms for the Purposes of Patent Procedure *(1)* regulates patent deposits in an international frame. The most important feature of the treaty is that a patent deposit that has been made with one International Depositary Authority (IDA) is sufficient and valid for all other states that signed the Budapest Treaty (Budapest Union). This means that any member state recognizes the deposit made with any IDA. Non-member countries may also accept deposits according to the Budapest Treaty, but often these countries wish an additional deposit in the purview of their own patent law. To date, 59 states worldwide are party to the Treaty.

An IDA is impartial, objective, and available to any depositor on the same terms. It must comply with the requirements of secrecy. Therefore, neither information on whether a sample of the biological material has been deposited with it nor information concerning the material itself will be given to anyone, except to parties entitled to obtain a sample of the said organism. The IDA tests the viability of a biological material promptly after receipt and stores the organisms for the period of time specified by the Treaty (5 yr after the most recent request for the furnishing of a sample of the strain, but at least for 30 yr) (*see* **Note 1**). It furnishes samples of the deposited material for trials and examinations only to persons who have proven their entitlement to receive a sample. Currently, 35 institutions worldwide have acquired the status of an IDA under the Budapest Treaty (*see* **Table 1**).

2. Practical Procedure of a Patent Deposit

Long before filing the patent application and before depositing an organism, it should be clarified with the help of a patent expert whether a patent deposit is indispensable in order to warrant reworkability of the invention (*see* **Note 2**). If the organism's abilities could be described sufficiently in writing (e.g., by sequence data), a deposit might not necessarily be needed. If, however, a sufficient written description is not possible, it should be deposited. To be on the safe side, it might even be advisable to care for the physical availability of the biological material in any case. If the biological material is already available by a deposit, it has to be clarified whether this deposit is acceptable according to patent regulations. When the organism has been deposited by the patent applicant according to the Budapest Treaty, the application can be filed right away (*see* **Fig. 1**). A deposit that is not in accordance with the Budapest Treaty should be converted into a deposit according to the Treaty. The original accession date may remain valid. If the organism has been deposited by a party different from the patent applicant, it has to be checked whether the patent applicant is authorized to refer to the biological material in his application (*see* **Fig. 2**).

Table 1
List of International Depository Authorities (as of April 2004)

IDA Acronym	IDA Full name	Country	Website	Biological material accepted	Risk level accepted
ABC	Advanced Biotechnology Center	Italy	http://www.iclc.it/iclceng.html	Human and animal cell lines	2
AGAL	Australian Government Analytical Laboratories	Australia	http://www.agal.gov.au	Bacteria, fungi, yeasts, bacteriophages, plasmids	2
ATCC	American Type Culture Collection	United States	http://www.atcc.org/	Algae, bacteria, fungi, yeasts, embryos, human, animal and plant cell cultures, bacteriophages, animal and plant viruses, mycoplasma, seeds, plasmids, RNA, protozoa	3
BCCM	Belgian Coordinated Collections of Microorganisms	Belgium	http://www.belspo.be/bccm	Bacteria, fungi, yeasts; animal and human cell lines, plasmids, RNA	2, 3
CBS	Centraalbureau voor Schimmelcultures	The Netherlands	http://www.cbs.knaw.nl/Services/depositi.htm	Bacteria, fungi, yeasts, bacteriopahages, plasmids	2
CCAP	Culture Collection of Algae and Protozoa	Great Britain	http://windermere.ceh.ac.uk/ccap/	Algae, free-living protozoa	1
CCM	Czech Collection of Microorganisms	Czech Republic	http://www.sci.muni.cz/ccm/ccmang.htm	Bacteria, fungi, yeasts	2
CCTCC	China Center for Type Culture Collection	People's Republic of China	http://www.cctcc.org/cctccgk.htm	Algae, bacteria, fungi, yeasts, animal, human and plant cell cultures, animal and plant viruses, bacteriophages, plasmids, seeds	2

(continued)

453

Table 1
List of International Depository Authorities (as of April 2004) *(continued)*

IDA Acronym	IDA Full name	Country	Website	Biological material accepted	Risk level accepted
CCY	Culture Collection of Yeasts	Slovakia	http://www.chem.sav.sk/intro.html	yeasts	1
CECT	Colección Española de Cultivos Tipo	Spain	http://www.uv.es/cect/	bacteria, fungi, yeasts	1
CGMCC	China General Microbiological Culture Collection Center	People's Republic of China	http://www1.im.ac.cn/typecc/junzhong/en.html	Algae, bacteria, fungi, yeasts mycoplasma, viruses, bacteriophages, plasmids	2
CNCM	Collection Nationale de Cultures de Micro-organismes	France	http://www.pasteur.fr/recherche/unites/Cncm/index-en.html	Human and animal cell cultures, bacteria, fungi, yeasts, viruses, bacteriphages	2
DBVPG	Collection of Industrial Yeasts	Italy	http://www.agr.unipg.it/dbvba	Fungi, yeasts	2
DSMZ	DSMZ-Deutsche Sammlung von Mikroorganismen und Zellkulturen GmbH	Germany	www.dsmz.de	Bacteria, fungi, yeasts, bacteriophages, plasmids, plant viruses, plant cell cultures, animal and human cell cultures, murine embryos	1
ECACC	European Collection of Cell Cultures	Great Britain	http://www.ecacc.org.uk/	Human and animal cell cultures, viruses, plant cell suspension cultures, plasmids, bacteria, yeasts, fungi, protozoa	4
IAFB	Collection of Industrial Microorganisms	Poland	http://www.ibprs.pl/	Bacteria, fungi, yeasts	1
IMI	CABI Bioscience, UK Centre	Great Britain	http://www.cabi-bioscience.ReferenceCollections.htm.org/html/	Bacteria, fungi, yeasts	2

454

IPOD	International Patent Organism Depositary	Japan	http://unit.aist.go.jp/ipod/contents_e/001%20ipod/ipod.html	Bacteria, fungi, yeasts, plasmids, animal and plant cell cultures, embryos, algae, seeds, protozoa	1
KCCM	Korean Culture Center of Microorganisms	Republic of Korea	http://www.kccm.or.kr/	Bacteria, fungi, yeasts, viruses, bacteriophages	1
KCLRF	Korean Cell Line Research Foundation	Republic of Korea	http://cellbank.snu.ac.kr/ENG/j_introductionKCLB.htm	Animal, human and plant cell cultures	1
KCTC	Korean Collection for Type Cultures	Republic of Korea	http://kctc.kribbre.kr/english/	Algae, bacteria, fungi, yeasts, animal embryos, human, animal and plant cell cultures, bacteriophages, animal and plant viruses, plasmids, protozoa	1
MSCL	Microbial Strain Collection of Latvia	Latvia	http://www.lu.lv/eng/dept/f_biology1.html	Bacteria, fungi, yeasts	2
MTCC	Microbial Type Culture Collection and Gene Bank	India	www.imtech.res.in	Bacteria, fungi, yeasts, bacteriophages, plasmids	2
NBIMCC	National Bank for Industrial Microorganisms and Cell Cultures	Bulgaria	http://nbimcc.cablebg.net/	Bacteria, fungi, yeasts, animal cell lines, animal viruses, plant viruses	2
NCAIM	National Collection of Agricultural and Industrial Microorganisms	Hungary	http://ncaim.kee.hu	Bacteria, fungi, yeasts,	2
NCIMB	National Collections of Industrial, Food and Marine Bacteria Ltd.	Great Britain	www.ncimb.co.uk/ncimb.htm/	Bacteria, yeasts, bacteriophages, plasmids, seeds	2

(continued)

Table 1
List of International Depository Authorities (as of April 2004) (*continued*)

IDA Acronym	IDA Full name	Country	Website	Biological material accepted	Risk level accepted
NCTC	National Collection of Type Cultures	Great Britain	http://www.phls.org.uk/labservices/nctc/index.htm	Bacteria	3
NCYC	National Collection of Yeast Cultures	Great Britain	http://www.ncyc.co.uk/	Yeasts	1
NMLHC	National Microbiology Laboratory, Health Canada	Canada	http://www.hc-sc.gc.ca/english/media/releases/2001/2001_110ebk1.htm	Bacteria, fungi, yeasts, bacteriophages, animal and human cell cultures, hybridomas, animal viruses (propagated in cell cultures), plasmids	3
NPMD	National Institute of Technology and Evaluation on Partial Microorganisms Depository	Japan	http://www.nbrc.nite.gv.jp/npmd/	Bacteria, fungi, yeasts, bacteriophages, plasmids	
NRCA	National Research Center of Antibiotics	Russian Federation		Bacteria, fungi, yeasts, bacteriophages, plasmids, plant viruses, animal, human and cell cultures	1
NRRL	Agricultural Research Service Culture Collection	United States	http://nrrl.ncaur.usda.gov/	Bacteria, yeasts	1
PCM	Polish Collection of Microorganisms	Poland	http://surfer.iitd.pan.wroc.pl/PCM/PCM_en.html	Bacteria,bacteriophages	1
VKM	Russian Collection of Microorganisms	Russian Federation	http://www.vkm.ru/	Bacteria, fungi, yeasts	1
VKPM	Russian National Collection of Industrial Microorganisms (VKPM) GNII Genetika	Russian Federation	http://www.vkm.ru/Collections/vkpm.htm	Bacteria, fungi, yeasts, plasmids, animal, human and plant cell cultures	1

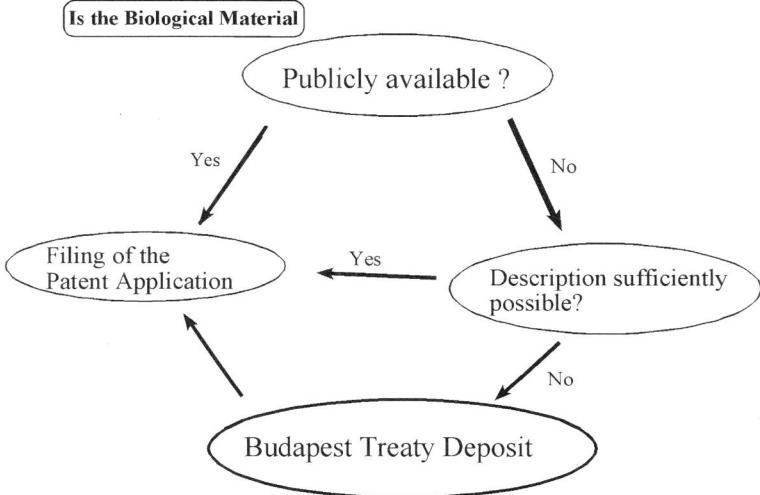

Fig. 1. Assistance for the decision whether biological material has to be deposited.

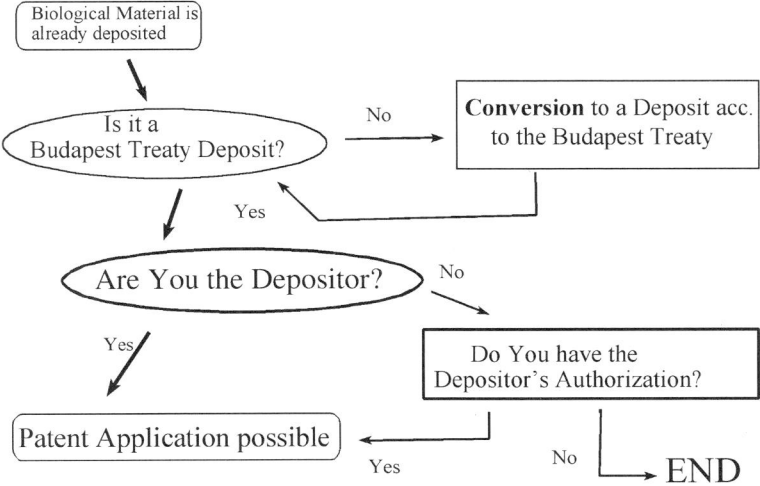

Fig. 2. Assistance for the decision whether the correct kind of deposit has been chosen.

2.1. Choice of an IDA

If a patent deposit is advisable or necessary, the future depositor first has to look for the suitable IDA. Each IDA decides individually on the kinds of microorganism it accepts. In the Budapest Treaty, the term "microorganism" is used without further definition. It is interpreted in a broad sense: Isolated DNA and cell cultures are regarded as microorganisms as well as bacteria or bacteriophages. The actual definition that is suitable for all kinds of biological material that is accepted as deposit for patent purposes states that the biological material must contain genetic information and must be capable of self-reproducing (e.g., bacteria, fungi) or of being reproduced in a biological system (e.g., plasmids, viruses) (2). Consequently, the range of biological material accepted varies between the individual IDAs. Some accept only one or two kinds of organism, whereas others might accept a broader range of different kinds of biological material (see Table 1).

In addition, the IDA decides about the risk level of the material it accepts. Organisms exceeding its limits will and have to be refused by the IDA. Therefore, material sent to an IDA at least should be characterized to a degree that allows its taxonomic and risk classification according to the applicable laws and regulations. Within the European Community, for example, the Council Directive 2000/54/EEC on the protection of workers from risks related to exposure to biological agents at work applies (3). In the case of genetically modified organisms, these must be classified accordingly (4).

Conclusively, first the future depositor and patent applicant has to know which kind of biological material he plans to deposit before he starts to find out where he can do so. The information about which kinds of biological material are accepted by an individual IDA can be retrieved directly from the IDA (in most cases, by consulting their respective websites) (see Table 1) or by looking into the Guide to the Deposit of Microorganisms under the Budapest Treaty (www.wipo) (5). The choice of the depositary should be done taking into account (1) the kind of biological material involved in the patent application, (2) the language(s) spoken and accepted by the IDA (to facilitate correspondence), and (3) the availability of the IDA concerning possible import and export restrictions.

2.2. Date of Patent Deposit

Because the patent deposit has to be interpreted as an essential part of a patent application, it is extremely important that both patent application and strain deposit be performed before any publication of relevant results (lecture, poster, publication). Otherwise, patent protection would no longer be grantable because the invention would not be new, would no longer be more than the state of the art. It is desirable to file the patent application including the deposition documents, the deposit should be performed in good time and before the prospective filing date. The very last date for depositing the biological material is the date of the filing of the patent application.

2.3. Depositing Procedure

In order to deposit a patent culture, it must be accompanied by an accession form that has been duly filled in and signed by an authorized official of the depositing organization. The depositor receives this form directly from the depositary. It is a deposition contract between the depositor and the IDA. To allow the immediate processing of a deposit, it is advisable to inform the IDA beforehand of an intended deposit and to specify which kind of biological material will be involved.

The culture should be supplied to the IDA in the amount and form requested by the culture collection in order to guarantee a smooth deposition procedure. The relevant information might be obtained directly from the IDA or the WIPO Guide to Deposit of Microorganisms under the Budapest Treaty *(5)*. The DSMZ (German Collection of Microorganisms and Cell Cultures), as an example of an IDA, asks the biological material to be deposited in the following forms and amounts:

1. Bacteria and fungi—strains without plasmids or plasmid bearing strains—should, where possible, be deposited in the form of actively growing cultures (two separate preparations, preferably in the form of screw-capped test tubes).
2. Plasmids as isolated DNA preparations should be deposited in a minimum quantity of 2×20 µg (two vials containing 20 µg each).
3. Bacteriophages should be deposited in minimum quantities of 2×5 mL (two vials containing 5 mL each) having a minimal titer of 1×10^9 pfu/mL.
4. Plant viruses should be deposited in the form of dried or frozen material along with the host's seeds, unless the host is generally available. One hundred µL of serum suitable for immunoelectron microscopy should be deposited in addition for the purity and identity test. When hybridomas for antibody testing of plants are deposited, the antigen (not pathogenic) necessary for the specificity test should be deposited at the same time.
5. Plant cell cultures should be deposited as active cultures in the form of a callus (4 Petri dishes), suspension (4 culture vessels) or, preferably, as frozen cultures in dry ice (18 ampoules).
6. Human and animal cell cultures should be deposited frozen in dry ice in a quantity of 12 cryoampoules containing 5×10^6 cells per ampoule, all prepared from the same batch of the culture.
7. Murine embryos should be deposited in 12 ampoules (with each 15–20 embryos at the 8- to 16-cell stage). Before preservation of the embryos by the depositor and subsequent dispatch to the DSMZ, information concerning the method to be used must be obtained from the DSMZ.

The technical and administrative procedure of a deposition is as follows (*see* **Fig. 3**):

1. The depositor prepares the culture in the form and amount requested by the IDA.
2. The accession form is filled in and signed.

Choose the suitable IDA

↓

Inform the IDA about the deposit to be expected

↓

Prepare the biological material in the form and amount requested by the IDA

↓

Fill-in and sign the accession form

↓

Wrap-up the biological material safely and forward it to the IDA in good time before the patent application

↓

Be prepared to provide missing information to the IDA

↓

After a positive viability and purity testing at the IDA you will receive the respective deposition statements

↓

Hand the statements of receipt and viability to the patent office (together with the application)

↓

Check the material preserved by the IDA for identity -when requested-

↓

Inform the IDA about the results of the identity check

Fig. 3. Procedure for a patent deposit; checklist for a depositor.

3. The IDA is informed about the deposit to be expected (by phone, e-mail, fax, or letter).
4. The culture is packaged in a manner such that it will reach the IDA in good condition. The relevant regulations for the shipping of biological material have to be taken into consideration (6).

5. Documents are checked for completeness at the IDA. If necessary, missing data will be demanded from the depositor.
6. After viability and purity of the culture have been checked at the IDA, the accession number will be assigned to the culture. Some collections assign provisional accession numbers when the culture reaches the IDA.
7. The depositor receives a statement of receipt and a statement of viability. These statements are important for the patenting procedure and should be submitted to the patent office involved when filing the application.
8. The biological material will be preserved and/or stored at the IDA.
9. After preservation at the IDA, the depositor will receive samples for identity checking (in dependence of the biological material deposited and the IDA chosen).
10. After checking the identity of the material, the depositor informs the IDA about the results of the check (OK/not OK).

According to the Budapest Treaty, an IDA is requested to perform a viability test before assigning an accession number to the culture. Therefore, it is in the interest of the depositor to send the cultures to the IDA as early as possible to provide sufficient time for the cultivation of the organism before the date of the filing of the patent application. This applies to all kinds of biological material. In general, the testing for viability and purity of bacterial and fungal cultures takes from one to several days. In the case of human and animal cell cultures viability testing, including the essential testing for contaminations with mycoplasmas, about 2 wk are needed; the cultivation of plant cell cultures needs approx 4 wk. The individual time needed not only depends on the kind of biological material but can also vary considerably according to the species of organisms handled.

A patent culture cannot be claimed back after it has been deposited. Patent strains will be preserved by sufficient safety measures to minimize the risk of losing deposits. If, however, a strain deposited under the Budapest Treaty should die or be destroyed during the period of storage, it would be the responsibility of the depositor to replace it by a viable culture of the same organism. Thus, the patent can be kept alive. Therefore, the depositor is recommended to keep samples of the culture for the same period of time: in the case that the culture is for any reason no longer available from the Depositary Authority, the stock can be replenished by the depositor.

3. Request for the Furnishing of Samples of Patent Organisms

The main reason for the deposit of biological material that is of relevance for a patent application is to render it available for entitled parties for trials and examinations, thus allowing reworkability of a protected invention. The accessibility to a deposited organism depends on the status of a patent. According to the Budapest Treaty, samples might be furnished either to interested industrial property offices, the depositor, third parties that obtained the authorization of the depositor, or to parties legally entitled.

A request for the furnishing of a patent culture by a third party can be sent either to the IDA harboring the strain or directly to the respective patent office. The IDA checks its files for information on whether or under which conditions the culture is to be furnished. If no information is available on whether the IDA is authorized to release a sample of the strain, the requestor will be asked to provide the IDA with a copy of the first page of the patent document in question and a copy of the page where the accession number of the collection of the strain in question is cited. After receipt of the requested documents, the IDA will be able to provide the relevant official forms (request for the release of a patent strain). In the case of a national or a European patent, the form must be filled in and sent to the relevant patent office for confirmation. The European Patent Office offers the necessary form 1140/1141 online (http://www.european-patent-office.org/epo/formul/epc_e.htm). If a US patent had been granted, the strain is available without further examination by the patent office.

After receiving the pertinent confirmation from a patent office, the IDA will ship the sample under consideration of other national and international relevant regulations [e.g., postal regulations *(7)* or biological weapons conventions like the EU dual-use directive *(8)*] (*see* **Fig. 4** and **Note 3**).

4. Notes

1. Material deposited for patent purposes according to the Budapest Treaty is stored at the IDA for at least 30 yr (plus an additional 5 yr after the most recent release). What will happen with the deposited patent organisms after the prescribed period of storage? Are they to be transferred into the public section of the collection to guarantee the state of the art? Are they to be destroyed, as they might represent a whole "factory" that is available so easily? Should they be handed back to the depositor who is the original owner of the material? As there is no jurisdiction concerning the fate of the organisms, some IDAs offer the possibility that the depositor himself decides what will happen with the organisms after 30 yr (+ 5 yr). The depositor's declaration will be valid until a relevant jurisdiction exists.

2. A conflict might arise from the interest of a researcher to obtain both patent protection and scientific merit. This conflict is based on the demand for the free availability of taxonomic reference strains that is in conflict with the patent practice of restrictively handled patent deposits. Many of the patented microorganisms isolated from unusual habitats are not only of industrial interest but are also new to systematists. Thus, the scientists might also be interested in validly describing a new species. The consequence in these cases of type strains is that agreed reference strains for the new species might be covered by patent protection. From a scientific point of view, however, type strains should be readily accessible for the user and no additional restrictions should lengthen the request procedure. The problem can be solved by keeping both activities separate and depositing a parallel subculture

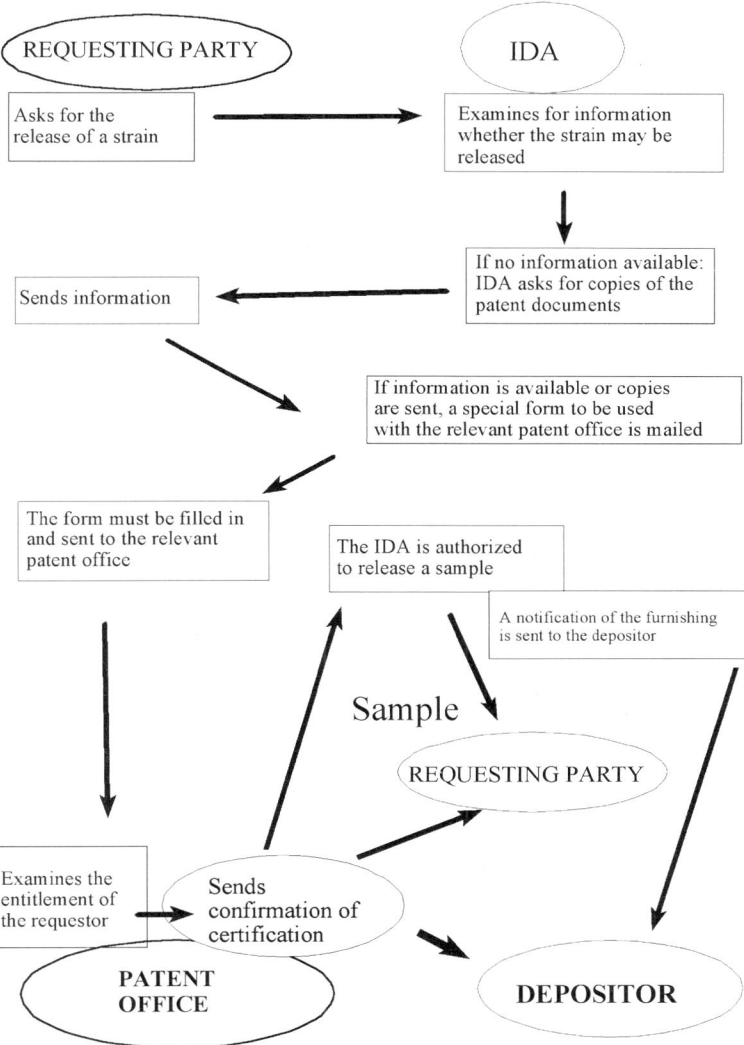

Fig. 4. Procedure for the furnishing of samples according to Rule 11.3 of the Budapest Treaty.

of the patent strain under a different accession number in the public part of a collection. From there, it will be available to anyone without delay resulting from patent regulations.

3. Patent applicants might fear that third parties will abuse the biological material deposited, especially when a patent has been withdrawn. To eliminate this fear, the

IDA promptly notifies the depositor in writing of each furnishing of a sample of the deposited organism. In addition, Rule 28 EPC *(9)* offers the possibility of choosing the "expert solution" in connection with patent deposits. Following this rule, after publication, the grant of the patent samples of the deposited organism will only be provided to a recognized expert. Experts in the field of research are scientists nominated by the European Patent Office. Alternatively, a scientist might be recommended as an expert by the requesting party. It would then be up to the depositor to accept this person as an independent expert. If a patent is withdrawn, the expert solution is applicable for 20 yr.

References

1. World Intellectual Property Organization (1989) Budapest Treaty on the International recognition of the deposit of microorganisms for the purposes of patent procedure. Done at Budapest 1977, as amended in 1980, and Regulations (as in force since 31.01.1981), Geneva.
2. European Parliament (1998) Council Directive 98/44/EC on the legal protection of biotechnological inventions. OJ No. L 213, pp. 13–21.
3. European Parliament (2000) Council Directive 2000/54/EC on the protection of workers from risks related to exposure to biological agents at work. OJ No. L 262, pp. 21–45.
4. European Parliament (1998) Council Directive 98/81/EC amending Directive 90/219/EEC on the contained use of genetically modified microorganisms. OJ No. L 330, pp. 13–31.
5. World Intellectual Property Organization (2002, yearly updates) *Guide to the Deposit of Microorganisms under the Budapest Treaty.* WIPO, Geneva.
6. Rohde, C. (1999) *Shipping of Infectious, Non-infectious and Genetically Modified Biological Materials, International Regulations.* DSMZ–Deutsche Sammlung von Mikroorganismen und Zellkulturen, Braunschweig.
7. Universal Postal Union (2000) *Official compendium of information of general interest concerning the implementation of the Letter Post Services* (1999 Beijing Convention and Letter Post Regulations).
8. European Parliament (1994) Council Directive 3381/94 setting up a Community regime for the control of exports of goods with dual use. OJ No. L 367, pp. 1–12.
9. European Patent Office (1998) Convention on the grant of European patents (European Patent Convention, EPC) of 5 October 1973, last amended by the decision of the Administrative Council of the European Patent Organisation of 10 December 1998.

28

Laws and Regulations for the Protection of Biotechnological Inventions

Uwe Fitzner

Summary

Microbiological processes and the products thereof are patentable. In this connection, "products" means the metabolic products of the microorganisms, the microorganisms themselves, and ingredients of the microorganisms, particularly genetic material. In addition to fulfilling the general requirements for obtaining patent protection, it is necessary in the field of microbiology to deposit the microorganism. It is undisputed currently that depositing is essential in many cases as a means for identification, disclosure, and the possibility of copying inventions in the microbiological field. Apart from this, modern genetic engineering now makes it possible to identify microorganisms with increased accuracy. Particularly in cases of genetic modification, an exact description can be provided by giving the genetic code, so that it is no longer necessary in such cases for the organisms to be deposited.

Key Words: Patent protection; invention; novelty; inventive step; microbiological process; microorganisms; genetic engineering; deposit; release; expert solution; plants; animals; cells; European Patent Convention; Patent Cooperation Treaty; Paris Convention.

1. Introduction

The rapid development of technology in the last 200 yr has brought about a revolution in the industrial sector. Biological processes became increasingly important in the second half of the 20th century. They open up new prospects in many areas of life, genetic technology, above all, playing a substantial role. Patent protection for biological processes and products is therefore increasingly important for industry and research *(1)*.

In the biotechnology sector, however, the case law has up to now been confronted by a series of unresolved issues as a result of the special features connected with the use of living organisms. The same applies to the special features of genetic engineering.

From: *Methods in Biotechnology, Vol. 18: Microbial Processes and Products*
Edited by: J. L. Barredo © Humana Press Inc., Totowa, NJ

2. Special Requirements in Respect of Biological Inventions
2.1. Historical Development

Nineteenth-century legislators only considered the patenting of inventions that covered inanimate nature and were subject to the laws of physics and chemistry. It was, however, only after long disputes in the literature and in case law that inventions in the biological field were recognized as forming part of the concept of technical inventions and as being applicable in the industrial field.

With increasing advances in science, a new approach became established, particularly after World War II. In 1950, the German Patent Office, in a prudent further development of the view taken by the German Imperial Patent Office, allowed patent protection for a technical method serving purely cosmetic purposes and relied in this respect on a decision by the Imperial Patent Office in 1935, according to which a diagnostic method that was equivalent to the analytical methods in chemistry and that was of a technical nature was not a non-patentable healing method.

The German Patent Office and the German Federal Patent Court have also taken the view in subsequent decisions that methods in which the living human body with its biological functions is involved could be technical inventions. Human involvement in the processes of animate nature (plant, animal, human) was only considered to be patentable where they could be controlled with certainty; it was recognized that even the human body, with all its biological functions, could be the subject matter of a technical invention. This led to a series of patents being granted in respect of diagnoses and absorbable surgical material. The fact that there were physiological effects on the body contributing to bringing about the effects of bodily functions no longer prevented patents from being granted. However, measures to restore and maintain human health, such as treatments and surgical interventions, were still excluded from patentability. The German Federal Supreme Court decided in 1967 that inventions relating to courses of treatment were excluded from patent protection, as they were not susceptible of industrial application. The medical profession was not an industry; the use of treatments should, therefore, be absolutely unrestricted.

Inventions relating to methods for treating the human or animal body for reasons other than therapeutic reasons are, however, patentable because they are susceptible of industrial application. These include, in particular, cosmetic methods. Since January 1, 1978, the German Patent Act has also contained corresponding provisions, just like the Patent Cooperation Treaty (PCT), the European Patent Convention (EPC), and most national laws.

Following the practice of the Imperial Patent Office, the German Patent Office granted substantive protection in 1957, for the first time, in respect of a plant variety, because the applicable patent law did not exclude product claims in the field of plant varieties. In 1962, the Federal Supreme Court allowed the granting of a patent in respect of a cultivation method for a rose, although the cultivation described was not repeatable with certainty because of the genetic changes.

In 1965, the Federal Patent Court again held that animal breeding methods were not patentable. However, in 1969, the Federal Supreme Court recognized the results of animal breeding as being patentable and gave detailed consideration in this case to the concept of technical inventions. It clarified that the planned exploitation of natural biological forces and phenomena could not, in principle, be excluded from patent protection, because biological processes were now also capable of being calculated and controlled.

The result of this new approach was the acknowledgment in principle that biological inventions were patentable. Any "teaching about planned action using controllable natural forces to achieve a casually clear result," to use the wording of the Federal Supreme Court, was available for patent protection. Following this decision, the Federal Patent Court commented on the requirements for the patenting of a plant cultivation method and emphasized that the principles developed by the Federal Supreme Court were also applicable to plant varieties. Only patent claims relating to the vegetative propagation of plants without any other inventive involvement were said to be inadmissible.

Microbiological processes were recognized at an early stage as being patentable. Patents were granted by the Imperial Patent Office in respect of inventions in which microorganisms were used (e.g., methods of making bread, brewing beer, and manufacturing vinegar and for producing butyl alcohol and acetone by fermentation). Additionally, biological methods are important for the production of organic acids (oxalic acid, glutamic acid), for obtaining vitamins and medicinal sera, and for producing antibiotics. There was, from the outset, no doubt that such microbiological processes are patentable. The main novelty in these processes was the choice of the microorganism. With the progress in science, however, the extent of the problem and the difficulties that arose in connection with the patenting of microbiological processes became clear.

In 1975, the Federal Supreme Court commented in detail on the question of the patentability of inventions in the microbiological field. The question to be decided was whether a baking yeast, its method of cultivation, and its use as a dry yeast could be subject to patent protection. The answer given was that microorganisms could, in principle, be protected if a repeatable method for producing them always in the same manner were given.

On the other hand, organisms occurring in nature that were not obtained by a special breeding process were not patentable. Consequently, the patenting of natural substances normally continued to be refused in the case law. The case law on patent protection for microbiological processes was consolidated in subsequent years and substantive protection for a microorganism culture was finally granted in Germany for the first time in 1978 *(2,3)*. The historical case law in Germany had substantial influence on international development, particularly on international conventions.

2.2. International Conventions

2.2.1. European Convention

2.2.1.1. SURVEY

The European Patent Convention (EPC) has created a uniform grant procedure for a number of European countries (*see* **Fig. 1**). Patent applications can be submitted to the European Patent Office (EPO) in Munich.

The application is examined to see how far the formal and substantive requirements for the grant of patent protection have been met. If the examination shows that the subject matter of the application is patentable, a European patent is granted.

European patents shall be granted for any inventions that are susceptible of industrial application, which are new and which involve an inventive step (*see* **Fig. 2**), whereby an invention must relate to the technology sector and contain a technical rule, as well as being repeatable, practicable, finished, and the solution to a technical problem (*see* **Fig. 3**).

According to Article 53 EPC, European patents shall not be granted in respect of plant or animal varieties or essentially biological processes for the production of plants or animals; this provision does not apply to microbiological processes or the products thereof. Reference is also made to Rule 23 (b)–(e) of the Implementing Regulations to the European Patent Convention. By these provisions, the EPO has implemented the regulations of Directive 98/44 of the European Parliament.

Consequently, no patent can be granted in respect of plant or animal varieties or "essentially biological processes for the production of plants or animals". Contrary to the ban on patenting essentially biological processes for the production of plants or animals, microbiological processes and the products thereof are exempted from the said exclusion provisions. In other words, Article 53(b) represents an exception to the exceptions in the first half-sentences of the relevant provisions and thus guarantees that microbiological processes and the products thereof do not come within the ban on patenting (*see* **Fig. 4**).

With the explosive development in genetic engineering, Article 53(b) EPC has, in the meantime, acquired significance for patent purposes that the legislators did not attach to these provisions when drafting them and that was also unforeseeable.

The terms "plant varieties," "animal varieties," "essentially biological processes for the production of plants or animals," and "microbiological processes or the products thereof" are not defined in the EPC. They must, therefore, be interpret-

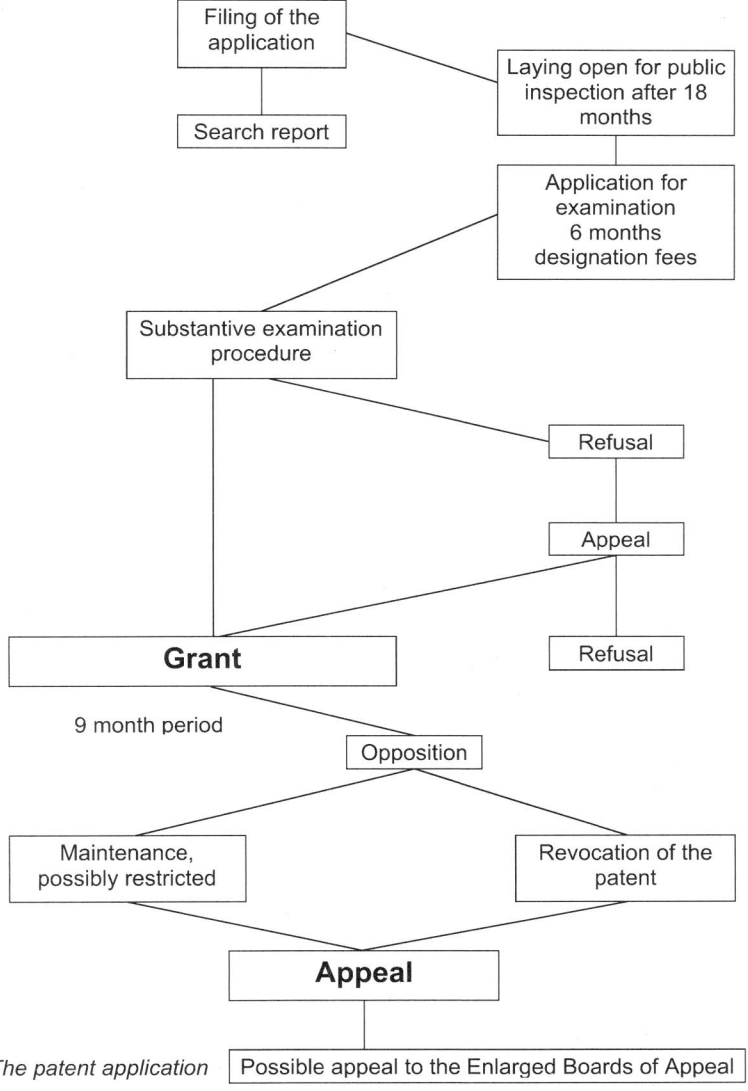

Fig. 1. The grant procedure in the European Patent Office.

ed in case law. In recent years, the national courts and the Boards of Appeal of the European Patent Office have made a series of interpretative decisions that all have an influence on the extent to which biotechnological inventions can be patented in Europe with regard to plants, animals, and biological processes *(4)*.

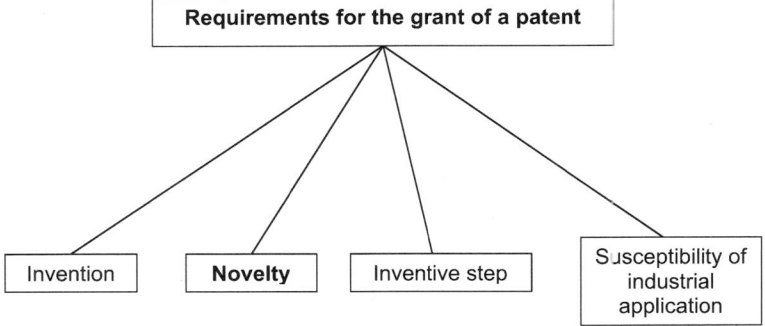

Fig. 2. General requirements for patent protection.

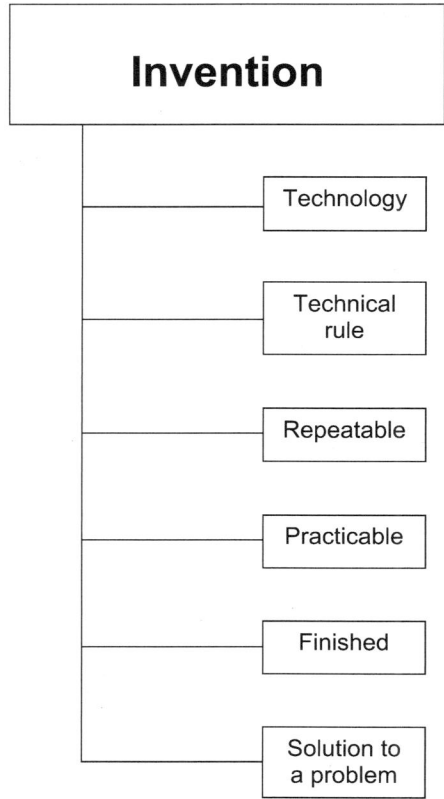

Fig. 3. Features determining the existence of an invention.

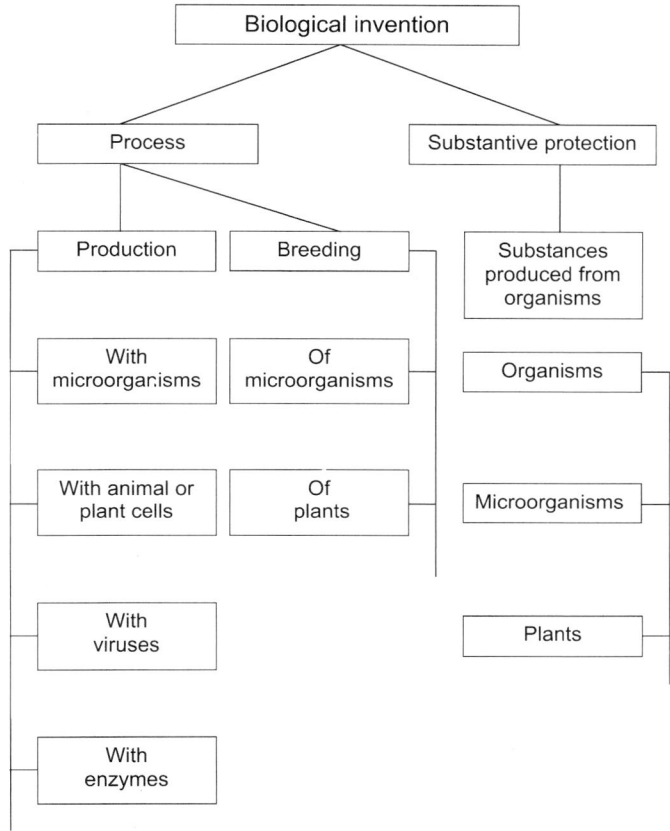

Fig. 4. Options for patent protection for biological inventions.

2.2.1.2. PLANT VARIETIES

In a decision in 1983, the European Patent Office defined the term "plant variety." In the context of this decision, the Board of Appeal had to decide on an application in which the applicant sought patent protection for improved propagating material, particularly seeds, for cultivated plants that had been chemically treated (5).

It was stated in the decision that the skilled person would understand the term "plant varieties" to mean a multiplicity of plants that are largely the same in their characteristics and remain the same within certain tolerances after every propagation or every propagation cycle and that this definition is reflected in the International Convention for the Protection of New Varieties of Plants of December 2, 1961. The Board of Appeal took the view that the legislator did not wish to afford patent protection under the EPC to plant varieties of this kind, whether in the form of propagating material or of the plant itself.

This decision must be understood in the light of the historical documentation relating to Article 53(b) EPC. The ban on patenting plant varieties in this Article originates from Article 2 of the Strasbourg Patent Convention, which allows countries to exclude plant varieties from patent protection. The provisions of Article 2 were adopted in the Strasbourg Convention in early 1960—in other words, at the same time as the International Convention for the Protection of New Varieties of Plants (the UPOV Convention) governing the protection of plant varieties came into existence.

The legislator decided at that time that living organisms are not suitable for patenting. The main reason preventing this was the identical reproduction of the invention required in traditional patent law; however, plant varieties could be protected under the provisions of the UPOV Convention; thus, the EPC excluded them from patent protection.

In the aforesaid decision, the Board also stated: "The subject-matter of Claims 13 and 14 (chemically treated seeds) is not an individual plant variety distinguishable from any other variety, rather the claims relate to any cultivated plants in the form of their propagating material which have been chemically treated in a certain way. However, Article 53(b) EPC prohibits only the patenting of plants or their propagating material in the genetically fixed form of the plant variety."

At the same time, the Board held that the chemical treatment of improved propagating material was not the result of an essentially biological process and that plant varieties and plants were not the same thing.

It may, thus, be inferred from the statements of the EPO that Article 53(b) EPC only excludes plant varieties in the case of an individual plant variety that is essentially different from other plant varieties. All in all, the scope of Article 53(b) EPC has been very narrowly interpreted *(6)*.

Plant breeding today is largely concerned with the genetic modification of plants. According to what has been said, Article 53(b) EPC only prohibits the patenting of plants or their propagating material in the genetically fixed form of the plant variety. The exceptions to patentability under Article 53(b) EPC before the semicolon do not therefore apply to plants as such and, in particular, do not apply to hybrid plants, which are not stable and thus do not meet a characteristic criterion of a plant variety *(7)*.

Another decision from 1993 *(8)* related to an invention whose objective was to develop plants and seeds that are resistant to a particular category of herbicides and that are, in this way, selectively protected against weeds and fungal diseases. This was achieved by the stable integration of a DNA sequence in the genome of the plants encoding a protein that is capable of inactivating or neutralizing this herbicide.

The Board assessed patentability in a restrictive manner in view of Article 53(b) EPC. The product claim for a "plant, nonbiologically transformed, which possesses, stably integrated in the genome of its cells, a foreign DNA nucleotide sequence encoding a protein capable of inactivating or neutralizing the particular herbicide" was not allowed. The Board took the view that such a claim comprised plant varieties, contrary to Article 53(b) EPC before the semicolon. Admittedly, the claim was not drafted in terms of a variety description because there was no reference to a single botanical taxon of the lowest known rank, but rather it was, in general, directed to a plant that possessed, integrated in its genome in a stable manner, a foreign DNA nucleotide sequence. However, the claim defined the distinctive feature common to all plants covered by this claim that was transmitted in a stable manner to the progeny. The exemplary embodiments showed that the practical forms of realization of the invention were genetically transformed plant varieties. Consequently, the claim encompassed genetically transformed plant varieties with the relevant distinctive feature. Even with regard to Article 53(b) EPC after the semicolon, according to which, products obtained with the aid of microbiological processes can be patented, it was not possible to grant this claim, because the multistep process whereby the plant was produced comprised not only microbiological but also significant agrotechnical and biological steps.

However, the claim for plant cells was granted, as these did not come within the definition of "plant" or "plant variety" and could be regarded as a microbiological process *(9)*. The view was expressed in the decision that plant cells as such that, thanks to modern technology, can be cultivated almost like bacteria and yeasts cannot be subsumed within the definition of a plant or a plant variety. On the contrary, according to the current practice of the EPO, they are regarded as "microbiological products" in the broader sense *(10)*.

The problem of the concept of the plant variety, in view of modern developments in genetic engineering, has repeatedly occupied case law. The following questions, *inter alia,* have been referred to the Enlarged Board of Appeal of the European Patent Office *(11)*: (1) Does a claim that relates to plant varieties but wherein specific plant varieties are not individually claimed *ipso facto* avoid the prohibition on patenting in Article 53(b) EPC even though it embraces plant varieties? (2) Does a plant variety, in which each individual plant of that variety contains at least one specific gene introduced into an ancestral plant by recombinant gene technology, fall within the provision of Article 53(b) EPC that patents shall not be granted in respect of plant varieties or essentially biological processes for the production of plants, which provision does not apply to microbiological processes or the products thereof?

2.2.1.3. ANIMAL VARIETIES

With regard to the term "animal variety," the three versions of Article 53(b) EPC before the semicolon are different from one another *(10)*. This is because the German term "Tierarten" is broader than the English term "animal varieties" and the French term "races animales." As, pursuant to Article 177(1) EPC, the German, English, and French versions of the EPC are equally binding, the common meaning of these three versions must be ascertained when interpreting the EPC, in order to be able to establish the extent to which animals are excluded from patentability under Article 53(b) EPC before the semicolon.

It is stated in a 1990 decision of the EPO Board of Appeal *(12)* that it is the EPO's duty to resolve the issue of the interpretation of Article 53(b) EPC before the semicolon in respect of the term "animal varieties" so that a middle way is found between; on the one hand, inventors' interests in having reasonable protection for their work in this area and, on the other hand, the public interest in excluding particular categories of animals from patent protection. The view was also expressed in this decision that account needed to be taken of the fact that no other industrial property rights were provisionally available in respect of animals, unlike in the case of plant varieties.

Animal breeding is increasingly concerned currently days with genetically engineered modification of animals. Article 53(b) EPC before the semicolon only prohibits the patenting of animals in the genetically fixed form of the animal variety. It is stated in this respect in the aforementioned EPO decision that the exclusion from patentability pursuant to Article 53(b) EPC before the semicolon does not apply to animals as such, irrespective of the meaning of the terms "Tierarten," "animal varieties," and "races animales."

Animal cells as such that, thanks to modern technology, can be cultivated almost like bacteria and yeasts cannot be subsumed under the definition of an animal or an animal breed. Like plant cells, they are, instead, regarded in accordance with current practice as "microbiological products" in the broader sense.

2.2.1.4. BIOLOGICAL PROCESSES

Inventions relating to processes other than essentially biological processes for breeding animals and the products thereof (viz. the animals produced with the aid of technical processes) are, however, not excluded from patenting. These include, for example, genetic engineering methods *(12)*. Likewise, essentially biological processes using animals to produce products are patentable. According to a decision by the European Patent Office, animals can also be protected, but not animal varieties *(12)*.

A biological process must primarily be understood as being distinct from a technical process in the narrower sense, in which the course of events is influenced using means other than animate material (e.g., using chemical or physical means) (Benkard-Bruchhausen, Section 2 of the German Patent Act, paragraph 18). According to the guidelines of the European Patent Office (EPO Guidelines Part C, Chapter IV, p. 70 et seq.), the question whether a process is "essentially biological" depends on the extent of any human technological involvement in the process: If such involvement plays a significant role in influencing the desired result, the process must not be excluded from patentability.

A number of examples are now discussed.

A method for cross-breeding, for mixing breeds, or a selective breeding method (e.g., for horses), in which only those animals that have particular characteristics are selected for breeding and for bringing together would be essentially biological and thus not patentable. However, a method for treating plants or animals to improve their characteristics or their yield or to promote or suppress their growth, regardless of whether it is a mechanical, physical, or chemical method (e.g., a method of pruning trees), would not be essentially biological because the invention is essentially technical in nature although it also contains a biological method; the same would apply to a method of treating plants that is characterized in that it uses a growth-promoting material or growth-promoting radiation. Treating the soil by technical means to suppress or promote the growth of plants is also not excluded from patentability.

The European Patent Office has, until recently, commented repeatedly on the patentability of "biological processes." For example, the following was stated in a decision of February 15, 1993 by an Opposition Division *(13)*:

The fact that particular subsequent generations of plants represent the direct product of a biological process does not, in principle, lead in itself to exclusion from patenting under Article 52 EPC. The question whether a process is to be regarded as essentially biological within the meaning of Article 53(b) EPC must be assessed on the basis of what constitutes the essence of the invention.

In another decision of 10 November 1988 *(5)*, a Technical Board of Appeal stated the following: "Whether or not a (nonmicrobiological) process is to be considered as 'essentially biological' within the meaning of Article 53(b) EPC has to be judged on the basis of the essence of the invention taking into account the totality of human intervention and its impact on the result achieved (cf. point 6 of the reasons). Hybrid seed and plants from such seed, lacking stability in some trait of the whole generation population, cannot be classified as plant varieties within the meaning of Article 53(b) EPC (cf. point 14 of the reasons)."

A patent can accordingly be granted in respect of essentially nonbiological processes for breeding plants and in respect of breeding processes that are generally applicable and not used merely for breeding a particular variety. This also applies to all inventions of plant varieties and to processes for breeding such plant varieties. Biological processes not used for breeding plants are also patentable (i.e., processes for producing products using higher plants with the exception of the plant itself). Plant and animal production processes are accordingly capable of being protected under patent law if they are not "essentially biological processes."

Finally, another decision of a Technical Board of Appeal of the EPO *(8)* states that a process for producing a plant that comprises the step of transforming the plant cell or the plant tissue with a recombinant DNA and the following steps of regenerating and copying plants or seeds is not an essentially biological process within the meaning of Article 53(b) EPC before the semicolon; that is, because the step of transforming the plant cell or the plant tissue with a recombinant DNA represents an essential technical step having a decisive influence on the desired final result, regardless of the extent to which the outcome of that step is a matter of chance.

2.2.1.5. MICROBIOLOGICAL PROCESSES

Patent protection can also be granted for microbiological processes. In the opinion of the German Federal Supreme Court *(14)*, microbiological processes can relate to producing valuable products or changing conditions by means of the metabolic activity of microorganisms. Additionally, such processes could relate to the production of new microorganisms under human control and the use of such microorganisms. In a 1993 decision *(8)*, the European Patent Office took the view that the term "microbiological processes" related to processes in which microorganisms (or their parts) were used to make or to modify products or in which new microorganisms were developed for specific purposes.

Traditional microbiology was primarily concerned with the production, by means of fermentation processes, of primary and secondary metabolites and with biotransformations. Modern microbiology, however, combined the traditional techniques with the genetic engineering techniques and made use of experimental approaches that were widely applicable to animal and plant cells that could be maintained and grown in culture much like bacteria and yeasts. In view of the significant developments that had recently taken place in the field of microbiology, what mattered when ascertaining the meaning of the statutory provision ([Article 53(b) EPC]) was not only what the legislator actually intended at the time when this provision was made but also what he might have intended, taking into account the changed circumstances that have subsequently occurred.

Article 53(b) EPC after the semicolon was, therefore, not restricted to processes in which microorganisms in the traditional sense, such as bacteria and yeasts, were used but also related to processes including fungi, algae, protozoa, and animal and plant cells—in other words all generally single-cell organisms invisible to the naked eye that could be propagated and manipulated in the laboratory.

Additionally, according to the current EPO practice, plasmids and viruses were included in this definition of the term "microorganism."

2.2.1.6. MICROBIOLOGICAL PRODUCTS

Microbiological products include substances produced or modified using microorganisms and also microorganisms as such.

It is stated in the 1993 decision of the European Patent Office cited earlier *(8)* that "microorganisms" mean not only bacteria and yeasts but also fungi, algae, protozoa, animal and plant cells, plasmids, and viruses. It is also stated that products produced or modified using microorganisms within the defined meaning and new microorganisms as such would be regarded as products of microbiological processes within the meaning of this provision.

2.2.1.7. DEVELOPMENTS IN GENETIC ENGINEERING

The patenting of human genes, in particular, has caused public concern. Apart from ethical considerations, it was asserted by environmental organizations in opposition proceedings that genes, whether they are human, plant, or animal genes, are substances that already existed in nature. They could, therefore, certainly not be invented but, at best, discovered *(15)*.

The starting point for considerations of this kind is the exclusion under Article 53 EPC whereby a patent shall not be granted if the publication or exploitation of an invention would be contrary to "ordre public" or morality.

The European Patent Office has had to comment on a number of occasions in recent years on the issue of breaches of morality and law. The following is stated, *inter alia,* in a decision of February 5, 1993 of the Opposition Division: "The exclusion from patentability, under Article 53(a) EPC, of inventions which are contrary to 'ordre public' or morality only relates to those extreme exceptional cases which are generally regarded as particularly abhorrent. Opposition proceedings are not the appropriate way of deciding on borderline cases for which there are no generally valid principles."

The following was stated in a decision by the Examination Division of April 3, 1992 (Harvard oncomouse): "(I) Article 52(1) EPC contains a general principle of patentability which is inapplicable only if other statutory provisions exclude particular subject-matters from patentability. (II) In view of the requirements of Article 53(a) EPC, every individual invention must be examined as to whether it is contrary to morality; possible adverse effects and risks must be assessed and weighed up against the benefits and the advantages of the invention. (III) A patent does not confer a positive right to use the invention but merely the right to exclude others from using it for a limited period. It is for the legislator to determine the conditions under which particular technical knowledge involving dealing with hazardous materials may be used."

The reason why it is difficult to answer the questions that have been thrown out is that the ethical assessment cannot cover technology as a pure potential resource but only the use to which such technology is put. However, it is only as mere potential that technology is the subject matter of patent protection *(16)*. This arises from the following.

First, the patenting of an invention does not in any way mean that the patent holder is even entitled to apply it, let alone use it on an industrial scale. This is only decided by general legislation. In the field of biotechnology, the legislation is primarily the German Genetic Engineering Act (Gentechnikgesetz), the German Protection of Embryos Act (Embryonenschutzgesetz), and the German Medicinal Products Act (Arzneimittelrecht).

Additionally, the rejection of a patent application—in other words, the refusal of patent protection—does not mean that exploitation of the invention is excluded. On the contrary, the refusal to grant a patent means that it is freely available for anybody to use, again provided that this is permissible in accordance with general legislation. Patent law is, therefore, not a suitable instrument for preventing or precluding abuses or even risks that might be associated with the use of any technology.

Finally, patent protection only affords an advantage for the inventor if the invention is, in fact, capable of exploitation on the market. Inventions, whose the exploitation would be contrary to "ordre public" or morality are for this very reason excluded from commercial trading. There is, therefore, no reasonable ground for the inventor of such an invention to apply to patent it. This might also explain why Article 53(a) EPC practically never comes up.

The question that therefore arises is, what purpose does this provision serve? The answer to this question is difficult to explain to laymen in patent matters but is nevertheless correct both from an historical and a substantive point of view *(16)*. The granting of a patent is a sovereign act; in the case of the EPO, it is a sovereign act by an international authority. The purpose of Article 53(a) EPC is to prevent an invention, the exploitation of which would be contrary to the principles of legal order or the feelings of propriety of all reasonable and right-thinking people, from being given the appearance of governmental approval.

It must, therefore, be stated that the act that is to be measured in accordance with the standards of "ordre public" and morality is not the patenting of the invention but the publication or exploitation thereof *(17)*.

In the "oncomouse decision" *(18)*, the Board of Appeal of the EPO referred the case back to the initial forum with the instruction to weigh the suffering for animals and possible risks to the environment connected with the exploitation of the invention against the benefit of the invention for humanity.

In the 1993 decision that has already been cited *(8)*, the Board of Appeal finally proceeded on the basis that there is no uniform concept of "ordre public" and morality for all Contracting States. It concludes from this that these terms must be defined specifically for the purpose of the EPC. "Ordre public" consisted of the protection of public safety and the physical integrity of members of society. This also encompassed the protection of the environment. Inventions whose the exploitation seriously endangers the environment must, therefore, be excluded from patenting.

Corresponding statements were made on the concept of morality. It would be contrary to morality if the exploitation of the invention were not in conformity with the generally recognized standards of conduct pertaining to European culture.

On the basis of these definitions, the Board of Appeal undertook an extensive examination of environmental acceptability but finally came to the conclusion that serious risk to the environment was not sufficiently documented in the case that has to be decided.

2.2.2. EU Law

In this respect, Directive 98/44/EC of the European Parliament and of the Council of July 6, 1998 on the legal protection of biotechnological inventions must, in particular, be complied with *(19)*. This Directive must be implemented by the Member States of the European Union. However, only a few states have carried out this implementation. By contrast, the European Parliament has adopted the provisions of the US Directive in the EPC implementation regulation although it was not obliged to implement them.

Various terms are defined to start with in Article 2 of the Directive. For the purposes of the Directive, "biological material" means any material containing genetic information and capable of reproducing itself or being reproduced in a biological system.

A "microbiological process," according to the definition in the Directive, means any process involving or performed upon or resulting in microbiological material. Additionally, the definition in Article 2(2) makes it clear that a process for the production of plants or animals is essentially biological if it consists entirely of natural phenomena such as crossing or selection. With regard to the concept of "plant variety", reference is made to Article 5 of Regulation (EC) No. 2100/94.

According to the Directive, inventions that are new, that involve an inventive step, and that are susceptible of industrial application shall be patentable even if they concern a product consisting of or containing biological material or a process by means of which biological material is produced, processed, or used (Article 3).

Biological material that is isolated from its natural environment or produced by means of a technical process could be the subject of an invention even if it previously occurred in nature.

The Directive defines exceptions to patentability in accordance with those that have already been discussed in the context of the European Patent Convention. In other words, plant and animal varieties and essentially biological processes for the production of plants or animals are not patentable. It is also made clear that inventions that concern plants or animals shall be patentable if the technical feasibility of the invention is not confined to a particular plant or animal variety. It is expressly made clear that the patentability of the inventions that concern a microbiological or other technical process or a product obtained by means of such a process is not affected.

A particular feature of the EU Directive is that it contains provisions relating to the human body and the related issues of patent protection. Article 5 of the Directive states that the human body, at the various stages of its formation and development, and the simple discovery of one of its elements, including the sequence or partial sequence of a gene, cannot constitute patentable inventions. However, an element isolated from the human body or otherwise produced by means of a technical process, including the sequence or partial sequence of a gene, might constitute a patentable invention, even if the structure of that element is identical to that of a natural element. In this connection, it is a special feature not known from other patent systems that the industrial application of a sequence or partial sequence of a gene must be specifically described in the patent application.

The EU Directive also contains a provision whereby inventions shall be excluded from patentability where their commercial exploitation would be contrary to "ordre public" or morality. Under Article 6, processes for cloning human beings, processes for modifying the germline genetic identity of human beings, the use of human embryos for industrial or commercial purposes, and processes for modifying the genetic identity of animals that are likely to cause them suffering without any substantial medical benefit to man or animal, and also animals resulting from such processes, are prohibited.

The protection conferred by a patent on a biological material possessing specific characteristics as a result of the invention shall extend, according to the Directive (Article 8) to any biological material derived from that biological material through propagation or multiplication in an identical or divergent form and possessing those same characteristics.

The protection conferred by a patent on a process that enables a biological material to be produced possessing specific characteristics as a result of the invention shall extend to biological material directly obtained through that process and to any other biological material derived from the directly obtained biological material through propagation or multiplication in an identical or divergent form and possessing those same characteristics.

In Article 9, the Directive also contains a special provision regarding the protection of products with genetic information. According to this Article, the protection conferred by a patent on a product containing or consisting of genetic information shall extend to all material in which this product is incorporated and in which the genetic information is contained and performs its function.

The patent protection referred to shall not extend to biological material obtained from the propagation or multiplication of biological material placed on the market in the territory of a Member State by the holder of the patent or with his consent, where the multiplication or propagation necessarily results from the application for which the biological material was marketed, provided that the material obtained is not subsequently used for other propagation or multiplication.

Finally, Article 11 contains exceptional provisions for farmers. According to this Article, the sale of plant-propagating material to a farmer by the holder of the patent or with his consent implies authorization for the farmer to use the product of his harvest for propagation or multiplication by him on his own farm. A similar provision is implied in Article 11(2) in respect of the sale of breeding stock.

As set forth in the above, the Directive distinguishes between plants and animals that are patentable and plant and animal varieties that are not. The reason for this differentiation lies in the means of achieving the product concerned: A plant or animal variety is generally obtained by essentially biological processes (sexual reproduction observable in nature), whereas transgenic plants and animals are obtained through nonbiological processes forming part of genetic engineering.

By virtue of the leeway provided by Article 27(3)(b) of the TRIPS Agreement, the Directive did not make use of the possibility afforded to the Members to exclude plants and animals from protection through patents.

The Directive reiterates that, although plants are patentable, plant varieties are excluded from patentability and are protected by plant variety rights. This right complies with the *sui generis* protection provided for by the TRIPS Agreement.

Article 5(2) of Council Regulation (EC) 2100/94 of 27 July 1994 defines a plant variety as a plant grouping within a single botanical taxon of the lowest known rank. The relevant provisions of the Directive are to be found in Article 4 and recitals 29 to 32.

In the action for annulment of Directive 98/44/EC, the applicants considered that the provisions relating to the patentability of plants and animals were unclear and ambiguous, and, hence, a source of legal uncertainty that justified an annulment of the Directive.

The Court rejected those arguments. It referred to the substance of Article 4 of the Directive, which lays down that a patent cannot be granted for a plant variety, but can be for an invention if its technical feasibility is not confined to a particular plant variety.

On the basis of recitals 29 to 32 of the Directive, therefore, it reiterated that plants varieties are defined by their whole genome and are protected by plant variety rights. However, plant groupings of a higher taxonomic level than the variety, defined by a single gene and not by the whole genome, can be protected by patent if the relevant invention incorporates only one gene and concerns a grouping wider than a single plant variety.

The Court concluded that a genetic modification of a specific plant variety is not patentable, but a modification of wider scope, concerning, for example, a species, can be protected by a patent.

It should be noted that this distinction does not apply in the United States. The Supreme Court, in a Decision of 10 December 2001, judged that a patent could be granted for an invention relating to a plant variety if it met the conditions required (novelty, nonobvious matter, utility, sufficient description, and deposit of biological material accessible to the public).

Article 12 of the Directive establishes a system of cross-licences between plant variety rights and patents where a breeder cannot obtain or exploit a plant variety right without infringing a prior patent, and vice versa.

Applicants for licences must demonstrate that they have applied unsuccessfully to the holder of the patent or the plant variety right to obtain a contractual licence and that the plant variety or the invention constitutes significant technical progress of considerable economic interest compared with the invention claimed in the patent or the protected plant variety.

Paragraph 4 of that Article lays down that where a licence for a plant variety can be granted only by the Community Plant Variety Office, Article 29 of Regulation (EC) No 2100/94 shall apply.

Member States cannot be expected not to transpose into national law a provision that would have to be amended by the Commission under the Regulation cited above.

Article 29 lays down that the Community Plant Variety Office shall grant such licences only on grounds of public interest.

Moreover, pursuant to Article 29(7) of Regulation (EC) 2100/94, only the Community Plant Variety Office is authorized to grant compulsory licences. However, under the applicable national law, that Office cannot be responsible for granting compulsory licences for national patents.

The Commission has examined the impact of Article 12 of the Directive on Article 29 of Regulation 2100/94. It has already taken the necessary steps to submit to the Council any suitable proposal for overcoming this difficulty.

This question was not broached in the judgment of the Court. There is no legal definition of an animal variety. This can be defined as a taxonomical grouping ranking next below a subspecies (where present) or species, whose members differ from others of the same species or subspecies in minor but permanent or heritable characters.

The relevant provisions of the Directive are essentially Articles 4 and 6(2)(d). It should also be noted that there is no protection of animal varieties in Community law.

Under Article 4(1)(a), animal varieties are not patentable. However, inventions relating to animals are patentable if the technical feasibility of the invention is not confined to a particular plant or animal variety. If an animal can be obtained only through genetic engineering, to the exclusion of any natural breeding, the invention relating to such an animal can be protected by patent.

This question has been debated many times in the context of the patent for the Harvard oncomouse. This patent relates to a mammal modified by genetic transfer. Thanks to this manipulation, the animal can, under certain conditions, develop tumors that can be used for cancer research.

After more than 16 yr of proceedings, the Opposition Division of the EPO responsible for this case decided on November 7, 2001 to limit this patent to transgenic rodents with the cancerous gene and, hence, not to authorize its extension to all mammals with the introduced gene. In the United States, this patent was granted in its initial form (i.e., it covers any nonhuman transgenic mammal).

It should also be noted that in a judgment of August 3, 2000, the Canadian Federal Court of Appeal accepted that this patent had the same scope as that granted by the American Patent Office (USPTO).

To be exhaustive on this question, it should be pointed out that essentially biological processes for obtaining animals and plants are not patentable. To the contrary, an essentially nonbiological process will be patentable. It is for the courts to decide, in each case, whether a process is essentially biological or not.

Article 27(1) of the TRIPS Agreement lays down a general principle of patentability in all fields of technology. However, under Article 27(3)(b), Members could exclude plants and animals from patentability even when the inventions relating to them meet the classic conditions for patentability. However, the same Article states that its Members must provide for patent protection of nonbiological processes.

The same applies to microbiological processes. Moreover, again according to the TRIPS Agreement, microorganisms must be patentable if the patentability conditions are met.

That is why Article 4(3) lays down that inventions that concern a microbiological or other technical process or a product obtained by means of such a process are not *per se* excluded from patentability.

The human body, at the various stages of its formation and development, is not patentable, as it involves a simple discovery. The same applies to the simple decoding of one of its elements. This exclusion also covers the discovery of a sequence or partial sequence of a gene.

However, an element isolated from the human body, including a sequence or partial sequence of a gene, by techniques of identification, purification, characterization, and multiplication, can constitute a patentable invention, even if the structure of that element is identical to that of a natural element. The same reasoning can obviously be applied to any element produced otherwise synthetically by a technical process. This type of invention, which is eligible for patent protection, must nevertheless fulfil the classic conditions for patentability (i.e. novelty, inventive step and industrial application).

The Directive allows a degree of flexibility as to the extent of the protection to be conferred on inventions relating to elements isolated from the human body (Commission of the European Communities Report from the Commission to the European Parliament and the Council of 07.10.2002).

2.2.3. Budapest Treaty

In patent law, all inventions are treated the same with regard to their patentability, including those relating to microbiological processes or the products thereof. In particular, every invention must be disclosed sufficiently clearly and fully for a skilled person to be able to implement it. Whereas in other technological areas the subject matter of the invention can be explained by means of a description, formulas and drawings in such a manner that it is identifiable with the aid of these information carriers, this is not automatically the case when microorganisms are the subject matter of the invention. These cannot normally be described and figuratively depicted in the patent application in such a manner that it would be possible to identify them solely by those means. Additionally, it is not normally possible for a skilled person to copy microbiological processes without inventive input unless the organism is available. In such cases, the organism must, therefore, be deposited at the same time as the application, at the latest, in a suitable place for that purpose and the depository must be stated in the application (*see* **Fig. 5**).

The details of the depository and the deposit number cannot, however, replace the description. It is merely a stopgap measure intended to make it easier for the inventor to identify beyond doubt, as required for patenting purposes, an organism that has not previously been described in the literature and that also cannot be described by means of morphologic or physiologic features or biochemical parameters. The prior deposit and notification thereof does not furthermore discharge the applicant from the obligation to do everything in his power to characterize the organism as exhaustively as possible using conventional means.

According to the principles in respect of the deposit that have been developed by case law, the organism must be deposited at a scientifically recognized place, irrespective of whether this is located in this country or abroad. As the

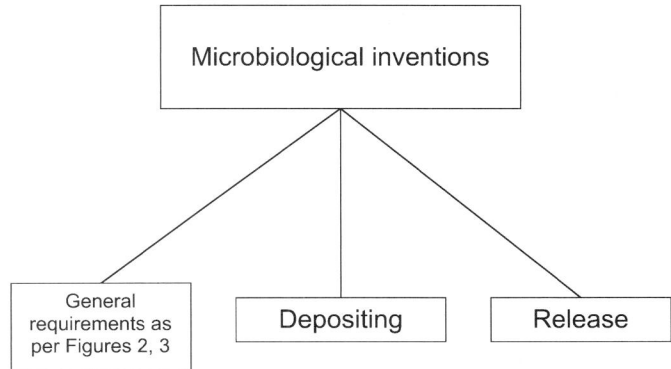

Fig. 5. Special requirements for protection for microbiological inventions.

depositing of the organism supplements or takes the place of the description of the invention, the details must be orientated toward the purpose pursued by the patent claims and the description of the invention. It is necessary to ensure that the deposited organism is available to the granting authorities at any time; an irrevocable declaration to the depository is prescribed for this purpose. The depositing must take place no later than the priority date or filing date in order to guarantee that the applicant was in possession of the complete invention, so that the invention was ready.

Provision must be made at the depository to ensure that the microbiological material is available to experts also for a reasonable period after the term of the patent being claimed, in order for the protected invention to be copied, even in the event of no patent being granted after the application is published.

The Budapest Treaty *(20)* contains more detailed provisions about the international recognition of the depositing of microorganisms for the purposes of patent grant proceedings. This Treaty governs the powers of the internationally recognized depositories, what happens in the event of interruption, and how long the deposit is to last. It obliges the Member States to recognize international depositories, with the effect that the applicant, on making a deposit at one of these depositories, satisfies the conditions with regard to all Member States. It also creates the possibility of a new deposit if the initial deposit is interrupted for technical reasons. Implementation regulations lay down the procedure for the deposit and any new deposit, including the retention of an organism for 30 yr. These regulations are in accordance with German legal practice, according to which a reasonable period for the deposit is regarded as being 5 yr after the last request for a sample to be handed over, but at least 30 yr in any event.

In some cases, however, no deposit is necessary—for example in the case of a known organism that is generally available. The relevant procedure is then sufficiently specified with regard to the initial material and the intended end product and is thus adequately disclosed if the nature of the process to be performed using the organism is described and the relevant nomenclature of the organism is stated unambiguously.

It is also unnecessary for microbiological processes the implementability of which is not restricted to one species to be secured by deposit, provided that at least one known type of microorganism can be used. The same applies to inventions relating to facilities for conserving, lyophilizing, storing, and so forth, microbiological processes. Such processes are applicable to various organisms. In this case, the essence of the invention is not the selection or the specific characteristic of a particular organism. Additionally, it is unnecessary for organisms to be deposited if they are analogous to organisms that have already been deposited. Analogous means that two similar strains of the same kind lead to the same result by producing the same metabolic product in a fermentation process under the same conditions.

However, the deposit alone is not sufficient. The culture must furthermore be freely available to third parties (release). The applicant must ensure that the relevant strain is generally available—in other words freely available, at the time of the disclosure at the latest. This is because, in spite of the publication, it is only possible for others to use the teaching if the strain itself is available to the public.

The depositing of the strain is an essential component of the disclosure of the invention. This guarantees that the organism is permanently available and not subject to any further influence by the applicant. By the deposit and release, the applicant has irrevocably relinquished the organism. The technical teaching and the organism provided by him make it directly possible for the technology to be further developed and for the patented teaching to be used. In particular, the organism becomes unrestricted public property because of the possibility of obtaining samples and propagating them naturally. According to more recent case law, the possibility of repeating a microbiological invention is guaranteed by means of the deposit. It is, therefore, no longer necessary in such a case for the possibility of repeating the cultivation process to be proved separately.

Further developments of microorganisms must be based on known organisms. These are primarily available to the skilled person in the scientifically recognized depositories. The material that is kept there and that is freely available is the state of the art on which microbiology is constructed by means of mutation and selection or by genetic engineering methods. It will, therefore, only be possible to copy repeatable processes for producing new organisms by using known deposited organisms. Substantive protection for organisms produced in a repeatable manner can, therefore, only be appropriately guaranteed if it is permissible to make the initial material available by depositing.

Inventions that make use of a new organism are placed at substantially greater risk by the disclosure and handing out of a culture than any other inventions. The devices and containers in which microbiological processes are performed are possessed by many companies, as are the necessary nutrient solutions and the necessary experience in dealing with such cultures. The culture that is handed out can, therefore, normally be used without any special outlay. It is rightly felt to be deplorable that the deposited organism is handed over to interested third parties as early as the disclosure of the filing documents. The option is therefore provided in European patent grant proceedings for the applicant to be able to restrict the release of a deposited organism to the provision of samples to an expert (the "expert solution") until notification of the grant of the patent. This is intended to prevent abuse by persons taking samples for the period of the grant proceedings. The expert acts on the applicant's instructions, as it were as a representative of the public, and conducts investigations and examinations of the invention. However, there is no corresponding provision in German law *(2)*.

2.2.4. UPOV International Convention for the Protecton of New Varieties of Plants

The Office of the International Union for the Protection of New Varieties of Plants (UPOV) has announced that the 1991 UPOV Convention came into force on April 24, 1998. The new convention is the result of a diplomatic conference that was held in Geneva in 1991 *(21)*.

The UPOV convention lays down the international rules of the system whereby the countries grant intellectual property rights to individuals or organizations that cultivate new plant varieties. The breeding process is time-consuming and costly, but plant varieties can frequently be readily propagated after they have been released. The breeding of new plant varieties does not normally attract private funds unless there is appropriate protection for the intellectual property.

The UPOV Convention was originally established in 1961. Noteworthy scientific and technical developments took place between 1961 and 1991, resulting *inter alia* in the emergence of genetic engineering and the advances in in vitro propagation. These developments, in conjunction with the experience of dealing with the 1961 Convention, constituted strong reasons for reviewing the Convention in 1991.

The 1991 Convention strengthened the position of breeders. It requires UPOV Member States to grant a right in respect of the entire production of seeds or plants to the breeder but allows each country to exclude from the breeder's right the use of seed obtained and used in a farm (the "farmer's privilege"). Member States must additionally grant a breeder certain rights in the final product of his variety (the "harvest"), subject to certain requirements (the harvest must be derived from seed or plant material used without the breeder's consent and the breeder must not have been given any reasonable opportunity to exercise his right in relation to this seed or plant material).

The most important point might be that a genetic engineer who uses a protected variety as a medium for his innovation (e.g., pest resistance or herbicide resistance) might not be in a position to use the genetically modified variety without the consent of the proprietor of the protected variety, namely if the genetically modified variety is regarded as being "essentially derived" from the protected variety. Prior to the 1991 revision, it would have been possible to use the genetically modified variety without recognizing the contribution made by the breeder of the protected variety to the final product.

Article 27(3)(b) of the Agreement on Trade-Related Aspects of Intellectual Property Rights (TRIPS Agreement) *(22,23)* requires members of the World Trade Organization (WTO) *(24)* to protect plant varieties under the patent system or by an effective *sui generis* plant variety protection system or by a combination of these systems. The UPOV Convention provides the only internationally recognized *sui generis* system for protecting plant varieties. The requirement in the TRIPS Agreement to protect the rights of the breeders of new plant varieties is already applicable to all industrialized countries that are members of the WTO and came into force on January 1, 2000 in respect of numerous developing countries.

2.2.5. TRIPS

Article 8(1) of the TRIPS Agreement *(22,23)* allows measures that are necessary to protect public health and nutrition provided that such measures are consistent with the provisions of this Agreement. This wording implies, first of all, that restrictions of patent protection to protect public health and nutrition might indeed be necessary, as is also stated in the Doha declaration. It is not, however, possible to predict with scientific accuracy the situations in which this is the case or which specific patent measures have a positive effect on the protection of public health and nutrition. It is therefore necessary to make a prognosis decision in circumstances of uncertainty. Recourse may, however, be had to the experience of other countries in broadly comparable positions, positive and negative examples again being found in many cases. Ultimately, WTO members can, therefore, only be required to make those patent policy decisions that they consider necessary without it being possible to require evidence from them as to the necessity thereof. The limit is reached when it is no longer possible to expect that the patent measure can have the desired result.

The extent of public health and nutrition that WTO members can select under Article 8(1) of the TRIPS Agreement as a target requirement is not specified in the TRIPS Agreement.

It must be noted in all this that Articles 7 and 8(1) of the TRIPS Agreement must not be regarded as independent legal bases for exceptional provisions in favor of protecting public health or nutrition. The wording of Article 8(1) of the TRIPS Agreement only permits measures "provided that such measures are consistent with the provisions of this Agreement." This wording makes it clear that patent law measures are only permissible in the context of the requirements and exceptional provisions of Article 27 et seq. of the TRIPS Agreement.

The objectives and principles of Articles 7 and 8 of the TRIPS Agreement therefore require points of entry into the patent provisions of Article 27 et seq. of the TRIPS Agreement. Aspects of nutrition and health are expressly mentioned in Articles 27(2) and 27(3) of the TRIPS Agreement relating to the exclusion of inventions from patentability. Article 27(2) of the TRIPS Agreement provides a possibility of exclusion where it is necessary to prohibit the commercial exploitation of an invention to protect ordre public "including to protect human life or health." Article 27(3)(a) of the TRIPS Agreement relates to diagnostic, therapeutic and surgical methods for the treatment of humans or animals, whereas Article 27(3)(b) of the TRIPS Agreement covers patent protection for plants and animals, which developing countries in particular regard as being highly important for feeding their inhabitants.

Article 27(1) of the TRIPS Agreement deprives WTO members of the most potent weapon, namely the exclusion from patentability of inventions that it is desired for political reasons to keep free from patents. The requirement to protect inventions both of products and also of processes in all technological areas that are new, based on an inventive step, and susceptible of industrial application is regarded as the outstanding innovation of the TRIPS Agreement in the patent law area, in view of the fact that the exclusion of medicinal and chemical products from patentability by a large number of developing countries was an important reason for including intellectual property protection in the negotiations in the Uruguay round. WTO members still have some flexibility in that the differentiation between inventions and discoveries and the concepts of novelty and the inventive step and nonobvious inventions are not defined in the TRIPS Agreement. This enables WTO members to assess the discovery of substances existing in nature that can only be used in isolated and purified form, including human proteins, as a discovery and so to exclude them from patentability. WTO members can go a very long way in assessing novelty, as they provide no territorial or formal restrictions or periods of grace for the inventor and additionally regard earlier suppositions or speculations (e.g., about the potential effects of medicinal products) as prejudicial to novelty. High requirements in respect of the inventive step might exclude the patentability of synthetically produced medicinal products based on known effects of natural substances. Finally, Article 27(1) of the TRIPS Agreement also says nothing about the patent law treatment of the discovery of new areas of application for substances that are already known, which is far more important than the discovery of new substances, particularly in pharmaceutical research, and is also highly important in genetic engineering. The TRIPS Agreement does not, therefore, require the granting of product patents that is provided for in the EPC and in the United States but allows a new process to be assessed as an invention.

A further possibility of exclusion, which is based both on grounds of food safety and on ethical considerations, exists under Article 27(3)(b) of the TRIPS Agreement in respect of "Plants and animals other than microorganisms and essentially biological processes for the production of plants or animals other than non-biological and microbiological processes." The WTO members must, however, introduce an "effective *sui generis* system" to protect plant varieties if they wish to exclude patent protection, whereas they are not required to provide any protection for breeds of animals. The reference to an effective *sui generis* protection system, which is not specified in greater detail and does not refer to any of the existing protection systems, particularly the Union for the Protection of New Varieties of Plants (UPOV) again allows the WTO members considerable flexibility that they can use by orientating themselves in particular to the objective of food safety under Article 8(1) of the TRIPS Agreement.

In addition to these exceptions under Article 27(3) of the TRIPS Agreement, which are not dependent on further conditions being fulfilled, Article 27(2) of the TRIPS Agreement permits the excluding of inventions from patentability where the prevention within Members' territory of the commercial exploitation of such inventions is necessary to protect "ordre public" or morality, including to protect human, animal, or plant life or health. This provision makes it possible not to patent inventions that are socially undesirable and dangerous and the use of which is prohibited. However, it provides no further assistance to WTO members that wish to use an invention without patenting it, as was originally the intention of a group of developing countries. Admittedly, a ban on commercial exploitation could be announced while permitting noncommercial exploitation—for example by medicinal products being distributed free of charge by the public authorities or charitable organizations. It is easy to imagine in such circumstances that noncommercial exploitation will lead to greater availability, which benefits public health. It would, however, be necessary to explain further why a ban on commercial exploitation was required, instead of permitting such exploitation in addition to noncommercial exploitation. The latter is hardly conceivable in the case of inventions that it is desirable to exploit. Preventing patenting while exploiting the invention on a noncommercial basis can, therefore, only be considered if the invention must be completely withdrawn from commercialization on ethical grounds or must be reserved for the State for reasons of safety *(23)*.

2.3. National Laws

The national laws are largely aligned with the international provisions. In the individual countries, there are formal differences and differences resulting from the development of case law. In principle, however, it is possible to obtain patent protection for biotechnological processes and products in practically all countries. This applies both in the case of conventional biotechnology and in the case of modern genetic engineering.

3. Final Comment

Microbiological processes and the products thereof are patentable. In this connection, "products" means the metabolic products of the microorganisms, the microorganisms themselves, and ingredients of the microorganisms, particularly genetic material. In addition to fulfilling the general requirements for obtaining patent protection, it is necessary in the field of microbiology to deposit the microorganism. It is undisputed nowadays that depositing is essential in many cases as a means for identification, disclosure, and the possibility of copying inventions in the microbiological field, although there is no guarantee that the deposited organism will retain its characteristics in the long term. However, as a result of modern storage methods, there is at least sufficient likelihood that the organism will remain unchanged. The problem of the release of microorganisms is disputed. This is connected with the fact that a disclosed invention does not provide the opportunities of claims for injunctions and damages that the applicant has in the case of a patent that has been granted. The applicant is, therefore, exposed to the substantial risk of third parties using his invention without the law providing him with effective countermeasures. This is because the applicant, as it were, hands over the whole works when he hands over the organism. This could deter him from disclosing or filing his invention or application, because the risk of misuse is extremely great.

European patent law provides wider-reaching protection at the time of disclosure. It would be welcome if it were made possible on a uniform worldwide basis for the organism to be handed over to only one neutral expert, so that the organism was only indirectly available to the public (the "expert solution"). This provision could prevent abuses and encourage potential applicants to achieve a higher level of application.

Apart from this, modern genetic engineering now makes it possible to identify microorganisms with increased accuracy. Particularly in cases of genetic modification, an exact description can be provided by giving the genetic code, so that it is no longer necessary in such cases for the organisms to be deposited.

Overall, in the current legal position, it is necessary for the applicant to consider in a specific case whether a deposit is, in fact, absolutely necessary. If a deposit can be avoided, it should, for the abovementioned reasons, be dispensed with. However, in the event of an application without depositing the microorganism, the national law of the countries must be checked in each individual case. This is because it is entirely possible that the omission of a deposit will be accepted in one country and the depositing of microorganisms is normally unavoidable in other countries (e.g., Taiwan).

References

1. Fitzner, U. (1988) Der patentrechtliche Schutz von Mikroorganismen [Patent protection of microorganisms], *CLB,* pp. 181.
2. Fitzner, U. (1980) Der Patentrechtliche Schutz mikrobiologischer Erfindungen [The patent protection of microbiological inventions], dissertation, Free University of Berlin.
3. Kurz, P. (2000) *Weltgeschichte des Erfindungsschutzes [World History of the Protection of Inventions].* Carl-Heymanns-Verlag, Cologne.
4. Fitzner, U. (1999) Der Patentrechtliche Schutz von Erfindungen auf dem Gebiet der Biologie [Patent protection of inventions in the field of biology], in *Festschrift Nordemann,* (Zollner, B., and Fitzner, U., eds.), Baden-Baden, p. 51.
5. OJ EPO 1984, p. 112.
6. Thomsen, H. (1998) GRUR Int., p. 213 et seq.
7. OJ EPO (1990), p. 71.
8. OJ EPO (1995), p. 545.
9. Lancon, P. (1998) GRUR Int., p. 227.
10. Moser, W. (1998) GRUR Int., p. 209 et seq.
11. EPO GRUR Int. (1998), p. 251.
12. OJ EPO (1990), p. 476.
13. EPO GRUR Int. 1993, p. 865.
14. BGHZ 64, 101.
15. OJ EPO 1995, p. 388.
16. Schatz, U. (1997) GRUR Int., p. 588.
17. OJ EPO (1975), p. 388.
18. OJ EPO (1990), p. 476.
19. Directive 98/44/EC of the European Parliament and of the Council of 6 July 1998 on the legal protection of biotechnological inventions, OJ 213 of 30 July 1998, p. 13 GRUR Int. 98, 675 = p. 98, 458 = OJ 99, 101.
20. BGBl. II/97, p. 1095
21. BGBl. II/98, p. 259
22. Agreement on Trade-Related Aspects of Intellectual Property Rights of 15 April 1994, *Bundesgesetzblatt* (BGBl.) 1994 II 1438, 1730 = p. 95, 23.

23. Rott, P. (2003) TRIPS Abkommen, Menschenrechte, Sozialpolitik u. Entwicklungsländer [TRIPS Agreement, human rights, social policy and developing countries], GRUR Int., pp. 103–198.

24. Agreement to establish the World Trade Organization (WTO) of 15 April 1994, BGBl. 1994 II 1438 = p. 95, 19.

Index

(δ-L-α-aminoadipyl)-L-cysteinyl-D-
    valine) synthetase, *see* ACVS
(δ-L-α-aminoadipyl)-L-cysteinyl-D-
    valine), *see* ACV
μBondapak C$_{18}$, 135, 145
[^{13}C$_6$] glucose, *see* ^{13}C tracer substrate
1,4 butandiol-diglycidylether, 169
1,4 dithio-DL-threitol, *see* DTT
^{13}C tracer substrate, 193, 194
16S rDNA analysis
    polymerase chain reaction-denaturing
        gradient gel electrophoresis
        (PCR-DGGE), 382
    polymerase chain reaction-temperature
        gradient gel electrophoresis
        (PCR-TGGE), 382
2,3-dehydratase, 138
2-hydroxymethylclavam, 143
2xTY, 135
3-ketoreductase, 138
3-HO-echinone, 263
4-ketoreductase, 138, 142
4,6-dehydratase, 138
5-bromo-4-chloro-3-indolyl-β-D-
    thiogalactopyranoside, *see* X-gal
7-ACA, 43, 44
7-ADCA, 42, 44
    bioconversion from DAOC, 45, 57
    chemical expansion, 44
    chemical extraction, 60
    production, 45, 46
7-aminocephalosporanic acid, *see* 7-ACA
7-aminodeacetoxycephalosporanic acid,
    see 7-ADCA
8-demethyl-tetracenomycin C, 144
8-DMTC, 144, 145

A

Acetate buffer, 318
Acid phosphatase, 413, 414, 417, 423
Acid rock drainage, 405
Acidianus brierleyi, 406
Acidimicrobium ferrooxidans, 405
Acidithiobacillus, 405, 408
Acidithiobacillus caldus, 405
Acidithiobacillus ferrooxidans, 405, 408
Acidithiobacillus thiooxidans, 405, 408
Acidophilic Fe(II) oxidizers, 406, 408
Acidophilic Fe(II) oxidizing bacteria,
    see Acidophilic Fe(II) oxidizers
Acidophilic sulfur oxidizers, 407, 408
Acidophilic sulfur oxidizing bacteria,
    see Acidophilic sulfur oxidizers
Acremonium chrysogenum
    DAOC-producing strain, 45, 48
        construction, 45
        selection by bioassay, 48, 54, 55
    flask fermentation, 48, 55
    maintenance and preservation, 55
    PN-producing strain, 45, 48
        construction, 45
        selection by bioassay, 48, 54
    protoplasts formation, 53
    transformation, 47, 53, 54
        hygromycin, 48
        phleomycin, 48
Acridine orange, *see* AODC
Acridine orange direct counting, *see*
    AODC
Actinomycetes, 65, 70, 131
Activity stain, 219
ACV, 42
ACVS, 42

Adipyl-7-ADCA, 44, 45
AEC, 188
Air-lift fermentor, 414, 416, 417, 418
Alanylclavam
    mode of action, 152
    quantification by bioassay, 157
    quantification by HPLC, 156
    structure, 153
Alkaline lysis, 165, 166, 167, 169
Allelic replacement, 183, 184
Ammonium sulfate
    elution of the pDNA, 167, 170
    equilibration of the column, 170
    precipitation, 165, 167, 169, 170,
        173, 174, 175, 177
    separation from pDNA, 170
Ampicillin, 47, 135, 136, 139, 141, 142,
        144, 303, 305, 306
Analog-resistant mutants, 8, 9
Antibacterials, *see* Antibiotics
Antibiotic resistance, 13
Antibiotic-resistance markers, 15
Antibiotics
    7-ACA, 43, 44
    7-ADCA, 42, 44
    ampicillin, 135, 136, 139, 141, 142,
        144
    apramycin, 135–137, 144, 146
    azithromycin, 71
    bacteriocins, 370
    cephalosporin, 41
    cephamycin, 143
    chemical class, 66
    chloramphenicol, 47
    clarithromycin, 71
    clavulanic acid, 143
    elloramycin, 132, 136, 144
    erythromycin, 65, 71
    G418, 260
    geneticin, 260
    holomycin, 143
    holothin, 143
    hygromycin, 48
    kanamycin, 135, 136, 139, 141

oleandomycin, 138, 139
    penicillin, 48, 116
    phleomycin, 48
    roxithromycin, 71
    target, 66
    tetracenomycin, 132
    thiostrepton, 135, 136, 139, 144, 145
    tobramycin, 135, 136, 146
    tunicamycin, 153
Antifoam Mazu DF204, 73
Antiparasites, 131
Antitumor, 131
AODC, 406, 408
Apramycin, 135–137, 144, 146
Arabidopsis thaliana, 334
ARD, *see* Acid rock drainage
Arrhenius equation, 410
Ashbya gossypii, 283
    isolation of high-molecular-weight
        DNA, 292, 293
    media, 285, 286
        MA2, 285
        SPA, 286
    mutagenized library, 294, 295
    riboflavin production, 283, 284
    strain, 285
    transformation, 296
        AT buffer, 273–282
        heterokaryotic mutant, 296
        homokaryotic mutant, 297
        STM buffer, 286
    vitamin B2, 283
Aspartate-family amino acids, 9
Aspartokinase, 180
Aspergillus, 225
Aspergillus awamori, 225, 226
Aspergillus awamori TGDTh-4, 227, 234
Astaxanthin, 248, 263–268
    isomers, 265, 266
    natural sources, 258
    precursors, 263
    xanthophyll, 258
Automation of high-throughput
        screening, 5. 6

Auxotrophic mutants, 9
Avidin, *see* Streptavidin
Azithromycin
   structure, 71
   synthesis, 83

B

Bacillus megaterium
   expression plasmids, 207
      free-replication, 207, 209, 222
      integration, 210, 222
   protein production, 210
   protoplasts, 208, 220
   secretion, 212
   strains, 222
   vitamin B_{12} production, 211
Bacteriocins
   bioassay, 374, 377
      indicator plates, 372, 374, 378
   Listeria indicator, 376
   modeling of activity, *see* Modeling
   poor activity, 370
Batch culture
   preparation of cells, 415, 417
   vessel, 417
BCA
   working reagents A and B, 168
   working solution, 168, 175
Bennett, 135, 137
Bioaccumulation of metals, 413, 414,
      419, 429
Bioactive, 131
Biotransformation, 132, 144, 145
Bioassay, 45, 48
Bioleaching, 405, 407, 409
Bioleaching activity, 407, 409
Biolegislation, 437
Biological invention, 451, 466, 467,
      471, 479
Biological methods, 467, 475
Biological material
   definition, 458
   deposit, 451, 452, 458–461
   diagnostic specimens, 447

UN 3373, 448
   genetic information, 458
   genetically modified organisms, 446
   infectious substances, 437, 441,
      445–448
      UN 2814, 445–448
      UN 2900, 445–448
   microorganism, 457, 458
   perishable biological substances, 441
   reproducibility, 458
   risk level, 458
Biological substances, *see* Biological
      material
Biological process
   animal, 465, 475, 480, 482
   animal breed, 474
   animal cells, 474, 476
   bacteria, 474, 477
   human, 465
   plants
      hybrid plants, 465, 472, 473, 475,
         480
      herbicides, 472
      plant cells, 474, 476
      cross-breeding, 475
      mixing breeds, 475
      selective breeding, 475
   process, 465, 468, 469, 472–474,
      476, 477, 483, 490
   yeast, 474, 476, 477
Biological weapons
   Biological and Toxin Weapons
      Convention (BTWC), 439
   Aspergillus fumigatus, 439
   Bacillus anthracis, 439
   Clostridium botulinum, 439
   EEC Dual Use Directive, 437
   end-user certificate, 440
   Staphylococcus aureus, 440
   Yersinia pestis, 439
Bioplastics, 336
Bioremediation
   heavy metals, 413, 414
      cobalt, 413, 414

nickel, 413, 414
    uranium, 413, 414
Biosafety precaution, 309
Biotechnological invention, *see* Invention
Biotechnology, 478
Biotin, 378
Blakeslea trispora, 240, 273
Blocking competing pathways, 12
Brain–heart infusion (BHI), 371, 372, 378
Brevibacterium linens
    cheese, 400
    growth, 399
BSA standards, 168, 175
BTWC, *see* Biological Weapons
Budapest Treaty
    Budapest Union, 452
    conversion, 452
    storage, period of, 462
    treaty, 452, 457

C

Calcofluor white, 244, 249
Canthaxanthin, 263
Carotene
    β-carotene, 274, 278
    γ-carotene, 274, 278
    ζ-carotene, 274, 278
Carotenoids
    absorption spectra, 274, 278
    analysis by HPLC, 245, 266
    biosynthesis, 274
    extraction, 245, 266, 276, 277
        DMSO, 266
        polar organic solvents, 266
    hyperproducing strains, 244
    lability, 282
    physical properties, 263
        β-carotene, 263
        3-HO-4-keto-β-ionone group, 265
        3-HO-4-keto-torulene, 265
        3-HO-canthaxanthin, 265
        absorbance spectra, 263
        absorption chromatography, 265
        astaxanthin, 263–268

isomers, 265
    mobility, 265
    monocyclic carotenoids, 263, 265
    polarity, 265
    retention time, 265
production in *E. coli*, 258
    3-HO-4-keto-torulene, 268
    3-HO-canthaxanthin, 268
purification, 277-279
quantitation, 279, 280
Caylase, 48
cDNA, 338
cefE
    expression in *A. chrysogenum*, 45
    expression plasmid, 51
    expression in *P. chrysogenum*, 44, 45
    gene, 42, 44
cefEF
    disruption plasmid, 50
    expression in *A. chrysogenum*, 44, 45
    gene, 42, 50
    inactivation, 45, 50
Cells immobilization
    methods, 413, 424
        biofilm, 415, 422
        covalent, 418, 425
        entrapment, 417, 424
        packed-bed reactor, 413, 414, 419,
            420, 429
Cells preparation
    biofilm, 415, 422
    planktonic, 415, 422
Cellular debris, 169
Central metabolism perturbing, 11, 12
Cephalosporins
    biosynthetic pathway, 42
    chemical structure, 42
    clinical spectrum, 41
    quantification by HPLC, 49, 56, 145
    semisynthetic derivatives, 41
Cephamycin C
    quantification by bioassay, 154
    quantification by HPLC, 155
    structure, 153

CFW, *see* Calcofluor white
Chemiluminescent, 339
Cheddar cheese, 336
Cheese slurry, 400
Chitin deposition, 249
Chloramphenicol, 47
Chromatography
  alumina column, 277
  thin layer, 279
    alumina, 279
    Rf values, 278
    silica gel, 279
  HPLC, 275
    columns, 275
    diode array detector, 279
    instrument, 275
    Rt values, 278
    solvents, 278
Chromogenic agents, 7, 8
Citrobacter sp., *see Serratia* sp.
Clarithromycin
  structure, 71
  synthesis, 83
Class 9, 448
Clavam-2-carboxylate, 153
Clavulanic acid
  mode of action, 152
  quantification by bioassay, 156
  quantification by HPLC, 156
  structure, 153
Cloning in humans, 480
Cobalt
  assay, 421, 430
  bioremediation, 413, 414
    intercalation in HUP, 419, 429
Combinatorial biosynthesis, 132
Community Plant Variety Office, 482
Conjugation and DNA transfer, 14
Continuous culture
  air-lift fermentor, 414, 418
  preparation of cells, 415, 422
Corynebacterium glutamicum
  amino acid production, 13, 151
    glutamate, 192

lysine, 192
  genome sequence, 187
  L-lysine-producing mutant, 180
    double mutant, 185
    hom gene, 180–183
    lysC gene, 180–183
    L-lysine production, 186, 187
    mutant generation, 183–185
    single mutant, 184, 185
  minimal medium, 194
  protein hydrolysis, 196
  shake flask cultivation, 195
Corn steep liquor, *see* CSL
Cos16F4, 135, 136, 144
CosAB61, 135, 138, 139, 141
Crab-tree, 243
CSL, 71, 73, 87
Customs inspection, 444
CYC1, 321
Cytokines
  interferon-γ (IFN-γ), 389, 390
  interleukin 1-β (IL-β), 388, 390
  interleukin 10 (IL-10), 388, 390
  interleukin 4 (IL-4), 390
  interleukin 8 (IL-8), 389
  monocyte chemotactic protein-1
    (MCP-1), 389
  transforming growth factor-β (TGF-β),
    388
  tumor necrosis factor-α (TNF-α),
    388, 390

D

D-amino acid oxidase, *see* DAO
DAC
  quantification by HPLC, 49, 56, 155
  structure, 43, 153
Dangerous Goods
  Class 6, Division 6.2, 445, 447
  packing instruction 602, 447, 448
  UN 2814, 445–448
  UN 2900, 445–44
Dangerous Goods Regulations (DGR)
  Class, 445

packing instruction, 445
proper shipping name, 445
UN number, 445
DGR, *see* Dangerous Goods Regulations
DAO
    assay, 58
    source, 57
    use, 45
DAOC
    bioconversion into 7-ADCA, 49, 57, 60
    bioconversion into glutaryl-7-ADCA,
        58
    quantification by bioassay, 54
    quantification by HPLC, 49, 56, 155
    production in *A. chrysogenum*, 45, 48
    production in *P. chrysogenum* , 44
    structure, 46, 153
DAOCS, 42
Deacetoxycephalosporin C synthase,
        see DAOCS
Deacetoxycephalosporin C, *see* DAOC
Deacetylcephalosporin C, *see* DAC
Deoxysugar
    D-desosamine, 138
    D-olivose, 142
    L-oleandrose, 138, 139
    L-olivose, 138, 139
    L-rhamnose, 139, 142
    L-rhodinose, 142
Deposit for patent purposes, *see* Patent
        Deposit
Detachment of cells, *see* Detachment of
        microorganisms
Detachment of microorganisms, 406, 407
Dextransucrase, 213, 216, 219
Dextrose, 316
Diagnostic culture media, 7
Diagnostic specimens, *see* Biological
        material
Dimethyl disulfide, 397
Dimethyl sulfide, 397
Dimethyl trisulfide, 397
Dimethylallyl-pyrophosphate, *see* DMAPP
Dimorphism, 241

Disclosure, *see* Manifestation
Dispatch, 437, 440
DMAPP, 260
DNA
    foreign, 14, 15
    shuffling, 16
    uptake, 13, 14
DNA microarray, 16
DNA sequence, 472, 473
DNA vaccination, 165
DOC, 168, 175
Dry ice, 448
dsrS, see Dextransucrase
DTT, 48
Dual Use, *see* Biological Weapons

E

Echinone, 263
EDTA, 167, 317
Elution buffer
    buffer 1 for HIC, 167
    buffer 1 for HPLC, 168
    buffer 2 for HIC, 167
    buffer 2 for HPLC, 168
Electroporation, 14, 262
Elliker's broth, 399
Elloramycin, 132, 136, 144
ElmGT, 132, 136, 137, 144
Endotoxin
    contamination of pDNA, 166, 176, 177
    determination by the kinetic- QCL
        assay, 165, 167, 176
    reduction, 167
Enrichment methods, 7–9
Enzymes
    animal-derived, 166
    single-cutting restriction, 168
Epifluorescence microscopy, *see*
        Fluorescence microscopy
Epimerase, 138
EPO Board of Appeal, 474
ERM, *see* Methodology of experimental
        research
ermEp promoter, 136, 139, 141, 142, 146

Erythromycin
  activity assay, 82
  base preparation, 81
  biosynthesis, 71
  genes, 71
  isolation, 81
  isothiocyanate, 81
  market, 71
  producer, *see Saccharopolyspora erythraea*
  production, 71, 77, 78, 80
  purity, 82
  structure, 65, 67, 71
Erythromycin A, 71, 74, 80, 82, 86
Erythromycin B, 71, 74, 80, 82, 87
Erythromycin C, 71, 74, 80, 82, 87
Erythromycin D, 71
Escherichia coli
  ATCC 9637/pJC200, 49, 57, 146
    fermentation, 49, 59
    GLA activity, 57
    maintenance and preservation, 58
  BL21 (DE3), 287, 316
  DH10B, 135, 289
  DH5α, 47, 167, 168, 176
    gDNA, 168, 175, 176
    transformation and bacterial culture, 168
  ESS22-31, 45
    bioassay, 54, 55
  HB101, 316
Esterase LipA, 217
Ethanol precipitation, 166
Ethidium bromide, 168, 174, 176
Ethylenediaminetetraacetic acid, *see* EDTA
Ethylmethane sulfonate, 303, 304, 309
European Patent Convention (EPC), 465, 466, 468, 472–475
European Patent Office (EPO), 462, 468, 469, 471–477
Evaporation of medium, 6
Exoenzyme, 205, 212, 219
Experimental design

  mixtures, 36
  particular, 36
Experimental domain, 27
Experimental Research Methodology, *see* Methodology of experimental research
Experimental response, 29, 38
Expert solution, 464
Exploratory research, 27
Expandase, *see* DAOCS
Export, *see* Secretion
Export restrictions, 439
Expression array, *see* Gene expression macroarray
Expression vector, *see* Plasmid

F

Factors, 27, 38
Family gene shuffling, 16
Federal Patent Court, 466, 467
Federal Supreme Court, 466, 467
Feedback
  elimination of regulation, 10
  inhibition, 8
  regulation, 8
  repression, 8
Fermentation
  analysis, 6, 7
  optimization, 1–3, 17, 18
  scale-up, 17, 18
Fermentation medium, 17
Fermented sausage, *see* Meat fermentation
Fermentor, 18, 371, 372
Ferroplasma, 406
Ferroplasma acidarmanus, 406
Ferroplasma acidiphilium, 406
FeS$_2$, *see* Pyrite
Filter enrichment, 247
Filtration
  0.45-μm filters, 303, 307
  syringe filters, 303, 305
Flasks in screening, 6
Fluorescence microscopy, 406, 408

Food
  cheese, 397
  VSC, 397
Form 1140/1141, 462
Freight, *see* Transport

G

G418, 260
Galactose, 318, 325
Gas chromatography, 400
GC/MS
  ^{13}C analysis, 194
    mass isotopomer, 192
    positional isotopomer, 192
    summed fractional labeling, 201
  analysis of protein hydrolysates, 196
    derivatization, 197
    fragmentation patterns, 199
    measurement conditions, 197
    sample preparation, 197
Gellan
  composition, 301
  filtration, 303, 307
    precipitate, 307, 309
  gravimetric determination, 302, 303
  overproduction, 305, 306
  production, 302, 305, 306
  structure, 301
Gellan determination
  filtration, 307
  gravimetric, 303
    ethanol, 307
    precipitate, 307
  treatment, 306, 307, 309
    acid, 307
    alkali, 307
    heat, 306
    prolonged acid exposure, 309
Gene
  encoding rate-limiting step, 10
  expression analysis, 15, 16
  expression macroarray, 333–343
  identification, 298, 299
  manipulating regulatory, 11

  probe, 333–343
  replacement, 183, 184, 187
  technology
    animal gene, 477
    human gene, 477
    plant gene, 477
  therapy, 165, 166
  transfer methods, 13, 14
Genetic engineering, 2, 9–16, 466–468,
    474, 483
Genetic material, 491
Genetic modification, 482
Genetic transfer, 483
Genetically modified organisms, *see*
    Biological Material
Genetically modified variety, 488
Geneticin, *see* G418
Genome, 179, 180, 335
Genome breeding
  L-lysine-producing mutant in *C.*
    glutamicum, 179, 180
    double mutant, 185
    single mutant, 184, 185
Genome sequence, 187
Genome shuffling, 16
Genomic DNA, *see* gDNA
Genomic integration, *see* Integration
Genomics, 333
German Federal Patent Court, 466
German Federal Supreme Court, 466,
    475, 476
German Genetic Engineering Act, 478
German Patent Office, 466
German Patent Act, 466
German Patent Office, 467
German Medical Products Act, 478
German Protection of Embryos Act, 478
gDNA
  breaking, 169
  co-elution with pDNA, 166
  contamination of pDNA, 166, 175, 177
  denaturation, 169
  determination by real-time PCR,
    165, 167

determination by Southern Slot Blot assay, 177
hydrophobic character of, 167
single-stranded fragments, 177
GFX polymerase chain reaction DNA, 135
GLA
  assay, 59
  bioconversion of glutaryl-7-ADCA, 45, 59
  source, 57, 58
Global regulators, 11
Glucose, 167, 302, 304
Glucosyltransferase (*see also* Dextransucrase), 179
Glutaryl acylase, 45, 57, 58
Glutaryl acylase, *see* GLA
Glutaryl-7-ADCA
  bioconversion into 7-ADCA, 58, 59
  chemical extraction, 58
  production from DAOC, 57
  structure, 46
Glycosyltransferase, 132, 144
GNA33 signal peptide sequence
  construction of mutants, 319
  fusion with *phoC*, 319
  sequence, 314
Growth promoters, 131

H

Head space analysis, 400
hemA, 211
Heme genes, 211
Herbicides, 131
HIC
  gel
    packing, 170
    preparation, 169
  HPLC technique, 177
  matrix, 166, 167
  purification, 165, 170
High-performance liquid chromatography, *see* HPLC
Himar1 transposase

activity, 289
coding region, 286
  pBADMar1, 285
expression plasmid, 287
  pJR2888, 287
purification, 288
  lysis buffer, 285, 288
Himar1 transposon, 289
Histidine, 316
Holomycin
  bioautography, 159
  extraction, TLC, 159
  quantification by HPLC, 158
  structure, 153
Holothin
  quantification by HPLC, 158
  structure, 153
hom gene, 180–183
Homologous recombination, 183, 184, 187
Homoserine dehydrogenase, 180
HPLC, 121, 135, 145, 263, 266, 371, 372, 374
Human embryos, 480
HUP, 413, 419, 429
Hybridization, 339
Hydrogen uranyl phosphate, *see* HUP
Hydrophobic interaction chromatography, *see* HIC
Hydrolyzed soybean meal, 302
Hygromycin
  resistance cassette, 50
  transformation marker in *A. chrysogenum*, 53, 60, 61

I

IATA, *see* International Air Transport Association
IDA, *see* International Depositary Authority
idi gene, 258
Immobilization and cross-linking reagents
  glutaraldehyde, 426

silane, 426
Immobilization methods
    biofilm, 415, 422
    covalent, 418, 425
    entrapment, 417, 424
Immobilization supports
    ceramic Raschig rings, 413, 415
        application, 429
        methods, 422, 426
        product specification, 416, 418
    polyurethane, 413
        application, 429
        prepolymer, 417
        reticulated foam, 416
Immune system
    antibodies, 384
        IgA, 384, 392
        IgG1, 392
    innate mucosal defenses, 383
        innate epithelial response, 384
    peripheral blood mononuclear cells,
        385, 389
    in vitro models, 387
        co-culture model, 386, 387
Imperial Patent Office, 465–467
Import restrictions, 440
    Chlamydia psittaci, 440
    Erwinia amylovora, 440
    import permit, 440
    quarantine regulations, 440
    Synchytrium endobioticum, 440
    Zymonema farcinimosus, 440
Impurities
    binding, 16
    determination, 165
    washing, 166, 167
Induction by xylose, 207, 210
Integration, 210, 222
Intestinal epithelial cells
    Caco-2 human cell lines, 384–386,
        389
    HT-29 human cell lines, 384
    mucins, 382, 383
        sialomucins, 382

sulfomucins, 382
Intestinal microbiota
    Commensals (non-pathogenic
        bacteria), 382, 383, 386, 393
    neonatal colonization, 381
        Bifidobacterium, 382
        Clostridium, 382
        Enterococci, 382
        Lactobacillus, 382, 385, 388, 390,
            392
        Ruminococcus, 382
        Streptococci, 382
Isopropanol
    concentration, 165
    precipitation, 166, 169, 194
In vitro transposition, 294
Infectious substances, *see* Biological
    material
Insecticides, 131
Insertional mutagenesis, 294
International Air Transport Association
    (IATA), 437, 446–448
International Convention for the
    Protection of New Varieties of
    Plants (UPOV)
    animal varieties, 468, 481–483
    genetically transformed plant
        varieties, 473
    multiplicity of plants, 468, 471
    plant varieties, 468, 471–475, 481,
        482, 487
International Depositary Authority (IDA)
    acronym, 453
    choice of, 457
    full name, 453
    impartiality, 452
    risk level, 453
    secrecy, 452
Invention
    commercial use, 451
    industrial value, 451
    novelty, 451
    reworkability, 451, 461
    state of the art, 458, 462

IPN, 42
IPNE, 42
IPNS, 42
IPP, 257
IPTG, 47, 317, 325
Isoleucine, 10
Isopenicillin N epimerase, *see* IPNE
Isopenicillin N synthase, *see* IPNS
Isopenicillin N, *see* IPN
Isopentenyl-pyrophosphate, *see* IPP
Isopentenyl-pyrophosphate isomerase-
 encoding gene, *see idi* gene
Isopropyl-β-D-thiogalactopyranoside,
 see IPTG

K

Kanamycin, 135, 136, 139, 141

L

Lactic acid bacteria
 Enterococcus faecium, 376
 Lactobacillus curvatus, 369
 Lactobacillus sakei , 369, 371
 maintenance and incubation, 372
 Pediococcus, 369
Lactococcus lactis
 cheese, 397
 growth, 399
LB, 47, 74, 167, 168, 176, 316
LB-ampicillin, 305
LB-ampicillin-kanamycin, 285
Leader peptide, (*see also* Signal
 peptide), 212
Leathen medium, 406, 408
Legal protection of biotechnological
 inventions, 479
Leptospirillum, 405, 408
Leptospirillum ferriphilum, 405
Leptospirillum ferrooxidans, 405
Leptospirillum thermoferrooxidans, 405
Letter Post Compendium, 441
Leuconostoc mesenteroides, 213, 216
Levansucrase gene, *see sacB*
LiCl, 316

Lipid A, 167
Lipopolysaccharide, *see* LPS
Listeria, 367–377
Lithium chloride, *see* LiCl
Loading buffer, 174
Localization of exported protein, 219
LPS
 co-elution with pDNA, 166
 contamination of pDNA, 166
 hydrophobic character, 167
Luria-Bertani, *see* LB
Lycopene, 257, 274, 278, 282
Lysate
 concentration, 166
 precipitation with isopropanol or
  ethanol, 166
lysC, 180–183
Lysine, 139
 analysis, 187
 flask fermentation, 186
 jar fermentor cultivation, 186, 187
 production in *C. glutamicum*, 186,
  187
Lysis buffer, 167, 169

M

Macrolide, 66–69
Mail, *see* Transport
MALDI-TOF, 364
Manifestation, 451
Mass spectrometry, *see* GC/MS
MBDSTFA, 197
MDFA, 226, 227, 236
Meat fermentation
 in situ simulation, 370, 376, 377
 in vitro simulation, 370, 372–374
  meat simulation medium (MSM),
   371, 372
  pH profile, 372, 373
  sampling, 373, 374
  temperature control, 373
  water activity profile, 373
 pilot scale, 370, 371
 sausage batter, 371

MECHM
    cocrystallization with HUP, 419, 429
    intercalation in HUP, 419, 429
Media optimization. 17
Metabolic activity profiling
    analysis of protein formation, 193
        de novo synthesis, 193
        synthesis from building blocks, 193
    ^{13}C metabolic flux analysis, 201
        computational tools, 201
        experimental tools, 201
Metabolic engineering
    bioconversion into 7-ADCA, 49, 57
    in *A. chrysogenum*, 44
    in *P. chrysogenum*, 44, 45
    in *X. dendrorhous*, 258
Metabolic flux analysis, 16, 201
Metabolite
    production, 1, 2, 8–13
    quantification, 6, 7
Metal dissolution, 407, 409
Metal sulfide, 405, 409
Metal sulfide oxidation, 409
Metallosphaera sedula, 406
Methanthiol, 397
Methodology of experimental research,
    26. 38
Methyltransferase, 138
Methyl green, *see* MG
MG, 317
MIC, 8
Microbial activity, *see* Bioleaching activity
Microbial enhanced chemisorption of
    heavy metals, *see* MECHM
Microbiological invention
    deposit, 451, 466, 486, 491
    expert solution, 464, 465
    microbiological material, 458, 485
    microorganism culture, 466–468,
        473, 476, 477, 479, 483–487,
        490–492
    release, 461, 465, 486
Microbiological product
    biotransformations, 476

    fermentation process, 476
    with genetic information, 480
Microcalorimetry, 409
Micrococcaceae, 369
Micrococcus luteus ATCC9341, 74, 82
Microorganism, *see* Biological material
Microtiter plates, 6
Mine waste, 405, 407
Miniaturization of fermentation vessels,
    6
Minimal media
    carbon source, 304
    glycerol, 415
    lactose, 415
    salt solution, 302, 304
Minimum inhibitory concentration, *see*
    MIC
Minitransposon
    minitransposon, 291
        ColE1, 291
        pJR2585, 291
        minitransposon A, 289–291
        pJR2125, 291
Modeling
    bacteriocin activity, 376
    cell growth, 374, 375
        inhibition exponent, 374
        logistic equation, 374
        Monod equation, 375
        nutrient-depletion model, 375
    Euler integration, 376
    lactic acid production, 375
    sugar depletion, 375
Molecular-weight marker, 168
Most-probable-number, *see* MPN
Mouse model
    colonization, 390
        Peyer's patches, 391
    germ-free mice, 390
MPN, 408
MRS medium, 371, 372
Mucor circinelloides
    batch culture, 250
        ethanol production, 250

batch spore preparation, 250
carotene source, 273
flask fermentation, 249
hyperproducing strains, 244
    isolation and selection, 245
monomorphic mutants, 246, 247
    isolation procedure, 206, 207
mutagenesis, 204
Mutagenesis, 3, 4
Mutagenesis of *Pseudomonas* sp.
aeration, 305
cell survival rate, 304, 309
    treated cells, 304
    untreated cells, 304
ethylmethane sulfonate, 303, 304
nutrient broth, 302, 304
sodium thiosulfate, 303, 309
Mutagens, 3, 4
Mutant
analog-resistant, 8, 9
auxotrophic, 9
isolation, 2, 3, 5, 8, 9
Mutant isolation of *Pseudomonas* sp.
ampicillin, 303, 305
identification, 305
    strain EGP-2, 305
mucoid, 305
    polysaccharide formation, 305
screening, 305, 306
    biomass production, 305
    gellan production, 305
Mutation
characterization and reconstitution,
    180
    single mutant, 184, 185
    double mutant, 185
    L-lysine-producing mutant, 179,
    180

N

N-aceylclavaminic acid, 153
N-methyl-N-t-butyl-dimethylsilyl-
    trifluoroacetamide, *see*
    MBDSTFA

N-Propionylholotine, 153
Near-infrared (NIR) spectroscopy, 7
Negative regulators, 11
Neurosporene, 274
Neutralization buffer, 167, 169
NG, *see* Nitrosoguanidine
Nickel
assay, 413, 414
bioremediation
    cocrystallisation with HUP,
        419, 429
    intercalation in HUP,
        419, 429
Nitrosoguanidine, 244, 245
Non-biological processes, 483, 490
Nutrient agar, 135
Nutrient broth, 302, 304
Nylon membrane, 336

O

Office of the International Union
        for the Protection of New
        Varietiesof Plants (UPOV),
        487, 488, 490
oleE, 138, 139, 146
oleL, 138, 139, 146
oleS, 138, 139
oleU, 138, 139, 141, 142, 146, 147
oleV, 138, 141, 142, 146
oleW, 138, 141, 142
oleY, 141
Oligonucleotide
cloning with, 217
printing, 335
probe design, 167, 335
One-step gene disruption, 45
Open reading frame, *see* ORF
Optimal design
first-order model, 31
second-order model, 31, 32
ORF, 15
Orange Book, 444
Overmutagenesis, 4
Oxidation rate, 407, 409

P

p-nitrophenol, *see* *p*NP
p-nitrophenyl phosphate, *see* *p*NPP
Packaging
    EN 829, 441
    packing instruction, 445, 447, 448
    triple packaging, 437, 441
        absorbent material, 441
        labels, 441
        outer packaging, 441
        primary receptacle, 441
        secondary packaging, 441
    UN certified, 441, 445
Packed-bed reactor, 419, 429
Palcam agar, 372, 377
Paris Convention, 465, 476
Patent deposit
    date of deposit, 458
    depositing procedure, 459
        accession form, 459
        accession number, 461
        checklist, 460
        identity checking, 461
        preservation, 461
        viability check, 461
    practical procedure, 452
    statement of receipt, 461
    statement of viability, 461
Patent protection
    for microbiological processes,
        451, 467
    grant of the patent, 487
    human body, 480
    invention, 465, 470
    inventive step, 470
    novelty, 465, 470
Patent strain, *see* Biological
    material
Patenting teaching, 486
PBS
    elution, 170
    equilibration of the column, 170
    phosphate-buffered saline, 167, 177,
        226, 231

pDNA
    assessment of the purity, 168
    assessment of the quality, 165, 167,
        173
    assessment of the transformation
        efficiency, 168, 176
    binding, 166
    biological active, 176
    biological activity, 165
    concentration by precipitation with
        isopropanol, 167
    denatured, 166, 167
    desalting, 165, 170, 173–175, 177
    determination of HPLC purity, 174,
        177
    determination of non-supercoiled
        isoforms by agarose gel
        electrophoresis, 174
    determination of concentration with
        anion-exchange HPLC, 165,
        167, 171, 177
    double-stranded, 167
    elution, 166, 167
    identity, 175
    isolation, 168
    loading, 170
    precipitation, 166
    purification, 165–167, 170
    quantification, 168
    standards, 168, 172
    supercoiled, 175
    therapeutic, 176
    yield, 165, 172, 173
PDP, 316
PDP-MG technique, 323
PEG 4000, 316
Penicillin N, *see* PN
Penicillins
    enzymatic synthesis of penicillins, 116
    chemical synthesis of natural
        penicillins, 116
    penicillin G, 48
Perishable biological substances,
    see Biological material

pH electrode, 371, 376
PhAc-CoA ligase, 110, 111
Phaffia rhodozyma, see
 Xanthophyllomyces
 dendrorhous
Phenolphthalein diphosphate, *see* PDP
Phenolphthalein diphosphate-Methyl
 Green technique, *see* PDP-MG
 technique
Phenotype, 1
Phenylacetyl-CoA catabolon
 analytical procedures, 121
  identification of PhAc, 121
  HPLC analysis of penicillins,
   121
  separation of PhAc derivatives
   by thin tayer thromatography,
   122
 bioplastics, 114
  PHA polymerase assay, 114
  isolation of PHA granules, 114
  PHA granule-associated proteins,
   114
  synthesis of 3-hydroxyacyl-CoA,
   115
 biotechnological applications, 116
  bioplastic production, 119
  chemical synthesis of natural
   penicillins, 116
  enzymatic synthesis of penicillins,
   116
  extraction of PHAs, 120
  genetic manipulation of
   Penicillium chrysogenum, 117
  overproduction of plastic
   polymers, 119
  production of 2-hydroxy-PhAc,
   119
 hydroxylation complex, 112
 PhAc-CoA ligase assays, 110, 111
  colorimetric, 110
  fluorimetric, 111
 PhAc uptake, 109
 structure, 95

styrene monooxygenase, 112
 phenylacetaldehyde
  dehydrogenase assay, 113
 upper pathways, 112
Phleomycin, 48
phoC gene
 reporter in *E. coli*, 314
 reporter in *S. cerevisiae*, 315
PhoC phosphatase
 qualitative assay, 323
  plate screening in *E. coli*, 317, 323
  plate screening in *S. cerevisiae*,
   317, 325
 quantitative assay, 318
  induction of cultures, 317, 325
  *p*NPP assays in *E. coli*, 327
  *p*NPP assays in *S. cerevisiae*, 328
PhoN
 Escherichia coli, 413
 Citrobacter sp., *see Serratia* sp.
 Salmonella typhimurium, 413
 Serratia sp., 413
Phosphate assay, 421, 430
Phycomyces blakesleeanus
 carB, carS, carRA, 274
 carotene source, 240, 243
 commercial value, 273
 growth and maintenance, 274
  harvesting mycelia, 276
  lyophilization, 276, 280
  master stocks, 280
  media, 274
  shake cultures, 276
  spores, 276
 mutant strains, 274
Physiological saline solution, 372, 377
Phytoene, 257, 274, 278, 279
Phytofluene, 274
Phoenicoxanthin, 263
Phosphate-buffered saline, *see* PBS
Plasmid
 construction in *E. coli*, 315, 318
 construction in *S. cerevisiae*, 315,
  321

construction with oligonucleotides, 217
expression in *B. megaterium*, 205
GNA33Muts*phoC*, 319
instability, 14
pALfleo7, 47
pAG100, 285
pAN7.1, 47
pBADmar1, *see Himar1* transposase coding region
pBC KS (+), 47
pBluescript I KS (+), 47
pBluescript I KS (+), 285, 289
pEM4, 135, 139, 141, 146
pEM4V, 141, 146
pEM4VWUY, 141
pET12-a, *see Himar1* transposase, expression vector
pET21b(+), 319
pET33phoC, 319
pETphoC$_{wt}$, 319
pGB15, 137
pGB16, 137
pHBIntE, 208, 211
pHSG299, 180
pIAGO, 135, 137, 139
pIJ2925, 135, 137
pJR2125, *see* minitransposon A
pJR2585, *see* minitransposon R
pKC796, 135, 137
pLITMUS29, 135, 141
pLITMUS29V, 141
pLN1b, 140, 141
pLN2, 138, 139, 142, 144, 146
pLNR, 142, 144, 146
pLNRHO, 142, 144, 147
pLNZ3, 144, 147
pLND, 142, 144, 147
pLR2347, 139, 141, 146
pLR2347D7, 141, 146
pMM1520, 208, 213
pMM1522, 217
pMM1533, 217
pmMar1, 289, 290

pOLV, 138, 144, 146
pQE, 94
pRHAM, 138, 139, 144, 146
profile analysis, 378
pUC18, 135
pUC18W, 141
pUK21, 135, 139
rescue, 298
secretion vector, 217
stability, 14
uptake, 13, 14
YEpphoC, 321
YEpphoCDP yeast promoter probe vector, 321
YEpsecI yeast expression vector, 321
Plasmid DNA, *see* pDNA
PN
    bioassay, 54
    quantification by HPLC, 49, 56
    structure, 43
*p*NP, 327, 423
*p*NPP, 318, 417, 423
Polyinosine, 335
Polyketide
    biosynthesis, 70
    type I, 70
        macrolide, *see* Macrolide
    type II, 70
Polysaccharide
    applications, 301
    heteropolysaccharide, 301
        tetrasaccharide, 301
    production, 302, 306
    separation, 309
Positive regulators, 11
Predictive microbiology, 25
Primer, 134, 141, 142, 147, 168, 175
Product transport enhancing, 13
Production
    animals, 468, 480
    plants, 468, 480
Protease protection assay, 220
Protein
    analysis, 349, 353

contamination of pDNA, 175
denaturation, 169
determination with the micro-BCA
assay, 165, 167, 175
extraction, 348, 352
precipitation with TCA, 135
production with *Bacillus
megaterium*, 165
Protein synthesis, 16
Protein transport, *see* Secretion
Protoplasts, 137, 144, 208, 220
Pseudomonas putida
directed mutagenesis, 95
gene deletion, 101
succinyl-CoA, 92
transposon mutagenesis, 98
Tn5, 98
Pseudomonas sp. ATCC 31461
biomass determination, 307
filtration, 307
gravimetric, 303
culture medium, 302, 304
flask fermentation, 304, 305
inoculation, 304, 305
maintenance and preservation, 306,
309
long-term, 306, 309
short-term, 306, 309
mutant strain, 302, 305
ampicillin resistance, 302
mutagenesis, 303–305
screening, 305, 306
optical density, 304
strains, 302, 304, 305
Pseudomonas sp. EGP-2, 305, 306
PW, 226, 227
Pyrite, 405, 407, 409
Pyrite oxidation, 407, 409
Pyrolysis mass spectroscopy (PyMS), 7

Q

Quantitative studies of factors, 30
Quantitative studies of responses, 30

R

R5, 135, 136, 144
R5A, 135, 144, 145
Random screening, 2–5
Rate-limiting steps debottlenecking, 10
Recombinant DNA, 205, 476
Regulatory agencies, 166
Regulatory genes manipulation, 11
rep-PCR, 378
Response surface methodology, 31
Resuspension buffer, 167, 169
Rhodotorula gracilis
ATCC 26217, 49, 57
DAO activity, 58
fermentation, 49, 57, 58
maintenance and preservation, 57
Riboflavin, 283
Ribosome binding site, 215
Ribosome binding site optimization,
215
Risk Groups, WHO, 439
RNA
binding to the HIC matrix, 167
cleavage with RNase , 166, 177
co-elution with pDNA, 166
contamination of pDNA, 175
determination by agarose gel
electrophoresis, 165, 167,
174
hydrophobic character, 167
isolation, 337
mRNA, 337
reduction, 167
16S rRNA gene, 175
RNase (DNase-free), 168, 171
Roxithromycin
synthesis, 84
structure, 71
Rule 28 EPC, 464

S

S-(2-aminoethyl)-L-cysteine, *see* AEC
sacB, 180, 184, 185

Saccharomyces
  cell breakage, 352
  growth conditions, 347, 351
    Triple M medium, 351
  proteases, 352
  stock cultures, 351
Saccharomyces cerevisiae KD53B, 316
Saccharopolyspora erythraea
  colony morphology, 76
  flask cultivation, 77
  fermentation, 71, 77, 78, 80
    feed, 78, 87
    productive stage, 78
    seed stage, 77
    viscosity, 80
    yield, 71, 80
  maintenance, 71, 77
    agar slants, 74, 77
  NRRL 2338, 70, 71
  recombinant strain, 71, 80
  sporulation, 76
  strain selection, 71,76
Samples, furnishing of
  entitled party, 461
  request for, 461
  third party, 462
  trial and examination, 461
Scale-up, 17, 18
Screening
  automation, 5, 6
  factors, 27
  random screening, 2–5
SD/H, *see* Synthetic Dextrose/Histidine
  medium
SDS, 128, 135
SEC, *see* Secretion
Secondary metabolite
  function, 65, 66
  production (*see also* Erythromycin
    production), 65, 70
Secretion, 212, 219
Seed culture preparation, 18
Sepharose CL-6B, 167, 169
Serratia sp.

acid phosphatase, 413, 414
batch culture, 415, 417
Citrobacter sp., 413, 415
continuous culture
  air-lift fermentor, 414, 418
  preparation of cells, 415, 422
immobilization methods
  biofilm, 415, 422
  covalent, 418, 425
  entrapment, 417, 424
MECHM, 413, 414
minimal media
  glycerol, 415
  lactose, 415
packed-bed reactor, 419, 429
PhoN, 413
  bioaccumulation of HUP,
    419, 429
  cocrystallization with HUP,
    419, 429
  intercalation with HUP, 419, 429
removal of heavy metals
  cobalt, 413, 414
  nickel, 413, 414
  uranium, 413, 414
Shear stress, 17
Shipper's Declaration for Dangerous
  Goods, 444, 446
Shipping, *see* Transport
Signal peptide
  evaluation, 213
  secretion vector, 217
SignalP, 215
Significance Analysis of Microarrays
  (SAM), 335, 341
Silver staining 2D-PAGE, 350, 364, 365
Single-stranded nucleic acids, 167
Site-directed mutagenesis, 182
Sodium borohydride, 167
Sodium deoxycholate, 168
Sodium phosphate buffer, 372
Sodium thiosulfate, 303, 309
Solvents
  chromatography, 277

column, 277
HPLC, 278, 282
thin layer, 279
extraction of carotenoids, 282
Soybean oil, *see* Soy oil
Soy oil, 73, 77, 87
Soytone, *see* Hydrolyzed soybean meal
SpotFinder, 335, 339
Starkey medium, 407, 408
Sterilization, 431
Stomacher, 371, 377
Strain breeding in *C. glutamicum*, 179, 180
Strain improvement, 1, 2
classical approach, 2, 3–7
genomic approach, 15, 16
targeted approach, 2, 3
use of genetic engineering, 9–16
Strain selection, 7–9
Streptavidin, 339
Streptomyces
S. albus, 132, 136, 146
S. albus GB16, 135, 137, 144, 145
S. albus GB16-LN2, 144
S. albus GB16-LNR, 144
S. albus GB16-LND, 144
S. albus GB16-OLV, 144
S. albus GB16-RHAM, 144
S. albus GB16-RHO, 144
S. antibioticus, 138
S. clavuligerus
culture conditions, 144
production of antibiotics, 142
structure, 153
S. fradiae Tü2717, 135, 142
S. lividans GB16F4, 135, 144
S. olivaceus Tü2353, 132, 136
Submerged leaching, 409
Sucrose, 303, 306
Sugar, *see* Deoxysugar
Sulfidic mine waste, *see* Mine waste
Sulfobacillus, 405, 406
Sulfobacillus acidophilus, 405
Sulfobacillus thermosulfidooxidans, 406
Sulfolobus metallicus, 406

Sulfur compounds, *see* Volatile sulfur
compounds
Superclean LC18, 144
Supor membranes, 135, 145
Sutter's IV medium, 274
Synthetic Dextrose/Histidine medium,
316

T

t-butyl-dimethylsilyl, *see* TBDMS
TAE buffer, 168, 174
Target gene identification, 2, 3
TBDMS, 197
TBS, 226
TCA, 168, 175, 176
TE buffer, 47
Technical Board of Appeal, 476
Technical invention, 465–493
Technical process, 480
Technical teaching, 486
Tetracenomycin C, 132
Thaumatin production process
downstream processing, 232
diafiltration and concentration,
232
ion-exchange chromatography,
233
ELISA, 231
immunoblotting, 231
lyophilization, 234
monitorization, 229
dry weight, 231
sugar and ammonium, 231
thaumatin, 231
pre-inoculum, 226, 235
purification, 233
scale-up, 227
ThermoPol buffer, 136
Thin-layer chromatography, *see* TLC
Thiobacillus, see Acidithiobacillus
Thiostrepton, 135, 136, 139, 144, 145
Throughput in screening, 4, 5
TLC, 122, 265, 267
Tobramycin, 135, 136, 146

Torulene, 263
Total cell counts, 406, 408
Total cell numbers, *see* Total cell
    counts
Toxicity overcoming, 12, 13
TPMG medium, 317
Transcriptome, 345
Transformation
    A. awamori, 225–238
    A. chrysogenum, 47, 53, 54
    A. gossypii, 296
    B. megaterium, 209
    C. glutamicum, 192
    E. coli, 313–331
    M. circinelloides, 239–256
    P. chrysogenum, 117
    Pseudomonas, 91–129
    S. albus, 136, 137
    S. cerevisiae, 313–331
    S. erythraea, 71
    X. dendrorhous, 261
Transformation and DNA uptake, 13, 14
Transport
    by freight, 444
        air (ICAO), 446
        inland waterways (ADN), 446
        rail (COTIF/RID), 446
        road (ADR), 446
        sea (IMDG), 446
    by mail, 441
    diagnostic specimens, 447
        packing instruction 650, 448
        UN 3373, 448
    genetically modified organisms, 446
    infectious substances, 437, 441, 445–448
        packing instruction 602, 447
        UN 2814, 445–448
        UN 2900, 445–448
    perishable biological substances, 441
    transport emergency card, 445
Transport across cell membrane, 13
Transposition reaction
    inverted terminal repeats
        amplification, 289–291

    3'ITR, 290, 291
    5'ITR, 290, 291
    pmMar1, 289, 290
    TA target, 289
    T buffer, 285
    yeast tRNA, 289
Trichloroacetic acid, 303, 307, 309
TRIPS Agreement, 488–490
Tris-buffer saline, *see* TBS
Tris-HCl
    buffer, 167
    elution, 170
Tryptic soy broth, *see* TSB
Tryptone, 167
Tryptose-phenolphtalein-methyl-green,
    see TPMG medium
TSB, 71, 82, 399
Tunicamycin, 153
Two-dimensional polyacrylamide gel
    electrophoresis (2D PAGE)
    gel drying, 365
    reproducibility, 346, 347
    first dimension gels, 354
        pouring, 354
        running, 358–361
    second dimension gels, 356
        loading, 361–363
        preparing, 356
        running, 363
Type strain availability, 462

U

UAS_{GAL}, 321
UN number, *see* Dangerous Goods
    Regulations
Undermutagenesis, 4
Universal Postal Union (UPU), 437, 441
Unmethylated CpG, 383
UPU, *see* Universal Postal Union
Uranium
    assay, 420
    bioremediation, 413, 414
        bioaccumulation, 419, 429
urdQ, 142, 147

urdR, 142, 146
urdZ3, 142, 144, 147

V

Vector, *see* Plasmid
Vent DNA polymerase, 135
Viable cell counts, 406, 408
Viable cell numbers, *see* Viable cell
    counts
Vitamin B2, 283
Vitamin B12, 211
Viscosity
    standard, 303, 309
    viscometer, 303, 309
    whole-culture , 307
        ATCC 31461, 307
        comparison, 307
        glucose minimal medium, 307
        strain EGP-2, 307
    crude polysaccharide, 307, 308
        acid treatment, 308
        alkali treatment, 308
        ATCC 31461, 308
        comparison, 308
        heat treatment, 308
        strain EGP-2, 308
Volatile sulfur compounds, 397

W

WIPO, 459
World Intellectual Property
    Organization, *see* WIPO
World Trade Organization, *see* WTO
WTO, 488–490

X

X-gal, 47
Xanthan, 302

Xanthophyllomyces dendrorhous

    astaxanthin source, 240, 241, 257,
        258
    doubling time, 270
    idi overexpressing strain, 258, 261
        carotenoid analysis, 263
        construction, 261
        Southern analysis, 268
    idi overexpression vector, 260
    isolation of chromosomal DNA,
        268
    transformation, 261
        efficiency, 269, 270
        electroporation, 262
        integration, 260
        preparation of cells, 262
        preparation of DNA, 262
        system, 260, 261
        vectors, 260
        yeast, 258
Xanthophylls, 258
Xanthomonas campestris, 302

Y, Z

Yeast extract, 167
Yeast extract peptone dextrose medium,
    see YPD
Yeast extract peptone galactose
    medium, *see* YPG
Yeast extract peptone medium, *see* YP
Yeast nitrogen base, *see* YNB
YNB, 316
YP medium, 317
YPD medium , 318
YPG medium, 317, 318
YPG/PDP medium, 317
Zone of inhibition, 7